An Introduction to the Statistical Analysis of Data

Houghton Mifflin Company / **Boston**

Dallas   Geneva, Illinois   Hopewell, New Jersey   Palo Alto   London

An Introduction to the

STATISTICAL ANALYSIS OF DATA

T. W. Anderson   Stanford University

Stanley L. Sclove   University of Illinois, Chicago Circle

Library of Congress Catalog Card Number: 77-78890
ISBN: 0-395-15045-0

To

Dorothy
    Robert
    Janet
    Jeanne

Suzan
    Sarabeth
    Benjamin

PREFACE

1 The Nature of the Book

This book is a text for a first course in statistical concepts and methods. It introduces the analysis of data and statistical inference and explains various methods in enough detail so that the student can apply them. Little mathematical background is required; high school algebra is used occasionally. No mathematical proof is given in the body of the text, although algebraic demonstrations are given in appendices at the ends of some chapters. The explanations are based on intuition, verbal arguments, figures, and numerical examples. The verbal and conceptual levels are higher than the mathematical level.

Many illustrations, from a number of fields, help the student understand the subject and present questions that can be answered by statistical analysis. Some examples are taken from daily life; many deal with the behavioral sciences, some with business, the life sciences, the physical sciences, and engineering. The exercises are of varying degrees of difficulty.

This book is suitable for undergraduates in various majors, for graduate students in the behavioral sciences and education, and for omnibus courses with various types of students. It has grown out of our experience over the past 20 years in teaching such courses at Columbia, Stanford, Carnegie-Mellon University, and the University of Illinois at Chicago Circle. Several drafts of this text have been tried out in these courses.

2 The Organization of the Book

Part One, "Introduction," shows how statistical methods are used in several substantive fields to answer interesting questions. Part Two, "Descriptive Statistics," considers the organization of data and summarization by means of descriptive statistics; the student learns how to approach information that comes in numerical form. Part Three, "Probability," develops the ideas of probability to form the basis of statistical inference. Part Four, "Statistical Inference," begins by considering the use of sample characteristics to estimate population characteristics; the idea of variability of sample quantities leads directly to confidence intervals. Part Four discusses tests of hypotheses and the allied concepts of significance level and power. This part treats some of the basic methods for means, proportions, and variances. Part Five, "Statistical Methods for Other Problems," includes techniques such as chi-square tests, analysis of variance, regression and correlation, and sample surveys.

3 The Use of the Book

The 16 chapters in the book provide enough material for a course of two semesters or three quarters. In some chapters there are starred sections which may be omitted without affecting the understanding of subsequent chapters; some such material is put into appendices in order not to burden the main development.

Chapter 1 presents examples of the use of statistics in studies of general interest and some basic concepts of statistics.

Chapters 2 through 5 are largely *descriptive statistics*. Some students need to begin their study with such relatively simple material. In other cases the instructor will allow the students to read this material for themselves or move through it quickly in the lectures, perhaps slowing down for the study of weighted averages in Chapter 3. Association is discussed at some length in Chapter 5; some instructors may wish to limit the study to the first two sections.

Chapters 6 and 7 on probability require more careful development and study. However, the somewhat formal early part of Chapter 6 may be given less emphasis. The ideas of estimation and testing hypotheses in Chapters 8 and 9 are more difficult, and the student needs time to reflect on them. If time is at a premium, the sections on nonparametric tests (Sections 9.5 and 9.6) can be omitted. Chapter 10, dealing with statistical inference based on two samples, presents useful methods. Again, the instructor may omit the material on nonparametric tests (Section 10.6) if time is short. Chapter 11 on variances in one and two populations presents chi-square and F-tests, which are based on an assumption of normal populations. Although sensitive to nonnormality, they are presented for pedagogical reasons. Chapter 12 presents statistical inference for "contingency" tables. Chapter 13 is on analysis of variance; the nonparametric material is optional. Chapters 14 and 15 are on simple and multiple regression, respectively, and Chapter 16 discusses sampling techniques.

The chapters are divided into sections (numbered 1.1 and 1.2 in Chapter 1, for example). Many sections are in turn divided into subsections. The effect of this hierarchy of chapters, sections, and subsections is to form an outline of the material. The summary at the end of each chapter reviews that chapter's contents and helps the student determine whether the important concepts have been learned.

The material of the book is organized so that the 16 chapters provide a basis for courses of varying lengths. Some guidelines for coverage are as follows.

A one-year course (two semesters or three quarters). All 16 chapters can be covered in a one-year course. Some teachers may wish to omit the starred sections and possibly de-emphasize other relatively advanced sections, such as Section 3.4 on standardized averages and Sections 5.3 and 5.4 on effects of a third variable on the relationship between two given variables.

A two-quarter course. One can go in the direction of either breadth or depth. A survey course covering all 16 chapters would mean (i) the omission of starred sections, (ii) de-emphasis of Chapters 5 and 6, and (iii) omission of peripheral topics in Chapters 12–15. An instructor can design a course which goes in the direction of depth rather than breadth by omitting one or more of Chapters 12–16.

A one-semester course. The first ten chapters comprise a text for a one-semester course. Some teachers may wish to include smatterings of later chapters, such as 11, 12, and 14.

A one-quarter course. The first ten chapters provide a basis for such a course. The instructor can omit starred sections and de-emphasize Chapters 5 and 6. Again, some teachers may wish to include parts of Chapters 11, 12, and 14.

4   Exercises

In order to learn statistical concepts and methods, one must do exercises. For convenience, exercises are grouped at the end of each chapter. In most cases, particularly in the second half of the book, the reader should work some exercises after studying one or several sections. (Most exercises have indications of the relevant sections.)

Some exercises directly apply the ideas presented. Others present the data and goals of studies that demonstrate the utility of statistical methods. Some instructors may wish to put more emphasis on the mathematical aspects. Certain problems, given under the heading of Mathematical Exercises, develop algebraic proofs of statements.

The student does not need these to understand the material in the text, but they give insights to readers with more mathematical background. Chapters 3 and 4 contain Exercises on Summation Notation, to give students practice in dealing with this notation. Starred exercises refer to starred sections and can be omitted.

Answers to Selected Exercises appear at the end of the book. The instructor has many options in assigning exercises of varying difficulty, either with or without answers available. The instructor can obtain a complete solutions manual from the publisher.

5 Acknowledgments

We are indebted to many colleagues who read various drafts, particularly Herman Chernoff (M.I.T.), Neil Henry (Virginia Commonwealth University), Gary Simon (S.U.N.Y. at Stony Brook), and Paul Switzer (Stanford). We benefited from our teaching assistants (then graduate students in statistics or sociology) who supplied illustrative material as well as reactions to pedagogy. They include Beth Gladen, Louis Gordon, James Inglis, Raymond Maurice, John Reed, Richard Roth, Barbara Rubin, Robert Somers, and Nira Herrmann Szatrowski.

Parts of the manuscript were reviewed by James Buckley (University of South Carolina), James Chumbley (University of Massachusetts), Ralph D'Agostino (Boston University), Kenneth Goldstein (Miami-Dade Junior College, North Campus), David Hildebrand (University of Pennsylvania), Robert Hogg (University of Iowa), Jerome Sacks (Northwestern), I. Richard Savage (Yale), Arnold Schroeder (Long Beach City College), Gerald Sievers (Western Michigan University), Fred Steier (University of Pennsylvania), and Charles Stone (U.C.L.A.). David Berengut and Hamparsum Bozdogan served as accuracy reviewers. These people made many helpful suggestions.

Many others helped in the preparation and publication of the manuscript. Judi Campbell Davis did the lion's share of the typing; Dorothy Brothers, Wanda Edminster, Carolyn Knutsen, and Karola Lof also performed many secretarial tasks.

T.W.A.
S.L.S.

CONTENTS

APPENDICES

PART ONE
INTRODUCTION

1 THE NATURE
OF STATISTICS

INTRODUCTION

Statistics enters into almost every phase of life in some way. A daily news broadcast may start with a weather forecast and end with an analysis of the stock market. In a newspaper at hand we see in the first five pages stories on an increase in the wholesale price index, an increase in the number of *habeas corpus* petitions filed, new findings on mothers who smoke, the urgent need for timber laws, a state board plan for evaluation of teachers, popularity of new commuting buses, a desegregation plan, and sex bias. Each article reports some information, proposal, or conclusion based on the organization and analysis of numerical data.

Statistics in systematic and penetrating ways provides bases for investigations in many fields of knowledge, such as social, physical, and biological sciences, engineering, education, business, medicine, and law. Information on a topic is acquired in the form of numbers; an analysis of these data is made in order to obtain a better understanding of the phenomenon of interest; and some conclusions may be drawn. Often generalizations are sought; their validity is assessed by further investigation. In the following section some examples of such studies are described.

A definition of statistics is *making sense out of figures*. Statistics is the methodology which scientists and mathematicians have developed for interpreting and drawing conclusions from data. This first chapter begins with a survey of some uses of statistics and concludes with a discussion of some of the general ideas underlying the subject of statistics.

1.1 SOME EXAMPLES OF
THE USE OF STATISTICS

The large-scale trial of the Salk polio vaccine, the Surgeon General's report on smoking and health, the many political polls conducted by elected officials and by candidates for office, the use of educational screening devices such as the College Board Entrance Examinations, and the use of the census, consumer price indices, unemployment rates and population projections in assessing human progress and future potentialities all attest to the importance of statistical methods in matters of everyday importance. We discuss several such applications of statistics.

1.1.1   Political Polls

Political polls illustrate some uses of statistics in the field of political science and political sociology. During the period of January 7–8, 1972, each person in a national sample of 1380 adults, 18 years of age and older, was interviewed. Each respondent was asked a question which has been asked about every president since Franklin Delano Roosevelt held office: "Do you approve or disapprove of the way the president is handling his job?" The response in January, 1972, regarding Richard Nixon's performance, was 49% approving and 39% disapproving, the remaining 12% having no opinion. The pollster [Gallup (1972)] noted that Mr. Nixon was somewhat less popular among college students than he was among the whole adult population.

Some of the early political polls were made in order to forecast outcomes of elections. In 1920, 1924, 1928, and 1932 *The Literary Digest* (a magazine now defunct) successfully predicted the winner of the United States presidential election, forecasting the popular vote with only a small percentage of error—less than 1% in 1932.

Then came 1936. The contest was between the incumbent, Franklin D. Roosevelt, and Alfred E. Landon. An October 31 headline read

Landon	**1,293,669**
Roosevelt	**972,897**

Final Returns in The Digest's Poll of Ten Million Voters

The poll results were listed by state. The Landon vote, Roosevelt vote, and votes for minor party candidates were given; the total number of such "votes" was 2,376,523. This information is partially summarized in Table 1.1.

TABLE 1.1
Summary of Results of The Literary Digest *Poll*

	LANDON	ROOSEVELT
Number of states "carried"	32	16
Number of electoral votes	370	161

SOURCE: *The Literary Digest* (1936).

The *Digest* reported the figures; they did not "weight, adjust, or interpret" them. However, it was hard for any reader not to take the poll results as a prediction of the actual vote.

The rest is history. Instead of being a Landon victory, the election was a landslide for Roosevelt. The *Digest* poll had 42.9% for Roosevelt; the actual vote was 62.5% for Roosevelt! Instead of 161 electoral votes, Roosevelt actually received 523; instead of carrying only 16 states, he carried 46.

What happened? Why was there such a great difference between the percentages in the poll and the percentages of actual vote? To answer this we examine *The Literary Digest*'s procedure. Ballots were mailed out to some ten million persons; each recipient was asked to fill out the ballot and mail it back. Every third registered voter in Chicago and every registered voter in Allentown, Pennsylvania, were polled; names were drawn from telephone books, from the rosters of clubs and associations, from city directories, lists of registered voters, and other lists.

A major fault with this procedure was that the lists used for mailing tended to have more people with higher incomes than voters as a whole and in this election—because of the New Deal—those with above-average incomes tended to vote Republican more than those with lower income. Another questionable feature was that only one-fourth of the recipients of the ballots returned them; maybe those who did not respond had political opinions somewhat different from those who responded.

Among the respondents to the poll, most had voted for Hoover in 1932. This was a sure sign that the sample was not representative of all voters, since Roosevelt had won in 1932. On November 14 the *Digest* said they "were willing to overlook this inconsistency." The final outcome was the demise of the *Literary Digest* in 1937.

Later Pre-election Polls. Even before the *Literary Digest* debacle, improved methods of polling had been developed for use in election prediction, market research, and for other purposes. Mail question-naires with their inherent uncertainty of response were replaced by direct interviews. Interviewers were sent out to question respondents met on the streets or in homes. Their selections were guided by *quotas;* for example, in a given day an interviewer might be asked to inter-view 25 men and 25 women, of whom 30 lived in the city and 20 in the suburbs. These polls were able to provide much useful information on behavior of consumers and opinions of the citizenry, but forecasting voting provided a more acid test.

In 1948 it was Dewey vs. Truman. Three leading pollsters were George Gallup (Institute of Public Opinion), Elmo Roper, and Archi-bald Crossley. As early as September a Dewey victory was predicted, and interest in the election waned.

It was a close election; so close, indeed, that an early *Chicago Tribune* headline read, "Dewey wins." Truman won, however. Table 1.2 shows the popular vote together with the reports of the various polls.* In calculating the percentages those who were undecided were omitted; and Gallup included only Thurmond and Wallace in "Other can-didates." Again the polls were wrong. What happened?

TABLE 1.2

1948 Poll Results and Actual National Vote

	Dewey	Truman	Other Candidates
Actual national vote	45.1%	49.5%	5.4%
Crossley	49.9%	44.8%	5.3%
Gallup	49.5%	44.5%	5.5%
Roper	52.2%	37.1%	10.7%

SOURCE: McCarthy (1949).

Subsequent analysis showed that in spite of the quotas some parts of the electorate were over- or underrepresented in the polls; for in-stance, too many of those interviewed were college educated, and there were too few with only grade school education. Another contributor to the error in forecasting was the time element. One-seventh of the voters made up their minds only during the last two weeks preceding the election, and three-quarters of these decided in favor of Truman. Roper, however, closed down his operation September 9, two months before the election; and Gallup and Crossley did not gather new information

*We should note that even if the popular vote is predicted well, the problem of pre-dicting the outcome of a U.S. presidential election is complicated by the system of electoral votes. Dewey could have won by carrying Ohio, California, and Illinois, each of which he lost by less than 1%!

during the last two weeks. In the latest poll 15% were undecided, and it was assumed that those would be split in the same proportions as those who had already decided; this contributed about 1.5% points to the error of the predictions.

In the case of quota sampling there may be biases due to the interviewer's selection of persons to interview; he or she may prefer well-dressed or pleasant-looking individuals; these are factors not controlled by the quotas. There is also a chance factor in whom the interviewer meets. The factor of chance, luck, or probability comes into almost every sampling method. It gives rise to what we call *sampling error.* "Sampling error" refers to the inherent variation between an estimate of some characteristic as computed from a sample and the actual value of that characteristic in the population from which the sample was drawn. (See Section 1.2.4.) The desirable aspect of sampling error due to a chance factor is that the laws of probability can be used to assess sampling error. In these pre-election polls the sample sizes were large enough so that the sampling error was relatively small. The biases introduced by underrepresentation of some portions of the electorate and by the fact that the pollsters completed their interviewing early caused the gross error. Interestingly enough, the public opinion polls were similarly unsuccessful in predicting the outcome of the 1970 general election in the United Kingdom, and for the same reasons [The Market Research Society (1972)].

Modern Refinements.   Social scientists, market researchers, and pollsters have developed better methods of selecting interviewees and more sophisticated modes of inquiry. Many political polls today do not rely simply on voters' answers to the question of whom they will vote for. The interviewer may ask the occupation of the head of the household, the number of dependents (both children and senior citizens), and information concerning union affiliations.

The voter may be asked to state what issue is most important in the current campaign or be asked to rate the candidates as to how closely they agree with the voter's own position on this issue. The questioning may relate to several issues, and then by weighting each issue a score can be obtained which indicates the voter's relation to each candidate. Strengths and weaknesses of candidates are assessed by asking a voter about ways to convince a neighbor to vote for the favored candidate. These indirect methods, based on scientific studies of the motives and attitudes that underlie choice behavior, provide more power to the researcher when the time comes to calculate a prediction.

A Panel Study of Voting Behavior.   Pre-election polls simply find the intentions of a sample of persons (that is, report opinion) and then predict the vote (that is, predict action). In principle there are two sources of discrepancy inherent in this procedure. One is the discrepancy between present and future—a prediction is being made; that is, present data are being brought to bear on the future. The other is the discrepancy between people's stated opinions and their actions: polls ascertain people's opinions or intentions, whereas what is being predicted is an action, actually voting.

Social scientists have undertaken to study voting on a more scientific basis in order to learn how voters behave and why. One such study was an investigation of the evolution of opinion regarding the 1940 election among persons in Erie County, Ohio. The purpose of the investigation was to study how opinions are formed and changed.

Every fourth house was sampled to obtain 3000 adults. From these, four groups of 600 persons each were selected by sampling in such a way that the groups were matched on important characteristics. The study was a *dynamic* study in the sense that it was carried out over time rather than at a single point in time. Each of three groups was interviewed once, one group in July, one in August, and the third in October. The fourth group was used for a *panel* study in which the same people were interviewed successively over time, on seven occasions, about one month apart. The last of these seven interviews was in November, after the election. Out of the original 600 persons in the panel 483 were available for the interviews in October and November. Some of the results of the next-to-last interview and the actual vote are reported in Table 1.3. Among the 65 persons who in October said that they would not vote ("No vote"), 6 voted Republican and 59 did not vote. The 6 who voted Republican said Republican party members persuaded them.

TABLE 1.3
Tabulation of Persons Interviewed After 1940 Election

	VOTE INTENTION IN OCTOBER	ACTUAL VOTE
Republican	229	232
Democrat	167	160
Don't know	22	—
No vote	65	91
	483	483

SOURCE: Lazarsfeld, Berelson, and Gaudet (1968), p. xxiii.

The data gathered in this study were much more complete than the information usually obtained by a political poll. They can be used for answering questions about voting behavior as well as suggesting problems to investigate. For example, the data indicated that Catholics tended to vote Democratic and upper-income people tended to vote Republican. Especially interesting were people subject to cross pressures. Would Democrats in the upper income bracket be more likely to change their intention than Democrats in the lower income bracket? With richer data such as these, it is possible to use more sophisticated techniques, such as the construction of indices, composite measures, relating the propensity to vote for one party or the other to various social, economic, and religious factors.

1.1.2   The Polio Vaccine Trial

The largest public health experiment to date took place in 1954. The amount of publicity given the evaluative report, produced by the Poliomyelitis Evaluation Center at the University of Michigan, was at that time unequalled for a scientific work. The cost of the study, involving nearly 2,000,000 children, has been estimated to be about $5,000,000.

The question at hand was how effective the Salk vaccine might be in preventing the occurrence of poliomyelitis. The original plan for this massive field trial was to administer the Salk polio vaccine to children in the second grade of school. First and third graders would not be inoculated, but would be kept under observation for the occurrence of poliomyelitis. The rate of occurrence among the second graders would then be compared with that among the first and third graders. The second graders constituted the *experimental group;* the first and third graders constituted the *control group.*

Use of a control group is efficacious only if the control group differs from the experimental group in no important way, except that it is not subjected to the procedure under investigation. In the original plan the control and experimental groups would be observed for the same time period and in the same geographical areas. However, the following questions were raised:

1. Might the act of inoculation itself increase or decrease the chances of contracting polio?
2. Might second graders (about 7 years old) differ from first and third graders (about 6 and 8 years old) in the incidence of polio?
3. Might knowledge by the children (and their parents) of whether they had been inoculated affect the risk?
4. Might the diagnosis of physicians be affected by knowledge of whether children had been inoculated?

Since these questions indicated possible defects in the plan, another plan was proposed in which children of the first, second, and third grades would be treated alike. One half of all the children in the study were to be inoculated with a placebo, an inert solution of appearance similar to the polio vaccine. The incidence of polio in the placebo group would be compared with that in the vaccine group.

In implementing this plan it was important that there be no biasing factors that would influence the choice of which children received the vaccine. Possibly some physicians practiced in areas of higher susceptibility to polio than others. Children in different areas may differ in regard to overall health. There was the possibility that some physicians might give the placebo to the healthier children.

These problems were overcome by placing in each one of a number of boxes 50 vials, 25 filled with placebo and 25 with vaccine. The vials were labelled with code numbers; only certain persons at the Evaluation Center knew which vials were placebos. The vials were positioned randomly within the box. In effect, a nurse or doctor picking out a vial was doing it blindfolded. This procedure served to eliminate possible bias in the reporting of apparent cases of polio, since the diagnosing doctor would have had no way of knowing whether the child had received placebo or vaccine.

Results.   Some results of the improved experiment are summarized in Table 1.4. (In some areas the original plan was followed.) While the number of children inoculated with the vaccine was about the same as the number inoculated with the placebo, the number of cases of paralytic polio among those vaccinated was about $\frac{1}{4}$ the number of cases among those not vaccinated. This was convincing evidence that the vaccine was effective.

On the basis of this statistical evaluation, public health authorities instituted a campaign to have every child vaccinated. In time, the original Salk vaccine (which had various defects) was replaced by the Sabin preparation. Now the fearsome polio is virtually unknown in the United States.

TABLE 1.4
Incidence of Paralytic Polio

	INOCULATED WITH VACCINE	INOCULATED WITH PLACEBO
Number of children inoculated	200,745	201,229
Number of cases of paralytic polio	33	110

SOURCE: Francis, *et al.* (1957).

1.1.3   Smoking and Health

In the early part of this century statisticians concerned with vital statistics noted an increase in lung cancer. In the last two decades a great many studies have been concerned with the possible effects of smoking on health. The primary focus of much of this research has been the relationship, if any, between lung cancer and smoking. In 1962 the Surgeon General of the United States Public Health Service, Dr. Luther Terry, formed an advisory committee to review and evaluate the relevant data. *Smoking and Health,* the 1964 report of that committee, has received an amount of publicity and discussion surpassing even the attention given the polio vaccine trial.

Some of the conclusions of the report are stated in such a way as to identify cigarette smoking as a *cause* of certain diseases; for example (p. 37), "Cigarette smoking is causally related to lung cancer in men; the magnitude of the effect of cigarette smoking far outweighs all other factors. The data for women, though less extensive, point in the same direction. The risk of developing lung cancer increases with duration of smoking and the number of cigarettes smoked per day, and is diminished by discontinuing smoking." The report also points out that there is a definite relationship between other diseases and smoking.

What methods are used to reach such conclusions? First it is noted whether or not there is an *association* between smoking and the incidence of a given disease. Table 1.5 illustrates this; Hammond and Horn (1958) observed (via volunteers from the American Cancer Society) 187,783 men over a period of 44 months, determining at the outset the smoking habits of the men studied. The death rate from lung cancer was

TABLE 1.5 *Death Rates from Lung Cancer*

RATES PER 100,000 MEN PER YEAR, BASED ON OBSERVATION OF 187,783 MEN 50–69 YEARS OLD OVER 44 MONTHS

NEVER SMOKED	CIGARS ONLY	PIPES ONLY	CIGARETTES AND PIPES OR CIGARS	CIGARETTES ONLY
13	13	39	98	127

SOURCE: Table 7 from Hammond and Horn, *Journal of the American Medical Association* 166: 1294–1308. Copyright 1958, American Medical Association.

TABLE 1.6 *Death Rates from Lung Cancer by Current Daily Cigarette Smoking*

RATES PER 100,000 MEN PER YEAR, BASED ON OBSERVATION OF 187,783 MEN 50–69 YEARS OLD OVER 44 MONTHS

LESS THAN $\frac{1}{2}$ PACK	$\frac{1}{2}$ TO 1 PACK	1 TO 2 PACKS	MORE THAN 2 PACKS
95	108	229	264

SOURCE: Table 8 from Hammond and Horn, *Journal of the American Medical Association* 166: 1294–1308. Copyright 1958, American Medical Association.

ten times as great for cigarette smokers as for nonsmokers. A next step corroborates this evidence by considering the death rates for those who smoke different amounts of cigarettes, as shown in Table 1.6.

Many questions remain to be answered. Might smoking and lung cancer be related to something else, something which causes an individual both to smoke and to contract lung cancer? For instance, many persons who smoke a great deal also drink heavily; alcohol consumption might be a common factor associated with both smoking and lung cancer. However, reports have shown that smokers who drink little are more likely to contract lung cancer than are nonsmokers who drink little, and the same was observed among those who drink a lot.

Other possible causes of lung cancer, such as air pollution, have also been studied. Death rates from lung cancer are indeed higher in urban than in rural areas. However, in both urban and rural areas the incidence of lung cancer is higher for smokers than for nonsmokers. Differences both in levels of air pollution and in smoking habits of people in urban and rural areas account for some of the difference in death rates due to lung cancer.

1.1.4   Project "Head Start"

In the 1960's a nationwide program for disadvantaged youngsters of preschool age, administered by the Office of Economic Opportunity through local Head Start agencies, was begun. Most of the youngsters who participated in the program were four-year-olds whose parents were earning less than $4000 per year. The programs provided were similar to those in nursery schools. The philosophy of the project was that, while with middle- and upper-class children teachers can *build upon* awareness developed in the home, with lower-class youngsters teachers must build that awareness itself; Head Start centers in some areas were working with children who did not know what an orange was, had never held a crayon or a fork.

A team of investigators undertook assessment of the program by comparing Head Start youngsters with their counterparts who had not been in Head Start. A sample of 104 Head Start Centers was selected for study from the total of 12,927 centers. A total of 1980 children who had gone on to the first, second, and third grades were compared with 1983 non-Head Start children in the same grades and schools. The children in both groups were given a series of six tests of ability and achievement in language and of self-concept and attitudes.

As an example of the results, consider the Metropolitan Readiness Test, given to 864 children in first grade; of these, 432 children had been in full-year Head Start programs in 27 centers and 432 children of similar backgrounds had not been in programs [Westinghouse Learning

Corporation/Ohio University (1969), p. 153]. The average school readiness score for Head Start children was 51.74 and for the other children was 48.46. These scores indicated that the Head Start children were a little more ready than similar children not in such programs. Although the difference in average scores was not very large, analysis indicated it was significant in the sense that although readiness scores varied from child to child it was unlikely that the average scores of two sets of 432 children picked at random would differ so much.

1.1.5 "A Minority of One Versus a Unanimous Majority"

Does group pressure influence individual behavior? The experiments conducted by Solomon Asch at Swarthmore in an effort to answer this question constitute a good example of *experimental method.*

In the first experiment subjects were asked to compare lengths of line segments. One experiment consisted of sequences of 18 trials. On each trial a test line was to be compared with 3 comparison lines by each of 8 persons in turn, the test line being exactly equal in length to one of the 3 comparison lines. In each sequence of trials, 7 persons had been instructed by the experimenter to give incorrect responses on a prescribed dozen of the 18 trials. The eighth person, who was the experimental subject, was not instructed and considered the activity bona fide comparisons of lengths. Imagine yourself as the eighth person in one of these tests, finding that your seven predecessors have unanimously made an "obviously" incorrect choice on a trial! On 12 of the 18 trials the subject turned out to be "a minority of one versus a unanimous majority." Fifty such series were carried out. In most of these experiments the subject under pressure conformed to some extent and made errors of comparison. There were another 37 series in each of which all 8 subjects made legitimate comparisons; that is, no such pressure to conform was exerted on the last person; this is called the "control" situation. Table 1.7 gives the frequency of the number of errors made by persons in the experimental and control situations.

There was a clear and strong effect of the "unanimous majority." The investigation showed that the phenomenon was worth studying and led to questions for further research. What variables affect this phenomenon? What will be the effect of variations in group composition, for example, different size groups, or the presence of a "partner"? In Table 1.7 great individual differences are apparent: even under pressure 13 of 50 subjects made no errors, whereas one subject made 11 errors; this suggested questions about what types of persons are most likely, or least likely, to conform.

TABLE 1.7
Numbers of Errors Made by Last Person in Each Set of Trials

No. of Errors	FREQUENCIES	
	Groups with Persons Under Group Pressure	Groups with Persons Not Under Group Pressure
0	13	35
1	4	1
2	5	1
3	6	0
4	3	0
5	4	0
6	1	0
7	2	0
8	5	0
9	3	0
10	3	0
11	1	0
12	0	0
	50	37

SOURCE: Asch (1951).

1.1.6   Deciding Authorship

The "Federalist Papers" played an important role in the history of the United States. Written in 1787–1788 by Alexander Hamilton, John Jay, and James Madison, the purpose of these 77 newspaper essays was to persuade the citizens of the State of New York to ratify the newly written Constitution of our emerging nation. These essays were signed with the pen name Publius, and published as a book which also contains eight essays by Hamilton. The question of whether Hamilton or Madison wrote twelve of the "Publius" essays has been a matter of dispute. Standard methods of historical research have not settled the problem.

Mosteller and Wallace (1964) applied statistical methods towards its solution. The problem is a difficult one, for Hamilton and Madison used the same style, standard phrases, and sentence structure, which were characteristic of most educated Americans of their time; for example, their average sentence lengths in the undisputed papers were 34.5 and 34.6 words, respectively. However, there were some subtle differences of style. Scholars had noticed, for instance, that Hamilton tended to use "while," and Madison, "whilst." This pair of "marker" words, however, was not decisive because in some papers neither word appeared.

After painstaking analysis Mosteller and Wallace found that 30 words differed substantially in the frequency of use by the two authors.

Table 1.8 gives the rate of use of nine words by the two authors: the rate 3.24 is obtained by taking the number of times "upon" was used by Hamilton divided by the number of 1000's of words in the Hamilton essays. Note that Madison used "upon" infrequently but used "on" more frequently than Hamilton.

The frequencies of occurrence of the 30 words were combined into an index in such a way that the index was large for papers known to have been written by Hamilton and small for Madison papers. In fact, the score ranged from 0.3120 to 1.3856 for Hamilton papers and from −0.8627 to 0.1462 for Madison papers. Except for one paper, the scores of the disputed papers went from −0.7557 to −0.0145. Thus, these disputed papers were assigned to Madison. The paper with a score of 0.3161 was also assigned to Madison on the basis of further investigation and with less assurance.

TABLE 1.8

Rates of Use of Certain Words by Hamilton and Madison

| | FREQUENCY PER 1000 WORDS | |
Word	Hamilton	Madison
Upon	3.24	0.23
Also	0.32	0.67
An	5.95	4.58
By	7.32	11.43
Of	64.51	57.89
On	3.38	7.75
There	3.20	1.33
This	7.77	6.00
To	40.79	35.21

SOURCE: Reprinted by special permission from Mosteller and Wallace, *Inference and Disputed Authorship: The Federalist*, 1964, Addison-Wesley, Reading, Mass.

1.2 BASIC CONCEPTS OF STATISTICS

1.2.1 Organizing and Summarizing Numerical Information

A set of statistical data consists of one or several measurements, scores, or values for each of a number of individuals, objects, or events;

for example, the scores made by students on verbal and mathematical scholastic aptitude tests. A basic feature of statistical data is that the values vary from one individual to another; if all students' scores were the same, if all values or measurements were the same, or if every person held the same opinion, statistical analysis would be unnecessary. It is this variability that creates the problems in making sense out of figures.

A statistician should always advise an investigator on how to collect and record data; too often the process of statistical analysis starts with organization and summarization of data already obtained. These operations replace the many data by a few numbers, which retain most of the relevant information; an average and a measure of variability are usually included. The methodology developed for organization and summarization of the data is called *descriptive statistics.*

1.2.2   Relating a Small Set to a Large
Set: Population and Sample

The set of individual persons or objects in which an investigator is primarily interested is the *population* of relevance for that study. Sometimes values for all individuals in the population of relevance are obtained, but often only a set of individuals which can be considered as representatives of that population are observed; such a set of individuals constitutes a *sample.*

Finite Populations.   In many cases the population under consideration is one which could be physically listed. The books in a library constitute such a population, and the card catalog provides the list. For *The Literary Digest* poll and other political polls the population of relevance consists of all potential voters; these people are registered on lists held by boards of elections. The population of the United States as of noon yesterday could be listed, at least in theory.

Hypothetical Populations.   From a batch of 1000 light bulbs produced by a factory, a number of bulbs are tested—they are put into operation, and the time it takes for each to burn out is recorded. These tested bulbs are a sample from a finite, concretely defined population— the batch of 1000 bulbs. Using the same equipment, materials, and method of production, the factory will continue to produce light bulbs; and the bulbs that *will be* produced constitute a *hypothetical* population —it does not yet exist. The lifetimes of the bulbs in this hypothetical, to-be-produced population are important and can be predicted from the experience with the tested bulbs. Similarly, the children inoculated with polio vaccine (Section 1.1.2) were considered as a sample from the population of all children who would be inoculated with the vaccine if it were put into general use.

Statistical Inference. When we observe only a sample, the attempt to draw conclusions about the total population involves uncertainties. These uncertainties are inherent in the process of generalizing from the few to the many. The methodology developed for passing from observation of the sample to statements about the population is called *statistical inference.*

Statistical Analysis. This is the whole process of organizing, processing, summarizing, and drawing conclusions from the data. In statistical analysis, methods of description and methods of inference are combined. The questions under investigation indicate what types

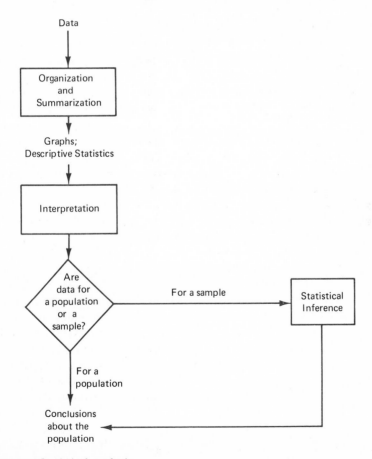

Figure 1.1 *Statistical analysis*

of inferences are to be made; these in turn indicate how the data should be summarized to extract the relevant information. Figure 1.1 is a diagram of these processes.

Generalization. The implications of a study often go far beyond the immediate material; the conclusions may be generalized to individuals and to situations in a framework larger than the population sampled. The effects of group pressure can be expected among individuals not at Swarthmore College and in situations other than the experimental one.

Necessity and Desirability of Sampling. We cannot test all the light bulbs manufactured in a factory because to test them is to destroy them. In any event, we are interested in the hypothetical population that includes light bulbs yet to be produced; those light bulbs cannot be tested. If the population of interest is a hypothetical one, it is necessary to regard the cases at hand as a sample.

In the case of a finite population other practical considerations make it desirable to sample instead of enumerate the entire population. For instance, the typical political poll is a sample because the cost of interviewing all of the voters in the United States would be exorbitant.

1.2.3   Random Sample: Use of Probability

Inferences about the population are necessarily based upon the information in the sample. Ideally, the sample should reflect as accurately as possible the relevant characteristics of the population; it should in a sense be representative. In practice we have to settle for a "fair" sample rather than insisting on perfect representation of the population. For, without complete knowledge of the population (with regard to the characteristics of interest), it is impossible to determine a sample that will be a miniature facsimile of the population.

In order to obtain a sample that is "fair," we sample randomly. If the population is finite, we figuratively put the names of all the members of the population into a bowl, shake the bowl, and draw out as many names as we need for the sample. (Actually this physical procedure is not a good way to draw a random sample; better ways will be described later in the use of random numbers.) A hand dealt in bridge is an example of a random sample of 13 cards from a population of 52 cards; the 10 birthdays with the highest priority numbers in the annual drawing for Selective Service was supposed to be a random sample from the population of birthdays.

The term "random" implies that every individual in the population has the same chance of being included in the sample. This feature helps to make it likely that the sample chosen will contain the salient features of the population in their proper proportions. It is possible to obtain a sample on a random basis which, by bad luck, is far from representative; but the probabilities of doing so are small if the sample is large enough. Random sampling is used, therefore, to ensure unbiased findings. It is possible to assess the chances that the results obtained from a random sample are misleading. Other methods of selecting members of the population, such as selection by an "expert," may result in samples subject to bias; and it may be impossible to evaluate the sampling error.

1.2.4 Using Sample Characteristics to Infer Population Characteristics: Estimation

The first paragraph of Section 1.1.1 reported that 49% of a national sample of 1380 adults said, when asked, that they approved of the way the president was handling his job in early 1972. We *estimate* the percentage of the voting population who would have answered "approve" to the same question if posed at the same time and the same way to be 49%. How accurate is this estimate? Probability theory tells us that when we take random samples of 1380 persons, for about 95 samples out of 100 the sample percentage for a given variable will not differ from the population percentage for that variable by more than 2.6%. Thus, assuming that the sampling procedure had inherent variability no greater than for random sampling, we can feel confident that 49% is a fairly good approximation of the percentage of voters who would have approved of the President's policy.

1.2.5 Using Sample Information to Answer Questions About the Population: Testing Hypotheses

In the group pressure experiment the first question was whether pressure to conform would affect behavior. An enormous difference between the performance of persons under pressure in the experimental situation and persons not under pressure in the control situation was observed. It seems clear that this difference could not be due to chance in the selection of subjects and in their behavior.

In another study it might not be so clear that an observed difference between the responses of individuals in an experimental group and those in a control group reflects a difference which holds also for finite or hypothetical populations of relevance. We test the *hypothesis* that the vaccine has no effect; that is, we ask the question whether or not inoculation by an inert liquid has the same effect on immunity to polio as the Salk vaccine. In the study reported, there was very little chance that the observed difference in rates of incidence could be due to the random assignment of inoculation to children.

Many questions that statistical data may answer are questions that can in principle be answered Yes or No; frequently such a question can be stated in terms of the truth of a hypothesis. It can happen, of course, that the investigator is led to an incorrect answer by chance in obtaining the sample; that is, by chance a sample that is quite different from the population may be drawn. One of the goals of this book is to show how to control the probability that this happens.

1.2.6   Explaining Variability

One purpose of statistical analysis is to assess the factors which cause a set of measurements to vary. If the heights of a number of children are recorded reasonably accurately, one can expect that not all of the heights will be equal. They will vary. This variability may be "explained" by virtue of the fact that the children's ages are different, the heights of their mothers and fathers vary, their diets are different, etc. When a researcher conducts a study in order to discover to what extent the amount of a vitamin in the diet is related to the height of the children, he or she must take these other factors into account in the analysis.

When a variable such as blood pressure is to be observed, the variability of the measurement for each individual becomes important. The difference between blood pressure readings for one person at two different times may be as large as the difference between blood pressure readings for two individuals. We must constantly remind ourselves that the observations we record as data are for a particular point in time, and that if we observed the same individual on another occasion we might see something very different.

The fact that there will be variability between measurements on members of a population and between measurements on one individual implies that a sample will differ from the population from which it comes. A precise study is one in which the effect of this difference, this sampling error, is small, so that there will be a good chance of detecting an experimental effect if one exists. A goal in designing an experiment is to make the experiment as precise as possible, within budgetary limits.

1.2.7   Planning Statistical Investigations

Tournaments, run according to specified plans, can be considered as "experiments" for determining who the best players are. The plan, or design, of a tournament should be such as to reduce to a minimum the chance that the player who is best in some overall sense may not win the tournament. This design will include, for example, the method of scoring, the length of play, and the number of tries.

In laboratory experiments investigators often systematically manipulate factors which may cause the results to vary. These factors are then assessed in terms of their effects on the variables under consideration.

A metallurgist trying to determine the best composition for an alloy may try several combinations of concentrations of copper and manganese, concentrations which would have a reasonable cost if adopted for large-scale production. The result may be that increasing the amount of copper included in the alloy increases the strength, but increasing the amount of manganese over the feasible range of cost has little effect on the strength.

When a survey of public opinion is to be made, whether the purpose is to determine voting intentions, preferences of consumers for various products, or sociological attitudes, the investigators must obtain considerable information while keeping the costs of the survey within the amount of money allocated. Above all, the procedures they use for drawing the samples and obtaining the data must be designed so that it is possible to use statistical methods to assess the precision of the results.

Quantification. General research topics must be formulated in terms of specific questions before scientific investigation can proceed. A study must then be planned in which objective phenomena can be measured, counted, and tabulated. To study the general problem of whether group pressure influences behavior, Asch planned an experimental situation to provide specific evidence. The general problem became a specific question of whether subjects would incorrectly compare the lengths of line segments when others in the group did so before them. The evaluation of the success of Project Head Start was made specific by posing such questions as, Will children who have participated in Head Start score better on tests than their counterparts who have not participated? Can such tests measure the effectiveness of the program? If so, what tests are appropriate? In some of the other studies presented in Section 1.1 the relevant quantities were the frequencies of the occurrence of certain phenomena, such as cases of paralytic polio or deaths from lung cancer.

It is sometimes difficult to put observations into numerical form. Such psychological quantities as pain and pleasure, for example, are hard to quantify. A physician, Dr. Harry Daniell, having studied 1104 people over a period of a year, reported [*Time* (1971)] a definite relationship between early and heavy face wrinkling and habitual cigarette smoking. He quantified wrinkling by studying each patient's face to arrive at a "wrinkle score." Then, knowing that age and amount of exposure to the sun are factors in causing wrinkles, he had his patients fill out a questionnaire about their smoking habits, exposure to the sun, and medical history. He found that in each age group cigarette smokers with indoor occupations had more wrinkles on the average than nonsmokers who worked outside.

1.2.8   Models

In considering random sampling from a finite population, it is convenient to think of a jar of beads, each bead representing an individual in the population. If the population consists of 10,000 students, the jar is to hold 10,000 beads of identical size, shape, and weight with a student's name on each bead. If a student would answer Yes to the question, "Do you favor eliminating letter grades?" the bead with that name might be red; the beads representing students not answering Yes might be white. A random sample of 25 students corresponds to drawing 25 beads from the jar after thorough mixing; the color of the bead represents the response of the student to the question.

The set-up with beads is a model of the information concerning students. A *model* is an abstract statement of a real situation or process. The relevant characteristics of the real situation are represented in the model—here the beads, color, and mixing. The jar of beads, its properties, and drawing need only exist in one's mind; it is an idealization or abstraction. Alternatively, the relevant characteristics can be stated mathematically; the mathematical statement is also a model.

Another model, that of the coin toss, is widely used when some phenomenon such as the determination of an unborn baby's sex is under discussion. Since a tossed coin is supposed to have a 50–50 chance of showing heads, and since boys and girls are born in roughly equal numbers, we can imagine that sex is determined by the toss of a coin, each side of which represents one of the sexes.

A model is useful because irrelevancies have been dropped and the logical consequences of what is considered relevant can be correctly deduced. Of course, if some important aspect of reality has been omitted from or incorrectly inserted into the model, conclusions can be misleading or wrong.

We shall see later that the formulation of a model is a necessary step in the process of statistical inference.

SUMMARY

Statistics is the methodology for collecting, analyzing, interpreting, and drawing conclusions from numerical information. The relevant *population* is the set of individuals, objects, or events in which the investigator is interested; it may be finite and concrete or it may be hypothetical. A *sample,* which is a segment of the relevant population, permits generalizing to the population. The process of generalization is part of *statistical analysis. Sampling error* refers to the inherent variation between a characteristic of a sample and the corresponding characteristic of the population. A random sample protects against biased generalizations about the population and permits assessment of the accuracy of the inference about the population. A *model* is a simplified representation of a real-world situation or process.

EXERCISES

1.1 Find a newspaper or magazine article or advertisement using statistics in a misleading way.

1.2 Find a newspaper or magazine article or advertisement using statistics in a helpful or illuminating way.

1.3 Each of the following studies may be regarded as an example of sampling. Discuss what an appropriate population might be for each.
  (a) Collection of prices (at various stores) for a certain brand of cereal
  (b) Collection of student ratings of a certain professor
  (c) Measurement of blood pressure reduction due to a certain medication in 100 patients of a physician

1.4 In the examples in Exercise 1.3, discuss some factors other than sampling which might cause the sample to be unrepresentative.

1.5 Give examples of hypothetical populations.

1.6 Give examples of finite populations and indicate how to list each one.

REFERENCES

Asch, Solomon (1940). "Studies in the principles of judgments and attitudes: II. Determination of judgments by group and ego standards." *Journal of Social Psychology* 12: 433–465.
Asch, Solomon (1955). "Opinions and social pressure." *Scientific American* 193: 31–35.
Asch, Solomon (1956). "Studies of independence and conformity. A minority of one against a unanimous majority." *Psychological Monographs* 70, no. 9 (whole no. 416).

Asch, Solomon (1951). "Effects of group pressure upon the modification and distortion of judgment." In H. Guetzkow (ed.), *Groups, Leadership, and Men,* Carnegie Press, Pittsburgh, 1951; and in Eleanor E. Maccoby, Theodore M. Newcomb, and Eugene L. Hartley (eds.), *Readings in Social Psychology.* Holt, Rinehart, and Winston, New York, 1958.

Berelson, Bernard R., Paul F. Lazarsfeld, and William N. McPhee (1954). *Voting.* University of Chicago Press, Chicago.

Brownlee, K. A. (1955). "Statistics of the 1954 polio vaccine trials." *Journal of the American Statistical Association* 50: 1005–1013.

Francis, Thomas, Jr., *et al.* (1957). *Evaluation of 1954 Field Trial of Poliomyelitis Vaccine: Final Report.* Poliomyelitis Vaccine Evaluation Center, University of Michigan, Ann Arbor, Michigan.

Gallup, George (1972). "Nixon's Rating at [State of the Union] Message Time," *San Francisco Chronicle,* January 20, 1972. Syndicated column, Field Enterprises, Inc. Courtesy of Field Newspaper Syndicate.

Hammond, E. Cuyler, and Daniel Horn (1958). "Smoking and death rates—report on forty-four months of follow-up of 187,783 men. II. Death rates by cause." *Journal of the American Medical Association* 166: 1294–1308.

Lazarsfeld, Paul F., Bernard Berelson, and Hazel Gaudet (1968). *The People's Choice.* 3rd ed. Columbia University Press, New York.

The Literary Digest (1936). Vol. 122: October 31, pp. 5–6, and November 14, pp. 7–8.

The Market Research Society (1972). *Public Opinion Polling in the 1970 Election.* London.

McCarthy, Philip J. (1949). "Election Predictions." In Frederick Mosteller, Herbert Hyman, Philip J. McCarthy, Eli S. Marks, and David B. Truman (eds.) (1949). *The Pre-Election Polls of 1948.* Social Science Research Council, New York.

Mosteller, Frederick, and David L. Wallace (1964). *Inference and Disputed Authorship: The Federalist.* Addison-Wesley, Reading, Mass.

Tanur, Judith M., *et al.* (eds.) (1972). *Statistics: A Guide to the Unknown.* Holden-Day, San Francisco.

Time (1971). December 27, p. 42.

U.S. Public Health Service (1964). *Smoking and Health.* Report of the Advisory Committee to the Surgeon General of the Public Health Service. Public Health Service Publication No. 1103. U.S. Department of Health, Education, and Welfare, Washington.

U.S. Public Health Service (1967). *The Health Consequences of Smoking; A Public Health Service Review.* Public Health Service Publication No. 1696. U.S. Department of Health, Education, and Welfare, Washington.

U.S. Public Health Service (1968). *The Health Consequences of Smoking; 1968 Supplement to the 1967 Public Health Service Review.* U.S. Department of Health, Education, and Welfare, Washington.

Westinghouse Learning Corporation/Ohio University (1969). *The Impact of Head Start: An Evaluation of the Effects of Head Start on Children's Cognitive and Affective Development.* Office of Economic Opportunity, Washington.

PART TWO
DESCRIPTIVE STATISTICS

2 ORGANIZATION
OF DATA

INTRODUCTION

In Chapter 2 we start with the statistical information as it is obtained by the investigator; this information might be an instructor's list of students and their grades, a record of the tax rates of counties in Florida, or the prices of Grade A large eggs in each of ten Chicago grocery stores averaged over the past 36 months. We refer to such statistical information as *data*, the recorded results of observation. After the collection of data, the next step in a statistical study is to organize the information in meaningful ways, often in the form of tables and graphs or charts. These displays or summaries are descriptions which help the investigator, as well as the eventual reader of the study, to understand the implications of the collected information. In later chapters we shall develop numerical descriptions that are more succinct than these tables and charts.

In considering methods for organizing, summarizing, and analyzing statistical data, it is necessary to distinguish two types of data. Data that consist of numerical measurement, such as weight, distance, or time, or counts, such as numbers of children or numbers of errors, are called *numerical*. On the other hand, *categorical* data result from the observation of characteristics—occupations, for example, or species, or physical states—defined by categories, such as clerical, professional, laborer, or canine, feline, bovine, or gas, liquid, solid. These distinctions among kinds of data will be made sharper in Section 2.1.

In Section 2.2 we outline the steps involved in the process of organizing categorical data. These organized data may be presented in a table or a graph.

Numerical data, treated in Section 2.3, permit more flexibility and detail in their organization and summarization. At the stage of first obtaining data there may arise the question of how accurately the information is to be recorded; for example, how well can or should height be measured? We shall see what insight into statistical problems can be obtained by tabulating numerical data and graphing them suitably.

2.1   KINDS OF VARIABLES; SCALES

2.1.1   Categorical Variables

People may be described as being blond, brunette, or redhead. Employed persons may be classified as white-collar workers, blue-collar workers, laborers, or farm workers. Plants may be classed as vines, trees, shrubs, etc. Rocks are grouped as sedimentary, igneous, or meta- morphic. In each of these examples, the members of the population under scrutiny (people, plants, rocks) are placed into *categories*.

A *variable* is any characteristic that varies from one individual mem- ber of the (statistical) population to another. Hair color, occupation, and plant type are examples of *categorical variables;* such a variable is defined by the classes or categories into which it falls. The categories may be natural, as in the case of sex, or rather arbitrary, as in the case of occupational classification.

We may refer to categorical variables as *qualitative* variables. The simplest type of such a variable has only two categories, such as male and female. The categorization may be the presence or absence of a given quality; for example, a person may or may not be employed, may or may not use a certain product; a plant may or may not bear flowers. In these instances, the information recorded is only whether an indi- vidual does or does not have the quality in question, that is, employ- ment, using the soap powder, or bearing flowers. The variable generated is thus of a Yes-No form and may be termed *dichotomous*. The literal meaning of this word, "divided into two parts," becomes especially appropriate when we have reduced a *quantitative* variable such as income to the two categories "low" and "high."

2.1.2   Numerical Variables

When the characteristic of interest may be expressed by a num- ber, we distinguish between a number that is obtained simply by *counting* and a number that requires *measurement*.

Discrete Variables. Some variables, such as the numbers of children in families, the numbers of accidents in a factory on different days, or the numbers of spades dealt in hands of thirteen cards from a deck of playing cards are the results of counting; these are discrete variables.

Typically, a discrete variable is a variable whose possible values are some or all of the ordinary counting numbers 1, 2, 3, . . . or the integers*

---

*We cannot write down all of these numbers because there is no end to them; the three dots in 1, 2, 3, . . . mean that the sequence continues indefinitely.

0, 1, 2, ... The number of spades dealt in a hand is a discrete variable having the possible values 0, 1, 2, 3, 4, 5, 6, 7, 8, 9, 10, 11, 12, 13. A shorter way of writing this set of values is 0, 1, 2, . . . , 12, 13; here the three dots signify the integers omitted, namely, 3, 4, 5, 6, 7, 8, 9, 10, 11. When it is not difficult for the reader to fill in the missing elements, we shall use these three dots (called "ellipsis").

As a definition, we can say that a variable is *discrete* if it has only a countable number of distinct possible values. That is, a variable is discrete if it can assume only a finite number of values or as many values as there are integers.

Continuous Variables. Quantities such as length, weight, or temperature can in principle be measured arbitrarily accurately. There is no indivisible unit. Weight may be measured to the nearest gram, but it could be measured more accurately, say to the tenth of a gram; the gram is divisible, as is the tenth of a gram. Such a variable, called *continuous*, is intrinsically different from a discrete variable.

2.1.3 Scales

Scales for Categorical Variables. Besides being described as either categorical or numerical, variables are typed according to the *scale* on which they are defined. (See Figure 2.1.) This typology of variables involves the amount of structure and meaning which variables have.

The categories into which a variable falls may or may not have a natural ordering. Occupational categories have no natural ordering. A variable, the categories of which are not ordered, is said to be defined

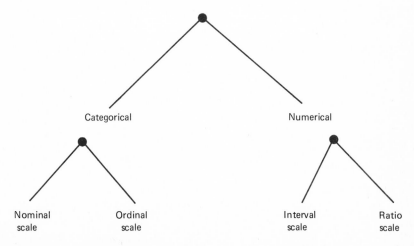

Categorical

Numerical

Nominal
scale

Ordinal
scale

Interval
scale

Ratio
scale

Figure 2.1 *Classification of variables*

on a *nominal* scale, the word "nominal" referring to the fact that the categories are merely names. If the categories can be put in order, the scale is called an *ordinal scale*. Examples of ordinal variables are "social status" (classified as low, medium, and high) and "strength of opinion" on some proposal (classified according to whether the individual favors the proposal, is indifferent toward it, or opposes it), and position at the end of a race (first, second, etc.). For some variables, the classification may be somewhat arbitrary: social status could as well be categorized simply as low and high, instead of low, medium, and high; strength of opinion could be categorized as strongly in favor, mildly in favor, indifferent, mildly opposed, strongly opposed.

Scales for Numerical Variables. Numerical variables, whether discrete or continuous, may be classified as to whether there is a natural, meaningful zero. If I say I have zero cents in my pocket, that statement makes sense; the statement that Mary has twice as many cents as Alice is also meaningful. On the other hand, 0° Fahrenheit is an artificial zero point; to say that an object whose temperature is 32° is twice as hot as one whose temperature is 16° is meaningless. However, the numbers can be used in other meaningful ways. An object that is half way in temperature between freezing and boiling (of water) is 50° in the Centigrade system $[(0° + 100°) \div 2]$ and is 122° in the Fahrenheit system $[(32° + 212°) \div 2]$. In either system the difference between freezing temperature and this temperature is equal to the difference between this temperature and boiling temperature. (The physical meaning is that it takes as much fuel to heat a liter of water from 0° C to 50° C as to heat it from 50° C to 100° C.) Thus, temperature is said to be measured on an *interval scale:* differences but not ratios can be compared. Money and weight are said to be measured on a *ratio scale* because we can sensibly consider ratios of values. For instance, 200 pounds is twice 100 pounds; that is, when weighed, two 100-pound objects will give the same value as one 200-pound object. A ratio scale is an interval scale with a meaningful zero.

A psychologist or a sociologist who wishes to evaluate a conceptual variable such as intelligence, strength of opinion, or degree of belief, which cannot be measured directly, may devise a test or questionnaire to measure the variable indirectly, the scores on the test or questionnaire supposedly being related to the variable of real interest. This nonobservable variable may be defined on an ordinal, interval, or ratio scale; the investigator must decide on what sort of scale the measurements of the observable variable shall be defined in order to know what kinds of comparisons can be made between measurements.

Even physical variables may be measured on different scales at different times. A two-point ordinal scale of height may be made by seeing who can walk under a bar without ducking; or an ordinal scaling of height can be made by having a set of individuals line up in order of increasing height. Any variable that can be measured on an interval or ratio scale could also be measured on the less stringent ordinal scale.

A detailed and authoritative discussion of scales is given by Stevens (1951).

2.1.4 Data and Observational Units

Sets of data result from observing the hair colors of children in a nursery, the major occupational categories of employed persons in St. Louis, the durations of drug effect in a group of experimental animals, and the fatigue points for I beams of a new alloy. *Data* are the values of a variable (or of several variables) recorded for a set of observational units.

The *observational units* may be individual persons; or they may be groups of people such as families, households, employees, or populations of cities, universities, and nations. They may be objects, such as light bulbs, blood cells, or counties. Or they may be events, such as strikes, wars, or volcanic eruptions. It is often convenient to refer to these units as "individuals," even when they are not people.

In the process of observation we record for each observational unit some particular (quantitative or qualitative) characteristic of that unit; this observation constitutes one datum.

2.2 ORGANIZATION OF CATEGORICAL DATA

2.2.1 Frequencies

The data can always be arranged in a table having two columns, the first column giving an identification of the observational unit and the second giving the datum for that unit. Table 2.1 is a hypothetical list of occupations of the heads of households collected by a student who interviewed heads of households in a particular metropolitan area. The household is identified by its address, and the occupational category of the household head is recorded. The variable observed is occupation; the observational units are households.

Perhaps the simplest kind of summary for this type of data is made by providing a list showing the number of times each category occurs in the list, that is, the *frequency* of each category.

TABLE 2.1

Occupations of Heads of Households

HOUSEHOLD IDENTIFICATION (ADDRESS)	OCCUPATIONAL CATEGORY OF HEAD OF HOUSEHOLD
515 Main Street	Clerical
314 Wilkins Avenue	Professional
212 Shady Avenue	Professional
519 Shady Avenue	Professional
917 Shady Avenue, Apt. 3	Clerical
812 Denniston Street	Professional
423 Denniston Street	Professional
1024 Beechwood Boulevard	Sales
531 Northumberland Street	Laborer
1313 Aylesboro Road	Sales
1057 Aylesboro Road	Sales
1016 Marlborough Drive	Sales
914 Marlborough Drive	Sales
1123 Beeler Street	Sales
13 Solway Place	Professional
597 Maple Street	Clerical
1212 Jones Street, Apt. 5	Sales
612 Smith Street	Clerical
1549 Mary Street	Sales
415 Edna Street	Laborer

(hypothetical data)

This is done by first writing down a list of the categories. Then as we come to each household in Table 2.1 we enter a tally for the category to which the household head belongs. The result is shown in Table 2.2. In passing from the original list to the table of frequencies we ignore all information except the categories and how often they were observed. Any two households for which the heads have the same occupation are now treated as equivalent. From the table of frequencies we know there are two households whose heads are laborers; we would no longer know that one is from 531 Northumberland Street and the other is from 415 Edna Street. This abstraction is a simplification because information which is regarded as irrelevant for considering the set as a whole is eliminated.

Frequency Distributions.   A table like Table 2.2, or a graph representing the same information, is called a *frequency distribution*, because it shows how the individuals are "distributed" over the categories, that is, how many individuals are associated with each category. When we focus attention on the variable observed, occupation here, we may simply say, "distribution of *occupation*" instead of the "distribution of *individuals* in occupational categories."

TABLE 2.2
Tally Sheet Showing Frequency of Occupational Categories

Category	Tally	Number of Household Heads in Category (Frequency)
Professional	⊔⊔⊔ │	6
Sales	⊔⊔⊔ │││	8
Clerical	│││││	4
Laborer	││	2
		20

SOURCE: Table 2.1.

Percentages. The *relative frequency* of a category, or the *proportion* of times that category has occurred, is the frequency of that category divided by the total number of observations. The relative frequencies in Table 2.3 are the frequencies in Table 2.2 divided by 20, the number of households observed. The sum of the relative frequencies in any table is 1, except for a possible discrepancy caused by rounding off the figures (for example, recording 1/3 as 0.33). Often we express the relative frequencies as *percentages* by multiplying the relative frequencies by 100. Table 2.3, like Table 2.2, is called a "frequency distribution."

TABLE 2.3
Relative Frequency of Various Occupations

Category	Relative Frequency	Percentage
Professional	0.3	30
Sales	0.4	40
Clerical	0.2	20
Laborer	0.1	10
	1.0	100

SOURCE: Table 2.2.

Comparing Distributions. Table 2.4 gives the numbers of employed persons in different occupational categories for 1950 and 1970. Of interest is the question of whether the composition of the labor force has changed much over the twenty-year span. It is difficult to make the necessary comparison from Table 2.4. The percentages in Table 2.5 tell the story better. There are relatively more white-collar workers in 1970 and relatively fewer farm workers. The percentage of blue-collar workers has decreased slightly, from 39% to 35%.

TABLE 2.4

Employed Persons by Major Occupation Groups in 1950 and 1970

| | NUMBERS OF PERSONS (IN THOUSANDS) | |
	1970	1950
White-collar workers[a]	38,068	22,373
Blue-collar workers[b]	27,452	23,336
Service workers[c]	9,724	6,535
Farm workers	3,164	7,408
Total	78,408	59,652

SOURCE: *Statistical Abstract of the United States* (1970), p. 225.
[a]Professional and technical workers; managers, officials, and proprietors; clerical workers; sales workers.
[b]Craftsmen and foremen; operatives; nonfarm laborers.
[c]Private household workers; other service workers.

TABLE 2.5

Percentage of Employed Persons in Major Occupation Groups in 1950 and 1970

	1970	1950
White-collar workers	49	38
Blue-collar workers	35	39
Service workers	12	11
Farm workers	4	12
Total	100	100
(Thousands of persons)	(78,408)	(59,652)

SOURCE: Table 2.4.

2.2.2   Frequency Graphs

A bar graph can be used to represent frequency or relative frequency. For example, each percentage in Table 2.3 is represented in

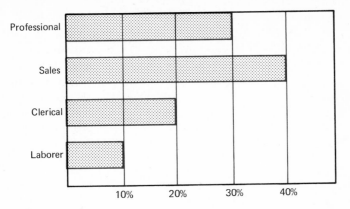

Figure 2.2 *Percentages of persons in various occupational categories (Table 2.3)*

Figure 2.2 as the length of a horizontal bar; the thicknesses of the bars are the same. Note that in Table 2.2 the tallies themselves form a bar graph. The numbers of persons in the United States belonging to various religions are shown in Figure 2.3 with heights representing frequencies. The bar graph indicates that Protestantism is most prevalent, with Catholicism next. The numbers of Eastern Orthodox and Jewish adherents are small, but not negligible.

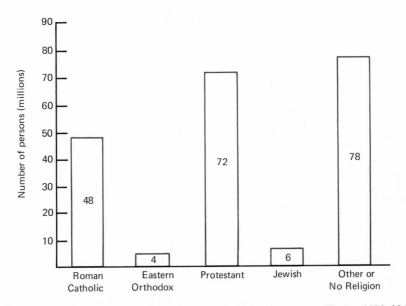

Figure 2.3 *1970 U.S. memberships in various religions, in millions. 1970 U.S. population: 208 million.* (Statistical Abstract of the United States (1972), p. 45.)

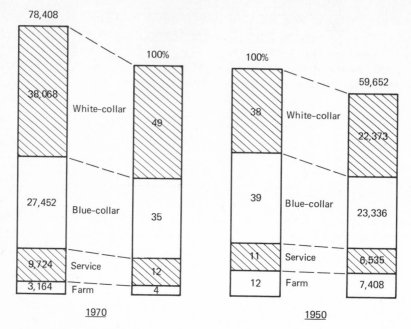

Figure 2.4 *Numbers (in thousands) and percentages of employed persons in major occupation groups, illustrating the change from numbers to percentages for comparative purposes (Tables 2.4 and 2.5)*

The purpose of such a visual presentation is to enable the reader to compare the different frequencies easily. Since occupational categories and religions form nominal scales, the orders of labels in Figures 2.2 and 2.3 are arbitrary. In the case of an ordinal scale, the categories are graphed in their intrinsic order.

Another graphical device for representing frequencies or percentages particularly appropriate for purposes of comparison is shown in Figure 2.4. The height of each segment of the left-hand bar is proportional to the frequency in that occupational category in 1970; the right-hand bar represents frequencies in 1950.

The two middle bars represent the percentages in the categories in the two years. Equalizing the base of both number columns to 100 and reducing the figures proportionately makes direct comparison easy. We can see that the blue-collar portion decreased from 39% to 35%, even though the number of blue-collar workers increased from 23,336 to 27,452.

Another context in which percentaging in this way is interesting is in business, where companies are interested not only in the change in gross sales, but also in the change in their *share of the market,* defined as a percentage of total sales. An automobile manufacturer might compare sales of Chrysler, Ford, General Motors, American Motors, and various foreign companies in successive years. He or she might want to study share-of-the-market figures in different metropolitan areas, different age groups, or different ethnic groups.

2.3 ORGANIZATION OF NUMERICAL DATA

2.3.1 Discrete Data

Frequency Distributions for Discrete Data. Quantitative data usually are recorded in a list, in a manner similar to the record for qualitative data. Suppose that for each household in Table 2.1 the number of children was recorded as in Table 2.6.

TABLE 2.6
Numbers of Children in 20 Households

HOUSEHOLD IDENTIFICATION (ADDRESS)	NUMBER OF CHILDREN
515 Main Street	2
314 Wilkins Avenue	0
212 Shady Avenue	1
519 Shady Avenue	0
917 Shady Avenue, Apt. 3	0
812 Denniston Street	1
423 Denniston Street	1
1024 Beechwood Boulevard	1
531 Northumberland Street	4
1313 Aylesboro Road	1
1057 Aylesboro Road	3
1016 Marlborough Drive	2
914 Marlborough Drive	2
1123 Beeler Street	1
13 Solway Place	1
597 Maple Street	2
1212 Jones Street, Apt. 5	0
612 Smith Street	3
1549 Mary Street	1
415 Edna Street	4

(hypothetical data)

As is the case with categorical data, the information can be summarized in a table of frequencies, ignoring all knowledge except what numerical values were obtained and how often each value was observed. Table 2.7 gives the frequencies derived from Table 2.6, as well as the relative frequencies.

TABLE 2.7

Tally Sheet Showing Frequency of Numbers of Children

NUMBER OF CHILDREN	TALLY	FREQUENCY	RELATIVE FREQUENCY	CUMULATIVE FREQUENCY	CUMULATIVE RELATIVE FREQUENCY
0	∣∣∣∣	4	0.2	4	0.2
1	⫽⫽⫽ ∣∣∣	8	0.4	12	0.6
2	∣∣∣∣	4	0.2	16	0.8
3	∣∣	2	0.1	18	0.9
4	∣∣	2	0.1	20	1.0
		20	1.0		

SOURCE: Table 2.6.

Frequencies can be displayed on a graph (Figure 2.5) by using the horizontal axis to denote the possible values of the variable and the vertical axis to indicate the relative frequency. In Figure 2.5 the dot above 1 and to the right of .40 shows that the relative frequency of families with one child is .40. Often the graph used to show frequencies is a *bar graph* (Figure 2.6); such a graph is called a *histogram*. The bars are centered at the values, in this case 0, 1, 2, 3, or 4 children per family.

Cumulative Frequency. The idea of *cumulative* frequency is relevant here because we are dealing with numerical data. The cumu-

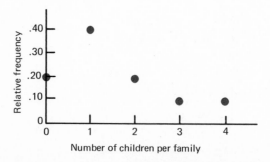

Figure 2.5 *Frequency distribution (dot graph) of number of children (Table 2.7)*

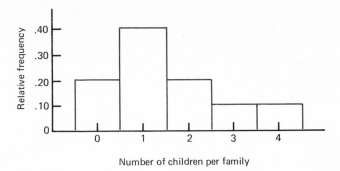

Figure 2.6 *Frequency distribution (bar graph or histogram) of number of children per family (Table 2.7)*

lative frequency at a given value is the sum of the frequencies of values less than and equal to that value; it is the number of observational units for which a value was obtained that is less than or equal to the given value. For example, Table 2.7 shows that 16 families had 2 or less than 2 children. *Cumulative relative frequency* at a value is the sum of the relative frequencies of values less than and equal to that value; this is equal to cumulative frequency divided by the total number of observational units. In the example, the cumulative relative frequency may be useful because one can read the relative frequency of small families, where "small" can be defined, for example, as families with 0, 1, or 2 children.

The cumulative relative frequency is graphed in Figure 2.7. The height above the horizontal axis at any value represents the proportion of families with that many children or less. At each integer the relevant height is the higher of the two line segments.

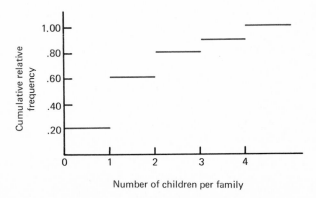

Figure 2.7 *Cumulative relative frequency of number of children (Table 2.7)*

Comparing Distributions.   Such visual presentations enable the investigator to receive an impression of some features of the data, such as their center and variability, that is, how scattered they are. They also help in comparing different sets of data; the investigator can see whether one set has generally larger values than another.

Consider again the example from the study by Asch discussed in Section 1.1.5. The frequencies and relative frequencies are given in Table 2.8. The maximum possible number of errors is 12 because there were 12 trials out of the 18 in which the "unanimous majority" matched lines incorrectly.

TABLE 2.8

Distribution of Errors Made by Last Person in Each Set of Trials

Number of Errors	FREQUENCY		RELATIVE FREQUENCY	
	Pressure Group	*No-Pressure Group*	*Pressure Group*	*No-Pressure Group*
0	13	35	.260	.946
1	4	1	.080	.027
2	5	1	.100	.027
3	6	0	.120	.000
4	3	0	.060	.000
5	4	0	.080	.000
6	1	0	.020	.000
7	2	0	.040	.000
8	5	0	.100	.000
9	3	0	.060	.000
10	3	0	.060	.000
11	1	0	.020	.000
12	0	0	.000	.000
	50	37	1.000	1.000

SOURCE: Table 1.7.

Figures 2.8 and 2.9 show the same information in graphical form. It can be seen from either graph that persons under group pressure tend to make more errors than persons not under pressure; that is, most persons in the control groups made no errors, but for the groups with persons under pressure as many as 11 errors were made.

Figure 2.9 is better for comparison purposes: using relative frequency adjusts for the fact that there were different numbers of subjects in the two groups.

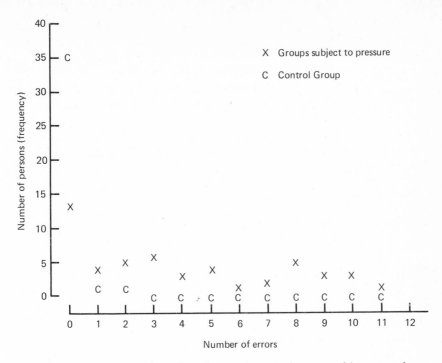

Figure 2.8 *Distributions of number of errors in control group and in group of persons subject to pressure (Table 2.8)*

2.3.2   Continuous Data

Accuracy.   In theory, continuous variables can be measured as accurately as desired. In practice, the accuracy of measurements is limited by the accuracy of measuring devices. In any particular case we may require only a certain level of accuracy. We might, for example, have a scale accurate enough to determine that a person's weight is 161.4376 pounds, but we would often be content to record it as 161 pounds. The extra accuracy may not be of real interest to our investigation since a person's weight can fluctuate several pounds during the course of the day and it is affected by the amount of clothing worn.

Rounding.   Reducing the accuracy of data measurement involves rounding. The rule for rounding to the nearest pound is to record the weight as 161 if the actual weight is more than $160\frac{1}{2}$ and less than $161\frac{1}{2}$ pounds, as 162 if the actual weight is between $161\frac{1}{2}$ and $162\frac{1}{2}$ pounds, etc. In effect, we are replacing any measurement within the interval $160\frac{1}{2}$ to $161\frac{1}{2}$ by the midpoint of the interval, that is, 161, etc.

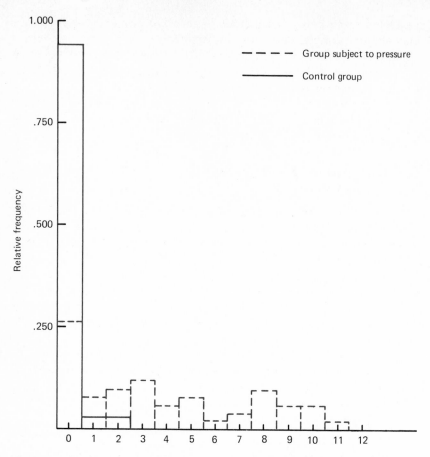

Figure 2.9 *Bar graphs for relative frequency of errors (Table 2.8)*

If we wish to round a measurement of $161\frac{1}{2}$ to the nearest pound, how do we decide whether to round it to 161 or 162? We could flip a coin, which is ambiguous and seems a foolish waste of time and effort, or we could always raise to the higher integer, which would result in an upward bias of our figures. The standard procedure is to round to the nearest *even* integer; this procedure is unambiguous and gives a result divisible by 2.

Sometimes we might wish to round to the nearest ten pounds, 160, 170, etc., or to the nearest tenth of a pound, 160.1, 160.2, etc. The same principles apply.

Rounding is not always done accurately in practice. In most countries it is the custom to give one's age, for example, as 28 years until the 29th birthday. (This is an example of rounding *down*.) Thus one's "age" is not *really* one's age, but rather the number of full years of life one has completed since being born. (This is reflected in some languages; in Spanish, for example, one does not ask "How old are you?" but rather "How many years have you completed?") If instead we rounded the years according to the rules of rounding—as is done for life insurance policies—a person having a birthday on November 25 would, during the 29th year, say he or she is 28 years old until May 25 and 29 thereafter. The ages of race horses are changed on January 1, regardless of the actual birthday.

2.3.3 Frequency Distributions
for Continuous Data

As with other statistical information, the data for continuous variables will originally be reported in a listing of individuals and the observed measurements. A list of the weights of workers in a factory is given in Table 2.9. Continuous data such as these can be displayed along a *continuum* or *axis* (Figure 2.10). The data of Table 2.9 are recorded to the nearest pound; in Figure 2.10 we mark a dot above the axis for each datum. The density of dots shows how the data cluster. Here one can see that there are especially many weights between 145 and 150 pounds.

Even when we record the weights to the nearest pound, there are still not very many individuals at any one recorded weight. In order to obtain a table of frequencies which gives a good summary of the data, we group observations into *classes* according to weights.

The smallest weight is 138 and the largest is 172; the *range* is $172 - 138 = 34$. We divide the interval containing all recorded values into seven separate *class intervals,* each 5 pounds wide, choosing the intervals so that their midpoints are convenient numbers, 140, 145, 150, etc. The choice of interval length and midpoints determines the *class limits* (the boundaries of the intervals) since they must lie halfway between the successive equally spaced midpoints. The class limits are thus 137.5, 142.5, 147.5, etc. Then, using a list of the class intervals, for each observation in Table 2.9 we enter a tally for the interval in which that observation lies. Thus, for 151 pounds, we enter a tally for the interval 147.5–152.5, for 141 pounds, we enter a tally in the interval 137.5–142.5, etc. The result is shown in Table 2.10.

Note that tabulating the data into these class intervals is equivalent to simply rounding each figure to the nearest five pounds.

The pattern of tallies provides a picture of how the data cluster. After all the tallies have been made, the tallies for each interval are counted. The number of observations in an interval is the *frequency* of that interval. The frequencies are given in the last column of Table 2.10 and are shown in the histogram of Figure 2.11. (The break in the horizontal

TABLE 2.9
Weights of 25 Employees

EMPLOYEE	WEIGHT (POUNDS)	EMPLOYEE	WEIGHT (POUNDS)
Adams	151	Martin	145
Allan	141	Newcombe	154
Brammer	166	Newsome	144
Brown	147	Peterson	162
Coles	172	Raines	157
Davidson	155	Richards	146
Jackson	153	Samson	149
Johnson	149	Smith	152
Jones	147	Taylor	142
Kellner	148	Tucker	138
Levy	150	Uhlman	161
Lewis	143	West	167
Margolis	160		

(hypothetical data)

axis of the figure indicates that the axis is not continuous from 0 to 140.) The pattern of weights is more easily discernible from Figure 2.11 than from Figure 2.10. Note that the tally column itself is a histogram when viewed from the side.

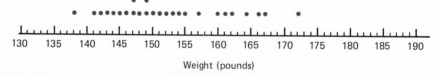

Figure 2.10 *Weights of 25 employees (Table 2.9)*

There are some guidelines to follow in forming frequency distributions. As a rule of thumb, it is generally satisfactory to use 5 to 15 class intervals. A smaller number of intervals is used if the number of measurements is relatively small; if the number of measurements is large, the number of intervals may be greater than 15. Secondly, note that the class limits all have an extra decimal digit 5. Thus, no recorded observation can be equal to a class limit. Each observation lies strictly *within* one class interval, never on the boundary between two intervals. Note what could have happened if we had used a class width of 10

TABLE 2.10
Tally Sheet and Frequency Distribution of Weights

CLASS INTERVAL	MIDPOINT	TALLY	NUMBER OF OBSERVATIONS IN INTERVAL (FREQUENCY)
137.5–142.5	140	|||	3
142.5–147.5	145	⊬⊤|	6
147.5–152.5	150	⊬⊤|	6
152.5–157.5	155	||||	4
157.5–162.5	160	|||	3
162.5–167.5	165	||	2
167.5–172.5	170			1
			25

SOURCE: Table 2.9.

pounds, with midpoints of 140, 150, 160, 170. Then the class limits would have been 135, 145, 155, 165, and 175. These numbers are possible recorded observations; so this choice of class width permits ambiguity. To avoid this pitfall, choose the midpoints to be possible recorded values, and choose the interval width to be an odd number of recorded units.

The data can also be represented by a dot graph (similar to Figure 2.5 for discrete data). Figure 2.12 represents the frequencies in Table 2.10. When the variable is continuous, sometimes the dots are connected by line segments in order to lead the eye from one point to another; then

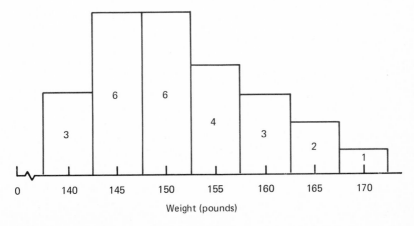

Figure 2.11 *Histogram of weights (Table 2.10)*

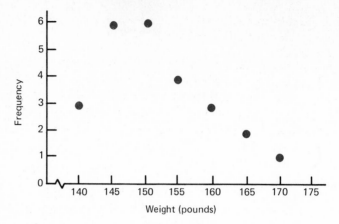

Figure 2.12 *Dot graph of frequency distribution of weights (Table 2.10)*

the graph is a *polygonal graph* or a *line graph*. A polygonal graph used to represent frequencies is a *frequency polygon.* The dots in Figure 2.12 are connected to yield Figure 2.13. The histogram (Figure 2.11) is usually preferred to the line graph (Figure 2.13); the histogram appropriately suggests that the observed measurements are spread out beyond the first and last midpoints, and a single large frequency is given its appropriate weight. In general it is desirable to have all classes the same width, including the first and last; it is better that the first interval have a left-hand endpoint and the last interval a right-hand endpoint. (The right-hand class interval should not be "167.5 and more," for example.)

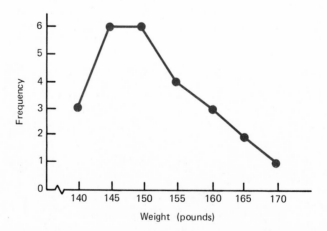

Figure 2.13 *Line graph of frequency distribution of weights (Table 2.10)*

Organization of quantitative data is discussed further in Section 5.5.

The histogram can be interpreted as representing frequency by *area*. This interpretation is important when we consider changing the width of the class intervals; we want to graph in such a way that the picture is affected as little as possible.

Table 2.11 reports the heights of 1293 11-year-old boys in one-inch intervals. These data are reduced by combining 5 intervals into one; Table 2.12 reports the heights in five-inch intervals. Both sets of frequencies are graphed in Figure 2.14 with area representing frequency. For example, the height of the bar for the interval 51.5–52.5 is proportional to the frequency 192, and the area is the width of the interval, namely 1, times the height, namely 192. The height of the dashed bar over the interval 47.5–52.5 is 109.8 and the area is $5 \times 109.8 = 549$, which is the area of the five bars over 47.5–48.5, 48.5–49.5, 49.5–50.5, 50.5–51.5, and 51.5–52.5.

TABLE 2.11
Heights of 11-Year-Old Boys

CLASS INTERVAL (INCHES)	MIDPOINT	FREQUENCY	RELATIVE FREQUENCY
44.5–45.5	45	2	.0015
45.5–46.5	46	8	.0062
46.5–47.5	47	6	.0046
47.5–48.5	48	25	.0193
48.5–49.5	49	50	.0387
49.5–50.5	50	119	.0920
50.5–51.5	51	163	.1261
51.5–52.5	52	192	.1485
52.5–53.5	53	204	.1578
53.5–54.5	54	187	.1446
54.5–55.5	55	156	.1206
55.5–56.5	56	96	.0742
56.5–57.5	57	41	.0317
57.5–58.5	58	23	.0178
58.5–59.5	59	12	.0092
59.5–60.5	60	5	.0039
60.5–61.5	61	1	.0008
61.5–62.5	62	1	.0008
62.5–63.5	63	2	.0015
		1293	.9998

SOURCE: Bowditch (1877).

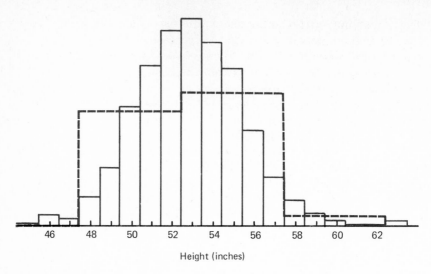

Height (inches)

Figure 2.14 *Histograms of heights of 11-year-old boys, with class-interval widths of 5 inches (dashed) and 1 inch (solid) (Tables 2.11 and 2.12)*

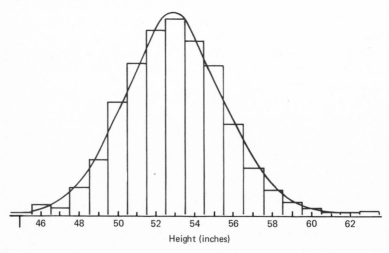

Height (inches)

Figure 2.15 *Histogram of heights of 11-year-old boys, with corresponding continuous frequency distribution (Table 2.11)*

Note that the smaller intervals produce a histogram that is fairly smooth. If the intervals were smaller than one inch, the histogram would be still smoother. Figure 2.15 has a smooth curve superimposed on the histogram. This is the shape we would expect if the interval were very small. It represents the frequency distribution of the continuous variable "height." In pictorial representation we shall often use smooth curves instead of histograms.

TABLE 2.12
Heights of 11-Year-Old Boys

CLASS INTERVAL (INCHES)	MIDPOINT	FREQUENCY	RELATIVE FREQUENCY
42.5–47.5	45	16	.0124
47.5–52.5	50	549	.4246
52.5–57.5	55	684	.5290
57.5–62.5	60	42	.0325
62.5–67.5	65	2	.0015
		1293	1.0000

SOURCE: Table 2.11.

SUMMARY

Statistical *data* are the recorded results of observation. Data arise from the observation of one or more characteristics of various *observational units*. These units, or *individuals*, may be persons, or groups of people, such as families or cities, or objects, such as lightbulbs being tested. The characteristics observed are called *variables*.

Variables are classified as being *categorical* or *numerical*.

The categories or classes of a categorical variable may be natural, as in the case of sex (male or female), or may be defined on other bases, such as occupational classes. A categorical variable with only two categories is called *dichotomous*.

Numerical variables are either *discrete* or *continuous*.

Variables may be characterized according to the *scale* on which they are defined; this typology involves the structure and meaning of the variables. Categorical variables are defined on a *nominal* or an *ordinal* scale. Numerical variables are defined on an *interval* or *ratio* scale. A variable that can be measured on a given scale could also be measured on any less stringent scale.

Categorical data are summarized in *frequency* or *relative frequency* tables and displayed in *bar graphs.* Discrete numerical data may be similarly summarized and displayed. Bar graphs of frequencies are called *histograms. Cumulative frequencies* and *cumulative relative frequencies* are the sums of frequencies and relative frequencies, respectively.

Continuous data may be graphed on a single axis. They can be grouped into class intervals; the frequencies are then tabulated and displayed as discrete numerical data.

APPENDIX 2A   Frequency as Area in a Histogram

Class intervals of equal width should be used when constructing a histogram. However, it is instructive to consider what happens when the intervals are not of equal width. It will then be seen that it is appropriate to *use area to represent frequency.*

Suppose in Table 2.10 we combine the two class intervals 137.5–142.5 and 142.5–147.5 into a single interval 137.5–147.5. This combined interval has a width of 10 instead of 5. The new frequency distribution is given in Table 2.13. The combined interval has a frequency of $3 + 6 = 9$ persons. These 9 persons will be represented by a bar ex-

TABLE 2.13

Revised Frequency Distribution of Weights

Class Interval	Width	Frequency
137.5–147.5	10	9
147.5–152.5	5	6
152.5–157.5	5	4
157.5–162.5	5	3
162.5–167.5	5	2
167.5–172.5	5	1
		25

source: Table 2.9.

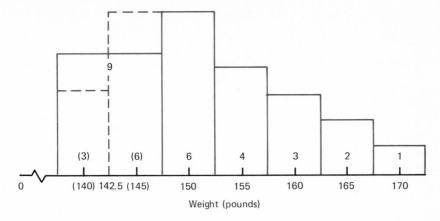

Figure 2.16 *Histogram of weights, with the first two class intervals of Figure 2.11 combined into a single class interval (Table 2.13)*

tending from 137.5 to 147.5. If the height of this bar is 9, it would be taller than all the other bars in Figure 2.11. Clearly this is inappropriate because the visual impression would be that the greatest *concentration* of weights is in that interval, which is not the case. Instead, we shall combine the two bars for 140 and 145 in Figure 2.11 in such a way that the total area remains the same. This means that the new bar, which is 10 units wide, has to be 4.5 units high. (See Figure 2.16.) Our reasoning is as follows. In Figure 2.11, the height corresponding to 137.5–142.5 is 3, the width is 5, and the area is therefore 15. Similarly, the area of the bar corresponding to 142.5–147.5 is $6 \times 5 = 30$. The total area of these two bars is $15 + 30 = 45$. Since, in Figure 2.16, this total area is spread out over a width of 10 units, the corresponding height must be 4.5, since $45 = 10 \times 4.5$.

EXERCISES

2.1 (Sec. 2.1) Indicate for each of the following variables on which scale (nominal, ordinal, interval, or ratio) it is usually measured.
  (a) strength of opinion
  (b) area
  (c) color
  (d) shade of red
  (e) Fahrenheit temperature

2.2 (Sec. 2.1) Indicate which of the following data are discrete and which are continuous.
  (a) student populations on the University of California campuses

(b) Dow-Jones weekly averages over the past year
(c) gross national products (GNP) of 15 countries
(d) murders in 1969 in each of 20 large U.S. cities
(e) murders in San Francisco for each of the last 20 years
(f) heights of 100 cows on a dairy farm
(g) lifetimes of lightbulbs

2.3 (Sec. 2.2)   Classify the members of your class according to their major subject in college. Find the distribution of majors by sex. Compare the two distributions. (For example, is there a higher proportion of political science majors among the men or among the women?)

2.4 (Sec. 2.3)   Show that in the Asch study (Table 2.8) 68% of the judgments of subjects in the experimental group were correct (despite the pressure of the majority). Consider only the 12 trials on which group pressure was exerted.

2.5 (Sec. 2.3)   List appropriate midpoints and intervals for a tally sheet using five intervals for the data in Table 2.9.

2.6 (Sec. 2.3)   Make a histogram of the heights of the males in your class. Do the same for the females. What is the proportion of males having heights greater than 5'6"? Of the females? What is the proportion of females having heights less than 5'4"? Of the males?

2.7 (continuation)   Using the data obtained for Exercise 2.6, make a histogram of the heights of males and females together. Compare it with the separate histograms for males and females.

2.8 (Sec. 2.3)   Consider the distribution in Table 2.14.

TABLE 2.14
Heights of 384 5-Year-Old Boys

HEIGHT (INCHES)	MIDPOINT OF CLASS INTERVAL	FREQUENCY
29.5–32.5	31	2
32.5–35.5	34	25
35.5–38.5	37	99
38.5–41.5	40	190
41.5–44.5	43	65
44.5–47.5	46	3
		384

SOURCE: *Isserlis* (1915), p. 62.

(a) Draw a histogram representing this distribution.

(b) Compute the relative frequencies, that is, the proportion of the sample in each interval of height.

2.9 (Sec. 2.3)   The weights and heights of the members of the 1970 Stanford football team are given in Table 2.15.*

(a) Mark the weights on a (single) horizontal axis.

(b) Tabulate the weights in a set of classes of weight. What considerations led you to your choice of class intervals?

(c) Graph the frequencies in a histogram.

(d) Do you think there is a tendency to report weight in round numbers ending in 5 or 0? Support your opinion.

TABLE 2.15
Weights and Heights of Stanford Football Players: 1970

Name	Weight (Pounds)	Height (Inches)	Name	Weight (Pounds)	Height (Inches)
Adams	225	74	Kehl	180	72
Alexander	216	75	Klippert	230	75
Alvarado	187	72	Kloos	178	72
Barnes	190	73	Koehn	183	73
Bleymaier	185	71	Lamanuzzi	185	71
Boryla	190	75	Lasater	195	74
Brown, J.	200	74	Lazetich	236	75
Brown, R.	200	72	Lightfoot	240	73
Butler	232	74	McCloud	193	71
Cowan	230	77	McClure	216	74
Cross	175	69	Margala	215	74
Ewing	180	70	Merrill	175	68
Fair	225	71	Meyers	236	76
Freitas	193	73	Moore, D.	205	76
Graves	215	72	Moore, P.	195	72
Grossi	213	71	Moore, R.	221	75
Hall	199	73	Murray	185	73
Hoftiezer	196	72	Perreault	216	75
Horowitz	218	74	Peterson	210	75
Jones	231	73	Plunkett	204	75
Jubb	260	77	Robnett	185	71
Kadzeil	220	76	Saibel	195	73
Kauffman	168	71	Sampson	235	77

*Several comparatively extensive data sets are given in these exercises; they will be used later in the book.

Table 2.15 (cont.)

Name	Weight (Pounds)	Height (Inches)	Name	Weight (Pounds)	Height (Inches)
Sanderson	195	70	Simone	195	72
Sande	226	74	Smiley	215	74
Satre	206	71	Smith	215	74
Schallich	247	77	Sones	206	73
Schultz	180	72	Tipton	230	78
Scott	200	75	Vataha	175	70
Sheehan	220	74	Waters	212	75
Shockley	220	73	Washington	181	71
Siemon	220	74	Zeisler	223	73

SOURCE: Official program.

2.10 (Sec. 2.3) From a study of purchasing intentions and actions of members of a consumers' organization Table 2.16 gives the annual family income reported by each member in a sample of members who are married, age 35–39, with one working adult in the family. The annual income is listed in $100's. (For example, 085 means $8500.)

TABLE 2.16
Incomes of 49 Families

Respondent Number	Income ($100's)	Respondent Number	Income ($100's)
31031	085	31807	065
31059	135	31867	147
31069	200	31927	084
31159	050	31933	120
31209	070	31949	092
31287	076	31951	090
31367	105	31957	100
31469	065	32247	074
31487	121	32249	070
31515	105	32317	110
31595	130	32365	100
31633	065	32475	075
31661	120	32545	069
31705	100	32561	067
31765	127	32685	060
31785	150	32273	120
31801	105	32829	075

Table 2.16 (cont.)

Respondent Number	Income ($100's)	Respondent Number	Income ($100's)
32963	083	36232	096
36000	092	36330	100
36118	060	36412	066
36120	105	36514	072
36140	088	36596	060
36164	055	36796	095
38198	085	36980	095
36230	110		

SOURCE: Anonymous.

(a) Tabulate the frequency distribution according to class intervals 37.5–62.5, 62.5–87.5, etc.

(b) Graph the corresponding histogram.

*2.11 (App. 2A)

(a) Suppose the three class intervals 137.5–162.5, 162.5–187.5, and 187.5–212.5 are combined into one interval 137.5–212.5. What is the number of families with incomes in that interval?

(b) Graph the histogram of the frequency distribution when the combined interval of (a) is used.

(c) Compare the histogram of Exercise 2.10(b) with that of (b) in this exercise. What are the advantages of using the three separate intervals, and what are the advantages of using the one combined interval?

2.12 (Sec. 2.3) Prepare a frequency table and construct a histogram for the data of Table 2.17.

TABLE 2.17
Weights of 58 Swine (Pounds)

73	139	143	195
72	126	256	126
89	156	117	82
166	106	97	160
103	214	139	151

Table 2.17 (cont.)

115	152	211	160
73	145	126	298
100	251	224	119
172	198	141	184
296	109	118	227
131	102	144	186
258	132	273	102
224	78	130	64
113	97	144	
115	78	179	

(hypothetical data)

2.13 (Sec. 2.3)   Prepare a frequency table and construct a histogram for the data of Table 2.18.

TABLE 2.18
State Gasoline Taxes

Alabama	7	Louisiana	8	Ohio	7
Alaska	8	Maine	8	Oklahoma	6.5
Arizona	7	Maryland	7	Oregon	7
Arkansas	7.5	Massachusetts	6.5	Pennsylvania	7
California	7	Michigan	7	Rhode Island	8
Colorado	7	Minnesota	7	South Carolina	7
Connecticut	8	Mississippi	7	South Dakota	7
Delaware	7	Missouri	5	Tennessee	7
Florida	7	Montana	7	Texas	5
Georgia	6.5	Nebraska	7.5	Utah	7
Hawaii	5	Nevada	6	Vermont	8
Idaho	7	New Hampshire	7	Virginia	7
Illinois	7.5	New Jersey	7	Washington	9
Indiana	8	New Mexico	7	West Virginia	7
Iowa	7	New York	7	Wisconsin	7
Kansas	7	North Carolina	9	Wyoming	7
Kentucky	7	North Dakota	7		

SOURCE: *Statistical Abstract of the United States* (1970), p. 542.
NOTE: Tax rates, in cents per gallon, in effect Dec. 31, 1969.

2.14 (Sec. 2.3) The total areas of the 50 states are given in Table 2.19.
(a) Mark the total areas on a horizontal axis.
(b) Tabulate the total areas in a set of classes. What considerations led to your choice of class intervals?
(c) Graph the frequencies in a histogram.

TABLE 2.19
Land Areas, by State

State	Total Area (Sq Mi)	Farm Area[a] (1000 Acres)	State	Total Area (Sq Mi)	Farm Area[a] (1000 Acres)
New England			*East South Central*		
Maine	33,215	2,590	Kentucky	40,395	16,265
New Hampshire	9,304	903	Tennessee	42,244	15,266
Vermont	9,609	2,524	Alabama	51,609	15,226
Massachusetts	8,257	902	Mississippi	47,716	17,752
Rhode Island	1,214	104			
Connecticut	5,009	721	*West South Central*		
			Arkansas	53,104	16,565
Middle Atlantic			Louisiana	48,523	10,411
New York	49,576	12,275	Oklahoma	69,919	36,077
New Jersey	7,836	1,156	Texas	267,339	141,706
Pennsylvania	45,333	10,804			
			Mountain		
East North Central			Montana	147,138	65,834
Ohio	41,222	17,619	Idaho	83,557	15,302
Indiana	36,291	17,933	Wyoming	97,914	37,053
Michigan	58,216	29,958	Colorado	104,247	38,259
Illinois	56,400	13,599	New Mexico	121,666	47,647
Wisconsin	56,154	20,378	Arizona	113,909	40,559
			Utah	84,916	12,867
West North Central			Nevada	110,540	10,483
Minnesota	84,068	30,805			
Iowa	56,290	33,758	*Pacific*		
Missouri	69,686	32,692	Washington	68,192	19,053
North Dakota	70,665	42,717	Oregon	96,981	20,509
South Dakota	77,047	45,567	California	158,693	37,011
Nebraska	77,227	47,793	Alaska	586,400	1,959
Kansas	82,264	50,271	Hawaii	6,424	2,354

Table 2.19 (cont.)

STATE	TOTAL AREA (SQ MI)	FARM AREA[a] (1000 ACRES)
South Atlantic		
Delaware	2,057	717
Maryland	10,577	3,181
Virginia	40,815	12,002
West Virginia	24,181	5,279
North Carolina	52,712	14,382
South Carolina	31,055	8,101
Georgia	58,876	17,887
Florida	58,560	15,411

SOURCE: *Statistical Abstract of the United States* (1970), p. 166 and p. 585.   [a]1964.

2.15 (Sec. 2.3)   The farm areas of the 50 states are given in Table 2.19.
(a) Mark the farm areas on a horizontal axis.
(b) Tabulate the farm areas in a set of classes. What considerations led to your choice of class intervals?
(c) Graph the frequencies in a histogram.

2.16 (Sec. 2.3)   Refer to Table 2.20.
(a) How many families had incomes less than $2,000?
(b) What percent of all families had incomes less than $2,000?
(c) How many families had incomes of at least $6,000?
(d) What percent of all families had incomes of at least $6,000?
(e) How many families had incomes between $3,000 and $5,999?
(f) What percent of all families had incomes between $3,000 and $5,999?

TABLE 2.20
Family Income in the United States: 1959

FAMILY INCOME	TOTAL NUMBER OF FAMILIES	PERCENT OF FAMILIES
Under $1,000	2,512,668	5.6
$1,000 to $1,999	3,373,813	7.5
$2,000 to $2,999	3,763,758	8.3
$3,000 to $3,999	4,282,945	9.5
$4,000 to $4,999	4,957,534	11.0
$5,000 to $5,999	5,563,516	12.3

Table 2.20 (cont.)

Family Income	Total Number of Families	Percent of Families
$6,000 to $6,999	4,826,563	10.7
$7,000 to $9,999	9,053,220	20.1
$10,000 and over	6,794,380	15.1
Total	45,128,397	100.1

SOURCE: *U.S. Census of Population: 1960.* Vol. 1, *Characteristics of the Population.*

2.17 (Sec. 2.3) Refer to Table 2.21.
(a) How many firms had receipts less than $250,000?
(b) What percent of all firms had receipts less than $250,000?
(c) How many firms had receipts of at least $500,000?
(d) What percent of all firms had receipts of at least $500,000?
(e) How many firms had receipts between $250,000 and $1,000,000?
(f) What percent of all firms had receipts between $250,000 and $1,000,000?

TABLE 2.21
Receipts of Architectural, Engineering, and Land Surveying Firms: 1967

Receipts	Number of Firms
Less than $100,000	2,270
$100,000–$249,999	2,816
$250,000–$499,999	1,208
$500,000–$999,999	687
$1,000,000 or more	437
Total	7,418

SOURCE: *Statistical Abstract of the United States* (1970), p. 749.

2.18 (Sec. 2.3) The prices of 10 long-playing records are $3.99, 5.99, 4.49, 3.49, 3.79, 4.99, 4.79, 4.59, 4.69, and 5.49.
(a) Plot these prices on an axis.
(b) Tabulate these prices in a frequency distribution.

2.19 (Sec. 2.3) Observations of several variables for a second-grade class of school children are given in Table 2.22.
(a) Mark the verbal IQ's on a (single) horizontal axis.

(b) Tabulate the verbal IQ's into a set of class intervals.

(c) Graph the frequencies in a histogram.

TABLE 2.22

IQ and Reading Achievement Scores of 23 Pupils

PUPIL	IQ Verbal	Math.	READING ACHV. Initial	Final	PUPIL	IQ Verbal	Math.	READING ACHV. Initial	Final
1	86	94	1.1	1.7	13	82	91	1.2	1.8
2	104	103	1.5	1.7	14	80	93	1.0	1.7
3	86	92	1.5	1.9	15	109	124	1.8	2.5
4	105	100	2.0	2.0	16	111	119	1.4	3.0
5	118	115	1.9	3.5	17	89	94	1.6	1.8
6	96	102	1.4	2.4	18	99	117	1.6	2.6
7	90	87	1.5	1.8	19	94	93	1.4	1.4
8	95	100	1.4	2.0	20	99	110	1.4	2.0
9	105	96	1.7	1.7	21	95	97	1.5	1.3
10	84	80	1.6	1.7	22	102	104	1.7	3.1
11	94	87	1.6	1.7	23	102	93	1.6	1.9
12	119	116	1.7	3.1					

SOURCE: Records of a second-grade class.

2.20 (Sec. 2.3)   Data on students in a graduate course on statistical methods are shown in Table 2.23.

TABLE 2.23

Data for 24 Students in a Statistics Course

STUDENT	MAJOR[a]	LOGIC[b]	SETS[c]	SEM. STAT.[d]	SEM. MATH. ANALYSIS[e]	FINAL GRADE[f]
1	EE	Yes	No	0	2	284
2	CE	Yes	Yes	0	3	292
3	S	Yes	Yes	1	6	237
4	P	Yes	Yes	0	5	260
5	PA	No	Yes	1	5	210
6	CE	Yes	No	1	3	238
7	PA	Yes	Yes	0	3	200
8	S	Yes	Yes	3	6	300

Table 2.23 (cont.)

Student	Major[a]	Logic[b]	Sets[c]	Sem. Stat.[d]	Sem. Math. Analysis[e]	Final Grade[f]
9	A	No	Yes	0	1	174
10	PA	Yes	Yes	1	4	266
11	PA	Yes	Yes	1	2	244
12	PA	Yes	Yes	0	5	207
13	PA	Yes	Yes	2	7	261
14	PA	Yes	Yes	2	2	293
15	PA	No	Yes	1	2	213
16	EE	Yes	Yes	1	3	262
17	CE	No	Yes	0	4	272
18	PA	Yes	Yes	0	4	196
19	PA	Yes	Yes	1	1	210
20	IA	Yes	Yes	2	5	267
21	A	Yes	No	1	1	226
22	PA	Yes	Yes	3	0	300
23	CE	No	No	0	4	288
24	M	Yes	Yes	0	2	274

SOURCE: Records of a class.

[a] *Major.* A: Architecture; CE: Chemical Engineering; EE: Electrical Engineering; IA: Industrial Administration; M: Materials Science; P: Psychology; PA: Public Administration; S: Statistics.

[b] *Logic.* "Are you familiar with the notions of elementary logic: logical 'and,' logical 'or,' 'implication,' and 'truth tables'?"

[c] *Sets.* "Are you familiar with the notions of 'union' and 'intersection' of sets?"

[d] *Sem. Stat.* "How many semesters of probability or statistics have you taken?"

[e] *Sem. Math. Analysis.* "How many semesters of analysis (calculus, advanced calculus, intermediate analysis, etc.) have you taken?"

[f] *Final Grade.* Sum of scores on three exams.

Plot the final grade on two parallel, horizontal axes, one for those who had taken no course in probability or statistics before, the other for those who had taken at least one course in probability or statistics before. Do you think a previous course in statistics or probability is an important factor in explaining the final grade?

2.21 The data in Table 2.24 constitute a "time series." The numbers 15.00, 16.54, . . . , 12.58 are arranged in order according to time. The "observational units" in a time series are points in time; in this case, decades. Note that this series increases to a maximum of 24.94 in 1940 and then decreases, there being only about half as many milk cows in

1970 as there were in 1940. Plot the data on graph paper, using the horizontal axis for years and the vertical axis for the number of cows.

TABLE 2.24
Number of Milk Cows in the United States (in millions)

1890	15.00
1900	16.54
1910	19.45
1920	21.46
1930	23.03
1940	24.94
1950	23.85
1960	19.53
1970	12.58

SOURCES: *Historical Statistics of the United States, Colonial Times to 1957,* pp. 292–293, and *Statistical Abstract of the United States,* 1972, p. 620.

2.22 Which of the data sets in Exercise 2.2 would be time series?

2.23 Plot the time series of Table 2.25 on graph paper.

TABLE 2.25
Number of Banks in the United States

1920	30,909
1925	29,052
1930	24,273
1935	16,047
1940	15,076
1945	14,660
1950	14,693
1955	14,285
1960	13,999
1965	14,324
1970	14,199

SOURCES: *Historical Statistics of the United States, Colonial Times to 1957,* pp. 623–624, and *Statistical Abstract of the United States,* 1972, p. 444.

2.24 Plot the data of Table 2.26 for Massachusetts and California on the same graph, plotting 1.783 and 0.865 above 1880, 2.239 and 1.213 above 1890, etc. Use one symbol for California and another for Massachusetts. (Connect the symbols, making a line graph for each state.)

TABLE 2.26
Populations of California and Massachusetts (millions)

YEAR	1880	1890	1900	1910	1920	1930	1940	1950	1960	1970
Mass.	1.783	2.239	2.805	3.366	3.852	4.250	4.317	4.691	5.149	5.689
Calif.	0.865	1.213	1.485	2.378	3.427	5.677	6.907	10.586	15.717	19.953

SOURCE: *Historical Statistics of the United States, Colonial Times to 1957*, p. 12, and *Statistical Abstract of the United States*, 1972, p. 12.

MATHEMATICAL EXERCISES

2.25 Using Table 2.27 write in symbols:
(a) The increase in number of employed persons.
(b) The percentage increase in number of employed persons.
(c) The increase in number of blue-collar workers.
(d) The percentage increase in number of blue-collar workers.
(e) The proportion of workers in 1970 who are blue-collar workers—that is, the blue-collar portion of the labor force in 1970.
(f) The proportion of workers in 1950 who are blue-collar workers.

TABLE 2.27
Employed Persons in 1950 and 1970

	1970	1950
White-collar	a'	a
Blue-collar	b'	b
Service	c'	c
Farm	d'	d
Total	n'	n

2.26 Show that the blue-collar portion of the population in Table 2.27 increases if and only if the percentage increase in number of blue-collar workers exceeds the percentage increase in the total number of workers.

REFERENCES

Anderson, T. W. (1971). *The Statistical Analysis of Time Series.* Wiley, New York.

Bowditch, H. P. (1877). "The growth of children." *Report of the Board of Health of Massachusetts,* VIII.

Isserlis, L. (1915). "On the partial correlation ratio. Part II: Numerical." *Biometrika* 11:50–66.

Stevens, S. S. (1951). "Mathematics, measurement, and psychophysics." Ch. 1 of *Handbook of Experimental Psychology,* S. S. Stevens (ed.), Wiley, New York.

U.S. Bureau of the Census (1961). *Historical Statistics of the United States, Colonial Times to 1957.* U.S. Department of Commerce, Washington.

U.S. Bureau of the Census (1964). *U.S. Census of Population: 1960,* Vol. 1, *Characteristics of the Population.* U.S. Department of Commerce, Washington.

U.S. Bureau of the Census (1970). *Statistical Abstract of the United States.* U.S. Department of Commerce, Washington.

U.S. Bureau of the Census (1972). *Statistical Abstract of the United States.* U.S. Department of Commerce, Washington.

3 MEASURES
OF LOCATION

INTRODUCTION

After a set of data has been collected, it must be organized and condensed or categorized for purposes of analysis. One summarizes the data in a *frequency table*. We can obtain a picture of quantitative data by graphing along an axis which represents the values of a variable. A dot along this axis shows the position, or *location*, of each datum. If we were to plot (mark) the location of each store along a "map" of Main Street, we could see on the map that the *middle* of the shopping district was located at some specific address. This address, so to speak, is analogous to the measure of *location* or the *central tendency* of a set of measurements. That is, we are speaking of a single number which specifies the center of the set. In this chapter we shall consider several measures of location and show how each of them tells where the data tend to be located.

3.1 THE MODE

3.1.1 Definition and Interpretation

The *mode* of a categorical or a discrete numerical variable is that value of the variable which occurs with the greatest frequency. The mode or *modal category* in Table 2.2 is "sales." The mode does not necessarily describe "most" (that is, more than 50%) of the individuals. Note, for example, only 40% of the household heads (Table 2.3) fall into the modal category. The modal occupational category in 1970 (Table 2.4) is "white-collar worker." The modal religious group in the United States (Figure 2.3) is Protestant.

Table 3.1 reports the numbers of children in a sample of 100 families. A family of 2 children occurs most frequently—46 times. The mode, therefore, is 2 children. We emphasize that the mode is a value of the variable, and the frequency of that value suggests its statistical importance.

In Table 3.1 families with 1 and 3 children are the next most frequent; there tends to be a grouping of the data around the modal value of 2. The distribution is like a mountain—there is a peak (maximum frequency), and the elevation (frequency) near the peak is also great. If the variable were the number of shoe stores per block on a shopping street, a shopper would probably begin shopping for a pair of shoes on the modal block, that is, the block with the most stores. The shopper might expect nearby blocks also to have many shoe stores, but not as many as the modal block.

The mode is usually a good indicator of the center of the data only if there is just one dominating frequency. For example, for another 100-family sample, shown in Table 3.2, the frequency increases as the number of children increases to 2, then decreases, then increases again for 4 children, and finally decreases. When the pattern of frequencies rises and falls twice like this, we say that the distribution is *bimodal* because there are two values (here, 2 and 4) which are more frequent than neighboring values. A bimodal distribution can arise from a mixture of two unimodal distributions. Thus, if the families of Table 3.2 could be identified as urban or rural the distributions might resemble those of Table 3.3. We can see that urban families are typified by a mode of 2 children per family, and rural families by a mode of 4 children per family. But neither 2 nor 4 is very useful for describing the entire set of 100 families. Figure 3.1 shows the histogram for all 100 families.

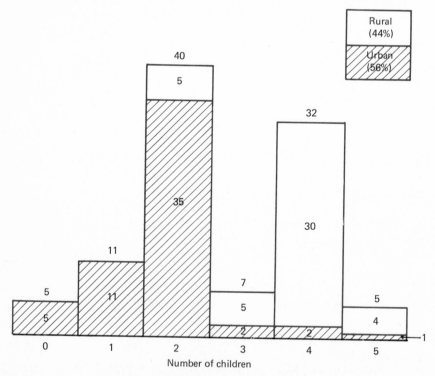

Figure 3.1 *Distribution of children among 100 families (Table 3.3)*

TABLE 3.1
Number of Children in 100 Families: I

Number	Frequency
0	10
1	23
2	46
3	11
4	5
5	5
	100

(hypothetical data)

TABLE 3.2
Number of Children in 100 Families: II

Number	Frequency
0	5
1	11
2	40
3	7
4	32
5	5
	100

(hypothetical data)

TABLE 3.3
Frequency Distributions of Number of Children in 100 Urban and Rural Families: II

Number of Children	Frequency		
	Urban	*Rural*	*Total*
0	5	0	5
1	11	0	11
2	35	5	40
3	2	5	7
4	2	30	32
5	1	4	5
	56	44	100

(hypothetical data)

3.1.2   Mode of Grouped Data

When we measure a continuous variable such as height or weight, all the measurements may be different. In such a case there is no mode because every observed value has frequency 1. However, the data can be grouped into class intervals and the mode can then be defined in terms of class frequencies. With grouped quantitative data, the *modal class* is the class interval with the highest frequency.

Data on income may be summarized by frequencies of class intervals as in Table 3.4. The *modal class* in Table 3.4 is $12,000–15,999, since more incomes are in this interval than in any other interval. The midpoint of the modal class interval is designated as the mode. (See Exercise 3.37 for another way of specifying the mode of grouped data.) In this example, then, the mode is $14,000.

When continuous data have been grouped, the value of the mode depends to some extent upon the grouping used. Suppose, for example, that in summarizing the income data we had used intervals such as $11,000–14,999 and $15,000–18,999 with midpoints $13,000 and $17,000. If there were many families with incomes of exactly $15,000, the modal class would be that with midpoint $17,000, which is quite different from the mode of $14,000 obtained when the data are grouped as in Table 3.4.

It was suggested in Chapter 2 that class intervals should be chosen so that the midpoints are convenient numbers and there are 5 to 15 classes. In order that the mode be reasonably defined, the class interval should be wide enough to contain many cases, but narrow enough to display

TABLE 3.4
Income of 64 Families

Income*	Midpoint	Frequency
$ 0–3,999	$ 2,000	4
4,000–7,999	6,000	6
8,000–11,999	10,000	15
12,000–15,999	14,000	28
16,000–19,999	18,000	7
20,000–23,999	22,000	2
24,000–27,999	26,000	2
		64

(hypothetical data)
*The interval $4,000–7,999 is meant to include all incomes of at least $4,000 and less than $8,000; for instance $7,999.81 is included. Sometimes, however, incomes are only reported to the nearest dollar. The U.S. Census reports data this way. (See Table 2.20.)

several different frequencies. A frequency table with a number of small frequencies and one large frequency would not ordinarily be informative. The intervals should be of equal length.

The mode has more meaning in the case of ordinal or numerical variables than it does with nominal variables because the existence of an ordering makes possible a grouping around the mode. Moreover, the definition of categories is often arbitrary. In Table 2.1, for example, "sales" could be divided into "salespeople" and "marketing."

Since the notion of *location* requires order it is not meaningful for nominal data. Consequently, the measures of central tendency discussed below are appropriate only for measurements made on ordinal, interval, or ratio scales. The median, to be discussed next, can be used with ordinal data; the mean requires an interval scale.

3.2 THE MEDIAN AND OTHER PERCENTILES

3.2.1 The Median

Median of Continuous Data. The median ("divider") of an interstate highway is a strip that divides the road in half. The median of a set of observations is a value that divides the set of observations in half, so that the observations in one half are less than or equal to the median value and the observations in the other half are greater than or equal to the median value.

In finding the median of a set of data it is often convenient to put the observations in order, so that the smallest is first, the next smallest is second, and so on. A set of observations arranged in order is called an *ordered sample*. If the number of observations is odd, the median is the middle observation. For example, if the weights at six weeks of five pigs are $59\frac{1}{2}$, 54, 55, 57, and $52\frac{1}{2}$ pounds, the ordered sample is $52\frac{1}{2}$, 54, 55, 57, $59\frac{1}{2}$; and the median is 55 pounds, since two of the measurements are larger than 55 and two are smaller than 55. Figure 3.2 shows how 55 divides the set of observations into halves; the median is the middle observation.

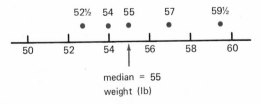

Figure 3.2 *The median of a set of observations*

If there were four measurements instead of five, say $52\frac{1}{2}$, 54, 55, and $56\frac{1}{2}$, there would not be a middle observation. Here, any number between 54 and 55 would serve as a median; but it is desirable to use a specific number for the median and we usually take this number to be halfway between the two middle measurements, for example $(54 + 55)/2 = 54\frac{1}{2}$.

The median is a "central" value—there are as many values greater than it as there are less than it. There are as many pigs heavier than the median weight as there are pigs lighter.

Median of Discrete Data.   In the case of a continuous variable one does not expect any two measurements to be equal, but in the case of a discrete variable, such as number of children in a family, one expects to obtain the same value several times. Even if the observations contain repeated values, they can still be put in order. If, for example, the numbers of children in five families are 2, 1, 5, 2, 2, the ordered sample is 1, 2, 2, 2, 5, and the median number of children is the middle observation, which is 2. If there were four observations, say 1, 2, 3, 5 children, 1 and 2 are the lower half and 3 and 5 the upper half of the set; any number between 2 and 3 divides the halves. To eliminate ambiguity, we take $2\frac{1}{2}$ as the median, even though $2\frac{1}{2}$ is not a possible value of the variable.

3.2.2   Quartiles

The median divides the set of observations into *halves*. Quartiles divide the set into *quarters*.

Quartiles of Continuous Data.   Suppose the weights of eight six-week-old pigs are arranged in order, $50\frac{1}{2}$, 51, 52, 54, 55, 56, 57, and 59 pounds. The median is $54\frac{1}{2}$; half (four) of the eight observations are less than $54\frac{1}{2}$, and half are greater than $54\frac{1}{2}$.

The quartiles are indicated on Figure 3.3. The value $51\frac{1}{2}$ inches is the *first quartile* of the pig weights: one-fourth (two) of the eight observations are less than $51\frac{1}{2}$ and three-fourths (six) of the eight observations are more than $51\frac{1}{2}$. The value $56\frac{1}{2}$ is the *third quartile* because three-fourths (six) of the observations are less than $56\frac{1}{2}$ and one-fourth (two) of the observations are greater than $56\frac{1}{2}$. The *second quartile* is the median, which is $54\frac{1}{2}$ here.

If the number of observations is a multiple of 4 and the observations are different, the quartiles are particularly easy to find. For example, if there are 8 observations, the first quartile is halfway between the second-ranking and the third-ranking observation, the second quartile is halfway between the fourth- and fifth-ranking observations, and the third quartile is halfway between the sixth- and seventh-ranking observations.

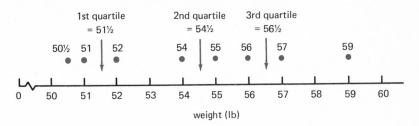

Figure 3.3 *Quartiles of a continuous variable: Weights of 8 pigs*

The general procedure of finding quartiles is carried out most easily in terms of cumulative relative frequencies. Suppose the weights of five pigs are 52, 54, 55, 57, and 59; the cumulative relative frequencies are graphed in Figure 3.4. The first quartile is 54 because the cumulative relative frequency is below 1/4 for values less than 54 and greater than 1/4 for values equal to and greater than 54.

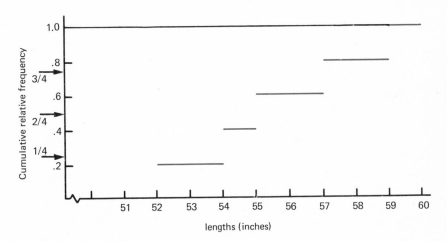

Figure 3.4 *Cumulative relative frequency*

Another way of saying the same thing is that less than 3/4 of the observations are above 54, and less than 1/4 of the observations are below 54. Hence the value 54 splits the set of observations into two sets, the lower set constituting about one-fourth of the total number of observations, the upper about three-fourths of the total number.

The second quartile (median) is 55 because the cumulative relative frequency is below 1/2 for values less than 55 and greater than 1/2 for values equal to and greater than 55. Similarly, the third quartile is 57.

It should be noted that "quartile" refers to a *value*, not to an individual or a set of individuals. It is a single value of the variable; it is not a set or range of values. We say, "John's test score is in the upper quarter," meaning his score exceeds the third (upper) quartile. We do not say, "John's score is in the upper quartile."

If continuous data have been grouped, the quartiles can be approximated by *interpolation*. This procedure is explained in Appendix 3B.

Quartiles of Discrete Data. Discrete data are first organized into a frequency distribution and then the quartiles are computed. Table 3.5 is a frequency distribution for the number of children in 100 families, and the cumulative relative frequency is graphed in Figure 3.5. We look for the classes in which the cumulative frequencies .25, .50, and .75 are achieved. The first quartile is 2 because the range of cumulative relative frequency corresponding to 2 is .22 to .47, which includes .25. The second quartile is 3 because the range of cumulative relative frequency corresponding to 3 is .47 to .74, which includes .50. The third quartile is 4 because the range of cumulative relative frequency corresponding to 4 is .74 to .95, which includes .75. The quartiles 2, 3, and 4 divide the 100 families into quarters, in that we can assign 25 of the 100 families to the first quarter with 0, 1, or 2 children per family, 25 to the second quarter with 2 or 3 children per family, 25 to the third quarter with 3 or 4 children per family, and 25 to the fourth quarter with 4, 5, or 6 children per family.

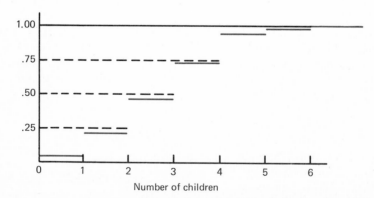

Figure 3.5 *Cumulative relative frequency: Number of children per family (Table 3.5)*

TABLE 3.5
Number of Children in 100 Families

NUMBER OF CHILDREN	FREQUENCY	RELATIVE FREQUENCY	CUMULATIVE RELATIVE FREQUENCY
0	3	.03	.03
1	19	.19	.22
2	25	.25	.47
3	27	.27	.74
4	21	.21	.95
5	4	.04	.99
6	1	.01	1.00
	100	1.00	

(hypothetical data)

3.2.3 Deciles, Percentiles, and Other Quantiles

Deciles, which divide the set of ordered observations into tenths, and *percentiles,* which divide it into hundredths, are defined analogously to quartiles.

If we want to compare colleges according to how "bright" their leading students are, we might use the ninth decile (ninetieth percentile) of scores on a Scholastic Aptitude Test. Telephone company scientists might compare systems of equipment for handling telephone calls by comparing the ninetieth percentiles of distributions of customers' waiting time under the various systems.

The 90th percentile in Table 3.5 is 4 children because the range of cumulative relative frequency corresponding to 4 children is .74 to .95, which includes .90.

Quantile is a general term that includes quartiles, deciles, and percentiles. The .999 quantile (99.9th percentile) is a value exceeded by only one-tenth of one percent (0.1%) of the individuals. Any quantile can be computed by reference to cumulative relative frequencies, in the manner already explained for quartiles, deciles, and percentiles.

3.3 THE MEAN

3.3.1 Definition and Interpretation

Definition. The *mean* of a set of numbers is the familiar *arithmetic average.* There are other kinds of "means," but the term "mean" by itself is understood as denoting the arithmetic average. It is the measure of central tendency most used for numerical variables.

The mean is the sum of observations divided by the number of observations. For example, the mean of the five observations 15, 12, 14, 17, and 19 is

$$\frac{15 + 12 + 14 + 17 + 19}{5} = (15 + 12 + 14 + 17 + 19) \div 5,$$

which is 77/5, or 15.4. Let these numbers represent the dollars that each of five persons has to spend on an outing; then, if the group pooled its resources and shared them equally, the *average* amount of $15.40 is what each person could spend.

Change of Units. If all the observations are multiplied by the same positive number, the measure of location for the derived set of observations is the measure of location for the original set multiplied by that number. In the above example the average amount of cents is 1540. Both the mean and the median have this property when computed from the original measurements (not from grouped data).

Another property necessary for a measure of location is that if a quantity is subtracted from every observation the measure of location for the derived observations should be the same as that obtained by subtracting the quantity from the measure of location for the original observations. Suppose that in the example above, each person on the outing needed $2 for return train fare. The individual amounts available for other spending are 13, 10, 12, 15 and 17; the average of these is $13.40, which is $2 less than the average money available before taking out train fare. These properties are proved for the mean in Appendix 3C.

The Mean and the Total. Ten families are labelled 1, 2, . . . , 10, and the number of children in each is recorded in Table 3.6. We can compute the mean number of children per family thus:

$$\frac{2 + 4 + 4 + 3 + 4 + 3 + 3 + 3 + 6 + 3}{10} = \frac{2 + 4 + \cdots + 3}{10} = \frac{35}{10} = 3\tfrac{1}{2}.$$

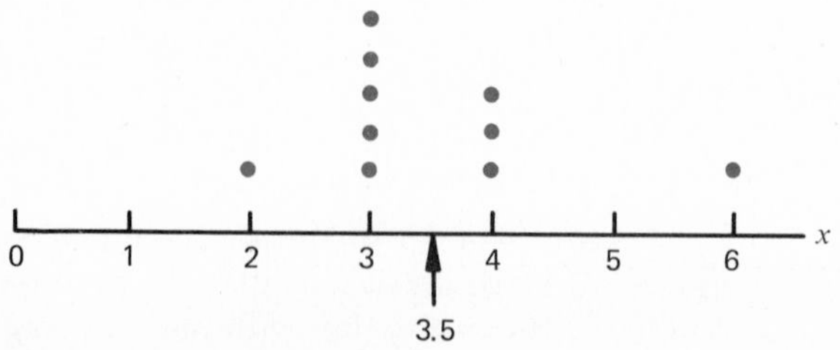

Figure 3.6 *Plotted observations (Table 3.6)*

Although there is no such thing as $3\frac{1}{2}$ children, it is useful to retain the fraction when reporting a mean, because the *total* number of children is the product of the mean and the number of families, and we would want this calculation to reproduce the total accurately: $10 \times 3\frac{1}{2} = 35$.

Center of Gravity. The data in Table 3.6 are plotted on an axis in Figure 3.6 and their mean is indicated. Think of the axis as a board and the data points as one-pound weights. If we place a wedge under the board, it will balance when the wedge is located at the mean, as in Figure 3.7. The mean is the "center of gravity."

TABLE 3.6
Number of Children in 10 Families

FAMILY	NUMBER OF CHILDREN
1	2
2	4
3	4
4	3
5	4
6	3
7	3
8	3
9	6
10	3
	35

(hypothetical data)

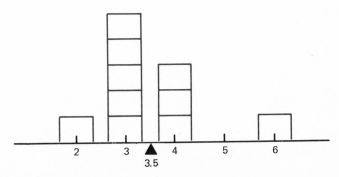

Figure 3.7 *The mean as the center of gravity (Table 3.6)*

3.3.2   Use of Notation

Understanding of arithmetic and other mathematical operations will be made considerably easier by the development of an appropriate system of notation. We denote the numbers in a set of five observations by x_1, x_2, x_3, x_4, and x_5. Their sum is written as

(3.1) $$x_1 + x_2 + x_3 + x_4 + x_5,$$

and their mean is

(3.2) $$\frac{x_1 + x_2 + x_3 + x_4 + x_5}{5}.$$

In the first example of Section 3.3.1, $x_1 = 15$, $x_2 = 12$, $x_3 = 14$, $x_4 = 17$, and $x_5 = 19$. The phrase "Take the five numbers as listed, add them, and divide by 5" is replaced by (3.2).

We do not always deal with *five* measurements, however, and must be able to express sums and means of an arbitrary number of observations. The number of observations is denoted by the letter n (or some other letter). Since n can be any integer, we cannot write all the observations explicitly. We write the n observations as x_1, x_2, ..., x_n. The three dots (ellipsis) indicate the omitted values; the number of omitted values is $n - 3$. The sum is then written as

$$x_1 + x_2 + \cdots + x_n,$$

and the mean, which we denote by $\bar{x}$, as

$$\frac{x_1 + x_2 + \cdots + x_n}{n} = \bar{x}.$$

The symbol for the arithmetic average, $\bar{x}$, is read "x bar." The mean of 2 observations ($n = 2$) is

(3.3) $$\frac{x_1 + x_2}{2};$$

the mean of 5 observations ($n = 5$) was given in (3.2).

The symbol x_i denotes the observation corresponding to the ith individual in the sample. If there are three individuals in the sample, the index (subscript) i can be 1, 2, or 3; we write $i = 1, 2, 3$. If there are n individuals, we write $i = 1, 2, \ldots, n$.

A more efficient way of indicating a sum is to use the "summation sign," which is Σ, the capital Greek letter sigma. The sum (3.1) is then written

(3.4)
$$\sum_{i=1}^{5} x_i.$$

This expression is read "the sum of x_i from i equals 1 to i equals 5" or "the sum of the x_i's from x_1 to x_5." The notation $i = 1$ under Σ says to begin the summation with x_1, and the 5 (short for $i = 5$) above Σ says to end the summation with x_5. It is understood that the sum is over consecutive integer values of i. Using (3.4), we may write (3.2) as

$$\bar{x} = \frac{\sum_{i=1}^{5} x_i}{5}.$$

Expression (3.3) is

$$\frac{\sum_{i=1}^{2} x_i}{2}.$$

The general formula for the mean of n observations is

$$\bar{x} = \frac{\sum_{i=1}^{n} x_i}{n}.$$

Summation need not start with x_1. The sum $x_4 + x_5 + x_6$ is indicated as

$$\sum_{i=4}^{6} x_i.$$

If x_i is dollars earned in the ith month of the year, then this sum is the total earnings for the second quarter (April, May, and June).

We shall use summation notation frequently. Some rules for working with summations are given in Appendix 3C. For convenience in writing, in the body of the text we shall write $\Sigma_{i=1}^{n} x_i$, where $i = 1$ is written as a subscript to Σ and n is written as a superscript. When the summation is not in the body of the text but is displayed as in expression (3.4), it is written as first described.

At times we shall need to distinguish between the measures of location for a sample and for a population because the measure for a sample may be used to make inferences about the measure for a population. In a *sample* we use n to denote the number of observations, $x_1, \ldots, x_n,$

and $\bar{x}$ for the arithmetic mean; we shall denote the number of individuals in a *population* by N and the mean of the values of a variable for all the individuals in the population by μ, which is the lower case mu, the Greek letter corresponding to the letter m (for mean).

3.3.3   Calculating the Mean of Discrete Data

When a set of observations contains repeated values, as does the data in Table 3.6, computation of the mean can be simplified. Let x_i be the number of children in the ith family, $i = 1, 2, \ldots, 10$. Then the computation of the mean is represented by the formula

$$\bar{x} = \frac{\sum_{i=1}^{10} x_i}{10}.$$

We can instead compute $\bar{x}$ from the frequency distribution given in Table 3.7. The sum is

$$2 + (3 + 3 + 3 + 3 + 3) + (4 + 4 + 4) + 6$$
$$= (1 \times 2) + (5 \times 3) + (3 \times 4) + (1 \times 6) = 2 + 15 + 12 + 6 = 35,$$

and the mean is this sum divided by $n = 10$. Here, we have multiplied each value by the frequency of that value, added the products, and divided the resulting sum by the number of observations.

Symbolic notation for these calculations is developed thus: We see from the frequency distribution that among the *observations* $x_1, x_2, \ldots,$ x_{10} there are only four distinct *values*. Call these $v_1 = 2, v_2 = 3, v_3 = 4,$ and $v_4 = 6$; the frequencies of these values are $f_1 = 1, f_2 = 5, f_3 = 3,$ and $f_4 = 1$, respectively. Symbolically, then, the above sum is computed as

$$f_1 v_1 + f_2 v_2 + f_3 v_3 + f_4 v_4.$$

The product $f_1 v_1$ is the contribution to the total of individuals having the value v_1; similarly for $f_2 v_2, f_3 v_3,$ and $f_4 v_4$. The mean is

(3.5)
$$\bar{x} = \frac{f_1 v_1 + f_2 v_2 + f_3 v_3 + f_4 v_4}{10}.$$

TABLE 3.7
Number of Children in 10 Families

j	NUMBER OF CHILDREN v_j	NUMBER OF FAMILIES HAVING v_j NUMBER OF CHILDREN f_j	TOTAL NUMBER OF CHILDREN IN FAMILIES HAVING v_j NUMBER OF CHILDREN $f_j v_j$
1	2	1	2
2	3	5	15
3	4	3	12
4	6	1	6
		$n = 10$	35

SOURCE: Table 3.6.

The general rule is that if there are m distinct values $v_1, v_2, \ldots, v_m$ with frequencies $f_1, f_2, \ldots, f_m$, respectively, then we can compute the mean as

$$(3.6) \qquad \bar{x} = \frac{\sum_{j=1}^{m} f_j v_j}{n}.$$

Note that the number of observations is equal to the sum of the frequencies: $n = \sum_{j=1}^{m} f_j$.

TABLE 3.8
Distribution of Families by Number of Children Under 18 and Residence in 1950 and 1969 (Number of Families, in 1000's)

NUMBER OF CHILDREN	1950 Nonfarm		1950 Farm		1969 Nonfarm		1969 Farm	
0	17,327	(.490)*	1,750	(.444)	20,971	(.438)	1,349	(.513)
1	7,674	(.217)	698	(.177)	8,762	(.183)	381	(.145)
2	5,941	(.168)	587	(.149)	8,187	(.171)	347	(.132)
3	2,581	(.073)	402	(.102)	5,075	(.106)	224	(.085)
4	919	(.026)	252	(.064)	2,442	(.051)	163	(.062)
5	460	(.013)	126	(.032)	1,245	(.026)	84	(.032)
6	460	(.013)	126	(.032)	1,198	(.025)	82	(.031)
	35,362	(1.000)	3,941	(1.000)	47,880	(1.000)	2,630	(1.000)

SOURCE: Proportions and totals from *Statistical Abstract of the United States* (1970). The proportion for "4 or more children" was allocated approximately 1/2 to 4 children, 1/4 to 5 children, and 1/4 to 6 children. The numbers of families were obtained by multiplying the proportions by the total.
*Proportion of group total is indicated in parentheses.

The utility of the mean in comparing several distributions is illustrated in Tables 3.8 and 3.9. Differences in numbers of children in the groups represented in Table 3.8 are hard to discern; there are just too many figures. By computing the average number of children per family in each of the four groups, we obtain Table 3.9.

TABLE 3.9
Mean Number of Children Per Family, by Residence: 1950 and 1969

	NONFARM	FARM
1950	1.019	1.388
1969	1.327	1.259

SOURCE: Table 3.8.

The averages show that while nonfarm families are larger than they used to be, farm families are smaller than they used to be. In fact, in 1969 farm families were smaller on the average than were nonfarm families, in spite of the fact that in 1950 the average number of children in farm families had been about 36% larger than in nonfarm families.

3.3.4   Calculating the Mean of Grouped Data

In Section 2.3 we saw that a set of continuous data may be grouped into a frequency table; the mean of *grouped* data is also computed according to formula (3.6). In this case the values v_j are the midpoints of the class intervals. We proceed as if the data in any one class interval were all equal to the midpoint of that interval; this approximation changes the results only slightly from the mean as computed from the original (ungrouped) data. We illustrate with the data in Table 3.4. The computation may be laid out as in Table 3.10. (On a desk calculator the products $f_j v_j$ can be cumulated; it is not necessary to record them individually.) The mean is

$$\bar{x} = \frac{\sum_{j=1}^{m} f_j v_j}{n} = \frac{808,000}{64} = \$12,625.$$

Sometimes data are available only in grouped form. If the data are given for all n individuals, the grouping is a convenience. When the calculations are done on an electronic computer, there is usually no advantage in grouping for calculating the mean, though a frequency table is desirable for visual presentation.

TABLE 3.10
Computation of Mean Income of 64 Families

j	v_j	f_j	$f_j v_j$
1	$ 2,000	4	8,000
2	6,000	6	36,000
3	10,000	15	150,000
4	14,000	28	392,000
5	18,000	7	126,000
6	22,000	2	44,000
7	26,000	2	52,000
		$n = 64$	$\sum_{j=1}^{7} f_j v_j = 808,000$

SOURCE: Table 3.4.

3.4 WEIGHTED AVERAGES AND STANDARDIZED AVERAGES

Instead of using every datum for calculation of an average, as described in Section 3.3, an investigator may often arrange the data so that the frequencies or relative frequencies of all obtained values can be used to compute an average. As an example, suppose a motor company reports its sales figures and prices as in Table 3.11.

TABLE 3.11
Auto Sales and Prices in 1969

MODEL	NO. OF CARS SOLD	PERCENT	PRICE
Tartan	3000	60	$1995
Voom	1500	30	2300
X 100	500	10	4500
Total	5000	100%	

(hypothetical data)

What was the average 1969 price? We proceed as in Section 3.3.3. The average price is

$$\frac{(3000 \times \$1995) + (1500 \times \$2300) + (500 \times \$4500)}{5000}$$

$$= \frac{3000}{5000} \times \$1995 + \frac{1500}{5000} \times \$2300 + \frac{500}{5000} \times \$4500$$

$$= .60 \times \$1995 + .30 \times \$2300 + .10 \times \$4500$$

$$= \$2337.$$

The relative frequencies .60, .30, and .10 are the *weights;* and the above average is known as a *weighted average.*

Suppose we are told that the prices are to be raised for 1970 according to Table 3.12. What is the average price increase? We can take a

TABLE 3.12
Auto Price Increases

MODEL	1969 PRICE	1970 PRICE	PRICE INCREASE
Tartan	$1995	$2000	$5
Voom	$2300	$2500	$200
X 100	$4500	$4900	$400

(hypothetical data)

weighted average of the price increase on the basis of 1969 sales to obtain

$$.60 \times \$5 + .30 \times \$200 + .10 \times \$400 = \$103.$$

At the time the price increases are announced, the relative sales of the three models in 1970 are unknown and the weighted average must be based on the 1969 weights.

Suppose that in 1970 the relative sales of the three models are .50, .35, and .15, respectively. Then the average 1970 price is

$$.50 \times \$2000 + .35 \times \$2500 + .15 \times \$4900 = \$2610.$$

This average price is higher than the 1969 average of $2337 not only because the price of each model was increased but also because there was a shift in sales away from the cheaper model toward the more expensive model. The average price increase of $103, based on 1969 weights, reflects the shift in prices but not in relative frequencies of models.

Standardized Averages. If the relative frequency of sales were the same in the two years the average price in 1970 would have been

$$.60 \times \$2000 + .30 \times \$2500 + .10 \times \$4900 = \$2440,$$

which is exactly $103 more than the 1969 average price. The figure of $2440 is called a *standardized average* because it is calculated on the basis of "standard weights." The year 1969 is the *base year*.

The Consumer Price Index. The *consumer price index* is a standardized average which measures the average change in prices of goods and services purchased by urban wage earners and by clerical families and single persons living alone. "The weights used in calculating the index are based on studies of actual expenditures by wage earners and clerical workers. The quantities and qualities of the items in the 'market basket' remain the same between consecutive pricing periods, so that the index measures the effect of *price change only* on the cost of living of these families" (*Statistical Abstract of the United States,* 1970, p. 337). The index is computed in such a way that its value for the period 1957–1959 is 100; that is, 1957–1959 is the *base period.* Values of the index for various years are given in Table 3.13.

TABLE 3.13
Consumer Price Index: 1960 to 1969 (1957–1959 = 100)

Year	1960	1963	1964	1965	1966	1967	1968	1969
Index	103.1	106.7	108.1	109.9	113.1	116.3	121.2	127.7

SOURCE: *Statistical Abstract of the United States* (1970), p. 344.

*Adjustment Using Standardized Averages. A standardized average can be computed so as to adjust for differences in the composition of groups to be compared.

Table 3.14 gives age-specific fertility ratios for rural nonfarm and rural farm women in Iowa. The fertility ratio is defined in this case as the number of children under 5 years old per 1000 women. From the tables one sees that the overall fertility ratio of 562 for rural nonfarm women is higher than that of 529 for rural farm women, in spite of the fact that the age-specific fertility ratios are higher for farm women in every age category except the youngest and the oldest. The difference between the two overall fertility ratios may be due at least in part to differences in the age distributions. For example, 14% of the nonfarm women but only 8.9% of the farm women are in the 20–24 age group. Standardized averages can be used to see whether the difference is accounted for by the age distributions.

What would be the fertility ratio of nonfarm women if they had the same age distribution as farm women? We answer by computing the weighted average for nonfarm women, using the weights from the farm group:

$$\frac{25{,}372 \times 114 + 12{,}275 \times 1{,}096 + \cdots + 21{,}357 \times 69}{137{,}825} = \frac{67{,}994{,}611}{137{,}825} = 493.$$

Similarly, using the nonfarm group as a base, the weighted average for farm women is computed, using the weights from the nonfarm group, as 604. If the farm women had the same age distribution as nonfarm women, their fertility ratio would be 605 per 1000, which is greater than the ratio of 562 per 1000 for nonfarm women. In fact, the fertility ratio is higher for farm women, regardless of which of the two sets of weights is used. (See Table 3.15.)

Here the difference is about 40 children per 1000 women, regardless of which weights are used. It is possible for the age distributions and age-specific fertility ratios to be such that the difference is quite different in the two cases. It would even be possible for the farm women to have the higher ratio when one set of weights is used and the non-farm women to have the higher ratio when another set is used.

TABLE 3.14

Own Children Under 5 Years Old for Iowa Women 15 to 49 Years Old in the Rural Nonfarm and Rural Farm Groups

	RURAL NONFARM		RURAL FARM	
Age (Years)	Frequency: Number (Proportion)	Children Under 5 Years Old Per 1000 Women	Frequency: Number (Proportion)	Children Under 5 Years Old Per 1000 Women
15–19	20,564 (.165)	114	25,372 (.184)	43
20–24	17,448 (.140)	1096	12,275 (.089)	1158
25–29	16,843 (.135)	1263	15,752 (.114)	1429
30–34	17,319 (.139)	800	18,898 (.137)	901
35–39	18,285 (.146)	478	21,963 (.159)	522
40–44	17,483 (.140)	210	22,208 (.161)	241
45–49	16,954 (.136)	69	21,357 (.155)	56
15–49	124,896	562	137,825	529

SOURCE: *U.S. Census of Population: 1950.* vol. IV, *Special Reports,* pt. 5, ch. 6, "Fertility."

TABLE 3.15
Fertility Ratios (Per 1000 Women), Computed Using Each Age Distribution

		FERTILITY RATIO	
		Nonfarm Women	*Farm Women*
BASE AGE DISTRIBUTION	*Nonfarm*	562	605
(WEIGHTS USED)	*Farm*	493	529

SOURCE: Table 3.14.

3.5   USES AND ABUSES OF AVERAGES

3.5.1   The Mean and the Total

A firm has 200 employees, and the mean annual salary is $9,000. The total annual payroll can be computed by multiplying the mean salary by the number of employees:

$$\text{total annual payroll} = 200 \times \$9{,}000 = \$1{,}800{,}000.$$

The total is the same as it would be if all 200 workers earned exactly $9,000 per year.

If x_i represents the salary of the ith employee, then

$$\text{total payroll} = \sum_{i=1}^{200} x_i,$$

and the preceding computation is

$$\sum_{i=1}^{200} x_i = 200 \times \$9{,}000.$$

This is a special case of the general formula

$$\sum_{i=1}^{n} x_i = n\bar{x},$$

which is simply the definition of $\bar{x}$, namely $\bar{x} = \sum_{i=1}^{n} x_i/n$, multiplied by n on both sides of the equals sign.

Suppose that the total, $\sum_{i=1}^{n} x_i$, represents *the total number of heads* obtained in n tosses of a coin. In this case, x_i is 1 or 0 according as the ith toss is a head or a tail. Then

$$\bar{x} = \frac{\sum\limits_{i=1}^{n} x_i}{n}$$

= total number of heads ÷ number of tosses

= proportion of heads.

That is, *a proportion is a mean.*

When one attempts to convey the idea that about half of all children are girls by using the phrase, "One in every two children is a girl," this phraseology imparts to the average properties which it does not possess. It is most assuredly *not* true that *every* pair of children contains exactly one girl. (Perhaps we should simply dismiss this use of the word "every" as idiomatic and not interpret it literally.)

3.5.2   Choosing an Appropriate Measure of Location

When you dial a phone number, there is a short delay before you hear either a ring or a busy signal. This delay varies from call to call; that is, it has a distribution. The shape of this distribution depends upon the pattern of telephone traffic through the exchange. It also depends upon the design of the equipment used at the exchange— different systems give different waiting-time distributions. Because it is annoying if the delay is more than momentary, telephone company engineers have paid some attention to evaluating systems on the basis of their associated waiting-time distributions.

Figures 3.8 and 3.9 represent waiting-time distributions under two alternative telephone systems. Statistics summarizing the two distributions are given in Table 3.16. Which statistics are most appropriate for comparing these distributions?

If human nature is such that a silence of one or one-and-a-half seconds is not noticeable, but a longer wait is noticeable and maybe even annoying, then it might be appropriate to evaluate a telephone system on the basis of *how many callers* have to wait more than 1.5 seconds—that is, on the basis of what proportion of the waiting-time distribution lies to the right of 1.5 seconds.

A similar idea is to compare the 90th percentiles of the waiting-time distributions, focusing attention on the 10% of calls that are delayed the longest.

Means of the distributions are not an adequate basis for comparison here because a distribution could have a relatively small mean, even though many customers might have to wait a long time. The distributions shown in Figures 3.8 and 3.9 have the same mean, 1.10 seconds, but the 90th percentile is 1.65 seconds in Figure 3.8, while it is 2.05 seconds in Figure 3.9. Obviously the system in which 10% of the calls

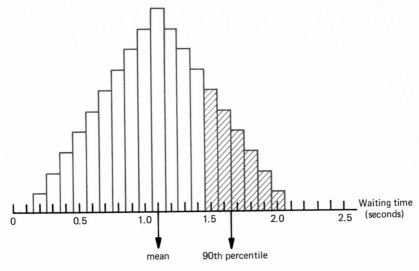

Figure 3.8 *Waiting-time distribution (18% of the calls wait more than 1.45 seconds)*

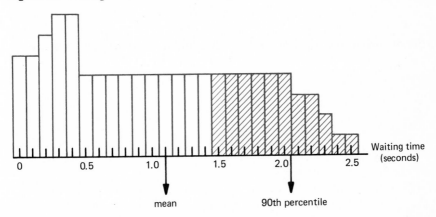

Figure 3.9 *Waiting-time distribution (32% of the calls wait more than 1.45 seconds)*

were delayed more than 1.65 seconds is more satisfactory than the one in which 10% were delayed more than 2.05 seconds. Comparison based on the proportion of calls delayed more than 1.45 seconds leads to the same conclusion. The distribution in Figure 3.9 is less satisfactory because 32% of the calls wait more than 1.45 seconds, compared to only 18% in Figure 3.8.

On the other hand, if the *total delay* experienced by *all callers* combined is felt to be of more importance, it would be appropriate to compare the means of the distributions because

total delay of all calls = mean delay × number of calls.

TABLE 3.16
Summary of Waiting-Time Distributions

	System 1 (Figure 3.8)	System 2 (Figure 3.9)
Proportion of calls delayed more than 1.45 sec.	18%	32%
90th percentile	1.65 sec	2.05 sec
Mean	1.10 sec	1.10 sec
Median	1.10 sec	1.05 sec

SOURCE: Figures 3.8 and 3.9. (hypothetical data)

3.5.3   Shape of Distributions

When the practical problem at hand does not dictate the choice of measure of location, the shape of the distributions may be of some help.

Bimodality.   If a firm employs two kinds of workers, 100 at a salary of $6000 and 100 at a salary of $12,000 per year, no one earns the mean salary of $9000 per year. When the distribution is bimodal, as is the case with these payroll data and with the distribution of number of children in Table 3.3, no one measure of location conveys useful information. Both modes should be stated. The median, on the other hand, at least tells us at what point the set of observations is divided in half.

Skewness.   Another situation in which the mean may not be descriptive is when the distribution is *skewed;* that is, when there are a few very high or a few very low values. A distribution of income, for example, may have a few very high incomes which can cause the mean to be much too large to be called "typical" or "central." Consider the five incomes $12,000, $14,000, $15,000, $17,000, and $1,000,000. The mean, $211,600, is not particularly close to any one of the incomes.

Some Advantages of the Median.   The median of these five incomes is $15,000. This figure is fairly close to most (4 out of 5) of the incomes.

In general the median is a better measure of location than the mean when the distribution is very skewed. The shape of a skewed distribution is shown in Figure 3.10, in which the location of the mean, median, and mode are indicated. A smooth curve is intended to be an idealization of a histogram with very small class intervals. (Alternatively, the reader can think of the smooth curve as an artist's rendering.)

Since the locations of the largest and smallest observations do not

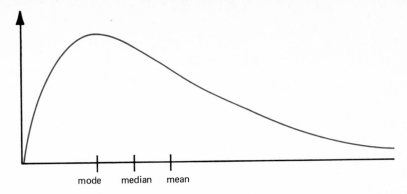

Figure 3.10 *Right-skewed distribution*

affect the median, it may be a more reliable measure of central tendency than the mean if it is possible that extreme values are due to errors of recording or other extraneous causes.

With small or medium-sized sets of observations, the median is more easily calculated than the mean. Calculation of the median entails only putting the observations in order, whereas calculation of the mean involves adding together all the observations. If, however, the number of observations is large, putting the sample in order is a tedious task. (In fact, even on a computer this takes a relatively long time.)

Symmetric Distributions. A symmetric distribution is a distribution that looks the same in a mirror. If it is drawn on translucent paper, the distribution or histogram will look the same from the back. Two symmetric distributions are sketched in Figures 3.11 and 3.12. The mean and median are equal for symmetric distributions. For symmetric *unimodal* distributions the mean, median, and mode all coincide.

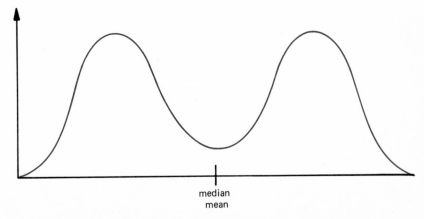

Figure 3.11 *Symmetric bimodal distribution*

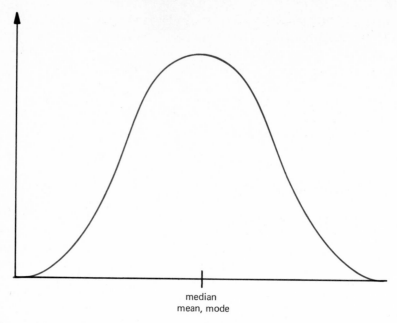

median
mean, mode

Figure 3.12 *Symmetric unimodal distribution*

3.5.4   Uses and Abuses of Weighted Averages

In comparing averages of different groups, it is often necessary to use standardized averages in order to make the groups comparable.

Tables 1.5 and 1.6 summarize death rates due to lung cancer in different groups. Age is known to affect the incidence of lung cancer. Therefore, to eliminate the effect of age in comparisons of lung cancer among groups of smokers standardized averages, that is, age-adjusted death rates, were used. A standard age distribution provided the weights. (Age-adjusted death rates are considered in Exercise 3.16.)

Failure to use appropriate weights, or failure to make any adjustment at all in such situations, can yield misleading averages.

SUMMARY

The *mode* of a distribution of observations is the value with the highest frequency and is valid for data defined on any scale, nominal, ordinal, interval, or ratio; continuous data must be grouped before a mode can be computed. *Bimodal* distributions can arise by mixing two unimodal distributions.

The *median* and other percentiles are valid for data defined on an ordinal scale. The median divides the ordered sample in half. *Quartiles* divide the ordered sample into quarters; the median is the second quartile.

The *mean* (arithmetic average) is valid for data defined on an interval scale.

Weighted averages can be used to adjust figures for comparative purposes. A weighted average computed using a set of standard weights is called a *standardized average*.

APPENDIX 3A   Coding

The mean of the incomes $10,000, $10,400, and $10,500 is $10,300. An easy way to compute this mean is to subtract $10,000 from each number, obtaining 0, 400, and 500, average these numbers, obtaining 300, and add in the $10,000 again, obtaining $10,300. This procedure is called *coding;* we "code" data by subtracting and/or dividing by convenient numbers, computing the mean of the resulting smaller numbers, and then uncoding the resulting mean. We illustrate the procedure in Table 3.17 with the grouped data on incomes in Table 3.10. Recall that the class intervals were determined to be equal in width and such that the midpoints are convenient numbers.

We start by subtracting from each midpoint the smallest midpoint, $2000 in this case. The result on the first line is 0 and the results on the other lines are divisible by the class interval width, here $4000. We divide each result by the width of the class interval to obtain the successive integers as recorded in the fifth column. The mean of the coded v_j' is

$$\frac{\sum_{j=1}^{7} f_j v_j'}{n} = \frac{170}{64} = 2.65625.$$

(Since the coded values in the fifth column designate the classes starting with 0, this is the mean class number.) To obtain the mean of the original values v_j we must decode this value. In coding we subtracted $2000 and then divided by $4000, therefore, to decode we must multiply by $4000 and then add $2000:

$$\bar{x} = \$4,000 \times 2.65625 + \$2,000$$

$$= 10,625 + 2,000$$

$$= \$12,625.$$

Note that this is the same as the result calculated directly from Table 3.10.

TABLE 3.17

Computation of the Mean of the Income Data of Table 3.10 by Coding

j	Frequency f_j	Midpoint v_j	$v_j - 2{,}000$	$v_j' = \dfrac{v_j - 2{,}000}{4{,}000}$	$f_j v_j'$
1	4	$ 2,000	0	0	0
2	6	6,000	4,000	1	6
3	15	10,000	8,000	2	30
4	28	14,000	12,000	3	84
5	7	18,000	16,000	4	28
6	2	22,000	20,000	5	10
7	2	26,000	24,000	6	12
	$n = 64$				$\displaystyle\sum_{j=1}^{7} f_j v_j' = 170$

SOURCE: Table 3.10.

$$\text{Mean of coded values} = \frac{170}{64} = 2.65625$$

Everyday arithmetic is done by coding. When we add 30 and 40, for example, we visualize the 3 and 4 in the tens column and add them mentally to get 7 in the tens column. In effect, we divide 30 and 40 by 10, add 3 and 4 to obtain 7, and multiply by 10. We are so accustomed to this procedure that we do not realize we are coding.

In formal terms, coding works because if we add a number to every value the mean of the resulting numbers is the same as what we obtain by adding that number to the average of the original values and because if we multiply every value by a number the mean of the resulting numbers is what we obtain by multiplying the average of the original values by that number. These ideas are developed in Appendix 3C.

When we use coding it is not necessary to have convenient midpoints for the class intervals, because they are replaced in the computation by the coded values. Hence, in grouping the data we can concentrate on obtaining a convenient grouping without worrying about the midpoints.

APPENDIX 3B   Computing the Median and Other Quantiles of Grouped Continuous Data

Computing Quantiles of Grouped Continuous Data.   When there are many observations we may find it convenient to group data instead of working with the raw data. Consider the frequency distribution of profits of nine businesses (in 1000's of dollars), shown in Table 3.18. (Since there are only 9 observations here, it would be easy to handle the original data. We purposely chose a small example to prevent the graphs below from being unwieldy.) The median profit is that of the firm which ranks fifth. The median therefore lies in the class interval 300–350. Within this interval lie the profits of 5 firms. Since we do not know the individual values for the 5 firms in this class, we proceed on the basis that they are spread uniformly over the interval (Figure 3.13). We allot to each of these 5 firms 1/5 of the interval width of 50, which is 10, and we assume that each of the firms falls at the midpoint of its interval. This gives a well-defined median, in this case 315.

If we make a graph showing the whole range of profits (Figure 3.14), we see that, in terms of the numbers 0 to 9 (on top of the axis), we can define the median as the profit which would correspond to the number $9/2 = 4.5$. It is this reasoning which gives rise to the formula for the median of grouped data as

$$l + \frac{\frac{n}{2} - b}{f} w,$$

where

n = number of observations
l = lower limit of interval containing the median
f = frequency of interval containing the median
b = sum of frequencies of intervals below that containing the median
w = class interval width

TABLE 3.18
Profits

PROFIT (1000's OF $)	FREQUENCY	CUMULATIVE FREQUENCY
200–250	1	1
250–300	2	3
300–350	5	8
350–400	1	9
	9	

(hypothetical data)

Figure 3.13 *Notion of the median of grouped data (Table 3.18)*

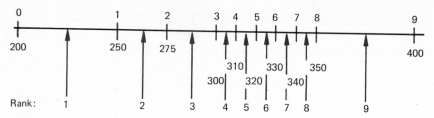

Figure 3.14 *Profits and their ranks (Table 3.18)*

Applying the formula to the distribution of profits, we approximate the median as

$$300 + \frac{\frac{9}{2} - 3}{5} \times 50 = 300 + \frac{1.5}{5} \times 50$$

$$= 300 + 15$$

$$= 315.$$

Turn back to the frequency distribution of weights of 25 employees given in Table 2.10. It is clear that the median lies in the class interval 147.5–152.5, as the 13th-ranking weight falls within that interval. We can apply the above formula, approximating the median as

$$147.5 + \frac{\frac{25}{2} - 9}{6} \times 5 = 147.5 + \frac{7}{12} \times 5$$

$$= 147.5 + 2.92$$

$$= 150.42.$$

Other quantiles are computed analogously, for example, with $n/4$, $3n/4$, and $9n/10$ replacing $n/2$ in the formula for the first quartile, third quartile, and ninth decile, respectively.

A Rule for Grouping.   Usually practical considerations will determine the choice of class intervals. However, a mathematical rule of thumb for the number of class intervals is given in Table 3.19. Roughly speaking, the number of observations is 2 raised to a power equal to one less than the number of intervals [Sturges (1926)]. In Table 3.4, for example, there are $64 = 2^6$ families, so we used 7 class intervals. In any case, the number of intervals may differ from that indicated in Table 3.19 in order to obtain midpoints and a class width that are convenient. As a rule no more than 15 class intervals are used.

TABLE 3.19
Suggested Number of Class Intervals

Total Frequency	Number of Class Intervals
10–25	5
25–50	6
50–100	7
100–200	8
200–400	9
400–800	10

APPENDIX 3C   Rules for Summation

Rules. We have already made use of some facts about summations in developing the idea of coding in order to calculate the mean. Since we use these facts often, we shall demonstrate them formally. They are stated as rules.

RULE 1 *Given a set of pairs of observations, adding the two members of each pair and then adding these sums produces the same result as adding the first members of the pairs, adding the second members of the pairs, and adding these two sums. In symbols:*

$$\sum_{i=1}^{n} (x_i + y_i) = \sum_{i=1}^{n} x_i + \sum_{i=1}^{n} y_i.$$

RULE 2 *The sum of a quantity a given number of times is the product of the quantity and the number of times. In symbols:*

$$\sum_{i=1}^{n} a = na.$$

RULE 3 *The sum of a set of observations, to each of which has been added the same quantity, is the sum of the original observations plus the product of the quantity and the number of summands. In symbols:*

$$\sum_{i=1}^{n} (x_i + a) = \sum_{i=1}^{n} x_i + na.$$

RULE 4   *The mean of a set of observations, to each of which has been added a number, is the mean of the original observations plus that number. In symbols:*

$$\overline{x + a} = \bar{x} + a,$$

where $\overline{x + a}$ denotes the mean of the quantities $x_i + a$, $i = 1, \ldots, n$.

RULE 5   *The sum of a set of observations, each of which has been multiplied by the same number, is the sum of the original observations multiplied by that number. In symbols:*

$$\sum_{i=1}^{n} cx_i = c \sum_{i=1}^{n} x_i.$$

RULE 6   *The mean of a set of observations, each of which has been multiplied by the same number, is the mean of the original observations multiplied by that number. In symbols:*

$$\overline{cx} = c\bar{x}.$$

RULE 7   *The sum of a set of observations, each of which has been multiplied by the same number and added to a given quantity, is the sum of the original observations multiplied by that number and added to the product of the given quantity and the number of summands. In symbols:*

$$\sum_{i=1}^{n} (cx_i + a) = c \sum_{i=1}^{n} x_i + na.$$

RULE 8   *The mean of a set of observations, each of which has been multiplied by the same number and added to a given quantity, is the mean of the original observations multiplied by that number and added to the given quantity. In symbols:*

$$\overline{cx + a} = c\bar{x} + a.$$

Rule 8 is the rule for computing the mean by coding. In coding the observations x_i, we replace them by quantities $(x_i - b)/d$, compute the mean $\overline{(x - b)/d}$ of these new quantities, then multiply by d and add b to obtain $\bar{x}$. Note that $(x_i - b)/d$ is the same as $cx_i + a$, where $c = 1/d$ and $a = -b/d$. To solve for $\bar{x}$ in terms of $\overline{cx + a}$, we use Rule 8 to obtain

$$\bar{x} = \frac{\overline{(cx + a)} - a}{c};$$

in terms of b and d this is

$$\bar{x} = d\,\overline{\frac{x - b}{d}} + b.$$

Proofs.

RULE 1 Both sides of the equation represent the sum of the $2n$ numbers $x_1, x_2, \ldots, x_n, y_1, y_2, \ldots, y_n$. The order of summation is immaterial.

RULE 2 We can write

$$\sum_{i=1}^{n} a = \sum_{i=1}^{n} x_i,$$

where x_i is simply a for each i. Thus we have

$$\sum_{i=1}^{n} a = \sum_{i=1}^{n} x_i = x_1 + x_2 + \cdots + x_n = a + a + \cdots + a = na.$$

RULE 3 From Rules 1 and 2 we obtain

$$\sum_{i=1}^{n} (x_i + a) = \sum_{i=1}^{n} x_i + \sum_{i=1}^{n} a = \sum_{i=1}^{n} x_i + na.$$

RULE 4 We have

$$\overline{x + a} = \frac{1}{n} \sum_{i=1}^{n} (x_i + a) = \frac{1}{n} \left(\sum_{i=1}^{n} x_i + na \right)$$

$$= \frac{1}{n} \sum_{i=1}^{n} x_i + \frac{1}{n}(na) = \bar{x} + a.$$

RULE 5 We have

$$\sum_{i=1}^{n} cx_i = cx_1 + cx_2 + \cdots + cx_n$$

$$= c(x_1 + x_2 + \cdots + x_n) = c \sum_{i=1}^{n} x_i.$$

RULE 6 We have

$$\overline{cx} = \frac{\sum\limits_{i=1}^{n} cx_i}{n} = \frac{c \sum\limits_{i=1}^{n} x_i}{n}$$

$$= c \frac{\sum\limits_{i=1}^{n} x_i}{n} = c\bar{x}.$$

RULE 7   This rule comes from combining Rules 3 and 5.

RULE 8   This rule comes from combining Rules 4 and 6 or alternatively from Rule 7.

APPENDIX 3D   Significant Digits

A measurement is usually recorded as accurately as it is measured. Thus, if a ruler can be read to the nearest tenth of a centimeter, one would record a reading to tenths of a centimeter, for example, 3.7 centimeters. Equivalently, this is 37 millimeters and .037 meter. In each case we say there are two *significant digits* (or two *significant figures*) in the number. That means that the measurement is considered valid to 1 part in 100. In .037 the 0 is for the purpose of indicating the decimal place and is not counted as a significant digit. If a measurement is recorded as 4.0 centimeters, the 0 after the decimal point indicates the reading is accurate to .1 centimeter; in this case, also, the number of significant digits is two.

The *significant digits* of a number include (a) all the nonzero digits, (b) the zeros that are included between nonzero digits, and (c) zeros that are on the right to signify accuracy.

In reporting the final results of a calculation, one should give only the *accurate significant digits*. If 1.2 and 1.23 are two recorded measurements, then their product, $1.2 \times 1.23 = 1.476$, should be rounded to 1.5, since the factor 1.2 contains only two significant digits.

On the other hand, the *mean* of a set of numbers is generally more precise than any one of the numbers alone. (This is explained in Chapter 7.) Hence it is reasonable to report the mean of 90, 100, 80, 72, and 109, as 90.2 rather than 90. (This is discussed further in the discussion of *confidence intervals* in Chapter 8.)

EXERCISES

3.1 (Sec. 3.3)   The 25 families of Little Town live on 4 blocks:

Block	A	B	C	D
Number of persons living on the block	20	50	20	10

(a) Compute the mean number of persons (arithmetic average) per block.

(b) Compute the mean number of persons (arithmetic average) per family.

(c) Compute the mean number of families (arithmetic average) per block.

3.2 There are 25 families in Rich Town. The distribution of number of automobiles is given.

Number of cars	0	1	2	3	4
Number of families having this number of cars	0	2	6	11	6

(a) What is the median number of cars per family?
(b) What is the modal number of cars per family?
(c) What is the mean number of cars per family?

3.3 Set A is 1, 1, 2, 2, 4 and Set B is 1, 1, 2, 2, 9; indicate which of the following is true of the two samples.
(a) Equal medians but unequal means
(b) Equal means but unequal medians
(c) Neither of these

3.4 We would like to know whether the majority of a junior high school class is eligible for a program for intellectually gifted children that accepts only children with IQ's over 120. Which would be more useful, the median or the mean?

3.5 (Sec. 3.2) Which observation is the median of a sample of 21 observations?

3.6 For each of the following hypothetical data sets state whether the mean, median, or mode is the most descriptive measure of location, give its numerical value, and tell why you chose that measure.
(a) Numbers of children in 12 families:
3, 2, 2, 2, 3, 2, 1, 2, 2, 2, 2, 2
(b) Incomes of 10 families (dollars per year):
8400, 8300, 8600, 7400, 7300, 9700, 8100, 17100, 9100, 9300
(c) Heights of 10 boys (inches):
39, 40, 39, 37, 38, 39, 38, 40, 41, 40
(d) Lifetimes of 10 lightbulbs (hours):
150, 110, 441, 2100, 1503, 1305, 257, 279, 215, 2536
(Plot the data to judge the amount of skewness.)
(e) Number of cars owned by each of 10 families in the city:
1, 1, 2, 1, 1, 1, 1, 1, 2, 1
(f) Number of cars owned by each of 12 families in the suburbs:
2, 2, 1, 3, 2, 2, 3, 1, 3, 2, 2, 3

3.7 Tabulate a frequency distribution, and find the mean and median for each variable of the data in Table 3.20.

TABLE 3.20
Data for 10 Farms

FARM i	NUMBER OF CHILDREN c_i	NUMBER OF TOYS t_i	NUMBER OF HOGS h_i	NUMBER OF CHICKS d_i
1	2	8	7	10
2	4	16	7	18
3	4	16	9	18
4	3	12	9	14
5	4	16	9	18
6	3	12	8	14
7	3	12	10	14
8	3	12	6	14
9	6	24	11	26
10	3	12	8	14

(hypothetical data)

3.8 In the "postural sway" test a blindfolded subject standing erect is told he or she is moving forward. The actual sway in reaction to this "information" is measured. These measurements, in inches, and rearranged in order, are as follows (a minus sign indicating a backward movement):

−2.0, −1.5, −1.4, −1.0, −.75, −.60, −.40, −.35, −.20, −.05, −.05, +.10, +.15, +.15, +.20, +.30, +.40, +.40, +.75, +1.00, +1.50, +2.00, +3.00, +3.25, +4.5, +5.5, +6.0, +8.2

(a) Make a histogram of these measurements.
(b) Compute the mean, median, and mode.
(c) What proportion of the subjects moved backwards?

***3.9** (App. 3B)   Find the median of the distribution of incomes of 64 families shown in Table 3.4.

3.10 In the United States, although the median duration of marriages ending in divorce is about 7 years, most divorces occur in the first 5 years of marriage or between the 20th and 25th years of marriage.
   Sketch a histogram that is consistent with these facts. Can you think of a hypothesis which would explain this bimodal distribution?

3.11 Data on reported family income in a study of purchasing intentions and actions of members of a consumers' organization were given in Table 2.16.

(a) Group the data of Table 2.16. Use class intervals $3,750–6,250, $6,250–8,750, etc.

(b) What is the mean of the grouped data? Note that, for example, $7,500 is taken as the midpoint of $6,250–8,750.

(c) What is the median for the grouped data?

(d) What is the mode?

(e) What is the mean of the (ungrouped) data?

(f) What is the median of the (ungrouped) data?

(g) Why is the mode of the (ungrouped) data of little interest or relevance?

3.12 (From the *San Francisco Chronicle*, June 30, 1971) "The 'average' Bay Area resident is a married, 28-year-old white woman, living with two others in a family-owned home worth $26,115."

What kind of "averages" are referred to in this statement?

3.13 (Sec. 3.3)   Plot the data of Table 3.21 on four different axes, each numbered from 1 to 10. Use one axis for each workbook and the fourth for all the data combined. Compute the mean corresponding to each workbook and the mean of the combined data.

TABLE 3.21
Reading Achievement Scores of 18 Children Using 3 Different Workbooks

Workbook 1:	2, 4, 3, 4, 5, 6
Workbook 2:	9, 10, 10, 7, 8, 10
Workbook 3:	4, 5, 6, 3, 7, 5

(hypothetical data)

3.14 (Sec. 3.3)   A study has related television viewing habits to the educational level of the viewer. Respondents (all being males over 25) were divided into four classifications according to formal education. Each respondent's viewing was monitored for three weeks, and his daily average number of hours spent watching TV is recorded in Table 3.22 (rounded to the nearest .5).

TABLE 3.22
Average Number of Hours of TV Viewing Per Day

Non-high school graduate	5, 6, 4
High school graduate only	5, 4, 3, 4
College graduate only	3, 4, 2
Post-College Study	1.5, .5

(hypothetical data)

Plot the data for each group on a different axis. Mark the mean for each group on the corresponding axis.

3.15 (Sec. 3.4)   In a study of television viewing habits in a community in which 20% of the household heads are college graduates, the average number of hours per week which household heads spend watching television is to be estimated. Fifty families are sampled. The average number of hours per week for the 30 households whose head is a college graduate is 15 hours. For the other 20 families in the sample, the average is 25 hours per week.

(a) Does the sample of 50 seem representative of the population?

(b) Estimate the average number of hours spent watching television by household heads in the community. [Hint: Take a weighted average using the weights .20 and .80.]

3.16 (Sec. 3.4)   Among the evidence considered in the Surgeon General's report on Smoking and Health were the findings of seven studies concerning the death rates among smokers and nonsmokers. In order to make the death rates comparable it was necessary to adjust for the differences in the ages of the men in the various studies. The age distribution of the 1960 white male population of the United States was used as a standard. Table 3.23 shows the age-adjusted death rates for ages 35 and over. The units are death rates per 1000 man-years, that is, per 1000 men per year.*

Explain what is meant here by "age-adjusted death rate."

*__3.17__ (App. 3A)   Compute the mean for Table 3.24 by coding. Note that the inconvenient midpoints do not matter, since in the computation they are replaced by coded values. Hence if you are going to use coding, you might as well choose convenient class intervals like these, which give the grouping of recorded weights as 120's, 130's, 140's, etc.

3.18   There are 30 students in each of the four sections of Statistics I at Stanu College. The sections are taught by different teachers. In order to learn the students' names (and to help ensure that the students come to class), the teachers call the roll for the first six class periods. For the second through sixth class periods each teacher records the number of names not yet learned. The data are given in Table 3.25.

---

*If there were 2000 persons in a given age group at the start of a 3-year study, and if 15 died by the end of the first year, 60 more died by the end of the second, and 20 more died by the end of the third year, then there would have been $15 + 60 + 20 = 95$ deaths. The number of person-years of exposure to risk of death would have been $2000 + (2000 - 15) + (1985 - 60) = 2000 + 1985 + 1925 = 5910$ person-years. The death rate for that age group would be $95/5.91 = 16.1$ deaths per 1000 person-years.

TABLE 3.23
*Age-Adjusted Death Rates per 1000 Man-Years for Current Smokers
of Cigarettes (Age 35 and over), by Amount Smoked, in Seven Studies*

| STUDY | NONSMOKERS | CURRENT SMOKERS OF CIGARETTES | |
		Less Than 1 Pack	*1 Pack or More*
British doctors	15.8	19.2	23.2
Men in 9 states	14.4[a]	22.4[a]	27.1[a]
U.S. Veterans	12.0	18.1	23.9
California occupational	10.5[a]	14.2[a]	18.0[a]
California Legion	11.3	16.4	16.3
Canadian Veterans	14.1	22.1	24.2
Men in 25 states	12.8[b]	18.5[b]	19.2[b]

SOURCE: U.S. Public Health Service (1964), Table 15, p. 95.
[a]Ages 50–69.
[b]These figures may be too low by about 1.7%, since the man-years used in the computation included some contribution by men who had not been fully traced.

TABLE 3.24
Weights of 100 Pigs (Pounds)

RECORDED WEIGHT	CLASS INTERVAL	MIDPOINT	NUMBER OF PIGS
120–129	119.5–129.5	124.5	13
130–139	129.5–139.5	134.5	28
140–149	139.5–149.5	144.5	35
150–159	149.5–159.5	154.5	12
160–169	159.5–169.5	164.5	9
170–179	169.5–179.5	174.5	3
			100

(hypothetical data)

TABLE 3.25
Number of Names Not Yet Learned

| TRIAL | TEACHER | | | |
	A	*B*	*C*	*D*
1	21	18	25	20
2	18	16	15	10
3	16	15	12	9
4	12	13	10	7
5	9	10	8	6

(hypothetical data)

(a) Plot the data using the horizontal axis for trials and the vertical axis for number of names not yet learned. Use A's to represent the points for Teacher A, B's for Teacher B, etc. Connect the five points for each teacher.

(b) Compute the mean for each trial.

(c) Plot the means against the trial numbers. Connect the five points.

3.19 Compute the mean, first decile, first quartile, and median for the weights of the Stanford football players given in Table 2.15. What positions do you think those in the lower tenth would be most likely to play?

3.20 Group the weights in Table 2.15 into a frequency distribution, and compute the measures mentioned in Exercise 3.19.

3.21 Compute the mean, median, third quartile, and ninth decile for the heights of the Stanford football players given in Table 2.15. What positions do you think those in the upper tenth would be most likely to play?

3.22 Group the heights in Table 2.15 into a frequency distribution and compute the measures mentioned in Exercise 3.21.

3.23 Table 3.26 gives the daily outputs of three bottle-cap machines. Plot the outputs of each machine on a separate axis. Compute the mean and median for each machine and mark these measures on the axis. Compare the outputs of the machines.

TABLE 3.26

Daily Outputs of Three Bottle-Cap Machines

Machine A:	340, 345, 330, 342, 338
Machine B:	339, 333, 344
Machine C:	347, 343, 349, 355

SOURCE: Kruskal and Wallis (1952).

3.24 (Sec. 3.3)   Table 3.27 gives the results of an experiment on reaction time. In such an experiment the subject stands on a low platform connected to an electric stopwatch. The experimenter closes a switch, which flashes a light and simultaneously starts the watch. When the light flashes, the subject jumps off the platform as soon as possible. In jumping off the platform, the subject breaks the circuit and stops the watch, automatically recording the "reaction time." The results of seven successive trials (which had been preceded by three preliminary trials) are given in the table.

(a) On three separate graphs, plot the data for each subject against the trial numbers.

(b) Compute the mean reaction time for each trial.

(c) On another piece of graph paper, mark off the trials on a horizontal axis. Plot the mean reaction times against the trial numbers.

(d) Compare the three separate graphs with the graph of means. Comment.

TABLE 3.27
Reaction Times (Hundredths of a Second)

TRIAL	SUBJECT		
	A	B	C
1	71	67	75
2	63	67	78
3	60	63	69
4	54	61	69
5	60	60	65
6	60	62	67
7	57	60	67

SOURCE: Laboratory notes of one of the authors.

3.25 (Sec. 3.3)   Ten individuals participated in a study on the effectiveness of two sedatives, A and B. Each individual was given A on some nights and B on other nights. The average number of hours slept after taking the first sedative is compared with the normal amount of sleep; a similar comparison is made with the second drug. Table 3.28

TABLE 3.28
Number of Hours Increase in Sleep Due to Use of Two Different Drugs, for Each of 10 Patients

PATIENT	DRUG A	DRUG B
1	1.9	0.7
2	0.8	1.6
3	1.1	−0.2
4	0.1	−1.2
5	−0.1	−0.1
6	4.4	3.4
7	5.5	3.7
8	1.6	0.8
9	4.6	0.0
10	3.4	2.0

SOURCE: Student (1908).

gives the increase in sleep due to each sedative for each individual. (A negative value indicates a decrease in sleep.)

(a) Compute the mean increase for drug A and the mean increase for drug B.

(b) For each individual, compute the difference (increase for drug A minus increase for drug B).

(c) Compute the mean of these differences.

(d) Verify that the mean of the differences is equal to the difference between the means.

3.26 (Sec. 3.3) For the data in Exercise 2.20 (Table 2.23) relating to students in a course in statistics, compute the mean final grade of

(a) Students who had taken at least one course on probability and statistics

(b) Students who had no previous course on probability and statistics

(c) All the students.

3.27 (Sec. 3.4) Compute the answer to Exercise 3.26(c) as the appropriately weighted average of the answers to Exercise 3.26(a) and (b).

3.28 (Sec. 3.4) We are interested in a certain region composed of three countries in different stages of development. Values of the standard of living index are given in Table 3.29.

Compute the average value of the standard of living index for 1970 and for 1960. How do you account for the fact that the 1970 average is *lower* than the 1960 average, even though the standard of living *increased* in each country?

TABLE 3.29

1960 and 1970 Population (Millions) and Values of Standard of Living Index for Countries A, B, and C

COUNTRY	1960		1970	
	Population	*Index*	*Population*	*Index*
A	10	100	12	110
B	20	80	24	88
C	10	30	40	36

(hypothetical data)

3.29 (Sec. 3.4) The rate of psychosis in Country A is 36.3 per 100,000 of population and in Country B is 30.1 per 100,000 of population.

Determine whether age influences these rates. Use the figures in Tables 3.30 and 3.31 to estimate what the rate would be in Country A if it had the same age distribution as Country B. (Weight the age-specific rates for Country A according to the age distribution of Country B.)

TABLE 3.30
Statistics on Psychosis for Country A

Age	Number of People		Number of Psychoses*	Rate** (Per 100,000)
10–19	1,000,000	(5%)	200	20
20–29	3,000,000	(15%)	150	5
30–39	4,000,000	(20%)	1840	46
40–49	3,000,000	(15%)	390	13
50–59	1,000,000	(5%)	120	12
60–	8,000,000	(40%)	4560	57
	20,000,000	(100%)	7260	

(hypothetical data)
*Number of first admissions to mental hospitals for psychosis.
**Overall rate = 7260/200 = 36.3 per 100,000.

TABLE 3.31
Age Distribution in Country B

Age	Percent	Age	Percent
10–19	5	40–49	40
20–29	10	50–59	20
30–39	15	60–	10

(hypothetical data)

3.30 (App. 3B) Compute the median 1959 U.S. family income from the distribution in Table 2.20. (The median computed from the original data is $5660.)

3.31 (Sec. 3.4) Verify that in Table 3.14 the numbers 562 and 529 could be computed as weighted averages of the numbers above them in the two respective columns.

EXERCISES ON SUMMATION NOTATION

3.32 Consider a business firm's monthly expenditures over a period of one calendar year. Let x_i represent the amount expended in the *i*th month of the year. Then x_3, for example, is the March expenditure. Using summation notation, give

(a) The total expenditure for the whole calendar year
(b) The expenditure for the last six months of the year
(c) The expenditure for the summer months of June, July, and August.

3.33 Let u_i be the amount expended for salaries in the ith month and v_i be the amount expended for office supplies in the ith month. The year's expenditure for both salaries and office supplies may then be computed either by adding for each month the two kinds of expenditures and then adding these 12 monthly totals, or by finding the year's expenditure for salaries and the year's expenditures for office supplies and then adding these two totals. State this fact symbolically using the u_i's and the v_i's.

3.34 Suppose we have a set of 100 persons who are to participate in an experiment. In some random manner we assign to each person a number between 1 and 100 so that different persons have different numbers. Then we assign those persons having an odd number to the "control" group and those having an even number to the "treatment" group. The experiment consists of making the numerical observation x_i on the ith person, $i = 1, 2, \ldots, 100$.

(a) Express symbolically the arithmetic mean of the observations on the persons in the treatment group.
(b) Do the same for the control group.

3.35 Let $a_j, j = 1, \ldots, m$, be positive numbers.
(a) Show that $(\sum_{j=1}^{m} a_j v_j / \sum_{j=1}^{m} a_j)$ is a weighted average of the values $v_1, v_2, \ldots, v_m$.
(b) Show that if we take $a_j = f_j, j = 1, \ldots, m$, where f_j is the frequency of occurrence of v_j, then $(\sum_{j=1}^{m} a_j v_j / \sum_{j=1}^{m} a_j) = \bar{x}$.

3.36 Let

$x_1 = 10$	$y_1 = 100$	$z_1 = 1$
$x_2 = 9$	$y_2 = 81$	$z_2 = -1$
$x_3 = 8$	$y_3 = 64$	$z_3 = 1$
$x_4 = 7$	$y_4 = 49$	$z_4 = -1$
$x_5 = 6$	$y_5 = 36$	$z_5 = 1$
$x_6 = 5$	$y_6 = 25$	$z_6 = -1$
$x_7 = 4$	$y_7 = 16$	$z_7 = 1$
$x_8 = 3$	$y_8 = 9$	$z_8 = -1$
$x_9 = 2$	$y_9 = 4$	$z_9 = 1$
$x_{10} = 1$	$y_{10} = 1$	$z_{10} = -1$

Compute each of the following expressions and show that it is equal to the number given on the right-hand side.

(a) $x_6 + x_7 + \cdots + x_{10}$ $\qquad = \sum\limits_{j=6}^{10} x_j \qquad = 15$

(b) $x_4 + x_5 + \cdots + x_8$ $\qquad = \sum\limits_{k=4}^{8} x_k \qquad = 25$

(c) $(x_4 - 2) + (x_5 - 2) + \cdots + (x_9 - 2) \quad = \sum\limits_{l=4}^{9} (x_l - 2) \qquad = 15$

(d) $y_2 + y_4 + \cdots + y_{10}$ $\qquad = \sum\limits_{m=1}^{5} y_{2m} \qquad = 165$

(e) $x_1 z_1 + x_2 z_2 + \cdots + x_{10} z_{10}$ $\qquad = \sum\limits_{g=1}^{10} x_g z_g \qquad = 5$

(f) $(y_1 - 9^2) + (y_2 - 8^2) + \cdots + (y_{10} - 0^2) = \sum\limits_{p=1}^{10} [y_p - (10 - p)^2] = 100$

(g) $(2^2 - 1^2) + (3^2 - 2^2) + \cdots + (7^2 - 6^2) = \sum\limits_{j=1}^{6} [(j+1)^2 - j^2] \quad = 48$

(h) $z_1 + z_3 + \cdots + z_9$ $\qquad = \sum\limits_{r=1}^{5} z_{(2r-1)} \qquad = 5$

(i) $(z_4 + 3) + (z_6 + 3) + \cdots + (z_{10} + 3) \quad = \sum\limits_{s=2}^{5} (z_{2s} + 3) \qquad = 8$

(j) $4y_7 + 4y_8 + \cdots + 4y_{10}$ $\qquad = \sum\limits_{a=7}^{10} 4y_a \qquad = 120$

MATHEMATICAL EXERCISES

3.37 (Sec. 3.1) An "interpolated mode" of grouped data may be computed as follows. Let

l = lower limit of modal interval

d_1 = difference between the modal frequency and the frequency of the next lower interval

d_2 = difference between the modal frequency and the frequency of the next higher interval

w = class interval width

Then

$$\text{Interpolated mode} = l + w \left(\frac{d_1}{d_1 + d_2} \right).$$

Compute the interpolated mode for the data of Table 3.4.

3.38 Prove that if a set of numbers is multiplied by a positive number the median of the new set is the median of the old set multiplied by that number.

REFERENCES

Kruskal, W. H., and W. Allen Wallis (1952). "Use of ranks in one-criterion analysis of variance." *Journal of the American Statistical Association* 47: 583–621.

Student (W. S. Gosset) (1908). "The probable error of a mean." *Biometrika* 6: 1–25.

Sturges, H. A. (1926). "The choice of class interval." *Journal of the American Statistical Association* 21: 65–66.

U.S. Bureau of the Census (1956). *U.S. Census of Population: 1950.* U.S. Department of Commerce, Washington.

U.S. Bureau of the Census (1970). *Statistical Abstract of the United States.* U.S. Department of Commerce, Washington.

U.S. Public Health Service (1964). *Smoking and Health.* Report of the Advisory Committee to the Surgeon General of the Public Health Service. Public Health Service Publication No. 1103. U.S. Department of Health, Education, and Welfare, Washington.

4 MEASURES
OF VARIABILITY

INTRODUCTION

Although for some purposes an average may be a sufficient description of a set of data, usually more information about the data is needed. An important feature of statistical data is their *variability*—how much the measurements differ from individual to individual. In this chapter we discuss the numerical evaluation of variability. Other terms for this concept are *spread*, *scatter*, and *dispersion*.

A measure of location is more representative of the whole set of observations if the variability of that set is small than if it is large. The three heights 71, 72, and 73 inches have the same mean, 72 inches, as the three heights 66, 72, and 78 inches. However, we have the feeling that 72 inches is more descriptive of the first set of heights than of the second. If two sets of observations have the same average, this average value is more descriptive of the set with smaller variability.

4.1 RANGES

4.1.1 The Range

A student can get an idea of the variability of the heights of men in the class by comparing the tallest man with the shortest, for if the difference is small the heights do not vary much. The difference between the largest measurement in a set (the *maximum*) and the smallest (the *minimum*) is called the *range*.

The basic shortcoming of the range as a measure of variability is that it is entirely dependent on these two extreme values. For example, if all of the males in a large college class are between 69 and 71 inches tall except for one child prodigy and one basketball player, we would feel that the variability is small and that 70 inches is quite descriptive of the heights of males in the class. If the numbers of children in eight families are 0, 1, 1, 2, 2, 3, 3, and 9, the range is 9, but all of the families except two have 1, 2, or 3 children; again the variability is small though the range is large.

A second disadvantage of using the range to characterize spread is that, because it depends directly on just the largest and smallest measurements, it depends indirectly on the *number* of measurements in the set. For instance, if two families have 1 and 4 children, respectively, the range is $4 - 1 = 3$; if we add a third family, the range continues to be 3 in case the number of children in this family is 1, 2, 3, or 4, and it is larger if the number of children is 0 or greater than 4. The point is that the range may increase with the size of the set of observations though it can never decrease, while ideally a measure of variability should be roughly independent of the number of measurements.

4.1.2   The Interquartile Range

A measure of variability that is not sensitive to the sample size and not affected by extremes is the *interquartile range*, which is the difference between the first and third quartiles. For the numbers of children in eight families, 0, 1, 1, 2, 2, 3, 3, 9 (shown in Figure 4.1), the first quartile is 1 and the third quartile is 3. The interquartile range is $3 - 1 = 2$.

Another set of eight observations with the same range is 0, 1, 1, 2, 3, 6, 6, 9. (See Figure 4.2.) Most of us would feel that this set exhibits more variability than the first set. The range does not reflect this fact, but the interquartile range does. The interquartile range of this new set is $6 - 1 = 5$, instead of 2.

The quartiles are measures of location; that is, they are positional indicators. The interquartile range may thus be called a *positional measure of variability*.

We have remarked that the range cannot decrease, but can increase, when additional observations are included in the set and that in this sense the range is overly sensitive to the number of observations. The interquartile range does not share this deficiency; it can either increase or decrease when further observations are added to the sample. In the set of observations 1, 2, 3, 4, 5, the range is $5 - 1 = 4$, the first quartile is 2, the third quartile is 4, and the interquartile range is $4 - 2 = 2$. If we obtain the four additional observations 6, 3, 3, and 4, the range has increased to $6 - 1 = 5$, but the interquartile range has *decreased* to 1, since the first quartile is now 3 and the third quartile is now 4.

Another positional measure of variability is the *interdecile range*, the difference between the ninth and first deciles; most (80%) of the observations lie between these deciles.

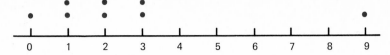

Figure 4.1 *Numbers of children in 8 families: Small variability (range is 9; interquartile range is 2)*

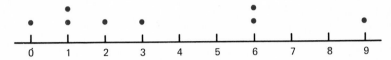

Figure 4.2 *Numbers of children in 8 families: Large variability (range is 9; interquartile range is 5)*

4.2 THE MEAN DEVIATION

The mean deviation is an intuitively appealing measure of variability. Table 4.1 gives the weights at birth of 5 babies. The mean weight is 7 pounds. The *deviation* of each weight from the mean is given in the third column of Table 4.1. The deviation is the measurement minus the mean. Note that some of the deviations are positive numbers and some are negative. Because we have subtracted the mean, the positive differences balance the negative ones in the sense that the (algebraic) sum (or the mean) of the deviations is 0. (This fact is proved in Appendix 4C.)

TABLE 4.1
Weights of 5 Babies

Baby	Weight (Pounds)	Deviation	Absolute Deviation
Ann	8	1	1
Beth	6	−1	1
Carl	7	0	0
Dean	5	−2	2
Ethel	9	2	2
Total	35	0	6
Arithmetic Average	7	0	1.2

(hypothetical data)

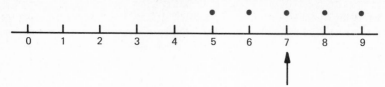

Figure 4.3 *Location of stores*

The *absolute value* of the deviation, that is, the numerical value without regard to sign, indicates how different a weight is from the mean. The average of these numbers provides a measure of variability. The absolute value of a quantity is indicated by placing the quantity between two vertical bars; for example, $|-2| = 2$. For the five weights in Table 4.1 the mean absolute deviation is

$$\frac{|-2| + |-1| + |0| + |1| + |2|}{5} = \frac{2 + 1 + 0 + 1 + 2}{5} = \frac{6}{5} = 1.2.$$

For a more graphic interpretation of these numbers, consider 5, 6, 7, 8, and 9 as the distances in blocks along a street of 5 stores from a bus transfer point. If one bus stop is to serve all of the stores, it should be located at the middle store (at block 7). (See Figure 4.3.) Then 1.2 is the average number of blocks walked by shoppers who go equally to all stores. If the variability in this sense were less, the shoppers would clearly be pleased.

In general terms we may refer to the measurements as $y_1, y_2, \ldots, y_N$, where N is the number of observational units in the population. The mean is

$$\mu = \frac{\sum_{i=1}^{N} y_i}{N};$$

the deviations are $y_i - \mu$, $i = 1, 2, \ldots, N$; and the absolute deviations are $|y_i - \mu|$, $i = 1, 2, \ldots, N$. In graphical terms the *mean deviation* (short for "mean absolute deviation") is the average distance of the points $y_1, y_2, \ldots, y_N$ from the mean, namely,

$$\frac{\sum_{i=1}^{N} |y_i - \mu|}{N}.$$

At this stage of our exposition we must again distinguish between *population* and *sample* measures. We shall often be drawing a sample from a population and using the sample characteristics to estimate the corresponding population characteristics. To make them match properly the sample measure of variability must be defined slightly differently from the population measure. Specifically, the *sample mean deviation* is defined to be

$$\frac{\sum_{i=1}^{n} |x_i - \bar{x}|}{n - 1},$$

where $x_1, x_2, \ldots, x_n$ constitute a sample of n and $\bar{x}$ is the sample mean. The divisor $n - 1$ is used here instead of n in view of the fact that the sample mean $\bar{x}$ is used instead of the population mean μ. The reason why this divisor should be less than n is explained by the following example.*

Figure 4.4 represents the results of a boy tossing 3 pennies at a line that is 100 inches away from him. The 3 coins land at distances x_1, x_2, and x_3 inches away from him. The sum of the distances of the three coins from the line (which defines presumably the mean of the infinite population of all possible tosses) is

$$|x_1 - 100| + |x_2 - 100| + |x_3 - 100|;$$

the average distance (this sum divided by 3) is a measure of scattering around the line of aim. However, the sum of the distances of the tosses from the dotted line established by $\bar{x}$ is

$$|x_1 - \bar{x}| + |x_2 - \bar{x}| + |x_3 - \bar{x}|,$$

which is usually smaller than the first sum because $\bar{x}$ is in the center of x_1, x_2, x_3 (the sample), while 100 (corresponding to the line of aim—the center of the population) may not be. To adjust for the fact that $\bar{x}$ tends to make the distance smaller, the sum is divided by 2 instead of 3.

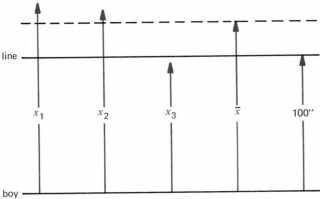

Figure 4.4 *Tosses of a coin*

*Many authors divide by n in defining the sample mean deviation. Dividing by $n - 1$ is consistent with what is done in the case of the sample standard deviation. (See Section 4.3.)

It is an interesting fact that the mean deviation from the median is less than (or equal to) the mean deviation from the mean. In fact, given a set of observations, the quantity $\Sigma_{i=1}^{n} |x_i - a|$ is made smallest by choosing a to be the median. (See Exercises 4.33 and 4.34.) Deviations could be defined as differences of observations from the *median*, and the mean deviation could be an average of these, but this is not customarily done.

4.3   THE STANDARD DEVIATION

4.3.1   Definitions

The *standard deviation* is the most frequently used measure of variability, although it is not as easily understood as ranges or the mean deviation. It can be considered as a kind of average (the "root mean square") of the absolute deviations of observations from the mean.*
In the case of the data in Table 4.1 we square each deviation to obtain $(-2)^2 = 4$, $(-1)^2 = 1$, $0^2 = 0$, $1^2 = 1$, $2^2 = 4$, and add these results to obtain 10. We find an average squared deviation by dividing by 5 to obtain 2. The (population) standard deviation is the (positive) square root of this number, $\sqrt{2} = 1.414$. For a population we denote this quantity by lower case *sigma*, σ, which is the Greek letter corresponding to s (for "standard deviation"). The computation can be summarized by

$$\sigma = \sqrt{\frac{(5-7)^2 + (6-7)^2 + (7-7)^2 + (8-7)^2 + (9-7)^2}{5}}.$$

The *squaring* of the numbers produces nonnegative numbers; one averages them and then takes the square root to return to the original units. We shall show shortly that summing squared deviations can be done quite expeditiously; in fact, the standard deviation is numerically and algebraically easier to use than the mean deviation.
For a population of measurements $y_1, y_2, \ldots, y_N$ the (population) *standard deviation* is

---

*The root mean square of a set of positive numbers $a_1, a_2, \ldots, a_N$ is a kind of average defined by

$$\sqrt{\frac{\sum_{i=1}^{N} a_i^2}{N}}.$$

In defining σ, $a_i = |y_i - \mu|$, $i = 1, 2, \ldots, N$.

$$\sigma = \sqrt{\frac{\sum\limits_{i=1}^{N}(y_i - \mu)^2}{N}}.$$

The average of the squared deviations itself is of importance; it is called the population *variance* and denoted by σ^2.

Now we turn to the definitions for a *sample*. The computation is the same as above except that we divide by one less than the number of observations. Thus if the 5 weights in Table 4.1 constitute a sample (for example, of the weights of all babies born in a certain hospital in a certain year), the standard deviation is $\sqrt{10/4} = \sqrt{2.5} = 1.581$. In general, the *sample standard deviation* is

$$s = \sqrt{\frac{\sum\limits_{i=1}^{n}(x_i - \bar{x})^2}{n-1}},$$

where $x_1, x_2, \ldots, x_n$ constitute the sample. The *sample variance* is

$$s^2 = \frac{\sum\limits_{i=1}^{n}(x_i - \bar{x})^2}{n-1}.$$

4.3.2 Reasons for Dividing by One Less than the Sample Size

It was noted in the discussion of the mean deviation that in defining a sample measure of variability in terms of deviations from the sample mean, it was appropriate to divide by a number less than the sample size. This idea can be made clearer in the case of the standard deviation. It is a fact that given n numbers $x_1, x_2, \ldots, x_n$, the quantity $\sum_{i=1}^{n}(x_i - a)^2$ is made smallest by choosing a to be $\bar{x}$. (This is proved in Appendix 4C.) The quantity $\sum_{i=1}^{n}(x_i - a)^2$ is called the "sum of squared deviations about a"; thus the sum of squared deviations about $\bar{x}$ is as small as the sum of squared deviations about any point a. In particular, the sum of squared deviations around the sample mean $\bar{x}$. $\sum_{i=1}^{n}(x_i - \bar{x})^2$, is as small as (or smaller than) the sum of squared deviations about the population mean μ, $\sum_{i=1}^{n}(x_i - \mu)^2$. If we know the population mean μ but do not know the population standard deviation σ, we should use deviations around μ to obtain

(4.1)
$$\sqrt{\frac{\sum_{i=1}^{n} (x_i - \mu)^2}{n}}$$

as the sample analogue of σ. When we do not know the value of μ, we could simply replace it by its sample analogue $\bar{x}$ to obtain

(4.2)
$$\sqrt{\frac{\sum_{i=1}^{n} (x_i - \bar{x})^2}{n}}.$$

However, because the sum of squared deviations around $\bar{x}$ is smaller than around μ, (4.2) is smaller than (4.1) and thus tends to underestimate σ; that is, (4.2) will tend to be too small. We can remedy this situation by dividing by something smaller than n (so that the result of the division will be larger). It turns out that dividing by $n - 1$ instead of n is just right. A hint of why this is so is gained by consideration of the deviations $x_1 - \bar{x}, x_2 - \bar{x}, \ldots, x_n - \bar{x}$. We noted at the beginning of Section 4.2 that the sum of deviations is zero; that is,

$$\sum_{i=1}^{n} (x_i - \bar{x}) = 0.$$

This means that the n deviations are not n independent quantities. Given the value of the deviations for $x_1, x_2, \ldots, x_{n-1}$, the value of the deviation of x_n is determined because the sum of all deviations must be zero. For example, with the $n = 3$ observations $x_1 = 4$, $x_2 = 2$, $x_3 = 6$, we have $\bar{x} = 4$, $x_1 - \bar{x} = 0$, and $x_2 - \bar{x} = -2$. Now, we know that

$$(x_1 - \bar{x}) + (x_2 - \bar{x}) + (x_3 - \bar{x}) = 0,$$

or

$$0 + (-2) + (x_3 - \bar{x}) = 0,$$

so $x_3 - \bar{x} = +2$. Thus, among the n quantities $x_1 - \bar{x}, x_2 - \bar{x}, \ldots, x_n - \bar{x}$ there are only $n - 1$ independent quantities, and this is one reason why we divide by $n - 1$ instead of n.

4.3.3   Interpretation

Pairwise Distances.   The variance has a natural interpretation in terms of *distances* (differences) between pairs of points (observations). The distances between stores A, B, C, D, and E at blocks 5, 6, 7, 8, and 9 along Center Street are given in Table 4.2. (The same distances are true for the set of observations in Table 4.1.) Roughly speaking, the larger the distances between pairs of observations the greater the variability, and a measure of variability should reflect this idea. Each number in Table 4.2 is a quantity of the form $|x_i - x_j|$, where x_i and x_j are pairs from 5, 6, 7, 8, and 9. The mean of these quantities is $20/10 = 2$. There is no direct relation between this mean and the mean of the absolute deviations $|x_i - \bar{x}|$ (which is 1.2). But there is a direct relation between the average of the *squares* of differences $(x_i - x_j)^2$ and the average of the squares of sample deviations $(x_i - \bar{x})^2$. The mean of the squared distances (given in Table 4.3) is $50/10 = 5$. The sample variance, which is the square of the sample standard deviation, is 2.5. This is one-half of the mean of the squared sample differences. *The variance of a sample is equal to one-half the mean of the squared pairwise distances.* The reader should try other examples to check this statement.

The interpretation is that the larger the variance the more scattered are the points in the sense of a large average of squared interpoint distances.

TABLE 4.2
Distances Between Stores

		STORE				
		A	B	C	D	E
	A	0	1	2	3	4
	B	1	0	1	2	3
STORE	C	2	1	0	1	2
	D	3	2	1	0	1
	E	4	3	2	1	0

TABLE 4.3
Squares of Distances Between Stores

		STORE				
		A	B	C	D	E
	A	0	1	4	9	16
	B	1	0	1	4	9
STORE	C	4	1	0	1	4
	D	9	4	1	0	1
	E	16	9	4	1	0

SOURCE: Table 4.2.

4.4   USES OF MEASURES OF VARIABILITY

4.4.1   Assessing the Amount of Information Contained in a Measure of Location

As indicated at the outset of this chapter, a measure of variability is often reported along with a measure of location as an indication of how typical of the population that measure of location is. Thus the fact that the median of a distribution of income is $9170 tells us more about the whole distribution of income when the interquartile range is $3760 than when it is $6000. Similarly, the standard deviation gives an indication of how typical of a whole distribution the mean is. In fact, the variance (the square of the standard deviation) is proportional to the sum of the squared distances of the measurements from the mean. If two distributions have large variabilities, any slight difference in their measures of location would be unimportant because many measurements in the population with the smaller measure of location would be larger than many measurements in the other.

4.4.2   Various Measures of Variability

The several measures of variability will usually compare different samples (or populations) differently. The means of both samples of size $n = 5$ in Figure 4.5 are 4 and the mean deviations are $6/4 = 1.5$. However, for Sample 1

$$s = \sqrt{\frac{10}{4}} = \sqrt{\frac{5}{2}} = 1.581,$$

while for Sample 2 we have

$$s = \sqrt{\frac{18}{4}} = \sqrt{\frac{9}{2}} = 2.121.$$

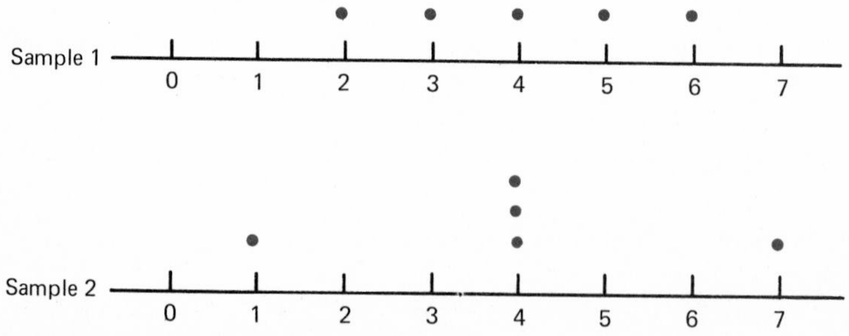

Figure 4.5 *Two samples of size* $n = 5$

Because it involves the *squares* of deviations from the mean, the standard deviation emphasizes extremes. It is, therefore, sometimes better to use another statistic, such as the interquartile range, to describe the variability of a distribution in which there is a small number of very extreme measurements.

4.4.3   Comparing Distributions

In comparing two distributions we often focus attention on their location and variability. If their locations are the same, they may still differ in variability. If their variabilities are the same, they may still differ in location. We shall give several examples.

Two countries having income distributions with the same mean but different standard deviations could have very different patterns of distribution of wealth. One country may have a large middle class; the other, a small middle class or no middle class at all.

Brand A tires are priced the same as brand B tires. There is a 20% discount on purchases of 4 or 5 tires of either brand. Thus there is a saving if the tires wear out at nearly the same time. If the lifetime of brand B tires has a standard deviation of 1000 miles and the lifetime of brand A tires a standard deviation of 5000 miles, one might prefer brand B tires; the total cost of tires might be smaller with brand B, even if the *average* lifetime of all brand B tires is shorter, because they are more likely to wear out at the same or nearly the same rate.

Given a choice between two farm locations where the rainfall had the same average but different variability, a farmer would buy the location with small variability so that he or she could plan for crops with less risk of drought and washouts. Consider, for example, that for the preceding 5 years the rainfall at farm A had been 40, 36, 25, 60, and 54 inches and at farm B it had been 38, 40, 35, 50, and 52 inches. Although the mean, 43, is the same for both locations, the (sample) standard deviation for rainfall at farm A is 14.07, while at farm B it is only 7.55. The relatively large standard deviation for farm A reflects extremely low (25 inches) and extremely high (60 inches) rainfall; farm B would be preferred because there the standard deviation is smaller.

In Section 3.5.2 we compared the waiting-time distributions of telephone calls for two alternative systems. To the already discussed characteristics of these two distributions we now add the fact that the standard deviation for the data in Figure 3.8 is 0.41 second and for the data in Figure 3.9 it is 0.69 second. The sizable difference between these standard deviations is a clue that it is not sufficient to compare only the means or medians (as reported in Table 3.16).

4.5 FORMULAS FOR THE STANDARD DEVIATION

4.5.1 Deviations from the Mean

The standard deviation and the variance have been defined in terms of the sums of squares of deviations from the mean. It is possible to calculate these quantities without explicitly obtaining the deviations; when the mean is carried to several significant figures it may be quite troublesome to find the deviations, and also difficult to maintain computational accuracy due to rounding error.

For the observations 5, 6, 7, 8, and 9, the squared deviations from the mean are 4, 1, 0, 1, and 4; these add to 10. An alternative way of obtaining this sum is to square the original numbers to obtain 25, 36, 49, 64, and 81, and add these squares to obtain 255. Then compute $n = 5$ times the square of the mean, 49, obtaining 245. Subtract this result from 255, ending up with 10. The rule is

(4.3)
$$\sum_{i=1}^{n} (x_i - \bar{x})^2 = \sum_{i=1}^{n} x_i^2 - n\bar{x}^2.$$

Since

$$\bar{x} = \frac{\sum_{i=1}^{n} x_i}{n},$$

we can also write this as

(4.4)
$$\sum_{i=1}^{n} (x_i - \bar{x})^2 = \sum_{i=1}^{n} x_i^2 - \frac{\left(\sum_{i=1}^{n} x_i\right)^2}{n}.$$

The identities (4.3) and (4.4) are proved as Rules 8 and 9 in Appendix 4C.

Note that in the right-hand side of equation (4.4) the first term is the *sum of the squares;* the second term involves the *square of the sum.* Equation (4.4) usually gives more accurate results than (4.3) because division is done at a later stage of the calculations, thereby eliminating some error due to rounding.

When using one of these formulas, care must be taken to carry enough digits, as digits are lost in the subtraction. The measurements 1020, 1012, 1021, 1017, 1015, 1010, and 1016, for example, might be recordings of atmospheric pressure, in millibars. If seven significant digits are retained in the intermediate steps of the computation, then the results are

$$\sum_{i=1}^{7} x_i^2 - \frac{\left(\sum_{i=1}^{7} x_i\right)^2}{7} = 7{,}223{,}855 - \frac{(7111)^2}{7}$$

$$= 7{,}223{,}855 - 7{,}223{,}760$$

$$= 95.$$

This result has only two significant digits.

The difficulty arises from the fact that all the data are large numbers, and their squares are even larger. An effective means of dealing with this problem is to subtract a "guessed" mean (*working mean*) from each datum and compute the variance from the resulting coded data. It is not necessary that the working mean be particularly close to the actual mean of the sample; in the above example 1000 could be used. Often it is convenient (especially with computers, when the data are read in order) to use the first observation, x_i, as the working mean. In the example, $x_1 = 1020$, and the coded observations would be

$$0, -8, 1, -3, -5, -10, -4.$$

It is easy to compute with these small numbers, and seven-digit accuracy will be more than sufficient. The sum of squares is 215; the sum is −29; the square of the sum is 841. Again retaining seven digits in the computation, we now have

$$215 - \frac{841}{7} = 215 - 120.1429 = 94.8571.$$

This result has six significant digits, more than sufficient for most purposes. Not only is this result the sum of squared deviations of the coded observations, *it is also the sum of squared deviations of the original measurements, because subtracting the same number from each observation changes only the location, not the spread, of the set of observations.* (See Rule 2, Appendix 4C, and Exercise 4.26.)

4.5.2 Calculations from Discrete Data

Just as the calculation of the mean can be simplified when some values are repeated, so can the computation of the variance and standard deviation. Table 4.4 uses again the data in Table 3.6, for which the mean was calculated as 3.5. To calculate the standard deviation we want the sum of squares of the numbers. Table 4.4 gives squares of the observations in the third column. Since many values are repeated, it is

TABLE 4.4

Number of Children in 10 Families

Family	Number of Children	Squares of Number of Children
1	2	4
2	4	16
3	4	16
4	3	9
5	4	16
6	3	9
7	3	9
8	3	9
9	6	36
10	3	9
	35	133

SOURCE: Table 3.6.

TABLE 4.5

Number of Children in 10 Families

j	Number of Children v_j	Number of Families Having v_j Children f_j	Squares of Number of Children v_j^2	$f_j v_j^2$
1	2	1	4	4
2	3	5	9	45
3	4	3	16	48
4	6	1	36	36
				133

SOURCE: Table 4.4.

convenient to use the frequency distribution in Table 4.5. Then the summing of squares indicated in Table 4.4 can be done by using the frequencies. The sum of squared deviations is

$$133 - 10(3.5)^2 = 133 - 122.5 = 10.5;$$

the sample variance is $10.5/9 = 1.1667$; and the sample standard deviation is $\sqrt{1.1667} = 1.0801$.

In symbols, the calculations are

$$\sum_{i=1}^{n} x_i = \sum_{j=1}^{m} f_j v_j, \qquad \bar{x} = \frac{\sum_{j=1}^{m} f_j v_j}{n}, \qquad \sum_{i=1}^{n} x_i^2 = \sum_{j=1}^{m} f_j v_j^2,$$

$$s = \sqrt{\frac{\sum_{j=1}^{m} f_j v_j^2 - n \left(\dfrac{\sum_{j=1}^{m} f_j v_j}{n} \right)^2}{n-1}}$$

by (4.3). Alternatively by (4.4)

$$s = \sqrt{\frac{\sum_{j=1}^{m} f_j v_j^2 - \dfrac{\left(\sum_{j=1}^{m} f_j v_j \right)^2}{n}}{n-1}} .$$

We have used

$$n\bar{x}^2 = n \left(\frac{\sum_{j=1}^{m} f_j v_j}{n} \right)^2 = \frac{\left(\sum_{j=1}^{m} f_j v_j \right)^2}{n} .$$

4.5.3 Calculations from Grouped Data

For *grouped continuous data* the pattern of the calculation is the same as for discrete data. Now the midpoints of the intervals play the role of the discrete values. In the first four columns of Table 4.6 we repeat the four columns of Table 3.10. The last three 0's of v_j are dropped for convenience (that is, the unit is $1000). The quantities $f_j v_j^2$ in the sixth column are computed by multiplying v_j by $f_j v_j$. The sum of squared deviations is

$$11{,}808 - \frac{(808)^2}{64} ,$$

which is $11{,}808 - 10{,}201 = 1{,}607$. The sample variance is

$$s^2 = \frac{1607}{63} = 25.50794,$$

and the sample standard deviation is

$$s = \sqrt{25.50794} = 5.05054$$

or $5051 to the nearest dollar.

TABLE 4.6

Computation of the Variance of Grouped Income Data

j	v_j	f_j	$f_j v_j$	$f_j v_j^2$
1	2	4	8	16
2	6	6	36	216
3	10	15	150	1,500
4	14	28	392	5,488
5	18	7	126	2,268
6	22	2	44	968
7	26	2	52	1,352
		$n = 64$	808	11,808

SOURCE: Table 3.10.
NOTE: v_j is in $1000's.

SUMMARY

A measure of location is more representative of a whole set of observations if the variability of that set is small than if it is large.

The *range* and *interquartile range* are *positional measures of variability.*

The mean deviation and standard deviation are interpreted as average deviations from the mean. The *mean deviation* is an arithmetic average of the absolute deviations; the *standard deviation* is their root mean square average. For the population the average involves division by the population size; for the sample it involves division by 1 less than the sample size.

The sample *variance* is half the average squared difference between observations.

When the same quantity is subtracted from each observation, the location changes but the spread does not. Hence, the sum of squared deviations, the variance, and the standard deviation are unchanged.

The standard deviation is in the same units as the individual measurements.

APPENDIX 4A   Coding for the Standard Deviation

If we have recorded observations in inches but want to do the arithmetic in terms of feet, so as to be working with smaller numbers, we divide each observation by 12. This division is an example of "coding." The computed standard deviation will then be in units of feet. To state the standard deviation in units of inches, we must "decode" by multiplying by 12.

We illustrate coding with the data in Table 4.6. These data were coded in Table 3.17. The sum of squared deviations in the coded units is (see Table 4.7)

$$552 - \frac{170^2}{64} = 552 - 451.5625 = 100.4375;$$

the variance is

$$\frac{100.4375}{63} = 1.594246;$$

and the standard deviation is

$$\sqrt{1.594246} = 1.26263.$$

The standard deviation in the original units is

$$1.26263 \times \$4000 = \$5051$$

to the nearest dollar, as obtained in Section 4.5.3. We note in passing that Table 4.1 is coded by using units of thousands of dollars.

TABLE 4.7
Computation of the Variance by Coding

j	Frequency f_j	Midpoint (1000's) v_j	Coded Value $v_j' = \dfrac{v_j - 2}{4}$	$f_j v_j'$	$f_j (v_j')^2$
1	4	2	0	0	0
2	6	6	1	6	6
3	15	10	2	30	60
4	28	14	3	84	252
5	7	18	4	28	112
6	2	22	5	10	50
7	2	26	6	12	72
	64			170	552

SOURCE: Table 4.6.

APPENDIX 4B   Interquartile Range of Grouped Continuous Data

For grouped continuous data the quartiles are obtained from the cumulative relative frequencies. But since the data are continuous, we interpolate to obtain a closer approximation to the true quartiles. We illustrate with the incomes of the subsample of 49 families of Table 2.16, which are tabulated into the frequency distribution of Table 4.8. The median lies in the class interval having $10,000 as its midpoint, since a cumulative relative frequency of .50 falls within that interval—44.9% of the families have incomes less than $8,750; 79.6% have incomes less than $11,250. To reach 50%, we need to penetrate a proportion $(.500 - .449)/(.796 - .449) = 51/347 = .147$ of the way into the interval $8,750 - 11,250$. Hence the median is approximately at $8,750 + .147 \times \$2,500 = \$8,750 + 368 = \$9,120.$* Similarly, we find

$$\text{first quartile} = \$6250 + \frac{.250 - .102}{.449 - .102} \times \$2500$$

$$= 6250 + .427 \times \$2500$$

and

$$= 6250 + 1070 = \$7320*$$

$$\text{third quartile} = \$8750 + \frac{.750 - .449}{.796 - .449} \times \$2500$$

$$= 8750 + .867 \times \$2500$$

$$= 8750 + 2170 = \$10,920.*$$

TABLE 4.8

Frequency Distribution of Income

INCOME INTERVAL	MIDPOINT OF INTERVAL	FREQUENCY	CUMULATIVE FREQUENCY	CUMULATIVE RELATIVE FREQUENCY
$ 3,750–6,250	$ 5,000	5	5	.102
6,250–8,750	7,500	17	22	.449
8,750–11,250	10,000	17	39	.796
11,250–13,750	12,500	7	46	.939
13,750–16,250	15,000	2	48	.979
16,250–18,750	17,500	0	48	.979
18,750–21,250	20,000	1	49	1.000

SOURCE: Table 2.16.

*Note that computations are rounded to an appropriate number of significant digits. See Appendix 3D.

The interquartile range is $\$10{,}925 - 7{,}325 = \$3{,}600$. That is, the half of the families having incomes in the middle of the distribution have incomes which span a range of approximately $\$3{,}600$.

APPENDIX 4C More Rules for Summation

Rules. These additional rules for summation verify the method of coding for calculation of variances as well as the method of summing squared deviations by summing squares and subtracting the product of the number of observations and the square of the mean. The proofs are given at the end of this Appendix.

RULE 1 *The sum of deviations around the sample mean is zero:*

$$\sum_{i=1}^{n} (x_i - \bar{x}) = 0.$$

RULE 2 *The sum of squared deviations around the mean is not changed by adding the same number to each observation:*

$$\sum_{i=1}^{n} [(x_i + a) - \overline{(x + a)}]^2 = \sum_{i=1}^{n} (x_i - \bar{x})^2.$$

RULE 3 *The standard deviation is not changed by adding the same number to each observation:*

$$s_{x+a} = s_x,$$

where s_{x+a} *denotes the sample standard deviation of* $x_1 + a, x_2 + a, \ldots, x_n + a$.

RULE 4 *If every observation is multiplied by the same number, the sum of squared deviations is multiplied by the square of that number:*

$$\sum_{i=1}^{n} [(cx_i) - \overline{(cx)}]^2 = c^2 \sum_{i=1}^{n} (x_i - \bar{x})^2.$$

RULE 5   *If every observation is multiplied by the same number, the standard deviation is multiplied by the absolute value of that number:*

$$s_{cx} = |c|s_x,$$

where s_{cx} denotes the sample standard deviation of $cx_1, cx_2, \ldots, cx_n$.

RULE 6   *If every observation is multiplied by the same number and added to a given quantity, the sum of the squared deviations is multiplied by the square of that number:*

$$\sum_{i=1}^{n} [(cy_i + a) - \overline{(cy + a)}]^2 = c^2 \sum_{i=1}^{n} (y_i - \bar{y})^2.$$

RULE 7   *If every observation is multiplied by the same number and added to a given quantity, the standard deviation is multiplied by the absolute value of that number:*

$$s_{cy+a} = |c|s_y.$$

If $y_1, y_2, \ldots, y_n$ are coded values, related to the original observations $x_1, x_2, \ldots, x_n$ by the formula $x = cy + a$, that is, $y = (x - a)/c$, then

$$s_x = s_{cy+a} = |c|s_y.$$

This is the mathematical justification for the method of coding used to calculate the standard deviation.

Rules 3, 5, and 7 hold also for σ_y, σ_{y+a}, σ_{cy}, and σ_{cy+a}, as is seen from the proofs.

RULE 8   *The sum of squared deviations around the sample mean is equal to the sum of squares minus the product of the sample size and the square of the mean:*

$$\sum_{i=1}^{n} (x_i - \bar{x})^2 = \sum_{i=1}^{n} x_i^2 - n\bar{x}^2.$$

This algebraic fact is the basis for an efficient method for calculation of the standard deviation.

RULE 9   *The sum of squared deviations can be computed by subtracting the quotient of the square of the sum and the number of observations from the sum of squares:*

$$\sum_{i=1}^{n} (x_i - \bar{x})^2 = \sum_{i=1}^{n} x_i^2 - \frac{\left(\sum_{i=1}^{n} x_i\right)^2}{n}.$$

This method of computing the sum of squared deviations has the least rounding error (but the subtraction on the right-hand side can lead to loss of significant digits).

RULE 10 *The sum of squared deviations around an arbitrary value is the sum of squared deviations around the sample mean plus the product of the sample size and the square of the difference between the mean and that value:*

$$\sum_{i=1}^{n} (x_i - a)^2 = \sum_{i=1}^{n} (x_i - \bar{x})^2 + n(\bar{x} - a)^2.$$

Note that Rule 8 is equivalent to Rule 10 for $a = 0$.

RULE 11 *The sum of squared deviations around the sample mean is as small as the sum of squared deviations around any other number.**

$$\sum_{i=1}^{n} (x_i - \bar{x})^2 \leq \sum_{i=1}^{n} (x_i - a)^2$$

for any number a.

This rule implies that $\sum_{i=1}^{n} (x_i - a)^2$ is minimized by choosing $a = \bar{x}$.

Proofs.
RULE 1 We have

$$\sum_{i=1}^{n} (x_i - \bar{x}) = \sum_{i=1}^{n} x_i - \sum_{i=1}^{n} \bar{x}$$

$$= \sum_{i=1}^{n} x_i - n\bar{x}$$

$$= 0.$$

*The notation $b \leq c$ means that b is less than or equal to c.

Rule 2   We have

$$\sum_{i=1}^{n} [(x_i + a) - \overline{(x+a)}]^2 = \sum_{i=1}^{n} [(x_i + a) - (\bar{x} + a)]^2$$

$$= \sum_{i=1}^{n} (x_i + a - \bar{x} - a)^2$$

$$= \sum_{i=1}^{n} (x_i - \bar{x})^2.$$

Rule 3   We have

$$s_{x+a} = \sqrt{\frac{\sum_{i=1}^{n} [(x_i + a) - \overline{(x+a)}]^2}{n - 1}} = \sqrt{\frac{\sum_{i=1}^{n} [x_i - \bar{x}]^2}{n - 1}} = s_x.$$

Rule 4   We have

$$\sum_{i=1}^{n} [(cx_i) - \overline{(cx_i)}]^2 = \sum_{i=1}^{n} (cx_i - c\bar{x})^2$$

$$= \sum_{i=1}^{n} [c(x_i - \bar{x})]^2$$

$$= \sum_{i=1}^{n} [c^2(x_i - \bar{x})^2]$$

$$= c^2 \sum_{i=1}^{n} (x_i - \bar{x})^2.$$

Rule 5   We have

$$s_{cx} = \sqrt{\frac{\sum_{i=1}^{n} [(cx_i) - \overline{(cx)}]^2}{n - 1}} = \sqrt{c^2 \frac{\sum_{i=1}^{n} (x_i - \bar{x})^2}{n - 1}} = \sqrt{c^2}\, s_x = |c|s_x.$$

Rule 6   From Rules 2 and 4 we obtain

$$\sum_{i=1}^{n} [(cy_i + a) - \overline{(cy + a)}]^2 = \sum_{i=1}^{n} (cy_i - \overline{cy})^2$$

$$= c^2 \sum_{i=1}^{n} (y_i - \bar{y})^2.$$

Rule 7   From Rules 3 and 5 we obtain

$$s_{cy+a} = s_{cy} = |c|s_y.$$

Rule 7 can also be obtained from Rule 6.

RULE 8   We have

$$\sum_{i=1}^{n} (x_i - \bar{x})^2 = (x_1 - \bar{x})^2 + \cdots + (x_n - \bar{x})^2$$

$$= x_1^2 - 2\bar{x}x_1 + \bar{x}^2 + \cdots + x_n^2 - 2\bar{x}x_n^2 + \bar{x}^2$$

$$= x_1^2 + \cdots + x_n^2 - 2\bar{x}x_1 - \cdots - 2\bar{x}x_n + \bar{x}^2 + \cdots + \bar{x}^2$$

$$= \sum_{i=1}^{n} x_i^2 - 2\bar{x}(x_1 + \cdots + x_n) + n\bar{x}^2$$

$$= \sum_{i=1}^{n} x_i^2 - 2\bar{x}(n\bar{x}) + n\bar{x}^2$$

$$= \sum_{i=1}^{n} x_i^2 - n\bar{x}^2.$$

RULE 9   Substitute

$$\frac{\sum_{i=1}^{n} x_i}{n}$$

for $\bar{x}$ in Rule 8.

RULE 10   We have

$$\sum_{i=1}^{n} (x_i - a)^2 = \sum_{i=1}^{n} [(x_i - \bar{x}) + (\bar{x} - a)]^2$$

$$= \sum_{i=1}^{n} (x_i - \bar{x})^2 + \sum_{i=1}^{n} (\bar{x} - a)^2 + 2\sum_{i=1}^{n} (x_i - \bar{x})(\bar{x} - a)$$

$$= \sum_{i=1}^{n} (x_i - \bar{x})^2 + n(\bar{x} - a)^2 + 0$$

because the cross product term gives

$$(\bar{x} - a) \sum_{i=1}^{n} (x_i - \bar{x}) = 0$$

by Rule 1.

RULE 11   The term $n(\bar{x} - a)^2$ in the right-hand side of Rule 10 is greater than or equal to zero, hence the left-hand side is greater than or equal to the first term in the right-hand side.

EXERCISES

4.1 (Sec. 4.3)   If sample A is 1, 2, 1, 2, 4 and sample B is 1, 2, 1, 2, 9, which of the following is true?
(a) Samples A and B have equal standard deviations.
(b) Sample A has a larger standard deviation than Sample B.
(c) Sample B has a larger standard deviation than Sample A.

4.2 (Sec. 4.2)   For a sample of three observations, the deviations of the first two observations from the sample mean are −3 and −1. What is the deviation of the third observation from the sample mean?

4.3 (Sec. 4.3)   Compute the standard deviation of the numbers 6, 6, 6, 10. (Treat these four numbers as a sample.)

*****4.4** (App. 4B)   Find the first and third quartiles and the interquartile range of the distribution of the heights of 5-year old boys as given in Table 2.14.

4.5 (Sec. 4.3)   There are five families on M Street. The numbers of children in these families are 3, 8, 2, 0, and 2.
(a) Graph the numbers as points on an axis.
(b) Find the mean deviation of the number of children in this sample of families.
(c) Find the standard deviation.

4.6 (continuation)   There are also five families on N Street. The numbers of children in these families are 2, 2, 3, 4, 4.
(a) Graph the numbers as points on a line.
(b) Find the mean deviation of the number of children in this sample.
(c) Find the standard deviation.

4.7 (Sec. 4.5)   Find the sample standard deviation of $n = 10$ observations with $\sum_{i=1}^{n} x_i = 20$ and $\sum_{i=1}^{n} x_i^2 = 161$.

4.8   Compute the (sample) variances of the data represented in Figures 4.1 and 4.2.

4.9
(a) Compute the variance for the number of children per family in the following samples.

    Sample C: 3, 2, 1, 5, 6
    Sample D: 3, 2, 1, 4, 6
    Sample E: 3, 2, 1, 4, 5

(b) Comment on the comparative variability of these three samples.

4.10 (Sec. 4.3)
(a) Compute the variance of the following samples.

   Sample F: 1, 1, 5, 5
   Sample G: 1, 3, 3, 5
   Sample H: 1, 1, 1, 5

(b) Comment on the comparative variability of these three samples.

4.11 (Sec. 4.5)   A cholesterol count was made on a blood sample from each of 9 30-year old males. The numbers of units counted were 273, 189, 253, 149, 201, 153, 249, 214, and 163. Compute the mean and the standard deviation of these cholesterol determinations.

4.12 (Sec. 4.1)   Compute the interquartile and interdecile ranges for the weights of the Stanford football players in Table 2.15.

4.13 (Sec. 4.1)   Compute the interquartile and interdecile ranges for the heights of the Stanford football players used in Table 2.15.

4.14 (Sec. 4.1)   The following data were collected from a study of professional communication among physicians. Nearly all the doctors in each of four cities were interviewed. It was then possible to divide the doctors into two groups on the basis of the following question(s): "When you need information or advice about questions of therapy where do you usually turn?" Then, if no names were given: "I see. Now, if you wish to discuss such matters with another physician, on whom are you most likely to call?"

As might be expected, some physicians were chosen as advisors many times while others were not chosen at all. The distribution, in fact, showed 61 doctors receiving no choices while 64 doctors received one or more choices as advisors on questions of therapy. The latter will be designated as advisors, the former as nonadvisors.

The researchers were concerned with the relationship between individual and social factors and the length of time between introduction and adoption of a new drug. They were able to obtain the prescription records of all the physicians in the sample ($n = 125$) and the date at which each first prescribed the drug.

The length of time from introduction of the drug to its adoption by doctors of each type (advisor, nonadvisor) is given in Table 4.9. (The period to adoption is shown in two-month intervals from the time the drug was first introduced on the market.)

TABLE 4.9

Length of Time from Introduction of Gammanym to Its Adoption by Doctors

NUMBER OF MONTHS	ADVISORS	NONADVISORS
0–2	13	6
2–4	9	11
4–6	17	6
6–8	11	9
8–10	3	2
10–12	2	6
12–14	3	4
14–16	1	5
16–17	0	1
>17	5	11
Total	64	61

SOURCE: Coleman, Katz, and Menzel (1966), Figure 22, p. 84. Copyright © by The Bobbs-Merrill Co., Inc. Used by permission.

(a) Find the first quartile, second quartile, and third quartile of the distribution for advisors according to Section 3.2. Do the same for the distribution for nonadvisors.

(b) Give the interquartile range for each frequency distribution.

(c) Plot the cumulative relative frequencies for advisors and for nonadvisors.

(d) Compare the salient features of the distribution of adoption times for advisors and nonadvisors.

(e) Is the answer to (a) or to (c) more informative in answering (d)? Why?

4.15 (Sec. 4.3)   Compute the (sample) standard deviation of the distribution of incomes of the 49 families given in Table 2.16. [The mean income was computed in Exercise 3.11(e).]

4.16 (Sec. 4.3)   Compute the standard deviation of the grouped frequency distribution having equal class intervals of $2500, which was given in Table 4.8. [The mean of this distribution was computed in Exercise 3.11(b).] Compare the result with that obtained in Exercise 4.15.

4.17   Consider the distribution of family incomes over the entire population of the United States who are married, age 35–39, with one working adult in the family, not just those families who belong to the consumers' organization (Table 2.16).

(a) What factors, if any, would tend to make the standard deviation of this income distribution greater than the standard deviation of the families in Table 2.16?

(b) What factors would tend to make it smaller?

(c) Does the number of cases per se make it any different?

4.18 For the weights of the members of the 1970 Stanford football team (Table 2.15) calculate:

(a) The sample standard deviation.

(b) The mean deviation.

(c) The interquartile range.

4.19 The length and breadth of the heads of 25 human males are given in Table 4.10. Of course the mean breadth will be less than the mean length. Is the variance of the breadths less than the variance of the lengths?

TABLE 4.10
Head Measurements (millimeters)

LENGTH	BREADTH	LENGTH	BREADTH
191	155	190	159
195	149	188	151
181	148	163	137
183	153	195	155
176	144	181	153
208	157	186	145
189	150	175	140
197	159	192	154
188	152	174	143
192	150	176	139
179	158	197	167
183	147	190	163
174	150		

SOURCE: Frets (1921).

4.20 (Sec. 4.5) Compute the means and sample variances of the cat weights summarized in Table 4.11.

TABLE 4.11
Weights of Cats (kilograms)

	MALES	FEMALES
Number, n	97	47
Sum, $\sum_{i=1}^{n} x_i$	281.3	110.9
Sum of squares, $\sum_{i=1}^{n} x_i^2$	836.75	265.13

SOURCE: R. A. Fisher, "The Analysis of Covariance Method for the Relation between a Part and the Whole," *Biometrics* 3: 65–68, 1947. With permission from The Biometric Society.

4.21 (App. 4A)   Compute the sample standard deviation of the distribution of heights in Table 2.14. Code the midpoints of the class intervals by subtracting 40 and then dividing by 3. Remember to decode at the end of the computation by multiplying the standard deviation of the coded values by 3 to obtain the standard deviation of the uncoded values.

4.22   The *coefficient of variation,* or *relative standard deviation,* is defined as the ratio of the standard deviation to the mean. Since the standard deviation and the mean are in the same units, the coefficient of variation is a dimensionless quantity. It is one index that is used in comparing distributions. Compare the following income distributions in Table 4.12 using the coefficient of variation.

TABLE 4.12
Mean and Standard Deviation of Annual Income of Persons in 3 Countries

COUNTRY	MEAN ANNUAL INCOME	STANDARD DEVIATION OF INCOME
A	10,000 dollars	5,000 dollars
B	3,000 pounds	1,500 pounds
C	36,000 francs	36,000 francs

(hypothetical data)

4.23 (Sec. 4.5)   Compute the standard deviation of the final grades given in Table 2.23. (See Exercise 3.26.)

EXERCISES ON SUMMATION NOTATION

4.24 Complete Table 4.13 by computing and adding column (3).

TABLE 4.13
Computation of Sum of Deviations

(1) i	(2) x_i	(3) $x_i - \bar{x}$
1	7	
2	11	
3	13	
4	17	
5	19	
6	29	
	$\sum_{i=1}^{6}(x_i - \bar{x}) =$	

4.25 Using Table 4.14, find

$$\sum_{i=1}^{5} a_i, \quad \sum_{i=1}^{5} a_i^2, \quad \sum_{i=1}^{5} (a_i + 1)^2,$$

and verify that

$$\sum_{i=1}^{5} (a_i + 1)^2 = \sum_{i=1}^{5} a_i^2 + 2 \sum_{i=1}^{5} a_i + 5.$$

TABLE 4.14
Computation of Sums of Squares

i	a_i	a_i^2	$a_i + 1$	$(a_i + 1)^2$
1	13			
2	17			
3	30			
4	27			
5	19			

4.26 Prove, using a rule for summation notation, that if x_1 is subtracted from each observation before computing the variance the answer is unchanged.

MATHEMATICAL EXERCISES

4.27 (Sec. 4.3)   Show that, when $n = 2$,

$$\sum_{i=1}^{n} (x_i - \bar{x})^2 = \frac{(x_1 - x_2)^2}{2}.$$

4.28   Suppose x is a dichotomous variable with values 0 and 1. Then each number in the sample $x_1, x_2, \ldots, x_n$ is either 0 or 1. Show that $\sum_{i=1}^{n} x_i$ is the number of 1's among $x_1, x_2, \ldots, x_n$.

4.29 (continuation)   Show that the mean $\bar{x}$ is the proportion of 1's among $x_1, x_2, \ldots, x_n$.

4.30 (continuation)   Show that

$$\sum_{i=1}^{n} (x_i - \bar{x})^2 = n\bar{x}(1 - \bar{x}).$$

[Hint: Since $x_i = 0$ or 1, $x_i^2 = x_i$.]

4.31   For the sample 4, 9, 3, 1, 3, compute

$$\sum_{i=1}^{n} |x_i - \bar{x}|, \qquad \sqrt{\sum_{i=1}^{n} |x_i - \bar{x}|^2},$$

$$\sqrt[3]{\sum_{i=1}^{n} |x_i - \bar{x}|^3}, \qquad \sqrt[4]{\sum_{i=1}^{n} |x_i - \bar{x}|^4},$$

and verify that as p increases, $\sqrt[p]{\sum_{i=1}^{n} |x_i - \bar{x}|^p}$ gets smaller and smaller, approaching the maximum absolute deviation, here 5.

4.32 (continuation)   Compute

$$\left(\frac{\sum_{i=1}^{n} |x_i - \bar{x}|^p}{n} \right)^{1/p}$$

for $p = 1, 2, 3, 4$, verifying that these increase with p.

4.33 (Sec. 4.2)   For $x_1 = 1$, $x_2 = 2$, $x_a = 4$, show that $\sum_{i=1}^{3} |x_i - a|$ is minimized by taking $a = 2$.

4.34 (Sec. 4.2) Show that $\Sigma_{i=1}^{n} |x_i - a|$ is minimized by taking a to be the median. [Hint: Arrange $x_1, x_2, \ldots, x_n$ in order as $x_{(1)} \leq x_{(2)} \leq \cdots \leq x_{(n)}$, so that $x_{(1)}$ denotes the minimum, $x_{(2)}$ the next larger observation, $\ldots, x_{(n)}$ denotes the maximum. The sum is made up of $\Sigma_i(x_{(i)} - a)$ over the $x_{(i)} \leq a$ and of $\Sigma_i(a - x_{(i)})$ over the $x_{(i)} \geq a$.]

REFERENCES

Coleman, James S., Elihu Katz, and Herbert Menzel (1966). *Medical Innovation: A Diffusion Study*. Bobbs-Merrill, Indianapolis.

Fisher, R. A. (1947). "The analysis of covariance method for the relation between a part and the whole." *Biometrics* 3: 65–68.

Frets, G. P. (1921). "Heredity of head form in man." *Genetica* 3: 193–384.

5 ORGANIZATION OF MULTIVARIATE DATA

INTRODUCTION

The organization and summarization of data concerned with *one* characteristic for each observed individual has been considered in Chapters 2, 3, and 4. In this chapter we discuss data based on the observation of *two or more* variables for each individual. For instance, age, income and occupation of each individual may be observed. Such data are termed *multivariate* data.

The first step in analysis of multivariate *categorical* data is tabulation of the data into a frequency table. Percentages are computed to determine the relationship, if any, between the variables.

A frequency table for two variables is a two-way table. The special, but important, case of two dichotomous variables is treated in Section 5.1. Section 5.2 treats larger two-way frequency tables.

The tabulation of data for three categorical variables results in a three-way frequency table (Section 5.3). Here, it is possible to consider the relationship between a pair of variables for different fixed values of a third variable (Section 5.4).

In Section 5.5 the discussion relates to *numerical* variables. Graphing, frequency tables, and their interpretations are discussed. Methods for pictorial representation of multivariate data are given.

5.1 TWO-BY-TWO FREQUENCY TABLES

5.1.1 Organization of Data

Bivariate Categorical Data. *Bivariate* categorical data result from the observation of two categorical variables for each individual. In the simplest case each variable has only two categories. Such data might result, for example, by classifying the 25 students in a statistics course by sex and university standing (graduate or undergraduate). Table 5.1 illustrates such data as originally recorded.

2 × 2 Frequency Tables. The data are summarized in a *two-by-two* (2 × 2) *frequency table.* Table 5.2 is a tally sheet and summarization of the data listed in Table 5.1.* (For example, "Adams M G" is recorded as the first tally in the lower left-hand box in Table 5.2.) This kind of summary is called a *cross-tabulation* or *cross-classification* of individuals, because they are categorized using two classifications simultaneously. The table indicates that 10 persons are male undergraduates, 4 are female undergraduates, 8 are female graduates, and 3 are male graduates.

TABLE 5.1

Data for a Double Dichotomy

	Name	Sex	University Division		Name	Sex	University Division
1	Adams	M	G	14	Sanchez	F	G
2	Brown	M	U	15	Schwartz	M	G
3	Clark	F	G	16	Thomas	M	U
4	Davis	M	G	17	Torres	M	U
5	Franz	M	U	18	Tucker	M	U
6	Humphreys	M	U	19	Tyler	M	U
7	Jones	M	U	20	Waters	F	G
8	Keller	M	U	21	York	F	U
9	Mann	M	U	22	Young	F	U
10	Martin	F	U	23	Youngman	F	G
11	Morgan	F	G	24	Yule	F	U
12	Rogers	F	G	25	Ziegler	F	G
13	Ruiz	F	G				

(hypothetical data)
U = undergraduate
G = graduate

Double Dichotomy. A variable with two categories, such as male/female or graduate/undergraduate, is called a *dichotomous* variable. A pair of dichotomous variables is called a *double dichotomy.* A double dichotomy arises, for example, when each of a number of persons is asked the same pair of yes-no questions. For example, Company X may ask each of 600 men whether or not they use the Brand X razor and whether or not they use Brand X blades. A physician may classify his patients according to whether they have been inoculated against a disease and whether they contract the disease.

*To illustrate procedures we have used small numbers in order to simplify presentation such as the listing of data and calculation.

TABLE 5.2

2 × 2 Frequency Table: Cross-Tabulation of 25 Students

		UNIVERSITY DIVISION	
		G	U
SEX	F	卌 ⏐⏐⏐ 8	⏐⏐⏐⏐ 4
	M	⏐⏐⏐ 3	卌 卌 10

SOURCE: Table 5.1.

TABLE 5.3

2 × 2 Table: General Notation for Frequencies

		QUESTION 2	
		Yes	No
QUESTION 1	Yes	a	b
	No	c	d

Because many statistical studies involve 2×2 tables and because many concepts of statistics can be presented in this form, we shall consider such tables in some detail. This discussion is simplified by the notation displayed in Table 5.3. The numbers $a, b, c,$ and d are *frequencies*, that is, numbers of persons falling into the four possible categories:

a is the number of persons answering "yes" to both questions.

b is the number of persons answering "yes" to Question 1 and "no" to Question 2.

c is the number of persons answering "no" to Question 1 and "yes" to Question 2.

d is the number of persons answering "no" to both questions.

We denote by n the total number of persons included in the table:

$$n = a + b + c + d.$$

Another kind of double dichotomy consists of variables of the high-low type. In Table 5.4, for example, the number b is the number of persons having high social status* but low income, c is the number having high income but low status, etc.

---

*"Social status" is a scale, involving education, occupation, etc., used by social scientists.

TABLE 5.4
2 × 2 Table for Ordinal Variables

		SOCIAL STATUS	
		Low	*High*
INCOME	*Low*	*a*	*b*
	High	*c*	*d*

Still another type of double dichotomy is shown by the two-by-two tabulation of the same dichotomous variable for the *same individuals* at *two different* times. Suppose in August we asked a number of people which of two presidential candidates they favored and then asked the same people the same question in September. We could tabulate the results as in Table 5.5. Here the number *c* would be the number of persons who switched from Democratic candidate McGovern to Republican candidate Nixon between August and September 1972. These data are change-in-time data and the table is called a "turnover" table. (Persons favoring candidates other than Nixon or McGovern are omitted from this tabulation.)

TABLE 5.5
2 × 2 Table for Change-in-Time Data ("Turnover" Table)

		SEPTEMBER	
		Nixon	*McGovern*
AUGUST	*Nixon*	*a*	*b*
	McGovern	*c*	*d*

5.1.2 Calculation of Percentages

Marginal Totals. From Table 5.2 we see that $8 + 4 = 12$ of the 25 students are females, while $3 + 10 = 13$ are males. Also, $8 + 3 = 11$ are graduates, while $4 + 10 = 14$ are undergraduates. These totals are appended to the margins of Table 5.2 to obtain Table 5.6 and therefore are called *marginal totals,* or marginals.

Percentages Based on Row Totals. Table 5.7 shows what percentage of the female students are graduates ($100\% \times 8/12 = 67\%$), what percentage of the female students are undergraduates (33%), what percentage of the male students are graduates ($100\% \times 3/13 = 23\%$), and what percentage of the male students are undergraduates

(77%). Here the *row* totals, 12 and 13, have been used as *bases* for the percentages. (A *row* is a horizontal line of the table.)

Percentages Based on Column Totals. Table 5.8 shows what percentage of the graduate students are females (100% × 8/11 = 73%), what percentage of the graduate students are males (27%), what percentage of the undergraduates are females (100% × 4/14 = 29%), and what percentage of the undergraduates are males (71%). (A *column* is a vertical line in the table.)

TABLE 5.6
2 × 2 Frequency Table, with Marginals

| | | UNIVERSITY DIVISION | | |
		Graduate	Undergraduate	Total
SEX	*Female*	8	4	12
	Male	3	10	13
	Total	11	14	25

SOURCE: Table 5.2.

TABLE 5.7
Percent Graduates and Percent Undergraduates, by Sex

| | | UNIVERSITY DIVISION | | |
		Graduate	Undergraduate	Total
SEX	*Female*	67%	33%	100%
	Male	23%	77%	100%
	Both Sexes	44%	56%	100%

SOURCE: Table 5.6.

TABLE 5.8
Percent Female and Percent Male, by University Division

| | | UNIVERSITY DIVISION | | BOTH DIVISIONS |
		Graduate	Undergraduate	
SEX	*Female*	73%	29%	48%
	Male	27%	71%	52%
	Total	100%	100%	100%

SOURCE: Table 5.6.

The computation of percentages is such a basic step that usually percentages are given right alongside the frequencies, as in Table 5.9. The information in this table can be summarized as in Table 5.10.

TABLE 5.9
2 × 2 Frequency Table, with Percentages Based on Column Totals

		UNIVERSITY DIVISION		Total
		Graduate	Undergraduate	
SEX	Female	8 (73%)	4 (29%)	12 (48%)
	Male	3 (27%)	10 (71%)	13 (52%)
	Total	11 (100%)	14 (100%)	25 (100%)

SOURCE: Tables 5.6 and 5.8.

TABLE 5.10
Percent Females, by University Division

	Graduate	Undergraduate	Both Divisions
Percent females	73%	29%	48%
(Number of students)	(11)	(14)	(25)

SOURCE: Table 5.9.

Percentages Based on the Grand Total. There is a total of 25 students in the class. Table 5.11 shows that 32% of the students in the class are female graduate students, 16% are female undergraduates, 12% are male graduates, and 40% are male undergraduates. The percentages in the margins show that 48% of all the students in the class are females, 52% are males, 44% are graduate students, and 56% are undergraduate students.

TABLE 5.11
Percentages Resulting from Cross-Classification of Students by Sex and University Division

	Graduate	Undergraduate	Total
Female	32%	16%	48%
Male	12%	40%	52%
Total	44%	56%	100%

SOURCE: Table 5.6.

5.1.3   Interpretation of Frequencies

Is there a tendency for those who use Brand X razors also to use Brand X blades? Company X made a market survey, interviewing a sample of 600 men. Each man was asked whether he uses the Brand X razor and whether he uses Brand X blades.

Such a large data set would usually not be processed by hand; cross-tabulation by making tallies manually as in Table 5.2 would be tedious and possibly error producing. The data would be punched onto IBM cards, and then run through a card sorter with an automatic counter to obtain the frequencies in Table 5.12. Alternatively, the data could be read into a computer programmed to make the cross-tabulation.

From Table 5.12 it can be seen that:

1. Of the 600 men, 245, or 41%, use Brand X blades.
2. Of the 279 men using Brand X razors, 186, or 67%, use Brand X blades.
3. Of the 321 men not using Brand X razors, 59, or 18%, use Brand X blades.

TABLE 5.12

Results of the X Company's Survey

	USE BRAND X BLADES	DO NOT USE BRAND X BLADES	TOTAL
USE BRAND X RAZOR	186 (67%)	93 (33%)	279 (100%)
DO NOT USE BRAND X RAZOR	59 (18%)	262 (82%)	321 (100%)
TOTAL	245 (41%)	355 (59%)	600 (100%)

(hypothetical data)

Association.   Most (67%) of the men using Brand X razors also use Brand X blades; only a few (18%) of the men not using Brand X razors use Brand X blades. Because the percentage of men using Brand X razors who use Brand X blades differs from that percentage among men not using Brand X razors, we say there is an association* between using the Brand X razor and using Brand X blades. Thus, from Table 5.13, we can say that men who use the Brand X razor are more likely to use Brand X blades than are men who do not use the razor. This difference may suggest that an advertising campaign for blades should be directed to men not using Brand X razors.

---

*Many authors use the term "correlation." We prefer to reserve that term for relationships among *quantitative* variables.

TABLE 5.13

Percent Using Brand X Blades, Among Those Who Use Brand X Razor and Among Those Who Do Not

	PERCENT USING BRAND X BLADES	(NUMBER OF MEN)
Men in Sample Who Use Brand X Razor	67%	(279)
Men in Sample Who Do Not Use Brand X Razor	18%	(321)
All Men in Sample	41%	(600)

SOURCE: Table 5.12

Independence. Table 5.14 is a cross tabulation of 350 persons according to whether they are Democrats or Republicans and whether they intend to vote in the next election. The proportion intending to vote in the next election is 80%, among both the Democrats and the Republicans. We say, therefore, that intending to vote in the next election is independent of whether one is a Democrat or a Republican. The percentages in the two groups are the same. Such a lack of association is called *independence.*

Since the percentage intending to vote is 80% in both columns of the table it is also 80% in the column labelled "Both Groups." That is, the marginal proportion 280/350 must also be equal to .8. In fact, it is an algebraic identity that if in a row the percentages in the two columns are equal, the same percentage holds for the groups combined. We demonstrate this arithmetically for Table 5.14. For "Both Groups" the denominator is the sum of the column totals $200 + 150 = 350$. The entries in the first row of the Democrat and Republican columns are $.8 \times 200 = 160$ and $.8 \times 150 = 120$, respectively. The entry in the first row of "Both Groups" is the sum $.8 \times 200 + .8 \times 150 = .8 \times (200 + 150) = .8 \times 350$. Thus the ratio of the first entry to the total for "Both Groups" is $.8 \times 350/350 = .8$.

The property of independence still holds when we use the *row* totals as bases for percentages, which results in Table 5.15. The percentage

TABLE 5.14

2 × 2 Table of Vote Intentions

		Democrat	Republican	Both Groups
INTEND TO VOTE IN	Yes	160 (80%)	120 (80%)	280 (80%)
NEXT ELECTION	No	40 (20%)	30 (20%)	70 (20%)
	Total	200 (100%)	150 (100%)	350 (100%)

(hypothetical data)

TABLE 5.15
2 × 2 Table: Percentages Relative to Row Totals

		Democrat	Republican	Total
INTEND TO VOTE IN	Yes	57.1%	42.9%	100%
NEXT ELECTION	No	57.1%	42.9%	100%
	Both Groups	57.1%	42.9%	100%

SOURCE: Table 5.14.

who are Democrats is 57.1%, whether one considers those persons who intend to vote, those who do not, or the entire group. The conclusion, that there is no association between the two variables, is the same. In fact, it is always the case that if the percentages based on column totals exhibit independence, then the percentages based on row totals also exhibit independence.

There is a third, very important, property of a table that exhibits independence, and that is that *any* frequency in the body of the table is equal to the product of the overall row proportion and the column total ($160 = .8 \times 200$, $120 = .8 \times 150$, $40 = .2 \times 200$, $30 = .2 \times 150$), and to the product of the overall column proportion and the row total ($160 = .571 \times 280$, $40 = .571 \times 70$, $120 = .429 \times 280$, $30 = .429 \times 70$).

Algebraic Condition for Independence.   We now state these properties using a general algebraic notation. The general notation for 2 × 2 tables with marginal totals is given in Table 5.16 where A_1 and A_2 denote the two values of one dichotomous variable, and B_1 and B_2 denote the two values of another dichotomous variable. The condition of independence is expressed algebraically in terms of the rows as

$$\frac{a}{a+b} = \frac{c}{c+d},$$

TABLE 5.16
2 × 2 Frequency Table: General Notation for Cell Frequencies, Marginal Totals, and Variables

		B		
		B_1	B_2	Total
A	A_1	a	b	$a + b$
	A_2	c	d	$c + d$
	Total	$a + c$	$b + d$	$n = a + b + c + d$

and it is expressed equivalently in terms of the columns as

$$\frac{a}{a+c} = \frac{b}{b+d}.$$

By "equivalently," we mean that if one of these algebraic statements is true, then the other is also true. To see this, note that by cross-multiplying, the first statement can be written as

$$a(c + d) = c(a + b),$$

or

$$ac + ad = ac + bc,$$

which simplifies to $ad = bc$. Similarly, the second can be written as

$$ab + ad = ab + bc,$$

which also simplifies to $ad = bc$. Thus, from Table 5.14 we have $ad = 160 \times 30 = 4800$ and $bc = 120 \times 40 = 4800$.

To summarize, we state the rule: *a* 2 × 2 *table* (written in the notation of Table 5.16) *exhibits independence if and only if* $ad = bc$. This equation provides an easy check for independence, namely, comparison of the product of frequencies on one diagonal with the product of frequencies on the other.

We have discussed the notion of independence in idealized terms; in our hypothetical numerical examples of independence corresponding percentages have been identical. In real-life situations one cannot expect to have percentages exactly the same. (In fact, the row or column totals may be such integers as 7 and 13 so that the possible equal percentages are only 0% and 100%.) In actual populations two variables would for practical purposes be considered independent if the percentages differed by very small amounts.

The Amount of Association. Suppose the 2 × 2 table looked like Table 5.17. Of those who voted in the last election, 90% intend to vote in the next election, while only two-thirds (67%) of those who did not vote in the last election intend to vote in the next one. Thus it appears that there is a strong association between having voted in the past election and intending to vote in the coming election. If we were to select a person at random from among the 350, that person is more likely to vote in the coming election if he or she voted in the last election than if he or she did not. However, even among those who did not vote in the last election, the proportion who intend to vote in the coming election is still two-thirds, considerably more than half. (In

TABLE 5.17

2 × 2 Table: Strong Association

		VOTED IN LAST ELECTION			
		Yes	No	Total	
INTEND TO VOTE IN	Yes	180 (90%)	100 (67%)	280 (80%)	
NEXT ELECTION	No	20 (10%)	50 (33%)	70 (20%)	
	Total	200 (100%)	150 (100%)	350 (100%)	

(hypothetical data)

this sense we can say that intending to vote is characteristic even of the group of people who did not vote in the last election.) Among all 350 persons the proportion intending to vote is 280/350, or 80%.

Suppose the data had appeared as in Table 5.18. Now the percentages intending to vote are 75% and 67%, compared to 90% and 67% before, so there is less of a difference between the two groups than before.

We need numerical measures of the *degree* of association in order to compare different tables and also to gauge how far the relationship in a given table differs from independence; one such measure is discussed below. When the table is obtained from a sample of individuals the association in the sample will probably differ from the association in the sampled population, and a measure of association in the table representing the sample may help the investigator to answer the question whether the variables are independent in the population.

TABLE 5.18

2 × 2 Table: Less Association Than in Table 5.17

		VOTED IN LAST ELECTION			
		Yes	No	Total	
INTEND TO VOTE IN	Yes	150 (75%)	100 (67%)	250 (71%)	
NEXT ELECTION	No	50 (25%)	50 (33%)	100 (29%)	
	Total	200 (100%)	150 (100%)	350 (100%)	

(hypothetical data)

Direction of Association of Low-High Variables. For low-high variables there is a notion of *direction* of association. Table 5.19 is a cross-classification of 250 persons according to income level and social status. Here it is evident that high social status and high income go together, and low status and low income go together, for 80% of those having low income have low social status, and 60% of those having high income have high social status. In this case we would say that the association between income and social status is *positive*.

TABLE 5.19
2 × 2 Table: Positive Association

		SOCIAL STATUS		
		Low	*High*	*Total*
INCOME	*Low*	100 (80%)	25 (20%)	125 (100%)
	High	50 (40%)	75 (60%)	125 (100%)
	Total	150 (60%)	100 (40%)	250 (100%)

(hypothetical data)

Consider, on the other hand, the hypothetical data summarized in Table 5.20 regarding the relationship between college grade-point average and the number of semesters of statistics courses taken. Four-fifths (80%) of those students having more than one semester of statistics have low grade-point averages, and most (56% = 100% × 250 ÷ 450) of those having taken only a little statistics have *high* grade-point averages. A *high* grade-point average is associated with a *low* number of semesters of statistics; we say that the association between the two variables, "grade-point average" and "number of semesters of statistics," is *negative*. Faced with such data, we would want to know whether the cause of the negative association is that the grades in statistics courses are low or whether the students who take statistics courses have lower averages to begin with. Association does not imply causality; if two variables are associated, one *may* "cause" the other, but they may both be associated with a third variable, a "common cause." This topic is examined in more detail in Section 5.4.

For a general definition of direction of association in 2 × 2 tables we use the notation of Table 5.16. If B_1 and B_2 are in the same order as A_1 and A_2 (for example, if A_1 and B_1 are "low" and A_2 and B_2 are "high"),

TABLE 5.20
2 × 2 Table: Negative Association

		NUMBER OF SEMESTERS OF STATISTICS		
		Low (0 or 1)	*High* (More than 1)	*Total*
GRADE-POINT AVERAGE	*Low* (0.0–2.9)	200 (44%)	40 (80%)	240 (48%)
	High (3.0–4.0)	250 (56%)	10 (20%)	260 (52%)
		450 (100%)	50 (100%)	500 (100%)

(hypothetical data)

we say there is positive association if $a/(a + b)$ is greater than $c/(c + d)$, independence if $a/(a + b)$ equals $c/(c + d)$, and negative association if $a/(a + b)$ is less than $c/(c + d)$. [These conditions are equivalent to $a/(a + c)$ greater than $b/(b + d)$, $a/(a + c) = b/(b + d)$, and $a/(a + c)$ less than $b/(b + d)$, respectively, or $ad > bc$, $ad = bc$, and $ad < bc$, respectively.]

Index of Association for 2 × 2 Frequency Tables. A measure of the strength and direction of association can be based on the quantity $ad - bc$ since a zero value of this indicates independence while large positive or negative values indicate strong positive or negative association, respectively. We divide the difference by a quantity such that the result will always be between −1 and +1. This divisor turns out to be the square root of the product of the four marginal totals. The resulting index of association is called ϕ (lower case Greek phi) or the phi-coefficient:

$$\phi = \frac{ad - bc}{\sqrt{(a + b)(c + d)(a + c)(b + d)}}.$$

Thus, for Table 5.17,

$$\phi = \frac{180 \times 50 - 100 \times 20}{\sqrt{280 \times 70 \times 200 \times 150}}$$

$$= \frac{7,000}{\sqrt{588,000,000}} = \frac{7,000}{24,249} = 0.289$$

and for Table 5.18,

$$\phi = \frac{150 \times 50 - 100 \times 50}{\sqrt{250 \times 100 \times 200 \times 150}}$$

$$= \frac{2,500}{\sqrt{750,000,000}} = \frac{2,500}{27,386} = 0.091.$$

Complete positive association occurs where $b = 0$ and $c = 0$; then $\phi = 1$. Complete negative association occurs when $a = 0$ and $d = 0$; then $\phi = -1$.

5.2   LARGER TWO-WAY FREQUENCY TABLES

5.2.1   Organization of Data for Two Categorical Variables

Larger Two-Way Tables. Two-way frequency tables are the result of classification of a number of individuals according to two sets of *categories* simultaneously. They may be 2 × 2 tables, 2 × 3 tables, 3 × 3 tables, etc. Thus a two-way frequency table is the *joint frequency distribution of two variables;* that is, a *bivariate distribution.*

Mode of a Bivariate Distribution. Table 5.21 is a 3 × 4 cross-classification of 6800 persons of Germanic stock according to colors of their hair and eyes. For example, 1768 is the number of persons with blue eyes and fair hair; it is the *joint frequency* of occurrence of blue eyes and fair hair. A comparison of the column totals 2829, 2632, 1223, and 116 shows that "fair" is the most frequent hair color; similarly, a comparison of the row totals 2811, 3132, and 857 shows that the modal eye color is "grey or green."

These facts might lead one to suspect that the most frequently occurring category in the joint distribution is the type with fair hair and grey or green eyes. However, this category is not the mode; it contains only 946 persons.

The modal category is fair hair and blue eyes, with a frequency of 1768 (underlined in the table). Perhaps ethnic stereotypes result from

TABLE 5.21

Cross-Classification of 6800 Males of Germanic Stock According to Eye and Hair Colors

		HAIR COLOR				All Hair Colors
		Fair	*Brown*	*Black*	*Red*	
EYE COLOR	Blue	1768	807	189	47	2811
	Grey or Green	946	1387	746	53	3132
	Brown	115	438	288	16	857
	All Eye Colors	2829	2632	1223	116	6800

SOURCE: Goodman and Kruskal (1954), who cite O. Ammon, "Zur Anthropologie der Badener."

TABLE 5.22

Percentages of Males in Each Eye-Hair Color Category

		HAIR COLOR				All Hair Colors
		Fair	*Brown*	*Black*	*Red*	
EYE COLOR	Blue	26.0%	11.9%	2.8%	0.7%	41.4%
	Grey or Green	13.9%	20.4%	11.0%	0.8%	46.1%
	Brown	1.7%	6.4%	4.2%	0.2%	12.5%
	All Eye Colors	41.6%	38.7%	18.0%	1.7%	100.0%

SOURCE: Table 5.21.

the assumption that the modal type is determined by the marginal modes. Sometimes this will in fact be the case, but sometimes not, as we have just seen.

The percentages of persons in each eye-hair category are given in Table 5.22. The percentage corresponding to the modal category is $100\% \times 1768/6800 = 26.0\%$. The percentage of persons in the grey/green-fair category is only 13.9%. Only 0.2% of the persons have brown eyes and red hair. These percentages are called *joint relative frequencies*.

5.2.2   Interpretation of Frequencies

2 × 3 Tables.   Table 5.23 is a 2 × 3 frequency table. The two variables, "level of education" and "level of political interest," are associated because the distributions are not the same in the two rows.

A 2 × 3 table for ordinal variables, such as the variables of Table 5.23, takes the form of Table 5.24. There is positive association if

(5.1) $a > b > c$ and $d < e < f,$

because low levels of variable B tend to go with the low level of variable A, and high levels of variable B tend to go with the high level of variable A. Condition (5.1) is just the simplest and most obvious case of positive association. Clearly Table 5.23 exhibits positive association. Although the frequencies do not satisfy condition (5.1), 20% of those with no high school have no political interest, compared to only 8% for those with some high school. And 31% of those with some high school have great interest in politics, compared to only 24% for those with no high school. The middle category, "moderate interest," gives no information about the direction of the association.

TABLE 5.23
Level of Education and Level of Political Interest

| | | LEVEL OF POLITICAL INTEREST | | | |
		No Interest	*Moderate Interest*	*Great Interest*	*Total*
LEVEL OF EDUCA-TION	*At Least Some High School*	132 (8%)	986 (61%)	495 (31%)	1613 (100%)
	No High School	245 (20%)	669 (56%)	285 (24%)	1199 (100%)
	Total	377 (13%)	1655 (59%)	780 (28%)	2812 (100%)

SOURCE: Lazarsfeld, Berelson, and Gaudet (1968).

TABLE 5.24
2 × 3 Frequency Table for Two Ordinal Variables: General Notation

		VARIABLE B		
		Low	Medium	High
VARIABLE A	Low	a	b	c
	High	d	e	f

3 × 3 Tables. The 3 × 3 cross-classification shown in Table 5.25 is from a study by Columbia University's Bureau of Applied Social Research. This study dealt with an international student exchange organization which sent young people abroad for a summer of living with a foreign family. Each group of ten students was accompanied by a leader. Part of the research focused on ways of evaluating participants in the program, and two possible ways of evaluation are shown in the cross-classification in Table 5.25. The columns of the table categorize group members according to the ratings given them by their leader in response to the following:

> Considering that one of the purposes of the Experiment is international good will, how would you rate the members of your group as ambassadors of America abroad? Please name the two who you think were the best and the two you think were the least successful as ambassadors.

Thus each leader designated two students from his group as "best ambassadors" and two as "worst ambassadors." The rows of the table distinguish group members according to the degree of change (that is, increase) in their "understanding of some of the simple cultural practices of their host family" as measured by a series of questions asked them both before and after the summer program.

TABLE 5.25
3 × 3 Table: Evaluation in Exchange Program

		LEADER'S RATING			
		Best Ambassadors	Not Mentioned	Worst Ambassadors	Total
DEGREE OF	Large	8	16	2	26
CHANGE IN	Small	7	30	9	46
UNDERSTANDING	None	2	8	4	14
	Total	17	54	15	86

SOURCE: Somers (1959).

The data shown in Table 5.25 are only for those groups whose leaders had previous experience as group leaders with this organization. The data are for a total of 86 group members (not a multiple of ten) because some groups actually had less than ten members, and in some of those cases the leader named only one "best ambassador" or one "worst ambassador." The column totals were nevertheless more or less fixed by the plan of the study.

The column totals 17, 54, and 15, and the row totals 26, 46, and 14 are shown in Table 5.25. We can see, for example, that 46 students showed a small change in understanding. Seven of these 46 were mentioned as "best ambassador," 30 were not mentioned, and 9 were mentioned as "worst ambassador."

TABLE 5.26

Distribution of Change in Understanding by Leader's Rating

| | | LEADER'S RATING | | |
		Best Ambassadors	*Not Mentioned*	*Worst Ambassadors*
DEGREE OF	*Large*	47%	30%	13%
CHANGE IN	*Small*	41%	56%	60%
UNDERSTANDING	*None*	12%	15%	27%
		100%	101%	100%
	(Number of persons)	(17)	(54)	(15)

SOURCE: Table 5.25.

Patterns of Percentages.   The percentages in Table 5.26 show a definite association between the variables. They decrease from left to right across the top row and increase from left to right across the middle and bottom rows:

1. 47% of those mentioned as best ambassador showed a large change in understanding, compared to only 30% for those not mentioned and 13% for those mentioned as worst ambassador.
2. Of those mentioned as best ambassador, only 41% showed a small change in understanding, compared to 56% for those not mentioned and 60% for those mentioned as worst ambassador.
3. Of those mentioned as best ambassador, only 12% showed no change in understanding, while 15% of those not mentioned showed no change in understanding, and 27% of those mentioned as worst ambassador showed no change in understanding.

The data provide strong evidence of a positive relationship between degree of change in understanding of cultural practices and rated effectiveness as an ambassador of America abroad.

TABLE 5.27
Best and Worst Ambassadors, by Degree of Change in Understanding

		LEADER'S RATING			
		Best Ambassadors	*Not Mentioned*	*Worst Ambassadors*	*Total*
DEGREE OF	*Large*	31%	62%	8%	101% (26)
CHANGE IN	*Small*	15%	65%	20%	100% (46)
UNDERSTANDING	*None*	14%	57%	29%	100% (14)

SOURCE: Table 5.25.

Notice what happens when we calculate percentages from Table 5.25 by using the *row* totals as bases, to obtain Table 5.27. This table shows, for example, that among those who showed a large change in understanding, 31% were mentioned as best ambassadors and 8% were mentioned as worst ambassadors. Those who showed a large change in understanding tended to be good or fair ambassadors but the 8% indicates two "surprises." The second and third rows are studied in the same way.

Association between the two variables is discovered by using either the column totals or the row totals to calculate percentages. However, one set of percentages may lead to more interesting or relevant interpretations than the other. For example, the percentages in Table 5.26 may suggest that the students who tried to be "good ambassadors" gained more understanding, while Table 5.27 might indicate that those who were affected more by the experience showed up better than fellow students.

The tables give a good summary of the information, but summaries necessarily hide some information. For example, the tables are based only on the *change* in understanding. One wonders, for example, if the two persons among the best ambassadors who had no change in understanding already had a high degree of understanding, which would certainly reduce the significance of the fact that their understanding did not increase. As a general rule, an analysis of *changes*, or *gains*, should include a simultaneous analysis of the initial scores, as well as a study of the relationship between the initial scores and the gains.

A different pattern of percentages is exhibited in Table 5.28, which is a cross-classification of pupils according to their scores on a test at the beginning of the semester and another test at the end of the semester.

TABLE 5.28

3 × 3 Table: Ratings by Initial Test and Final Test Scores

		SCORE ON FINAL TEST			Total
		Low	*Medium*	*High*	*Total*
SCORE	*Low*	30 (60%)	15 (30%)	5 (10%)	50 (100%)
INITIAL	*Medium*	4 (5%)	32 (40%)	44 (55%)	80 (100%)
TEST	*High*	4 (10%)	32 (80%)	4 (10%)	40 (100%)
	Total	38 (22%)	79 (46%)	53 (31%)	170 (99%)

(hypothetical data)

The modal frequency in each row is underlined. For those who scored low on the initial test, the modal final test score is low. For those who scored in the middle of the range of scores on the initial test, the modal final test score is high, and 95% (40% + 55%) of these pupils' scores were either high or medium.

From the results for those who scored low or medium on the initial test, one would expect those who scored high on the initial test to do well on the final test. Such was not the case. The modal initial test category for these students was medium, not high. Note that in cases like Table 5.28 we would almost always use the row totals to compute percentages because we take the *initial* state as the baseline for comparison and are interested in knowing how the distribution of outcomes (final states) vary given the initial conditions.

Independence. Suppose communities were cross-classified according to average income and crime rate, as in Table 5.29. (Note that in this table the "individuals," that is, the units of observation, are *communities*.) The table shows that for each income level most communities have a medium crime rate. In fact, the distribution of crime

TABLE 5.29

Distribution of Crime Rates, by Level of Average Income, for 80 Communities

		AVERAGE INCOME			All Income Levels
		Low	*Medium*	*High*	*Levels*
CRIME	*Low*	1 (10%)	4 (10%)	3 (10%)	8 (10%)
RATE	*Medium*	7 (70%)	28 (70%)	21 (70%)	56 (70%)
	High	2 (20%)	8 (20%)	6 (20%)	16 (20%)
	Total	10 (100%)	40 (100%)	30 (100%)	80 (100%)

(hypothetical data)

rates is the same for each level of average income; that is, crime rate is *independent* of level of average income. In this case, knowledge of a community's average income level would provide no information about its crime rate. Note that, since there is independence, the marginal distribution of crime rate has to be the same as the common distribution for each income level.

Now let us calculate percentages in the other direction, calculating the percentage of communities of different income levels within each category of crime rate to obtain Table 5.30. Note that the distribution of income levels is the same for all levels of crime rate and is equal to the marginal distribution. Thus, income level is *independent* of crime rate.

TABLE 5.30
Distribution of Income Levels, by Crime Rate

| | | LEVEL OF AVERAGE INCOME | | | |
		Low	Medium	High	Total
CRIME RATE	Low	1 (12%)	4 (50%)	3 (38%)	8 (100%)
	Medium	7 (12%)	28 (50%)	21 (38%)	56 (100%)
	High	2 (12%)	8 (50%)	6 (38%)	16 (100%)
	All Crime Levels	10 (12%)	40 (50%)	30 (38%)	80 (100%)

SOURCE: Table 5.29.

So it works both ways: income is independent of crime rate, and crime rate is independent of income level. In fact, it is *always* the case that if within the categories of the variable X the distributions of the variable Y are the same, then it is also true that within the categories of the variable Y the distributions of the variable X are the same. Because of this, we simply say, "X and Y are independent," or, in the present example, "crime rate and income level are independent."

If two variables are not independent they are said to be *dependent*, or *associated*, as we noted in the case of two dichotomies.

In many tables, the distribution in each line is *almost* the same. In such cases the percentages exhibit independence approximately; because the frequencies must be integers, exact independence may be impossible, given a specific set of marginals. When the individuals observed constitute a *sample* from a population, we do not expect exact independence in the sample even if there is exact independence in the population.

5.3 THREE CATEGORICAL VARIABLES

5.3.1 Organization of Data for Three Yes-No Variables

The Three-Way Table. We now consider situations in which interest centers on analyzing the interrelationships among three categorical variables.

In the simplest case each variable is dichotomous. The interviewers sent out by Company X to investigate razor and blade use also asked each of 600 men in the sample if he had seen the Brand X television commercial. The full set of data appeared as in Table 5.31.

Data for three dichotomous variables can be summarized as in Table 5.32. The rows indicate whether or not the Brand X razor is used. The columns are paired according to use of the blades, the first two columns for those who have not seen the commercial, the second two for those

TABLE 5.31

Data for Three Yes-No Variables

Person	Do You Use a Brand X Razor?	Do You Use Brand X Blades?	Have You Seen the Brand X Television Commercial?
1 Adams	No	Yes	Yes
2 Balch	No	No	No
⋮	⋮	⋮	⋮
600 Zuckerman	Yes	Yes	Yes

(hypothetical data)

TABLE 5.32

2 × 2 × 2 Table: Cross-Tabulation of 600 Men, by Use of Blades, Use of Razor, and Viewing of Commercial

	Have Seen Commercial		Have Not Seen Commercial	
	Use Brand X Blades	Do Not Use Brand X Blades	Use Brand X Blades	Do Not Use Brand X Blades
Use Brand X Razor	86	38	100	55
Do Not Use Brand X Razor	9	17	50	245

(hypothetical data)

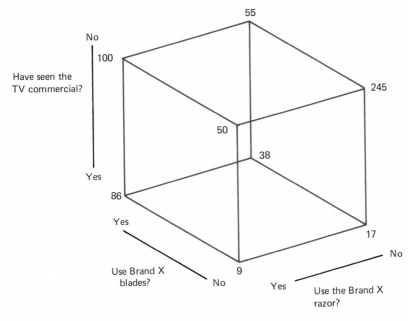

Figure 5.1 *A three-dimensional representation of a three-way table (Table 5.32)*

who have. For example, there were 86 men who answered Yes to all three of the questions: they have seen the commercial and use Brand X products, the blades and the razor.

The three-way table may be viewed as a set of two 2 × 2 tables, one for those who have not seen the commercial and the other for those who have. The effectiveness of the television commercial is assessed by comparing these two tables.

By thinking of one of these 2 × 2 subtables as being placed above the other, we can visualize this 2 × 2 × 2 table as a three-dimensional array of numbers. The table is represented in this way in Figure 5.1. Those who have not seen the commercial are represented at the top of the cube.

Three Two-Way Tables. By suppressing, that is, summing over the categories of, one variable at a time, we obtain three frequency tables for the three possible pairs of variables. (See Table 5.33.)

Interpretation. Of the 600 men, 150, or 25%, have seen the Brand X television commercial. Of these 150 men who have seen the commercial, 124, or 83%, use the razor.

Tables 5.32 and 5.33 can be used to study a number of questions. We can discover, for example, the relationship between popularity of the razor and the blades, both among those who have and among those

TABLE 5.33

2 × 2 Frequency Tables for Three Pairs of Variables

USE BRAND X BLADES?

		Yes	No	Total
USE BRAND X RAZOR?	Yes	186	93	279
	No	59	262	321
	Total	245	355	600

SEEN TV COMMERCIAL?

		Yes	No	Total
USE BRAND X RAZOR?	Yes	124	155	279
	No	26	295	321
	Total	150	450	600

USE BRAND X BLADES?

		Yes	No	Total
SEEN TV COMMERCIAL?	Yes	95	55	150
	No	150	300	450
	Total	245	355	600

SOURCE: Table 5.32.

TABLE 5.34

Percent Using Brand X Blades, by Use of Razor and Viewing of Commercial

		HAVE SEEN COMMERCIAL	HAVE NOT SEEN COMMERCIAL
USE BRAND X RAZOR	Percent	69%	65%
	(Number)	(124)	(155)
DO NOT USE BRAND X RAZOR	Percent	35%	17%
	(Number)	(26)	(295)

SOURCE: Tables 5.32 and 5.33.

who have not seen the commercial. Table 5.34 shows the percentage using the blades, in each of the four groups. (The base for each percentage is indicated below it in parentheses.) The percentages using the blades are higher for those who use the razor, both among those who have seen the TV commercial and among those who have not. Thus the association between using Brand X blades and using the Brand X razor exhibits itself in both groups.

Furthermore, we see that the third variable, viewing the commercial, has its own, separate effect: the percentage using the blades is higher among those who have seen the commercial, both among those who use the Brand X razor and among those who do not.

Various possible effects of a third variable are studied further in Section 5.4. Exercises 5.14 and 5.15 involve further analysis of the effects of Company X's television commercial.

5.3.2   Larger Three-Way Frequency Tables

Each dichotomous variable in the $2 \times 2 \times 2$ table could be replaced by a variable with more categories. We may refer to more general three-way tables as $r \times c \times t$ tables. Such a table can be considered as a set of t tables, each $r \times c$.

Table 5.35 is a $2 \times 3 \times 3$ table that was obtained by Lazarsfeld and Thielens (1958) in their study on the relationships between age, productivity, and party vote in 1952 for a sample of social scientists. Each of a number of social scientists was classified according to age, productivity score (an index constructed to measure professional productivity), and party vote in 1952.

TABLE 5.35

Classification of Social Scientists by Age, Productivity Score, and Party Vote in 1952

| | | PRODUCTIVITY SCORE | | | | |
| | *Low* | | *Medium* | | *High* | |
	DEMOCRATS	OTHERS	DEMOCRATS	OTHERS	DEMOCRATS	OTHERS
40 or younger	260	118	226	60	224	60
AGE *41 to 50*	60	60	78	46	231	91
51 or older	43	60	59	60	206	175

SOURCE: Lazarsfeld and Thielens (1958), Figure 1.7, p. 17.

These are data resulting from the observation of three variables for each social scientist. The three variables are categorical: age with the three categories, 40 or younger, 41 to 50, 51 or older; productivity score with the three categories Low, Medium, and High; and Party Vote with the two categories, Democrat and Others. This table can be considered as a series of three two-way tables for age and party, one for each productivity level.

The interrelationships among the three variables can be conveniently summarized by computing the percent Democratic in each of the nine age-productivity categories. This is part of Exercise 5.13.

5.4   EFFECTS OF A THIRD VARIABLE

5.4.1   Association and Interpretation

When analysis reveals association between two variables it is generally desirable to know more about the association. Does it persist under different conditions? Does one factor "cause" the other? Is there a third factor that causes both? Is there another link in the chain, a factor influenced by one variable and in turn influencing the other? Many of these problems can be studied by considering the interrelations between the two variables of interest and a third variable.

Do Storks Bring Babies?   In Scandinavian countries a positive association between the number of storks living in the area and the number of babies born in the area was noticed. Do storks bring the babies? Without shattering the illusions of the incurably romantic, we may suggest the following: Districts with large populations have a large number of births and also have many buildings, in the chimneys of which storks can nest. Figure 5.2 is a diagram representing the idea that the population factor explains both the number of births and the frequency with which storks are sighted. The three variables to study are populations of districts, numbers of births in districts, and numbers of storks seen in the districts.

When three variables are categorical, the interrelations are found from three-way tables. In this section we shall compare the overall association between two categorical variables with the association between them in the two-way tables defined by various values of a third variable. If two variables have a certain relationship overall, that association may be sustained in the component tables defined by values of the third variable, may be decreased, may be obliterated, or may be reversed. We shall study these cases.

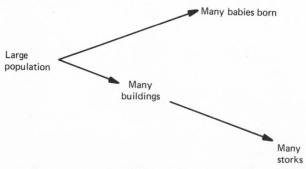

Figure 5.2 *Explanatory diagram indicating that high population accounts for many babies being born and for many storks being sighted*

There are several ways of interpreting the effect of a third variable. Several examples will be given.

5.4.2 Independence in Subtables

It frequently happens that two variables which appear to be associated are not when a third variable is taken into account. This idea can be illustrated most easily by a hypothetical example displaying *exact* independence.

Table 5.36 is a cross-classification of 6000 persons living in two suburbs of a city, according to income and residence (Suburb A or Suburb B). There is an association between residence and income level; 34% of the residents of Suburb A are in the low-income bracket, compared to only 24% for Suburb B, and 37% of the inhabitants of Suburb B are in the high-income bracket, compared to only 31% for Suburb A.

This study can be expanded by considering other factors. Residence and income may both be related to occupation. Table 5.37 is a three-way table. For each of three occupational groups, there is a 3 × 2 table of frequencies according to income level and residence. In each such subtable, income and residence are independent; that is, the income distributions are the same for suburbs A and B (though they differ from one occupational group to another).

TABLE 5.36
Income Distributions in Two Suburbs

| | | INCOME LEVEL | | | |
		Low	Medium	High	Total
SUBURB	A	1350 (34%)	1400 (35%)	1250 (31%)	4000 (100%)
	B	475 (24%)	775 (39%)	750 (37%)	2000 (100%)

(hypothetical data)

TABLE 5.37
Three-Way Frequency Table: Income and Occupation in Suburbs A and B

| | | OCCUPATIONAL GROUP | | | | | |
| | | Agricultural | | Blue Collar | | White Collar | |
		A	B	A	B	A	B
INCOME LEVEL	Low	1000 (50%)	250 (50%)	250 (25%)	125 (25%)	100 (10%)	100 (10%)
	Medium	500 (25%)	125 (25%)	500 (50%)	250 (50%)	400 (40%)	400 (40%).
	High	500 (25%)	125 (25%)	250 (25%)	125 (25%)	500 (50%)	500 (50%)
	Total	2000(100%)	500(100%)	1000(100%)	500(100%)	1000(100%)	1000(100%)

(hypothetical data)

How can it happen that income and residence are associated when all persons are considered, but are not associated within more homogeneous occupational groups? This can occur because occupation is related to both residence and income.

Table 5.38 shows a relationship between residence and occupation. [The modal occupational categories (with frequencies underlined) are agricultural for Suburb A and White Collar for Suburb B.] A little investigation reveals the reason: Suburb A is close to the farms where the agricultural workers are employed.

TABLE 5.38

Occupation Distributions in Suburbs A and B

| | | OCCUPATIONAL GROUP | | | |
		Agricultural	*Blue Collar*	*White Collar*	*Total*
SUBURB	A	2000 (50%)	1000 (25%)	1000 (25%)	4000 (100%)
	B	500 (25%)	500 (25%)	1000 (50%)	2000 (100%)

SOURCE: Table 5.37.

Table 5.39 is a cross-classification by occupation and income level. The modal income level is underlined for each occupational category. There is an association between occupation and income.

Figure 5.3 is a diagram of the pairwise associations among the three variables, occupation, income level, and residence. We see that the association between income level and residence is due to the fact that each of these variables is related to occupation. Figure 5.4 is a causal diagram representing this.

If we had restricted ourselves to observing only the association between income and residence, we might have suggested that persons with low incomes would prefer Suburb A because of certain public

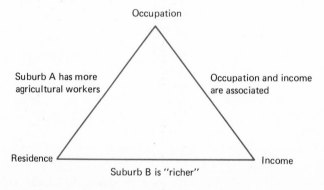

Figure 5.3 *Diagram of associations*

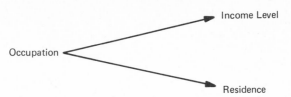

Figure 5.4 *Explanatory diagram: Occupation as a common cause, explaining both income level and residence*

facilities (for example, more public parks, child-care centers, etc.). If this were the case, then within occupational categories, we would expect the percent of residents with low income to be greater in A than in B. Such is not the case, however; the association between income and residence does not persist after occupation is taken into account.

Similarly, one would find that number of births and number of storks are independent in cities of roughly the same size.

Sometimes the term "spurious association" is used for the association between two variables which are dependent on a third. It is important to note that it is not the associations which are spurious; rather it is the *interpretations* people place on the associations. There is a real association here between income and residence. What would be "spurious" in this case is an interpretation that involves an assumption of some intrinsic or causal relationship between income and residence.

When two variables are associated, it is not necessarily true that one factor causes the other. Considering other factors, the investigator may find one which causes both of them.

TABLE 5.39
Cross-Classification of Persons in Suburbs A and B, by Occupation and Income Level

		OCCUPATIONAL GROUP		
		Agricultural	*Blue Collar*	*White Collar*
INCOME	*Low*	1250 (50%)	375 (25%)	200 (10%)
LEVEL	*Medium*	625 (25%)	750 (50%)	800 (40%)
	High	625 (25%)	375 (25%)	1000 (50%)
	Total	2500 (100%)	1500 (100%)	2000 (100%)

SOURCE: Table 5.37.

Education, Intent to Vote, and Political Interest.   Table 5.40 exhibits positive association between level of education and intent to vote, inasmuch as 92% of those with at least some high school education intended to vote, whereas only 86% of those with no high school education intend to vote. Table 5.41 is a three-way table in which the third factor, "level of political interest" (in terms of a scale based on the responses to certain questions), is introduced. Table 5.42 shows that there is a positive association between level of education and level of political interest, while Table 5.43 indicates that there is a strong association between level of political interest and intent to vote: 99% of those having great political interest intend to vote, compared to 92% for those with moderate political interest and only 58% for those with no political interest.

TABLE 5.40
Level of Education and Intention to Vote

		INTEND TO VOTE?		Total
		Yes	No	
LEVEL OF EDUCATION	At Least Some High School	1481 (92%)	132 (8%)	1613 (100%)
	No High School	1026 (86%)	173 (14%)	1199 (100%)

SOURCE: Lazarsfeld, Berelson, and Gaudet (1968). See Table 5.41.

TABLE 5.41
Level of Education, Intent to Vote, and Political Interest

	GREAT POLITICAL INTEREST Intent to Vote			MODERATE POLITICAL INTEREST Intent to Vote			NO POLITICAL INTEREST Intent to Vote		
	YES	NO	TOTAL	YES	NO	TOTAL	YES	NO	TOTAL
At least some high school	490	5	495	917	69	986	74	58	132
No high school	279	6	285	602	67	669	145	100	245
Total	769	11	780	1519	136	1655	219	158	377

SOURCE: Lazarsfeld, Berelson, and Gaudet (1968), p. 47, and Bureau of Applied Social Research.

TABLE 5.42
Distribution of Political Interest, by Education

	LEVEL OF POLITICAL INTEREST			Number of Persons
	Great	Moderate	No Interest	
At least some high school	31%	61%	8%	(1613)
No high school	24%	56%	20%	(1199)

SOURCE: Table 5.41.

TABLE 5.43

Percent Intending to Vote, by Level of Political Interest

| | LEVEL OF POLITICAL INTEREST | | |
	Great	*Moderate*	*No Interest*
Percent intending to vote	99%	92%	58%
(Number of persons)	(780)	(1655)	(377)

SOURCE: Table 5.41.

Table 5.44 shows that, within level-of-interest categories, there are only small differences in the percentages intending to vote. The differences between 99% and 98%, 93% and 90%, and 56% and 59% are small compared to the differences manifested in Table 5.40. We conclude that, within level-of-interest categories, there is essentially no association between education and intent to vote. The apparent association between these two variables in Table 5.40 is explained by their mutual associations with level of political interest. Figure 5.5 is a causal diagram consistent with this finding. It indicates that education increases political awareness, which in turn increases the likelihood of voting.

TABLE 5.44

Percent Intending to Vote, by Level of Education and Level of Political Interest

| | | LEVEL OF POLITICAL INTEREST | | |
		Great	*Moderate*	*No Interest*
LEVEL OF EDUCATION	*At Least Some High School*	99%	93%	56%
	No High School	98%	90%	59%

SOURCE: Table 5.41.

In the case where the third variable can be interpreted as a link in a chain from one variable to another, we call that third variable an *intervening variable*.

More Education ⟶ Greater Political Interest ⟶ Intent to Vote

Figure 5.5 *Causal diagram: Intervening variable*

5.4.3 Similar Association in Subtables

When an association between two variables persists even after other variables are introduced, we consider the interpretation that there is an intrinsic relationship to be better established. The associations in the subtables may be less than that in the overall table, but still in the same direction.

Population Density and Social Pathology. Galle, Gove, and McPherson (1972) made a study to determine the relationship, if any, between crowding (as measured by population density) and various "social pathologies" (high rates of juvenile delinquency, high admission rates to mental hospitals, high mortality ratios, etc.) in the 75 community areas of Chicago. We shall phrase our discussion in terms of one social pathology, juvenile delinquency. Table 5.45 shows a marked positive relationship between population density and juvenile delinquency.

TABLE 5.45
Population Density and Juvenile Delinquency

| | | POPULATION DENSITY | | | |
		Low	High	Total	
RATE OF	Low	30 (88%)	5 (12%)	35 (47%)	
JUVENILE DELINQUENCY	High	4 (12%)	36 (88%)	40 (53%)	
	Total	34 (100%)	41 (100%)	75 (100%)	

SOURCE: Hypothetical data suggested by the study by Galle, Gove, and McPherson (1972).

It is known that social pathologies are related to social class and that social class is a variable which should be considered when studying the relation between population density and social pathologies. Table 5.46

TABLE 5.46
Cross-Classification of 75 Community Areas of a Large City by Population Density, Rate of Juvenile Delinquency, and Social Class

| | | LOW SOCIAL CLASS | | | HIGH SOCIAL CLASS | | |
		Low Population Density	High Population Density	Total	Low Population Density	High Population Density	Total
RATE OF JUVENILE DELIN-QUENCY	Low	3	2	5	27	3	30
	High	2	33	35	2	3	5
	Total	5	35	40	29	6	35

SOURCE: Hypothetical data suggested by the study of Galle, Gove, and McPherson (1972).

is a cross-classification of 75 districts according to the three variables, population density, rate of juvenile delinquency, and social class. Examination of Table 5.46 shows that in each half of the table—the left half corresponding to low social class and the right corresponding to high social class—there is positive association between population density and juvenile delinquency. Table 5.47 gives the percentage of community areas having high rates of juvenile delinquency, in each of the four social-class–population-density combinations. In both low-social-class districts and high-social-class districts, the relationship between density and delinquency persists: for low-social-class areas, the 94% for high-density areas exceeds the 40% for low-density areas; and similarly for high-social-class areas, the 50% for high-density areas exceeds the 7% for low-density areas.

TABLE 5.47
Percentage of Community Areas Having High Rates of Juvenile Delinquency, by Social Class and Population Density

| | | SOCIAL CLASS | |
		Low	*High*
POPULATION DENSITY	*Low*	40%	7%
	High	94%	50%

SOURCE: Table 5.46.

Since the positive association between population density and juvenile delinquency is not so great in the social-class subtables as it is in Table 5.45, we next ask whether social class is associated with population density and whether it is associated with juvenile delinquency. The two-way tables relevant to these questions are given in Table 5.48. It will be seen that there is negative association in each table. Figure 5.6

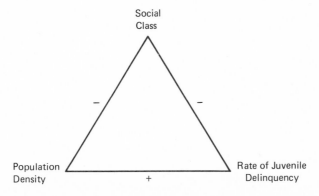

Figure 5.6 *Diagram of associations. Nature of each association is indicated.*

Figure 5.7 *A causal diagram representing low social class as a "common cause" of juvenile delinquency and population density*

is a triangular diagram representing the three pairwise associations.

We may ask whether social class is a common cause, totally explaining the association between population density and juvenile delinquency, as represented in Figure 5.7. However, this proposition is not consistent with the data of Table 5.46. It seems more plausible that social class affects both population density and juvenile delinquency; and, in addition, population density has its own, additional effect on delinquency (as represented in Figure 5.8).

TABLE 5.48

Relationships of Population Density and Juvenile Delinquency to Social Class

| | | SOCIAL CLASS | | |
		Low	High	Total
POPULATION	*Low*	5	29	34
DENSITY	*High*	35	6	41
	Total	40	35	75

| | | SOCIAL CLASS | | |
		Low	High	Total
JUVENILE	*Low*	5	30	35
DELINQUENCY	*High*	35	5	40
	Total	40	35	75

SOURCE: Table 5.46.

The association between population density and juvenile delinquency is "true," in the sense that it persists even after effects of a highly relevant third variable have been taken into account.

Drinking, Smoking, and Health. Since drinking and smoking are associated (see Table 5.60, for example), alcohol consumption was con-

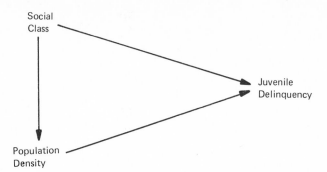

Figure 5.8 *Causal diagram: Density and social class each affect delinquency, and social class affects density*

sidered as a possible explanation of the association between smoking and lung cancer. But it was found that among both those who drink a great deal and those who drink little, the rate of lung cancer is higher for smokers than for nonsmokers.

    Urban Life, Smoking, and Health. Air pollution is greater in urban environments than in rural ones. Table 5.49 gives the death rate due to lung cancer in various urban and rural areas. It will be seen that in each type of area the incidence of lung cancer is higher for smokers than for nonsmokers. The relationship between smoking and lung cancer persists when environment is taken into account.

    Also, the relationship between environment and lung cancer persists when smoking habits are taken into account: urban smokers are more prone to lung cancer than are rural smokers, and urban nonsmokers are more prone to lung cancer than are rural nonsmokers.

    The conclusion is that both urban environment and smoking are contributing factors in lung cancer.

TABLE 5.49

Death Rates (Per 100,000 Man-Years) from Lung Cancer According to Residence and Smoking Habits

	CITIES OF OVER 50,000	10,000–50,000	SUBURB OR TOWN	RURAL
Cigarette smokers	85.2	70.9	71.7	65.2
Nonsmokers	14.7	9.3	4.7	0.0

SOURCE: Figure 6 from Hammond and Horn, *Journal of the American Medical Association* 166: 1294–1308. Copyright 1958, American Medical Association.

5.4.4   Reversal of Association in Subtables

So far we have considered examples in which a third factor "explained away" an association between two variables, and examples in which the association between two variables remained after taking the third factor into account. Now we consider an example in which taking a third factor into account changes the direction of association.

The Relation Between Weather and Crop Yield.   This example is suggested by the study of Hooker (1907), who studied the interrelationships among temperature, rainfall, and yield of hay and other crops in part of England. Table 5.50 is a three-way table giving a cross-classification of 40 years, by yield of hay, rainfall, and temperature.

TABLE 5.50
Cross-Classification of 40 years, by Yield of Hay, Rainfall, and Temperature

YEARS OF LIGHT RAINFALL
Yield

		LOW	HIGH	TOTAL
Temperature	LOW	4	1	5
	HIGH	11	4	15
	TOTAL	15	5	20

YEARS OF HEAVY RAINFALL
Yield

		LOW	HIGH	TOTAL
Temperature	LOW	4	11	15
	HIGH	1	4	5
	TOTAL	5	15	20

SOURCE: Hypothetical data suggested by the study by Hooker (1907).

The two-by-two tables derived from Table 5.50 are given in Table 5.51. The association between temperature and yield is negative, which is somewhat surprising because one would expect a warm summer to make for a good crop. The table suggests that high temperature tends to cause low yield. But looking at the other two pairwise associations, we see that rainfall and yield are positively associated, and rainfall and temperature are negatively associated. These relationships are summarized in the triangle diagram of Figure 5.9.

A hypothesis consistent with this pattern of association is that both

TABLE 5.51

Three 2 × 2 Tables for Rainfall, Temperature, and Yield of Hay in 40 Seasons

		YIELD		Total
		Low	High	
TEMPERATURE	Low	8	12	20
	High	12	8	20
	Total	20	20	40

		RAINFALL		Total
		Light	Heavy	
TEMPERATURE	Low	5	15	20
	High	15	5	20
	Total	20	20	40

		YIELD		Total
		Low	High	
RAINFALL	Light	15	5	20
	Heavy	5	15	20
	Total	20	20	40

SOURCE: Table 5.50.

heavy rainfall and high temperature increase yield, but usually the temperature is low when there is a good deal of rain. Figure 5.10 is a "causal diagram" corresponding to this hypothesis (the arrows are used to indicate causation).

Table 5.50 may be viewed as a cross-classification—by temperature and yield—of the 20 years in which the rainfall was light, together with the same cross-classification for the 20 years in which the rainfall was heavy. These tables are quite different from the portion of Table 5.51 giving the cross-classification of all 40 years according to temperature and yield. The tables in Table 5.50 show a weak positive association, if any, not a negative association, between temperature and rainfall: that is, in both tables

$$ad - bc = +5.$$

The conclusion is that, among years when the rainfall is light, the yield is apt to be greater when the temperature is high than when it is low, and the same is true among years when the rainfall is heavy.

Thus it would be a mistake to take the overall negative association

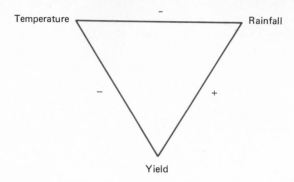

Figure 5.9 *Triangle diagram of observed associations*

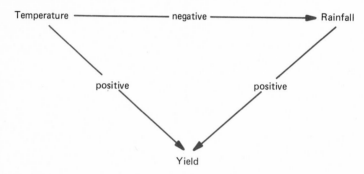

Figure 5.10 *Explanatory diagram: Rainfall as an intervening variable*

between temperature and yield at face value. In particular, it would be incorrect to deduce from the fact that the association is negative that decreasing the temperature would *cause* larger yield. It is just that low temperature *occurs naturally* with high rainfall, which gives a larger yield. In a greenhouse, we could achieve *both* high temperature *and* high humidity.

In summary: what we have seen in this example is that although the overall association was strongly negative, the associations in the sub-tables corresponding to fixed levels of rainfall are slightly positive. The overall negative relationship exists between temperature and yield because rainfall is an intervening variable between temperature and yield (Figure 5.10); rainfall is positively related to yield but negatively related to temperature.

5.4.5   Hidden Relationships

In Section 5.4.2 we saw that when a third variable is taken into account the association between two variables may disappear. The opposite can happen; two variables which appear independent may be found to be associated when the individuals are classified according to a third variable. This phenomenon can be expected when there is a reversal, that is, when the direction of the association in one group is different from that in the other.

In Table 5.52 there is little association between age and "listening to classical music." The percentages listening to classical music, among those below 40 and those 40 or older are 65% and 64%, respectively.

TABLE 5.52

Percentage Who Listen to Classical Music, by Age

	AGE	
	Below 40	*40 and Above*
Percentage who listen to classical music	65%	64%
(Number of persons)	(603)	(506)

SOURCE: Adapted from Lazarsfeld (1940), p. 98.

TABLE 5.53

Percentage Who Listen to Classical Music, by Age and Education

	AGE	
	Below 40	*40 and Above*
College	73%	76%
(Number of persons)	(224)	(251)
Below college	60%	52%
(Number of persons)	(379)	(255)

SOURCE: Same as Table 5.52.

But when the additional factor, education, is brought in, Table 5.53 results. Note the reversal: among college-educated persons, 76% of the "oldsters" listen to classical music, while 73% of the "youngsters" do. Among the noncollege persons, *fewer* of those 40 or older listen to classical music: 52% compared to 60% for those below 40.

Within the college group, the association between age and listening to classical music is small, but positive. Within the noncollege group, it is stronger but negative. When the groups are combined, these two opposite effects tend to cancel out, making the overall association small.

5.5   SEVERAL NUMERICAL VARIABLES

5.5.1   Scatter Plots

In Section 2.3 we considered the organization of data consisting of a single numerical measurement on each individual. In this section we study data consisting of *two* numerical measurements on each individual and the graphical and tabular methods of organizing such data.

The language and nonlanguage mental maturity scores (two kinds of intelligence quotients or "IQ's") of 23 school children from Table 2.22 are given at the right of Figure 5.11. These data are represented visually by making a graph on two axes, the horizontal, x, axis representing language IQ and the vertical, y, axis representing nonlanguage IQ. Such a graph is called a *scatter plot* (or *scatter diagram*). Each point in such a plot represents one individual.

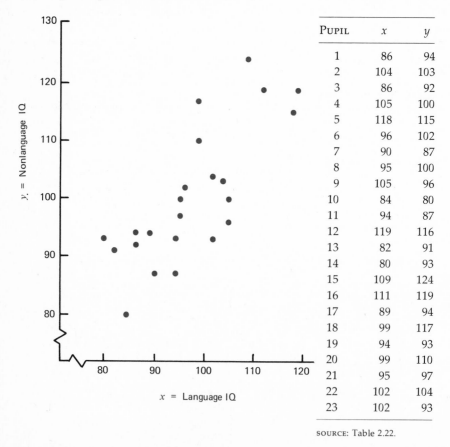

PUPIL	x	y
1	86	94
2	104	103
3	86	92
4	105	100
5	118	115
6	96	102
7	90	87
8	95	100
9	105	96
10	84	80
11	94	87
12	119	116
13	82	91
14	80	93
15	109	124
16	111	119
17	89	94
18	99	117
19	94	93
20	99	110
21	95	97
22	102	104
23	102	93

SOURCE: Table 2.22.

Figure 5.11 *Scatter plot of language and nonlanguage IQ scores of 23 children: Positive correlation (Table 2.22)*

The distances in the horizontal direction indicate language IQ; the points are above the axis at the respective IQ scores as in the graphing of a single numerical variable. (The axis is broken to indicate that the first interval represents 80 units rather than the 10 units that the other intervals represent.) The distances in the vertical direction indicate nonlanguage IQ; the points are to the right of the axis at the respective scores. Thus one point represents one pupil by indicating that pupil's two scores.

The scatter plot yields additional interesting information that is not provided by separate plots for the two variables. It gives a visual indication of the amount and direction of association, or *correlation*, as it is termed for numerical variables.

The swarm of points goes in the southwest-northeast direction. Large values of one variable tend to go with large values of the other; that is, the two IQ scores are *positively* correlated. This is not unexpected, as a pupil's general level of intelligence is a factor in both of the IQ scores.

The swarm of points in Figure 5.12, on the other hand, goes in the northwest-southeast direction. Large values of one variable tend to go with small values of the other. The two variables represented are negatively correlated.

For the points in Figure 5.13, knowing the value of one variable does little or nothing to locate the value of the other; the two variables represented have little or no association.

For the points in Figure 5.14, knowing the value of one variable does quite a bit to locate the other variable, but the relationship between y and x is not simple; that is, it is not true that y always increases as x increases, nor is it true that y always decreases as x increases.

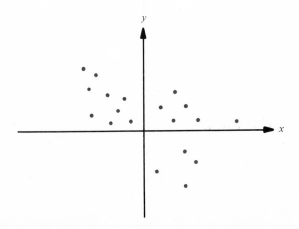

Figure 5.12 *Negative association*

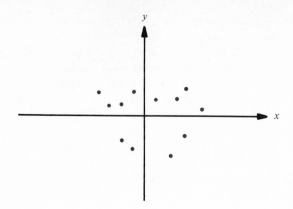

Figure 5.13 *No association*

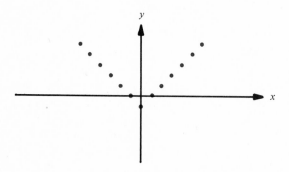

Figure 5.14 *Association*

5.5.2   Joint Frequency Table

Table 5.54 is a cross-classification of the 23 pupils represented in Figure 5.11. It is a joint frequency table for the two variables, language IQ and nonlanguage IQ. The procedure for making such a frequency table is similar to that discussed in Chapter 2 for a frequency table for data on a single variable, but now the categories for grouping are defined by class intervals for *two* variables simultaneously. A set of class intervals is selected for each variable. The two sets of class intervals define the two-way frequency table. The table tells how many pupils fall into each category. For example, the number 5 near the upper left means that 5 of the pupils had language IQ's in the 80's and nonlanguage IQ's in the 90's.

Such grouping is especially helpful when there are many observations, as is the case in Table 5.55, representing the joint frequency distribution of age and salary among 400 workers. As with categorical

TABLE 5.54

Cross-Classification of 23 Pupils, by Language and Nonlanguage IQ

		NONLANGUAGE IQ					
		80–89	*90–99*	*100–109*	*110–119*	*120–129*	*Total*
	80–89	| 1	⊥⊥⊤ 5	0	0	0	6
LANGUAGE IQ	*90–99*	|| 2	|| 2	|| 2	|| 2	0	8
	100–109	0	|| 2	||| 3	0	| 1	6
	110–119	0	0	0	||| 3	0	3
	Total	3	9	5	5	1	23

SOURCE: Table in Figure 5.11.

data, we can form percentages using the marginal totals. The percentages in the table show that the older workers tend to have higher salaries.

If we divide all the frequencies in the table by 400, we obtain Table 5.56, giving the *joint distribution* of age and income, which tells what percentage of the 400 workers fall into each salary-age category.

TABLE 5.55

Joint Frequency Table for Age and Salary, with Distribution of Salary, by Age

		AGE				*All Age Groups*
		20–29	*30–39*	*40–49*	*50–59*	
SALARY ($)	72.50–82.49	50 (50%)	20 (14%)	10 (8%)	4 (10%)	84 (21%)
	82.50–92.49	30 (30%)	70 (50%)	30 (25%)	6 (15%)	136 (34%)
	92.50–102.49	20 (20%)	50 (36%)	80 (67%)	30 (75%)	180 (45%)
	Total	100 (100%)	140 (100%)	120 (100%)	40 (100%)	400 (100%)

(hypothetical data)

We have discussed methods of organization and summarization of multivariate categorical data. These methods apply to numerical data as well, but with numerical data the categories are always ordered, since they came from quantitative variables.

TABLE 5.56

Joint Distribution of Age and Salary Among 400 Workers

		AGE				
		20–29	30–39	40–49	50–59	Total
SALARY ($)	72.50–82.49	12.5%	5.0%	2.5%	1.0%	21.0%
	82.50–92.49	7.5%	17.5%	7.5%	1.5%	34.0%
	92.50–102.49	5.0%	12.5%	20.0%	7.5%	45.0%
	Total	25.0%	35.0%	30.0%	10.0%	100.0%

SOURCE: Table 5.55.

5.5.3 Pictorial Representation

When there are only two variables we can use an ordinary scatter plot to represent individuals on a *plane*. (When there is only one variable, we can simply use a *line*.) If there are three variables, we could represent the data in *space*, say as points in a room. (For example, we might be able to use "tinker toys" to do this.)

Profiles. It is often desirable to be able to visualize data for more than three variables simultaneously. A simple method is the use of *profiles*. A profile is obtained by plotting the values of each variable for a given person vertically, but at different positions; the points are connected by lines. The scores of pupils 1, 15, and 16 from Table 2.22 are listed in Table 5.57 and the corresponding profiles are shown in Figure 5.15. The four variables are scaled for plotting on the same picture. The shape of the profile is dependent upon the scaling and the ordering of the variables along the axis. Here the reading achievement scores were multiplied by 50 so that they would fit with the IQ scores.

TABLE 5.57

Scores of Pupils 1, 15, and 16

PUPIL	LANGUAGE IQ	NONLANGUAGE IQ	INITIAL READING ACHIEVEMENT	FINAL READING ACHIEVEMENT
1	86	94	1.1	1.7
15	109	124	1.8	2.5
16	111	119	1.4	3.0

SOURCE: Table 2.22.

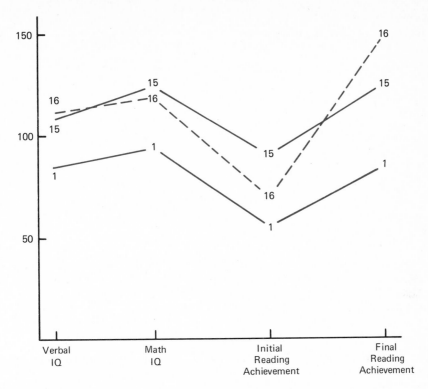

Figure 5.15 *Profiles of Pupils 1, 15, and 16 (Table 2.22)*

But once the ordering and the scaling are fixed, we can compare individuals by looking at their profiles.

It is evident that pupil 1 has less overall intellectual ability than pupils 15 and 16. The fact that pupil 16 had a lower initial reading score than pupil 15 but "passed him by" during the year is also made clear.

Polygons. Another way of plotting more than two positive variables is to plot them in different directions from a center [Daetz (1971)]; the points are connected to form a four-sided polygon. The four directions are North, East, South, and West. The three pupils are represented again in Figure 5.16. Figures 5.15 and 5.16 each show fairly clearly that pupils 15 and 16 have essentially equal IQ scores, but pupil 16 had a higher final reading achievement score, in spite of having a substantially lower initial reading achievement score.

We have illustrated the use of polygons when there are four variables so that the directions corresponding to variables are at right angles. If there were ten variables, the directions would be 36 degrees apart, since there are 360 degrees in a circle. The scale and ordering of variables is arbitrary.

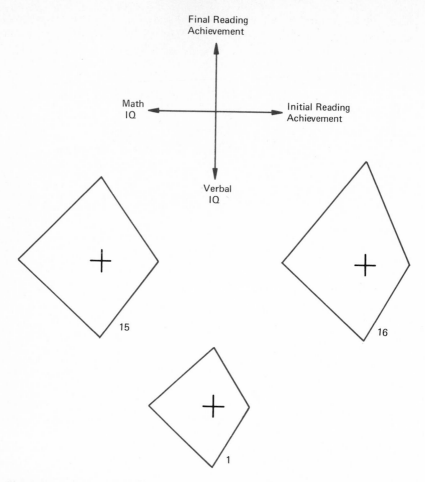

Figure 5.16 *Polygons representing Pupils 1, 15, and 16 (Table 2.22)*

Chernoff's Faces. The "face" technique [Chernoff (1971)] consists of representing an individual by a picture of a face, the characteristics of which are determined by the values of the several variables. Thus the curvature of the mouth, length of nose, size of eyes, etc., represent various variables. A set of data is represented by a collection of faces. In exploring the data the researcher can react to the *Gestalt* of the face.

The faces of Figure 5.17 represent measurements of six variables on each of 87 fossils* of a small sea animal from a yellow limestone formation in northwestern Jamaica [Wright and Switzer (1971)]. The faces

---

*There had been 88 specimens, but one (Number 34) was omitted because some of the measurements for that specimen had been incorrectly recorded.

were grouped and numbers were assigned to correspond to the grouping. It is clear from Figure 5.17 that faces 1–41 form a group, as do faces 42–75 and 76–88. The fact that this grouping is obvious is due

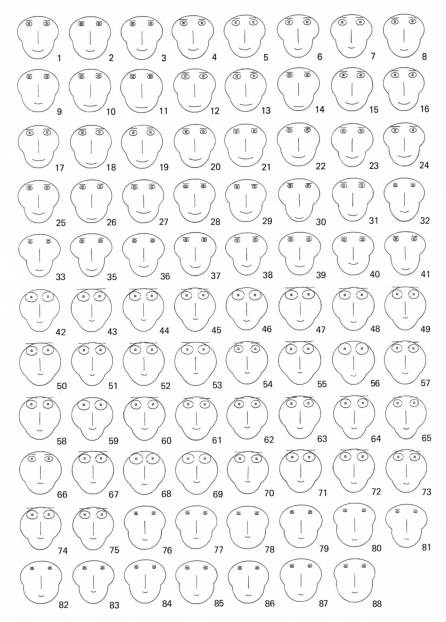

Figure 5.17 *"Faces" representing fossil specimens*

in part to the special arrangement of the faces. However, when copies of the faces were made (without the numbers), people sorted out the shuffled deck of faces into these same groups with only a few discrepancies.

5.5.4 Descriptive Statistics for Bivariate Numerical Data; Correlation Coefficient

A descriptive statistic for bivariate numerical data is the *correlation coefficient*

$$r = \frac{s_{xy}}{s_x s_y},$$

where

$$s_x = \sqrt{\frac{\sum\limits_{i=1}^{n}(x_i - \bar{x})^2}{n-1}}$$

and

$$s_y = \sqrt{\frac{\sum\limits_{i=1}^{n}(y_i - \bar{y})^2}{n-1}}$$

are simply the sample standard deviations of x and y and

$$s_{xy} = \frac{\sum\limits_{i=1}^{n}(x_i - \bar{x})(y_i - \bar{y})}{n-1}$$

is the *sample covariance* between x and y.

Let us examine the formula for the covariance, which is essentially an average of the quantities $(x_1 - \bar{x})(y_1 - \bar{y})$, $(x_2 - \bar{x})(y_2 - \bar{y})$, $\ldots$, $(x_n - \bar{x})(y_n - \bar{y})$. The ith individual in the sample is associated with the term $(x_i - \bar{x})(y_i - \bar{y})$.

If x_i exceeds $\bar{x}$, then individual i has a large x-score, and similarly if y_i exceeds $\bar{y}$, a large y-score. The algebraic sign of the product $(x_i - \bar{x})(y_i - \bar{y})$ tells whether individual i has a large score on both x and y or a small score on both x and y ($+$ sign), or a large score on one variable and a small score on the other ($-$ sign). The product $(x_i - \bar{x})(y_i - \bar{y})$ also reflects the *size* of the covariation in x and y for the ith individual. The covariations of all the individuals are combined by taking the sum of the individual products.

Dividing s_{xy} by the product of s_x and s_y produces a measure of as-

TABLE 5.58
Computation of the Correlation Coefficient

	Height x	Weight y	$x_i - \bar{x}$	$y_i - \bar{y}$	$(x_i - \bar{x})^2$	$(y_i - \bar{y})^2$	$(x_i - \bar{x})(y_i - \bar{y})$
Bill	65	162	−5	0	25	0	0
Ed	70	150	0	−12	0	144	0
Frank	75	174	5	+12	25	144	+60
Sum	210	486	0	0	50	288	+60
Mean	70	162					

(hypothetical data)

sociation r which is at least −1 and at most +1. If r is nearly zero, it is because s_{xy} is small, relative to the variability of x and y.

Table 5.58 illustrates briefly the computation of the correlation coefficient for the heights (in inches) and weights (in pounds) of only 3 individuals; the scatter plot is Figure 5.18.

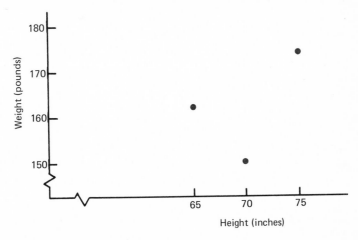

Figure 5.18 *Scatter plot for Table 5.58*

Note that Frank is tallest and also weighs the most. Bill is shortest and ranks second in weight. Ed weighs the least and ranks second in height.

The standard deviations are

$$s_x = \sqrt{\frac{50}{3-1}} = \sqrt{\frac{50}{2}} = \sqrt{25} = 5$$

and

$$s_y = \sqrt{\frac{288}{3-1}} = \sqrt{\frac{288}{2}} = \sqrt{144} = 12.$$

The covariance is

$$s_{xy} = \frac{60}{3-1} = \frac{60}{2} = 30.$$

The value of the correlation coefficient is

$$r = \frac{30}{5 \times 12} = \frac{30}{60} = \frac{1}{2} = 0.50.$$

This shows some positive association between the heights and weights of these 3 persons.

Note that the covariance, and hence the correlation coefficient, is based only on $x_i - \bar{x}$ and $y_i - \bar{y}$, $i = 1, \ldots, n$. It is possible for x and y to be related in a way that is not detected by r. Such is the case in Figure 5.14. There the correlation between y and x is 0 but the correlation between y and $|x|$ is +1.

SUMMARY

Bivariate categorical data are summarized in *two-way frequency tables*. In the case of two Yes-No variables, this is a 2×2 *table*. Two-by-two tables arise also from two Low-High variables, and from change-in-time data. A frequency table for change-in-time data is called a *turnover table*.

Interpreting tables involves observing patterns of percentages. If the distributions are the same in each row, the percentages exhibit *independence* of the variables defining the table. Otherwise, the variables are *associated*.

When the distributions are the same in each row, the distributions are also the same in each column.

Three-way frequency tables arise from *trivariate* data. A third variable may be the cause of association between two other variables. In sub-tables corresponding to fixed values of the third variable, there may be no relationship between the other two variables. The associations in subtables corresponding to fixed values of the third variable may differ from one another and from the overall association not only in size but even in direction. When these associations are nil but the overall association is not, the overall association is called "spurious."

Association must be supplemented by further evidence in order to establish causal relationship. Some evidence is supplied when the association persists after relevant third factors have been taken into account. The strongest evidence is supplied by a controlled experiment.

Bivariate numerical data are represented in a *scatter plot*. Two-way frequency tables can be made for numerical variables. In the case of continuous variables, the tables are made by categorizing each variable into class intervals and tabulating the corresponding joint distribution. A visual impression of *multivariate data* may be gained by *profiles* or by other pictorial representations.

EXERCISES

5.1 (Sec. 5.1.1)   Fill in the following 2×2 frequency tables in all possible ways.

(a)

		1
		1
1	1	2

(b)

		2
		1
1	2	3

5.2 (Sec. 5.1.1)   Fill in the following 2×2 frequency tables in all possible ways.

(a)

		2
		3
3	2	5

(b)

		2
5	1	

5.3 (Sec. 5.1.3)   Fill in the following 2×2 table in all possible ways. Rank the possible tables according to the degree and direction of association exhibited.

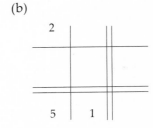

		B		
		Low	High	
	Low			4
A	High			4
		4	4	8

5.4 (Sec. 5.1.3) Fill in the following 2 × 2 frequency table in all possible ways. Rank the possible tables according to the degree and direction of association exhibited.

B

		Low	High		
Low					5
High					3
		4	4		8

(A on the left side)

5.5 (Sec. 5.1) The 1000 cars sold by a new-car dealer in a recent year were cross-classified as in Table 5.59. What do these data suggest about the relation between body type and type of transmission purchased?

TABLE 5.59

Cross-Classification of 1000 New Cars Sold in a Recent Year

		TRANSMISSION		
		Automatic	*Manual*	*Total*
BODY TYPE	2-door	250	225	475
	4-door	500	25	525
	Total	750	250	1000

(hypothetical data)

5.6 (Sec. 5.1) Table 5.60 gives the joint frequency distribution (in percentages) of white males in a single, long-range study, by smoking and drinking habits.
 (a) Compute the proportion of smokers who drink.
 (b) Compute the proportion of nonsmokers who drink.
 (c) For this group of men, are smoking and drinking associated?

TABLE 5.60

Joint Frequency Distribution of Men by Smoking and Drinking Habits (Percentages)

	SMOKER	NONSMOKER	TOTAL
DRINKER	73%	8%	81%
NONDRINKER	14%	5%	19%
TOTAL	87%	13%	100%

SOURCE: Marvin A. Kastenbaum, Director of Statistics, The Tobacco Institute, Inc., personal communication.

5.7 (Sec. 5.2) Summarize Table 5.61 using two different low-high categorizations of the variable, "Highest Proficiency Achieved."

(a) Combine "Average," "Better than Average," and "Extremely High" into the single category "High."

(b) Combine only "Better than Average" and "Extremely High" into the single category "High."

In each case, write down the 2 × 2 frequency table, and compute the percentages using languages after doctoral study in the Low and in the High groups.

TABLE 5.61

Cross-Classification of 247 Professional Geographers by Highest Proficiency Achieved in at Least One Language in Meeting the Language Requirement for the Doctoral Degree and Use of the Language After Doctoral Work

		Use The Language after Doctoral Study?		
		No	*Yes*	*Total*
Highest Proficiency	*Minimal*	32	23	55
Achieved in Meeting	*Average*	51	64	115
the Foreign Language	*Better than Average*	12	37	49
Requirement	*Extremely High*	1	27	28
	Total	96	151	247

SOURCE: Wiltsey, Robert G., *Doctoral Use of Foreign Languages: A Survey. Part II: Supplementary Tables*, p. 61. Copyright © 1972 by Educational Testing Service. All rights reserved. Reproduced by permission.

5.8 (Sec. 5.2.2) Each individual in a random sample of 200 freshmen, sophomores, and juniors was asked whether he or she favors greater representation of the freshman and sophomore classes in the student senate. The results are given in Table 5.62.

TABLE 5.62

College Political Poll

	Freshmen	Sophomores	Juniors	Total
Favor	45	30	30	105
Opposed	30	15	50	95
Total	75	45	80	200

(hypothetical data)

(a) (i) Compute percentages based on the column totals.

(ii) Compute percentages based on the row totals.

(b) (i)   Form a summary table having two columns: Lower classes (Freshmen and Sophomores) and Juniors.
   (ii)   Give percentages based on the column totals.
   (iii) Give percentages based on the row totals.
(c) (i)   Record the 2 × 2 subtable for the Freshmen and Sophomores only.
   (ii)   Compute percentages based on the row totals.

5.9 (Sec. 5.2.2)   Each owner in a random sample of 100 business firms was asked whether he or she favored a new tax proposal. The results are given in Table 5.63.

TABLE 5.63
Poll of Owners of Business Firms

		SIZE OF FIRM			
		Small	*Medium*	*Large*	*Total*
OPINION	*In Favor*	40	10	2	52
	Opposed	30	10	8	48
	Total	70	20	10	100

(hypothetical data)

(a) (i)   Compute percentages based on the column totals.
   (ii)   Compute percentages based on the row totals.
(b) (i)   Form a summary table having two columns:
          Small- or Medium-size Firms and Large-size Firms.
   (ii)   Give percentages based on the column totals.
   (iii) Give percentages based on the row totals.
(c) (i)   Record the 2 × 2 subtable for the small- and medium-size firms only.
   (ii)   Compute percentages based on the row totals.

5.10 (Sec. 5.2)   A study of reading errors made by second-grade pupils was carried out to help decide whether the use of different sorts of drills for pupils of different reading abilities is warranted. Errors were categorized as follows.
   DK: Did not know the word at all
   C: Substitution of a word of similar *configuration* (e.g., "bad" for "had")
   T: Substitution of a synonym suggested by the *context*
   OS: Other substitution
[These are combinations of categories that were suggested by Kottmeyer (1959), p. 30.] The children had been clustered into three relatively homogeneous reading groups on the basis of (1) their reading achievement scores at the end of the first grade, (2) their verbal IQ's,

and (3) the opinions of their first-grade teachers. The three groups of children each chose the name of an animal as their group name. It happened that the least able readers chose the name Squirrels; the most able readers chose the name Cats; and the middle group, the name Bears. There were five children in the Squirrels group, nine in the Bears, and eleven in the Cats. The numbers of errors of each type made by each child were added to obtain the group totals given in Table 5.64.

The *numbers* of errors are not directly comparable: the more able readers used more difficult texts and read more. For each group, compute the *relative frequency* of each type of error. Compare the three frequency distributions.

TABLE 5.64
Distribution of Reading Errors in Three Ability Groups

	T	C	OS	DK	TOTAL
SQUIRRELS	5	10	15	53	83
BEARS	28	34	72	172	306
CATS	8	10	15	36	69
TOTAL	41	54	102	261	458

SOURCE: Anonymous.

5.11 (Sec. 5.1) Despite the fact that 90% of the people in a community watch football on television and 90% smoke Tiparillos, watching football on television and smoking Tiparillos may not be associated in that community. Explain.

5.12 (Sec. 5.4) A researcher has collected certain data on all the students in a statistics course and claims that the summary in Table 5.65 shows that graduate students do better in the course. The following facts are also selected from the data: exactly 100 students in the course are male; exactly 60 are undergraduate females; exactly 25 of the females did poorly in the course; and exactly 35 undergraduate females did well in the course. Taking all this into account, do you think the graduate/undergraduate dichotomy is an important factor in determining performance in the course?

TABLE 5.65
Student Performance in a Statistics Course

	GRADUATE STUDENT	UNDERGRADUATE STUDENT
DID WELL IN THE COURSE	20	35
DID POORLY IN THE COURSE	40	85

(hypothetical data)

5.13 (Sec. 5.3.2)   Summarize the data given in Table 5.35 from the study, *The Academic Mind*, as follows:

(a) Calculate the indicated percentages (and show the steps in your calculations).

(i)   Find the percentage of social scientists who voted Democratic in 1952.

(ii) Among those social scientists in the 41–50 age group, what percentage are in the medium productivity group?

(iii) Among those social scientists 51 years of age or older having a low productivity score, what percentage voted Democratic in 1952?

(iv) Among those social scientists who did not vote Democratic in 1952, what percentage were 40 years or younger?

(v)   Among those social scientists 51 years or older having a low productivity score, what percentage did not vote Democratic in 1952?

(vi) Among those social scientists low on the productivity scale and Democratic in 1952, what percentage are in the 41–50 year age group?

(b) Compute the percent Democratic in each of the nine age-productivity categories. Arrange these percentages into a 3 × 3 table.

(c) Summarize the relationship between age, productivity, and vote for the Democrats in 1952.

5.14 (Sec. 5.4)   The X Company's commercial (Sec. 5.3) advertised Brand X products in general ("Brand X, a trusted name in men's toiletries," etc.), not the blades or the razor specifically. The advertising department is trying to decide whether it may be worthwhile to advertise specific products separately. Some background study is necessary.

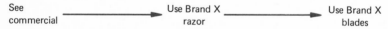

Figure 5.19 *Use of Brand X razor as an intervening variable*

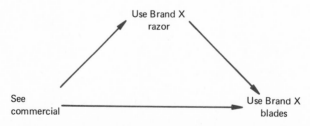

Figure 5.20 *Viewing the commercial has also an independent effect on use of Brand X blades*

Figure 5.19 shows use of the razor as an intervening variable, relating use of the blades to viewing of the commercial. On the other hand, Figure 5.20 indicates that the commercial affected use of the blades not only indirectly (through increased use of the razor), but also directly. Which causal diagram do you find more consistent with the data?

5.15 (continuation) Is Figure 5.21 consistent with the data concerning Brand X products? [Hint: Does taking use of the blades into account as a third factor eliminate the association between use of the razor and viewing of the commercial?]

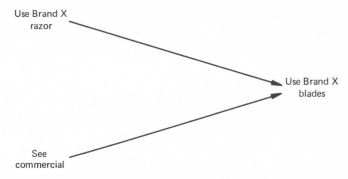

Figure 5.21 *Another possible causal diagram*

5.16 (Sec. 5.5.1) Blau (1955) compared scores of competitiveness and productivity of interviewers in two sections of a state employment agency. Circumstances are such that an interviewer who wants to make more job placements than other interviewers in the section is forced to monopolize the available openings so that the other interviewers are prevented from filling them. In Section A the norms of the group favor competition. The more competitive interviewers, then, hoard openings. In Section B the norms of the group favor cooperation, so that all job openings coming to the section are shared by the five interviewers in that section.

(a) Graph these data using the horizontal axis to represent competitiveness and the vertical axis to represent productivity. Mark each point representing an interviewer in Section A with the symbol A and each point representing one in Section B with B.

(b) Do the two sections seem to differ with respect to competitiveness? If so, how?

(c) Do the two sections seem to differ with respect to productivity? If so, how?

(d) Does there seem to be a relationship between competitiveness and productivity in Section A? If so, is it positive or negative?

(e) Does there seem to be a relationship between competitiveness and productivity in Section B? If so, is it positive or negative?

(f) What conclusions do you draw from this analysis? (Answer this question on the assumption that the differences observed are typical of such interviewers.)

TABLE 5.66
Competitiveness and Productivity in Two Sections of a State Employment Agency

SECTION A		SECTION B	
Competitiveness[a]	Productivity[b]	Competitiveness[a]	Productivity[b]
(X)	(Y)	(X)	(Y)
3.9	0.70	2.2	0.53
3.1	0.49	1.6	0.71
4.9	0.97	1.5	0.75
3.2	0.71	2.1	0.55
1.8	0.45	2.1	0.97
2.9	0.61		
2.1	0.39		

SOURCE: Blau (1955), Table 2, p. 53.
[a]Based on the degree to which each interviewer made more than the expected number of referrals of his or her own clients to job openings received personally, that is, the degree to which the interviewer hoarded job openings as they came in.
[b]The proportion: actual placements made (client actually gets job) divided by the number of openings per interviewer per section.

5.17 (Sec. 5.5.1)   Would you describe the data in Figure 5.22 by saying that X and Y are related, or would you say that there seem to be two groups of individuals and within groups there is no strong relationship between X and Y?

5.18 (Sec. 5.5.1)   Draw the scatter plot for the heights and weights given in Table 2.15.

5.19 (Sec. 5.5.1)   Table 3.28 (Exercise 3.25) presented data from a study on the effectiveness of two sedatives for each of ten individuals. Graph these pairs of values on a scatter plot. Comment on any pattern you observe.

5.20 (Sec. 5.5.1)   For the data for the second grade pupils in Table 2.22 (Exercise 2.19), do the following:

(a) Compute the gain in reading achievement for each child. Plot these against initial reading achievement. Does there appear to be a relation? If so, comment on it.

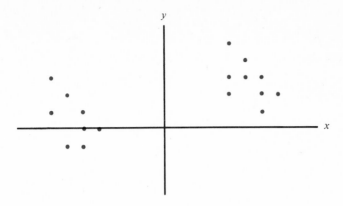

Figure 5.22 *Scatter plot*

(b) Plot final reading achievement score vs. verbal IQ.
(c) Plot initial reading achievement score vs. verbal IQ.
(d) Which appears to be more strongly related to verbal IQ, initial reading achievement or final reading achievement?
(e) Nonlanguage IQ is plotted against language IQ in Figure 5.11. Does there seem to be much of a correlation between these two variables?

5.21 (Sec. 5.5.1) Suppose the heights and weights of a football squad and a basketball squad were graphed together on one scatter plot, points representing football players being marked with F, points representing basketball players with B. Do you think the B's and F's would be separated, or scrambled together? Do you think F's representing ends would fall in with the B's?

5.22 (Sec. 5.5.1) In a traffic study the average speed of cars (in miles per hour) and the "acceleration noise" level were measured for each of 30 sections of roadway. In Table 5.67 the sections are listed in order of increasing average speed.
(a) Plot noise level, y, against average speed, x.
(b) What sort of policy on speed limits should be encouraged to decrease noise? (Consider lower and upper limits.) Does your anti-noise policy agree with or conflict with safety practices?

5.23 (continuation) Compute the values of x^2 from Table 5.67. Plot y against this new variable. Try to draw some conclusion about the mathematical relationship of y to x.

5.24 Discuss Exercise 3.12 further, in terms of what you know now about joint distributions.

TABLE 5.67

Average Speed and Noise Level in 30 Highway Sections

Section	$x =$ Average Speed (MPH)	$y =$ Noise Level	Section	$x =$ Average Speed (MPH)	$y =$ Noise Level
1	10.0	1.60	16	40.5	0.00
2	11.5	1.45	17	41.5	0.00
3	12.0	1.10	18	42.0	0.00
4	18.0	1.05	19	42.5	0.00
5	19.5	1.25	20	43.0	0.00
6	20.0	0.50	21	43.9	0.40
7	28.0	0.70	22	44.0	0.20
8	31.0	0.45	23	44.1	0.06
9	32.0	0.40	24	45.0	0.09
10	33.0	0.10	25	47.9	0.35
11	33.5	0.18	26	48.0	0.38
12	35.0	0.11	27	50.0	0.37
13	36.0	0.00	28	50.2	0.10
14	39.0	0.00	29	50.5	0.12
15	40.0	0.00	30	51.0	0.25

SOURCE: Hypothetical data suggested by Drew and Dudek (1965). Figure 20, p. 52.

5.25 (Sec. 5.4) Basic to the analysis in Lazarsfeld and Thielens' (1958) study, *The Academic Mind,* was a concept of "apprehensiveness," a mixture of worry and caution about the effects of infringements on academic freedom. The analysis based on Tables 5.68 and 5.69 is concerned with the impact of an individual's attitudes on his or her perception of the state of affairs at the college. Thus professors were asked questions designed to gauge their own apprehensiveness and to learn

TABLE 5.68

True Apprehension and Perceived Apprehension Among 1855 Faculty Members

		True Rate of Apprehension	
		Low	High
Perceived	Low	463	620
Apprehension	High	198	574
	Total	661	1194

SOURCE: Lazarsfeld and Thielens (1958).

how apprehensive they believed their colleagues were. In addition, colleges themselves were given a rating in terms of apprehensiveness—characterizing them by the (sampled) proportion of "apprehensive" faculty members at the school. It is with the perception of this "true apprehension rate" at the college that the analysis of these tables deals.

(a) From Table 5.68, compute the percent saying apprehension is high at their college, among those at low-apprehension colleges and among those at high-apprehension colleges.

(b) From Table 5.69 compute the percent saying apprehension is high, in each of the four groups.

(c) Interpret the results.

TABLE 5.69

True Apprehension, Perceived Apprehension, and Respondent's Own Level of Apprehension

		RESPONDENTS WITH LOW LEVEL OF APPREHENSION *True Rate of Apprehension*			RESPONDENTS WITH HIGH LEVEL OF APPREHENSION *True Rate of Apprehension*	
		LOW	HIGH		LOW	HIGH
Perceived	LOW	358	381	LOW	105	239
Apprehension	HIGH	84	148	HIGH	114	426
	TOTAL	442	529	TOTAL	219	665

SOURCE: Lazarsfeld and Thielens (1958).

5.26 (Sec. 5.2) Interpret Table 5.70 using percentages.

TABLE 5.70

Educational Level of Wife and Fertility Planning Status of Couple

		FERTILITY PLANNING STATUS OF COUPLE				
		A^a	B	C	D^b	Total
EDUCATIONAL	One Year or More of College	102	35	68	34	239
LEVEL OF	3 or 4 Years of High School	191	80	215	122	608
WIFE	Less Than 3 Years of High School	110	90	168	223	591
	Total	403	205	451	379	1458

SOURCE: Whelpton and Kiser (1950), Table 3, p. 373, and Table 6, p. 386.
[a]Most effective.
[b]Least effective.

5.27 (Sec. 5.5.2)   (*Cross-tabulation according to two discrete numerical variables.*) Table 5.71 is a cross-tabulation of 6000 household heads by number of children in the household and number of magazines read. Compute the distributions in each row, that is, the distributions of number of magazines read, by number of children. Is the number of magazines read associated with the number of children in the household? If so, is the association strong?

TABLE 5.71

Cross-Tabulation of Household Heads, by Number of Children in the Household and Number of Magazines Read

		NUMBER OF MAGAZINES READ					
		0	*1*	*2*	*3*	*4*	*Total*
	0	754	868	781	162	63	2628
NUMBER OF	*1*	297	363	330	87	23	1100
CHILDREN	*2*	256	358	307	71	32	1024
	3 or more	314	391	382	99	62	1248
	Total	1621	1980	1800	419	180	6000

(hypothetical data)

5.28 (Sec. 5.5.2)   On the other hand, Table 5.72 is a cross-classification of the same 6000 household heads by number of magazines read and income level. Is there an association between income level and number of magazines read? If so, what is its direction?

TABLE 5.72

Cross-Tabulation of Household Heads by Income Level and Number of Magazines Read

		NUMBER OF MAGAZINES READ					
		0	*1*	*2*	*3*	*4*	*Total*
	Less Than $8,000	1005	905	452	126	25	2513
INCOME	*$8,000 to $16,000*	536	930	1139	274	107	2986
	More Than $16,000	80	145	209	19	48	501
	Total	1621	1980	1800	419	180	6000

(hypothetical data)

5.29 (continuation)   Summarize Table 5.71 by computing the mean number of magazines read in each income group.

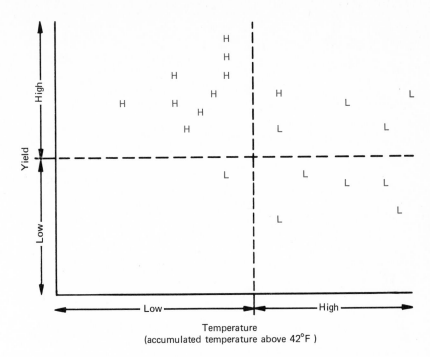

Figure 5.23 *Scatter plot of yield of hay and temperature for 20 years*

5.30 Figure 5.23 is a scatter plot consistent with Hooker's (1907) study. Points marked *H* are years in which the rainfall was heavy. Points marked *L* are years in which the rainfall was light.

(a) Cross-tabulate the 20 years into the $2 \times 2 \times 2$ frequency table which summarizes Figure 5.23.

(b) Make the three possible 2×2 frequency tables corresponding to the three-way table of (a).

Note that the table for yield and temperature corresponds to ignoring whether a point in the figure is an *H* or an *L*; that is, to ignoring whether the rainfall was heavy or light.

5.31 Hooker found a strong association between the size of the crop of peas (harvested in summer) and the warmth of the *subsequent* autumn. How is this possible?

5.32 Refer to Table 3.23, the age-adjusted death rates for nonsmokers, light smokers, and heavy smokers in seven studies.

(a) Within each category of smoking, rank the seven age-adjusted death rates; that is, replace the lowest rate with a 1 and the highest with a 7. Arrange these ranks in a two-way table, following the format of Table 3.23.

(b) Does a single study show consistently (that is, for each category of smoking) the highest death rate? If so, which study?

(c) What about the second highest death rates? The smallest death rates?

(d) Taking all this into account, would you say that there are systematic differences among the results for the different studies after adjusting for age?

5.33 (Sec. 5.5)   Hyman, Wright, and Hopkins (1962) made a study of 96 participants in a summer Encampment for Citizenship. Each participant was given a Civil Liberties test at the beginning and at the end of the summer. A decrease in score indicates improvement. The data are shown in Table 5.73.

(a) Make a scatter plot of the data, using the horizontal axis for the score before encampment, the vertical axis for the score at the end of encampment. Instead of using dots to make the plot, use numbers indicating the frequency (number of persons) at that point.

(b) Do you observe much correlation between the two scores?

TABLE 5.73

Individual Changes in Campers' Scores on a Civil Liberties Test

		SCORE AT BEGINNING OF SUMMER*											
		0	1	2	3	4	5	6	7	8	9	10	Total
	0	10	5	3	2	5		1		1			27
	1	2	2	2		4	1						11
	2	1	1	13	2	4		1	1				23
	3	1		1	2	1	3			1			9
SCORE AT	4	2		1		6	1	4	1				15
END OF	5						0	2	1				3
SUMMER	6					1		2					3
	7							2	0		1		3
	8						1		0	1			2
	9									0			0
	10											0	0
	Total	16	8	20	6	21	6	12	3	1	2	1	96

SOURCE: Hyman, Wright, and Hopkins (1962), Table D-6, p. 392. Copyright © 1962 by The Regents of the University of California; reprinted by permission of the University of California Press.
*A lower score indicates improvement.

5.34 (continuation)

(a) Make a table like Table 5.73 but for the two variables

x = score at beginning of summer

$d =$ difference score = score at end of summer − score at beginning of summer.
Again use frequencies instead of dots.
(b) Do the variables x and d appear to be correlated?

5.35 (continuation)   Do you think it would be adequate to focus on
the decrease in score when analyzing the data in Exercises 5.33 and
5.34?

5.36 (continuation)   Tabulate the frequency distribution of the difference in score obtained in Exercise 5.34.

5.37 (continuation)   For the scores reported in Table 5.73, compute
(a) the mean difference of the scores;
(b) the variance of this sample of scores.

TABLE 5.74
Relationship Between Voter Registration and Income, Santa Clara County

| GROUPS OF 21 CENSUS TRACTS | MEAN HOUSEHOLD INCOME, 1965[a] | | ADULTS[b] 1970 | | |
	Range	*Mean*	*Census Total*	*Known Registered*	*Percent Registered*
A[c]	$21,631–13,555	$16,005	55,584	45,461	81.79
B	13,373–11,558	12,412	68,663	53,037	77.24
C	11,518–10,585	11,135	86,016	58,321	67.80
D	10,559–9,818	10,150	71,271	48,665	68.28
E	9,785–9,181	9,417	75,518	49,696	65.81
F	9,178–8,552	8,891	74,718	46,496	62.23
G	8,540–8,173	8,403	71,010	39,716	55.93
H	8,076–7,273	7,706	64,236	33,845	52.69
I	7,269–6,136	6,681	51,815	24,540	47.36
J[c]	6,073–3,524	5,172	53,573	22,923	42.79

SOURCE: Professor Bradley Efron, Stanford University, personal communication (data obtained from
the 1970 U.S. Census and a 1965 Santa Clara County, California, Planning Commission report).
[a]Reported in 1966.
[b]Persons 18 years and over.
[c]20 census tracts.

5.38 (Sec. 5.5)   In Table 5.74 are given the mean income and percent
registered to vote for ten groups of census tracts in Santa Clara County,
California. Make a scatter plot using the horizontal axis for mean income
and the vertical axis for percent registered. Connect successive points,

to aid the eye. Does there appear to be a relationship between the two variables?*

MATHEMATICAL EXERCISES

5.39 (Sec. 5.2) Take a numerical example and verify that if the distributions across A-categories are the same for all B-categories then the marginal distribution across A-categories is the same as this distribution.

5.40 (Sec. 5.2.2) Write down a 3×3 frequency table in which the mode of the joint distribution corresponds to the modes of the marginal distributions.

5.41 (Sec. 5.2.2) Write down a 3×3 frequency table (different from that in the text) in which the mode of the joint distribution does not correspond to the modes of the marginal distributions.

5.42 (Sec. 5.2.2) In the 2×3 table (Table 5.75) show that if "Medium" is the mode of B for each level of A, then it is the mode of the marginal distribution of B.

TABLE 5.75
2 × 3 Frequency Table

		B		
		Low	*Medium*	*High*
A	*Low*	a	b	c
	High	d	e	f
	Total	$a+d$	$b+e$	$c+f$

5.43 (Sec. 5.4.5) Suppose Table 5.76 exhibits independence. Suppose further that when Z is low, X and Y have a *negative* association; that is,

---

*In early 1972, Miss Angela Davis, a left-wing activist and former UCLA philosophy instructor, was to be tried on charges of murder, kidnap, and conspiracy arising from an escape attempt in Marin County, California, in August of 1970, in which four persons, including a judge, were killed. The trial was to take place in San Jose, Santa Clara County, California. Miss Davis' defense argued that she could not receive a fair trial in San Jose because of Santa Clara county's procedure for jury selection, namely, choosing jurors from voter lists. Professor Bradley Efron of the Department of Statistics, Stanford University, testified he had studied data for the county's census tracts to determine the relationship of income to the percentage of people registered to vote (*San Francisco Chronicle*, February 4, 1972, p. 6). The data of Table 5.71 were given as a summary of his study. In spite of these arguments, and others by political scientists, the trial was not moved. Miss Davis was, however, subsequently acquitted.

a_1d_1 is less than b_1c_1. (See Table 5.77.) Does this imply that when Z is high, X and Y have a *positive* association?

TABLE 5.76
Overall 2 × 2 Table for X and Y

		Y	
		LOW	HIGH
X	LOW	$a = a_1 + a_2$	$b = b_1 + b_2$
	HIGH	$c = c_1 + c_2$	$d = d_1 + d_2$

TABLE 5.77
2 × 2 Tables for Variables X and Y, for Each Value of Variable Z

VARIABLE Z: LOW

		Y	
		LOW	HIGH
X	LOW	a_1	b_1
	HIGH	c_1	d_1

VARIABLE Z: HIGH

		Y	
		LOW	HIGH
X	LOW	a_2	b_2
	HIGH	c_2	d_2

REFERENCES

Anderson, T. W. (1958). *An Introduction to Multivariate Statistical Analysis.* Wiley, New York.

Blau, Peter Michael (1955). *The Dynamics of Bureaucracy.* University of Chicago Press, Chicago.

Chernoff, Herman (1971). "The use of faces to represent points in *n*-dimensional space graphically." Technical Report No. 71. Department of Statistics, Stanford University.

Daetz, D. (1971). "A graphical technique to assist in sensitivity analysis." Unpublished report. Department of Industrial Engineering, Stanford University.

Drew, Donald R., and Conrad L. Dudek (1965). "Investigation of an internal energy model for evaluating freeway level of service." Texas Transportation Institute, Texas A&M University.

Galle, Omer R., Walter R. Gove, and J. Miller McPherson (1972). "Population density and pathology: What are the relations for man?" *Science* 176: 23–30.

Goodman, Leo A., and William H. Kruskal (1954). "Measures of association for cross classifications." *Journal of the American Statistical Association* 49: 732–764.

Hammond, E. Cuyler, and Daniel Horn (1958). "Smoking and death rates—report on forty-four months of follow-up of 187,783 men. II. Death rates by cause." *Journal of the American Medical Association* 166: 1294–1308.

Hooker, R. H. (1907). "The correlation of the weather and crops." *Journal of the Royal Statistical Society* 70: 1–42.

Hyman, Herbert H., Charles R. Wright, and Terence K. Hopkins (1962). *Applications of Methods of Evaluation: Four Studies of the Encampment for Citizenship.* University of California Press, Berkeley and Los Angeles.

Kottmeyer, William (1959). *Teacher's Guide for Remedial Reading.* Webster Publishing, St. Louis.

Lazarsfeld, Paul F. (1940). *Radio and the Printed Page.* Duell, Sloan, and Pearce, New York. (Reprinted by Arno Press, New York, 1971.)

Lazarsfeld, Paul F., Bernard Berelson, and Hazel Gaudet (1968). *The People's Choice.* 3rd ed. Columbia University Press, New York.

Lazarsfeld, Paul F., and Wagner Thielens (1958). *The Academic Mind.* Free Press, Glencoe, Ill.

Somers, Robert H. (1959). "Young Americans Abroad: A Study of the Selection and Evaluation Procedure of the Experiment in International Living." Bureau of Applied Social Research, Columbia University, New York. Unpublished report.

Whelpton, P. K., and Clyde V. Kiser (1950). *Social and Psychological Factors Affecting Fertility. Vol. Two. The Intensive Study: Purpose, Scope, Methods, and Partial Results.* Milbank Memorial Fund, New York.

Wiltsey, Robert G. (1972). *Doctoral Use of Foreign Languages: A Survey. Part II. Supplementary Tables.* Graduate Record Examinations Board, Educational Testing Service, Princeton, New Jersey.

Wright, R. M., and P. Switzer (1971). "Numerical classification applied to certain Jamaican Eocene nummuliteds." *Mathematical Geology* 3: 297–311, Plenum Press.

PART THREE
PROBABILITY

6 BASIC IDEAS
OF PROBABILITY

INTRODUCTION

Each of us has some intuitive notion of what "probability" is. Everyday conversation is full of references to it: "He'll probably return on Saturday." "Maybe he won't." "The chances are she'll forget." "The odds on winning are small."

What do these vague statements mean? What are the ideas behind them?

In this chapter we discuss some of the basic notions of probability and make them precise. We show how to find probabilities of various events on the basis of certain given information.

6.1 INTUITIVE EXAMPLES OF PROBABILITY

6.1.1 Physical Devices Which Approximate Randomness

Some of our ideas of probability come from playing games in which chance is an essential part. In many card games a deck of 52 playing cards is used. There are two colors of cards, red and black, and four *suits:* clubs (black), diamonds (red), hearts (red), and spades (black). Within each suit there are the 13 cards Ace, 2, 3, . . . , 10, Jack, Queen, and King, the Jack, Queen, and King being "face cards" and bearing the letters J, Q, K, respectively, the other cards bearing their respective numbers. The deck is shuffled a few times, hopefully to put the cards in a *random order*—which means that all possible orderings of the cards are equally likely. In particular, we believe that when the cards are distributed (dealt), all of the 52 cards in the deck are equally likely to be the first card dealt. The terms "random" and "equally likely" are meaningful to a card player.

In bridge all of the cards are dealt out to four players, each receiving 13 cards. Bridge players have some ideas of the probabilities of receiving various combinations of cards, and their strategy of play involves applying a knowledge of these probabilities to the unseen combinations held by the other players. (See Exercises 7.27–7.32.)

There are some important factors underlying the determination of probabilities in card playing. For one thing, there is a finite, though very large, number of possibilities. Another aspect is that the players themselves produce the randomness; the player shuffling continues to shuffle until he or she feels the order is random. (We assume the players are honest.) The basis of calculating probabilities in card playing is that certain fundamental occurrences are "equally likely."

The shuffling of a deck of cards may be considered as a physical device for approximating randomness. Another physical device useful in describing probability is the flipping of a coin. Many games are played by flipping one or more coins. For a "fair" coin the probability of a head is considered to be 1/2. In other games one rolls *dice* (singular: *die*). A die is a cube with one dot on one face, two dots on a second face, etc. Our intuition suggests that for a balanced or fair die the probability of any specified one of the six faces coming up is 1/6.

In some games the player spins a pointer (Figure 6.1). When the pointer stops on blue one move is prescribed, on red another is prescribed, etc. The color areas can be designed so that the various moves have any desired probabilities.

As another example of probability, consider a jar with 1000 beads in it. The beads are spherical and of the same size and weight; 500 are red and 500 are black. The beads are stirred, or the lid is put on the jar and the jar is shaken. A blindfolded person takes out a bead. We would

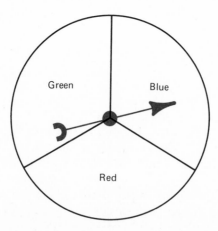

Figure 6.1 *A pointer for spinning, as in a game*

agree that all the beads are equally likely to be drawn and the probability that the bead drawn is red is 500 chances in 1000, or 1/2; that is, the first bead drawn is as likely to be red as it is to be black.

This example has shortcomings because in actuality it is hard to make the beads identical in size and weight and difficult to shake or stir them into random positions. In a somewhat similar situation slips of paper bearing the names of the 365 days of the year were put into capsules; the capsules were put into a drum which was rotated. The sequence of dates then drawn from the drum was supposed to be in random order, but there was considerable question whether this physical device in fact could come close to achieving randomness.

6.1.2   The Draft Lottery

In 1969 a draft lottery was instituted by the Selective Service System. Men born in 1950 would be called for military service according to a schedule determined by the lottery. Given the personnel requirements of the military services, those having birthdates near the top of the list would almost certainly be drafted and those having birthdates near the end of the list would almost certainly not be drafted. The priority list was determined by drawing capsules containing the dates from a rotated drum.

TABLE 6.1

1969 Draft Lottery: Month of Birthday and Priority Number

	PRIORITY NUMBERS			
Month	*1 to 122*	*123 to 244*	*245 to 366*	*Total*
January	9	12	10	31
February	7	12	10	29
March	5	10	16	31
April	8	8	14	30
May	9	7	15	31
June	11	7	12	30
July	12	7	12	31
August	13	7	11	31
September	10	15	5	30
October	9	15	7	31
November	12	12	6	30
December	17	10	4	31
	122	122	122	366

NOTE: Because some men born in a leap year were eligible, February 29 was included, making the total number of days 366.

Was the 1969 draft lottery really a random drawing? That is to say, was the *procedure* one in which all possible orders were equally likely? A more detailed account indicates that the capsules containing the dates of a particular month were put into one box. The 12 boxes were emptied into the drum one by one, first January, then February, etc.; and then the drum was rotated and capsules were drawn out, one by one. Perhaps the capsules put into the drum last were more likely to be drawn first. Table 6.1 gives evidence that this was the case.

The capsules for days in December tended to be drawn out of the drum early in the drawing: 17 December days have low numbers (between 1 and 122). Only 4 December days have high numbers. The strong trend manifests itself in the table in a striking manner when one looks at the clusters of single-digit frequencies. In the first column of figures, these are at the top; in the second column, they are in the middle; in the third column, they are at the bottom.

For the next year's lottery, deciding the order of call for men born in 1951, a procedure was used which could not be criticized so easily. Two drums were used, one for capsules containing the priority ranks 1 to 365, the other for capsules containing the numbers corresponding to birth dates. This time, each drum was loaded in a random order. The final drawing was by hand: a date drawn from one drum was paired with a number drawn from the second drum [Rosenblatt and Filliben (1971)].

6.1.3   Probability and Everyday Life

In the examples so far, the probabilities are fairly obvious. In many everyday situations things are not so simple. One may try to avoid driving an automobile on a holiday weekend because one thinks that the probability of an accident is greater on such a weekend; this idea may come from reading the newspaper reports on the frequency of accidents on weekends. A weather forecast in a newspaper may state that tomorrow the chance of rain is 1 in 10, but for another day it may be 8 in 10; one would not bother with a raincoat on the first day, but would on the second.

6.2   PROBABILITY AND STATISTICS

Often in statistical analysis the available data are obtained from individuals who constitute a *sample* from the population of interest.

A *random sample* of n individuals from a population of N is a sample drawn in such a way that all possible sets of n have the same chance of being chosen. One of the consequences of this is that all N members of the population have the same chance of being one of the n included in the sample. In many cases a sample is drawn from a population by a procedure that insures the property of randomness; in other cases there may be a natural physical mechanism operating to give a more or less random sample. In Chapter 1 examples of the use of statistics were given; let us see how notions of sampling and probability enter into some of these.

The procedure followed in political polls approximately achieves a random sample. Enumeration of the responses may lead the pollster to report that 40% of the respondents favor the Democratic candidate. How accurate is this as an estimate of the proportion of the entire population favoring the Democratic candidate? What is the probability that the pollster comes up with a percentage that differs from that of the entire population by as much as 5%? The pollster should be able to answer such questions and plan the sample so that the margin of probable error is sufficiently small.

In the polio vaccine trial the vials of vaccine and placebo were distributed in lots of 50; of the 50 one half were vaccine and one half were placebo. The positions of these vials in the container were random. There is some chance that the placebo was more often given to children susceptible to paralytic polio than was the vaccine and that therefore the higher incidence of polio among the children receiving the placebo could be due to chance and not to the vaccine. It is essential to be able to evaluate the probability of such an occurrence.

In the study of group pressure discussed in Section 1.1.5 the sets of measurements obtained are not strictly random samples from populations of direct interest. The persons in the experiment were students at Swarthmore College. The behavior of the students in the experimental situation is considered as the behavior of a sample of persons under group pressure, and the behavior of the students in the control situation as that of a sample of persons from a population not under group pressure. Could the large observed difference in behavior be reasonably expected if group pressure had no effect? What is the *probability* of such a large difference under the assumption of random sampling if in fact these populations are the same? If this probability is very small, then we cannot hold to the idea that the populations are the same. On the other hand, if the results of the experiment can be explained by random sampling from identical populations, they do not seem very important. In order to evaluate such statistical studies properly, we must be able to calculate the probabilities of certain results occurring under conditions of sampling from two identical populations. Thus, a study of probability is essential if we are to use samples to make inferences about populations.

There are times when, although samples are not drawn randomly from the corresponding populations, the investigator considers the method of selection close enough to random to justify statistical inference. The methods used in the selection of students for an experiment, the allocation to experimental or control situations, and the assignment of roles within the situation can all contribute to randomness.

6.3   PROBABILITY IN TERMS OF EQUALLY LIKELY CASES

Simple physical devices can be used, at least in principle, to draw a random sample from a given population. To choose at random one of two possibilities we can flip a coin, and to choose one of six possibilities we can roll a die. In this section we introduce the ideas of probability in terms of such simple operations. In the rest of this chapter and in Chapter 7 we extend these ideas to provide a background in probability for the study of statistical inference.

The various possible outcomes of such simple operations as tossing coins, drawing beads from a jar, throwing dice, and dealing cards are thought of as being *equally likely*. This notion of "equally likely" is not defined; it is anticipated that the reader will have an intuitive feeling for it. This notion goes along with what it means for the coin to be balanced or "fair," for the beads to be thoroughly mixed, for the dice to be balanced or "true," for the cards to be thoroughly shuffled.

Consider dealing from a shuffled deck of ordinary playing cards. Since there are 52 different cards in the deck, the first card dealt can be any one of 52 possibilities. We call these possibilities *outcomes;* they are the "atoms" or indivisible units under consideration.

We may be interested in the probability that the first card has some specific property, such as being a spade. The *event* that the first card dealt is a spade corresponds to a set of 13 possible outcomes: the ace of spades, the deuce of spades, the 3 of spades, . . . , the king of spades.

DEFINITION   An event *is a set of outcomes.*

Dealing a card which is a spade is an event. Dealing a card which is not a spade is an event. Dealing a face card is an event. Dealing a black card is an event.

If the individual outcomes are atoms, then events, being collections of outcomes, are molecules. Dealing a particular card, say the queen of hearts, is an event consisting of but a single outcome. It is a molecule, but a molecule which consists of a single "atom."

Typically an event is a set of outcomes with some interesting property in common.

What is the "probability" of dealing a spade? Since there are 13 spades and all cards are equally likely to be drawn, it seems reasonable to consider the probability of drawing a spade as $13/52 = 1/4$. This leads to the following definition.

DEFINITION *If there are n equally likely outcomes and an event consists of m outcomes, the* probability *of the event is m/n.*

In this situation, to find the probability of an event we count the number of possible outcomes to find n and we count the number of outcomes in the event to find m and calculate the ratio. Thus we obtain $13/52 = 1/4$ as the probability of a spade. The probability of an ace is $4/52 = 1/13$. The probability of a black card (spade or club) is $26/52 = 1/2$. The probability of a nonspade is $39/52 = 3/4$.

We want to be able to use probabilities of some events to obtain probabilities of new events. For example, the event of a black card being dealt is made up of the event of a spade being dealt together with the event of a club being dealt. (See Figure 6.2.) We shall show that we can calculate the probability of a black card from the probability of a spade and the probability of a club. First we note

$$\text{black cards} = \text{spades} + \text{clubs,}$$
$$26 \quad = \quad 13 \quad + \quad 13.$$

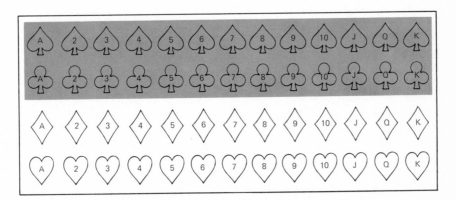

Figure 6.2 *A black card is a spade or a club*

By definition the probability of an event is the number of outcomes in the event divided by the total number of possible outcomes. If we divide the above equation by 52, the number of cards in the deck, we obtain*

$$\frac{\text{\# black cards}}{\text{\# cards}} = \frac{\text{\# spades}}{\text{\# cards}} + \frac{\text{\# clubs}}{\text{\# cards}},$$

$$\frac{26}{52} = \frac{13}{52} + \frac{13}{52},$$

$$\begin{matrix}\text{probability of} \\ \text{a black card}\end{matrix} = \begin{matrix}\text{probability} \\ \text{of a spade}\end{matrix} + \begin{matrix}\text{probability} \\ \text{of a club.}\end{matrix}$$

The general idea is that if we have two events with no outcomes in common, the first consisting of l outcomes and the second consisting of m outcomes, then the event which consists of all the outcomes in the two given events consists of $l + m$ outcomes. Similarly the two events have probabilities l/n and m/n, respectively, and the event of all outcomes has probability $(l + m)/n$. Thus if an event consists of all the outcomes of two events having no outcome in common, the probability of that event is the sum of the probabilities of the two events.

If we have three events with no outcome in more than one of the three events, the probability of an event consisting of all outcomes in the three events is the sum of the probabilities of the three events. For example, if the three events are a spade being dealt, a club being dealt, and a diamond being dealt, then

$$\frac{\begin{matrix}\text{\# spades, clubs, and}\\ \text{diamonds}\end{matrix}}{\text{\# cards}} = \frac{\text{\# spades}}{\text{\# cards}} + \frac{\text{\# clubs}}{\text{\# cards}} + \frac{\text{\# diamonds}}{\text{\# cards}},$$

$$\begin{matrix}\text{probability of a spade,}\\ \text{club, or diamond}\end{matrix} = \begin{matrix}\text{probability}\\ \text{of a spade}\end{matrix} + \begin{matrix}\text{probability}\\ \text{of a club}\end{matrix} + \begin{matrix}\text{probability}\\ \text{of a diamond.}\end{matrix}$$

What about the probability of an event consisting of all outcomes in two events which may have some outcomes in common? For example, consider the event of the card being *either* a spade *or* a face card (or both). The cards which are spade face cards are common to the two events. It is convenient to think of the large event as consisting of the outcomes in three events, no two of which have outcomes in common: spades that are face cards, spades that are not face cards, and face cards that are not spades. (See Figure 6.3.) Then we have

*The symbol # stands for "number of."

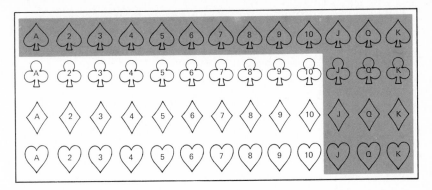

Figure 6.3 *Face cards, together with spades*

$$
\begin{array}{ccccc}
\dfrac{\substack{\text{\# cards that are} \\ \text{either spades or} \\ \text{face cards}}}{\text{\# cards}}
& = &
\dfrac{\substack{\text{\# spades that} \\ \text{are face cards}}}{\text{\# cards}}
& + &
\dfrac{\substack{\text{\# spades that} \\ \text{are not face} \\ \text{cards}}}{\text{\# cards}}
& + &
\dfrac{\substack{\text{\# face cards} \\ \text{that are not} \\ \text{spades}}}{\text{\# cards}}
\end{array}
$$

$$
\frac{22}{52} \quad = \quad \frac{3}{52} \quad + \quad \frac{10}{52} \quad + \quad \frac{9}{52},
$$

| probability of a card that is either a spade or a face card | = | probability of a spade that is a face card | + | probability of a spade that is not a face card | + | probability of a face card that is not a spade. |

An important event is the set of *all* cards. It has a probability of $52/52 = 1$; it is considered an event that happens for sure. Any event and the event consisting of all outcomes not in that event have no outcome in common; but the event consisting of the outcomes in these two events contains all the outcomes, and has a probability of 1. Hence

$$
\substack{\text{probability of a} \\ \text{given event}} + \substack{\text{probability of event} \\ \text{consisting of all} \\ \text{outcomes not in} \\ \text{given event}} = 1.
$$

It is convenient to define the absence of any outcome as the *empty* event, and it has probability $0/52 = 0$; it is certain not to happen.

Finally we note that if one event contains all the outcomes in another event (and possibly more), the probability of the first event is at least as great as the probability of the second. The probability of dealing a black card is 1/2, which is greater than the probability of dealing a club, which is 1/4.

Equally likely outcomes occur (in principle) in many games of chance. We consider that a head and a tail are equally likely in the toss of a coin. When we toss a penny and a nickel we consider the following outcomes equally likely: heads on both coins, head on the penny and tail on the nickel, tail on the penny and head on the nickel, and tails on both coins. In rolling a die we consider all of the six faces as equally likely.

The hand (set of 13 cards) dealt to one of four bridge players is a sample from a population of 52 cards. There are many possible hands; all are considered equally likely. Likewise, from any specific population there are many possible combinations of individuals. *Random sampling* is designed to make all combinations equally likely. The calculation of many probabilities can be based on the notion of equally likely outcomes. However, there are often situations in which a more general notion of probability is required; we develop this in the next section.

6.4 EVENTS AND PROBABILITIES IN GENERAL TERMS

6.4.1 Outcomes, Events, and Probabilities

There are two contexts in which the notion of a definite number of equally likely cases does not apply: (1) where the number of possible outcomes is finite but all outcomes are not equally likely, and (2) where the whole set of outcomes is not finite. If a coin is not fair, the probability of a head may not be 1/2; even if eyes are classified only as blue or brown, the probability of a blue-eyed baby is not 1/2. The possible states of weather are not finite. Outcomes which depend on the value of a continuous variable are not finite in number. For instance, in the spinner portrayed in Figure 6.1, an outcome is the position of the pointer measured by the angle between the pointer and the vertical; that angle can take on any of the infinite number of values between 0° and 360°. The events, red, blue, and green, are made up from these outcomes.

In this section we treat probability in a more general setting. The means used to calculate probabilities from general principles should agree with the methods that obtain for equally likely outcomes; the properties and rules of probability will be the same in the general case as in the special case of equally likely outcomes.

As before, we start with a set of outcomes and consider events, which are collections of outcomes.

DEFINITION   *An* event *is a set of outcomes.*

The *probability* of each event E is a number denoted as $\Pr(E)$. In Section 6.3, where the space of all outcomes, as well as each event, consists of a finite number of equally likely outcomes, $\Pr(E)$ is the ratio of the number of outcomes in the event to the total number of outcomes. Since probabilities in the general setting must have the same properties as these ratios, any probability is a number between 0 and 1 (inclusive).

PROPERTY 1

$$0 \le \Pr(E) \le 1.$$

The set of all outcomes is called the *space,* and the set containing no outcome is called the *empty event.* In agreement with Section 6.3 we require the following property:

PROPERTY 2

$$\Pr(\text{empty event}) = 0, \qquad \Pr(\text{space}) = 1.$$

6.4.2   Addition of Probabilities of Mutually Exclusive Events

Mutually Exclusive Events.   The events of a spade being dealt and of a club being dealt are incompatible, in the sense that they cannot occur simultaneously; they are *mutually exclusive.* Figure 6.4 denotes mutually exclusive events.

DEFINITION   *Two events are* mutually exclusive *if they have no outcome in common.*

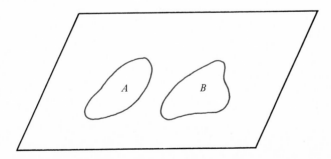

Figure 6.4 *Mutually exclusive events*

Probabilities of Mutually Exclusive Events. In terms of equally likely events we saw that the probability of the event made up of all outcomes of two mutually exclusive events (spades and clubs, for example) was the sum of the probabilities of the two events. We want this fact to be true in general. If A and B are two mutually exclusive events, we denote the event composed of the outcomes from both A and B as "A or B" because an outcome of this event is an outcome of A or an outcome of B.

PROPERTY 3 *If the events A and B are mutually exclusive, then*

$$\Pr(A \text{ or } B) = \Pr(A) + \Pr(B).$$

The probability of an event may be determined by a systematic procedure such as the proportion of "equally likely" outcomes contained in the event, or it may be determined on another basis. (See Section 6.5.) However determined, if the numbers $\Pr(E)$ have Properties 1, 2, and 3, then the *mathematics* of probability theory can be used.

If the three events A, B, and C are mutually exclusive, then

$$\Pr(A \text{ or } B \text{ or } C) = \Pr(A) + \Pr(B) + \Pr(C).$$

If $A_1, A_2, \ldots, A_m$ are an arbitrary number of mutually exclusive events, then

$$\Pr(A_1 \text{ or } A_2 \text{ or } \ldots \text{ or } A_m) = \Pr(A_1) + \Pr(A_2) + \cdots + \Pr(A_m)$$

$$= \sum_{j=1}^{m} \Pr(A_j).$$

DEFINITION *The* complement *of an event is the event consisting of all outcomes not in that event.*

If A is the given event we denote the complement of A by $\overline{A}$. By definition, A and $\overline{A}$ are mutually exclusive, but every outcome in the space is either in A or $\overline{A}$. If we substitute $\overline{A}$ for B in Property 3 and use the second part of Property 2, we obtain

$$1 = \Pr(A) + \Pr(\overline{A}),$$

or

$$\Pr(\overline{A}) = 1 - \Pr(A).$$

6.4.3 Addition of Probabilities

The Event "A and B." In Figure 6.5 we indicate the event consisting of outcomes that are in both A and B. We call the event "A and B." Another representation is similar to the 2×2 frequency table in Figure 6.6. The event A consists of the points in the square above the horizontal line; B consists of the points to the left of the vertical line; "A and B" consists of the points in the upper left-hand corner. In obtaining a corresponding 2×2 frequency table one counts the number of outcomes in A and B.

The probabilities of the events indicated in Figure 6.6 are given in Table 6.2. Since the event "A and B" and the event "A and $\bar{B}$" are mutually exclusive (because B and $\bar{B}$ are mutually exclusive),

$$\Pr(A) = \Pr(A \text{ and } B) + \Pr(A \text{ and } \bar{B}).$$

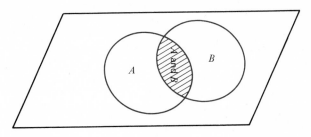

Figure 6.5 *A* and *B*

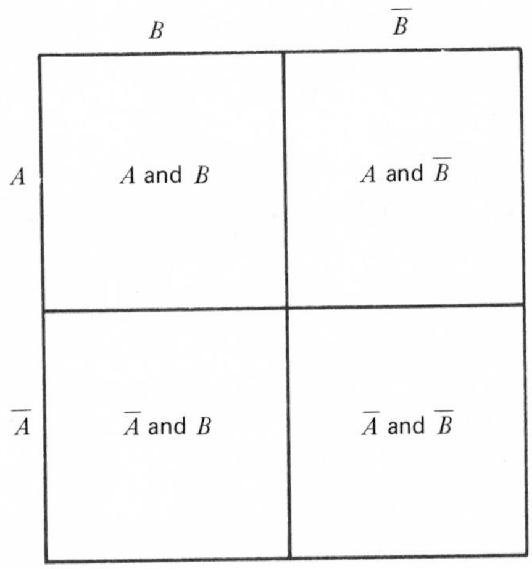

	B	$\bar{B}$
A	A and B	A and $\bar{B}$
$\bar{A}$	$\bar{A}$ and B	$\bar{A}$ and $\bar{B}$

Figure 6.6 *Events based on a two-way classification*

The other sums indicated in Table 6.2 are derived similarly.

TABLE 6.2
Probabilities Associated with Two Events

	B	$\overline{B}$	
A	$\Pr(A \text{ and } B)$	$\Pr(A \text{ and } \overline{B})$	$\Pr(A)$
$\overline{A}$	$\Pr(\overline{A} \text{ and } B)$	$\Pr(\overline{A} \text{ and } \overline{B})$	$\Pr(\overline{A})$
	$\Pr(B)$	$\Pr(\overline{B})$	1

The Event "A or B." Let us find the probability of "*A or B*," when the events A and B are not necessarily mutually exclusive. We know the three events "A and $\overline{B}$," "$\overline{A}$ and B," and "A and B" are mutually exclusive. Thus

$$\Pr(A \text{ or } B) = \Pr(A \text{ and } \overline{B}) + \Pr(\overline{A} \text{ and } B) + \Pr(A \text{ and } B).$$

This is the sum of all the probabilities in the interior of Table 6.2 except the probability in the lower right, namely, that of "$\overline{A}$ and $\overline{B}$" ("neither A nor B"). We can also see from Table 6.2 that

$$\Pr(A \text{ or } B) = \Pr(A) + \Pr(B) - \Pr(A \text{ and } B).$$

An event B is *contained* in an event A if every outcome of B is in A, as indicated in Figure 6.7. Then

$$\Pr(B) \leq \Pr(A)$$

because A is made up of the two mutually exclusive events, B and the event consisting of outcomes in A but not in B.

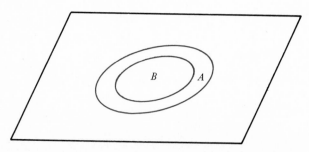

Figure 6.7 *B contained in A*

Sets. The notion of set used here is the same as that already encountered by students having some familiarity with the branch of mathematics known as "set theory." We have used the term "event" where "set" is used in set theory. The correspondence between the terms and notations of this book and of set theory are given in Table 6.3.

TABLE 6.3
Correspondence Between Terms and Notations

Term or Notation	
In This Book	*In Set Theory*
Event	Set
Outcome	Element, member, or point
Mutually exclusive	Disjoint
A or B	$A \cup B$ ("A union B")
A and B	$A \cap B$ ("A intersect B")
$\overline{A}$	$\overline{A}$, A^c, or $\widetilde{A}$ ("A complement")
empty event	null set

6.5   INTERPRETATION OF PROBABILITY: RELATION TO REAL LIFE

Probability theory, as presented in Section 6.4, can be developed as a formal theory; it is a branch of mathematics. How does this model relate to real life?

In Section 6.3 we saw that one way of assigning probabilities to events was to allocate *equal probabilities* to all possible outcomes and then define the probability of an event as the proportion of outcomes in the event. Although this seems a reasonable procedure in appropriate situations, these probabilities may be only approximations. In tossing a coin one side may be more likely to turn up than another; the shuffling of a deck of cards may not be random; the drawing of draft priority numbers from a rotated basket is not conceded to make all orderings equally likely. Another shortcoming of this approach is that in many cases all events cannot be expressed in terms of a finite number of equally likely outcomes. For example, it is hard to see how the probability of rain tomorrow can be given a meaning in that way.

A more satisfactory way of relating probabilities to the real world is in terms of *relative frequency*. If a coin were tossed unendingly, one would expect that after a while the ratio of heads to the total number of tosses would stay very near a certain number, which would be considered the probability of a head. In principle, the probability of an event can be thought of as the relative frequency with which it occurs in an indefinitely large number of trials. But the frequency approach to probability also has its problems. For one thing the determination is hypothetical (we spoke of tossing a coin "unendingly"). For another thing the notion of the relative frequency being approximately a constant for a large number of trials needs more careful development.

A third approach is *subjective* or *personal*. A person can assign a probability to an event according to how likely he or she feels the event is to occur. In this assignment he or she may be guided by mathematical analysis as well as other information. An advantage of this approach is that an individual can give probabilities to events when the other approaches cannot. He or she can, for example, provide a probability that extrasensory perception exists. A disadvantage is that different people may assign different subjective probabilities to the same event. Another disadvantage is that a given individual might assign probabilities in an inconsistent manner so that the rules of Section 6.4 would not hold.

For this book we shall adopt the point of view that a probability represents the relative frequency in a large number of trials. We shall, however, often appeal to the reader's intuition and use the "equally likely" approach in many examples.

6.6   CONDITIONAL PROBABILITY

The probability of one event given that another event occurs is a conditional probability. A conditional probability is defined in terms of the probabilities of the given events and combinations of them.

Consider a population of 100 individuals who have answered the questions, "Have you seen an advertisement of Spotless toothpaste in the last month?" and "Did you buy Spotless toothpaste in the last month?" Their responses are tabulated in Table 6.4.

If we were to draw one person at random from those who had seen the ad, the probability of obtaining a person who bought the toothpaste is $20/40 = 1/2$; that is, we have considered each of the 40 individuals who have seen the ad as equally likely to be drawn. We could as well have made the calculation from the table of proportions, Table 6.5, as

$$\frac{20/100}{40/100} = \frac{20}{40} = \frac{1}{2}.$$

The entries in Table 6.5 can also be considered as probabilities in terms of a random draw.

The probabilities of events A and B and derived events were given in Table 6.2. The conditional probability of B given A, denoted $\Pr(B|A)$, is $\Pr(A \text{ and } B)/\Pr(A)$.

Of course, this formula can be used only if $\Pr(A)$, the denominator, is different from zero.*

TABLE 6.4

2 × 2 Table of Frequencies

	BUY	NOT BUY	
SEEN AD	20 (50%)	20 (50%)	40 (100%)
NOT SEEN AD	10 (16.7%)	50 (83.3%)	60 (100%)
	30 (30%)	70 (70%)	100 (100%)

(hypothetical data)

TABLE 6.5

2 × 2 Table of Proportions

	B: BUY	$\bar{B}$: NOT BUY	
A: SEEN AD	$\Pr(A \text{ and } B) = \dfrac{20}{100}$	$\Pr(A \text{ and } \bar{B}) = \dfrac{20}{100}$	$\Pr(A) = \dfrac{40}{100}$
$\bar{A}$: NOT SEEN AD	$\Pr(\bar{A} \text{ and } B) = \dfrac{10}{100}$	$\Pr(\bar{A} \text{ and } \bar{B}) = \dfrac{50}{100}$	$\Pr(\bar{A}) = \dfrac{60}{100}$
	$\Pr(B) = \dfrac{30}{100}$	$\Pr(\bar{B}) = \dfrac{70}{100}$	$\dfrac{100}{100}$

SOURCE: Table 6.4.

DEFINITION   *The* conditional probability *of B given A when* $\Pr(A) > 0$ *is*

$$\Pr(B|A) = \frac{\Pr(A \text{ and } B)}{\Pr(A)} \, .$$

*If $\Pr(A) = 0$, then A cannot occur, and the conditional probability of B given A has no meaning. Mathematically, if $\Pr(A) = 0$ then $\Pr(A \text{ and } B) = 0$ and the ratio

$$\Pr(A \text{ and } B)/\Pr(A)$$

is not defined.

The calculation of $\Pr(B|A)$ from Table 6.2 and the calculation of percentages in 2×2 frequency tables in Chapter 5 are the same. The phrase, "given A," says to restrict attention to row A of the bivariate table, just as is done when calculating a percentage based on the total for row A. From Table 6.2 we can calculate other conditional probabilities, such as

$$\Pr(\overline{B}|A) = \frac{\Pr(A \text{ and } \overline{B})}{\Pr(A)}.$$

The results of these calculations are tabulated in Table 6.6.

TABLE 6.6
Conditional Probabilities

	B	$\overline{B}$	
A	$\Pr(B|A)$	$\Pr(\overline{B}|A)$	1
$\overline{A}$	$\Pr(B|\overline{A})$	$\Pr(\overline{B}|\overline{A})$	1

The conditional probabilities given one specified event A have all the properties of ordinary probabilities. They are greater than or equal to 0 (because the numerator is nonnegative and the denominator is positive) and they are less than or equal to 1 (because the numerator is less than or equal to the denominator since "A and B" is an event contained in A). If B and C are mutually exclusive, "A and B" and "A and C" are mutually exclusive and

$$\Pr(B \text{ or } C|A) = \Pr(B|A) + \Pr(C|A).$$

In particular,

$$\Pr(B|A) + \Pr(\overline{B}|A) = 1.$$

Conditional probabilities do not need to be calculated in terms of equally likely cases. For instance, at some season in some places the probability of clear skies is .6, the probability of cloudy skies without rain is .2, and the probability of cloudy skies with rain is .2. Then the conditional probability of rain given the skies are cloudy is $.2/(.2 + .2) = 1/2$.

The equation defining the conditional probability $\Pr(B|A)$ can be rewritten

$$\Pr(A \text{ and } B) = \Pr(A) \times \Pr(B|A).$$

This says that the probability of the event "A and B" can be calculated by multiplying the probability of A by the conditional probability of B given A.

A cautious man planning to fly to New York asked an actuary friend, "What is the chance that there will be a bomb on the plane? On hearing that the probability was 1/1000, he became anxious and puzzled. Then he asked his friend, "What is the chance that there will be *two* bombs on the plane?" His friend answered 1/1,000,000. "Aha," said the man, "I know what I'll do; I'll carry one bomb myself." The reader can use a knowledge of conditional probability to find the fallacy in this fictitious story.

An important use of the last equation is to calculate probabilities of events defined by sampling from finite populations. Dealing *n* cards from a shuffled deck of 52 cards can be thought of as drawing a random sample of *n* from a population of 52. The deck can be thought of as a model for any finite population of 52, such as the weeks in a year or 52 students in a class. When one deals *n* cards, the sample consists of *n* different cards. This is called *sampling without replacement*. Sampling *with replacement* would mean dealing one card, recording it, returning it to the deck, and shuffling again; the procedure would be repeated until *n* cards have been recorded. We shall always consider *dealing* as done without replacement.

Suppose the event A is dealing the spade ace as the first card and the event B is dealing the spade king as the second card. The probability of A is 1/52. Given that the spade ace has been dealt, there are 51 cards left, and one of these is the spade king. The conditional probability of the spade king given the spade ace has been dealt is 1/51. Thus the probability of dealing the spade ace and then the king is

$$\frac{1}{52} \times \frac{1}{51} = \frac{1}{52 \times 51}.$$

This probability can be calculated another way. There are 52 different cards that can be dealt first and for each of them there are 51 ways of dealing the next card. Thus there are 52×51 ways of dealing a pair of cards when we distinguish the first card dealt from the second. The number of "ordered" pairs is 52×51. Hence, the probability of dealing the particular ordered pair, spade ace followed by spade king, is $1/(52 \times 51)$.

The event of dealing the spade ace and king as the first two cards (regardless of order) is made up of the two mutually exclusive events, spade ace followed by spade king and spade king followed by spade ace, each having probability $1/(52 \times 51)$. Hence, the probability of the spade ace and king appearing as the first two cards in either order is $2/(52 \times 51)$.

As another example, consider the probability of dealing aces as the first two cards. The event A is an ace as the first card, having probability 4/52, and the event B is an ace as the second card. If an ace is dealt as the first card, there are 3 aces left in the remaining 51 cards; hence $\Pr(B|A) = 3/51$. The probability of aces as the first two cards dealt is

$$\Pr(A \text{ and } B) = \Pr(A) \times \Pr(B|A) = \frac{4}{52} \times \frac{3}{51}.$$

6.7   INDEPENDENCE

Independence of events in terms of probabilities is defined exactly like independence in frequency tables. Table 6.7 resembles Table 5.52 with rows and columns interchanged and percentages replaced by proportions. The responses in Table 5.52 were said to be independent because the proportion of individuals listening to classical music was about the same among those less than 40 and among those over 40.

If we sample one person at random from the 35 million persons enumerated in Table 6.7 we have the probabilities given in Table 6.8: each frequency has been divided by 35. The conditional probability of drawing a person listening to classical music given that the person was less than 40 is

$$\frac{12/35}{20/35} = \frac{12}{20} = .6,$$

and the conditional probability of drawing a person listening to classical music given that the person was 40 or over is

$$\frac{9/35}{15/35} = \frac{9}{15} = .6.$$

These two conditional probabilities are of necessity the proportions in the first column of Table 6.7. Since the conditional probability of listening to classical music does not depend on age, we say that listening to classical music is independent of age.

Table 6.2 gives the probabilities associated with two arbitrary events A and B. If we divide the entries in row A by $\Pr(A)$ and in row $\bar{A}$ by $\Pr(\bar{A})$ we obtain Table 6.9. If the corresponding conditional probabilities in the two rows are equal, we say that A and B are independent.

TABLE 6.7
2 × 2 Frequency Table
NUMBER OF PERSONS (in millions)

| | | LISTEN TO CLASSICAL MUSIC | | |
		Yes	No	Total
AGE	Less than 40	12 (.6)	8 (.4)	20 (1.0)
	40 or Above	9 (.6)	6 (.4)	15 (1.0)
	Total	21 (.6)	14 (.4)	35 (1.0)

(hypothetical data suggested by Table 5.52.)

TABLE 6.8
2 × 2 Probability Table

| | | LISTEN TO CLASSICAL MUSIC | | |
		Yes	No	Total
AGE	Less than 40	$\dfrac{12}{35}$	$\dfrac{8}{35}$	$\dfrac{20}{35}$
	40 or Above	$\dfrac{9}{35}$	$\dfrac{6}{35}$	$\dfrac{15}{35}$
	Total	$\dfrac{21}{35}$	$\dfrac{14}{35}$	$\dfrac{35}{35}$

SOURCE: Table 6.7.

TABLE 6.9
Conditional Probabilities

	B	$\overline{B}$	
A	$\Pr(B|A)$	$\Pr(\overline{B}|A)$	1
$\overline{A}$	$\Pr(B|\overline{A})$	$\Pr(\overline{B}|\overline{A})$	1
	$\Pr(B)$	$\Pr(\overline{B})$	1

DEFINITION   *The event B is* **independent** *of the event A if*

(6.1) $$\Pr(B|A) = \Pr(B|\overline{A}).$$

This notion of independence is analogous to the notion of independence in 2 × 2 frequency tables, as discussed in Chapter 5. There we noted that, if B is independent of A, it is also true that A is independent of B. The same is true in terms of probabilities; (6.1) implies that A is independent of B:

(6.2) $$\Pr(A|B) = \Pr(A|\overline{B}).$$

Just as the condition of independence in 2×2 tables can be expressed in different ways so can independence in the probability sense.

In Table 6.7 the marginal distribution is (.6, .4) because both of the distributions in the body of the table are (.6, .4). In the general notation of Table 6.9, this says that the common value of the conditional probabilities in (6.1) is $\Pr(B)$:

(6.3) $$\Pr(B) = \Pr(B|A) = \Pr(B|\overline{A}).$$

From (6.3) and the definition of $\Pr(B|A)$, we obtain

$$\Pr(B) = \frac{\Pr(A \text{ and } B)}{\Pr(A)},$$

or

$$\Pr(A \text{ and } B) = \Pr(A) \times \Pr(B).$$

Thus the definition of independence is equivalent to the following:
The events A and B are independent if

(6.4) $$\Pr(A \text{ and } B) = \Pr(A) \times \Pr(B).$$

It follows from (6.4) that every entry in Table 6.2 is the product of the corresponding row and column probabilities when A and B are independent. Thus, independence of A and B is equivalent to independence of A and $\overline{B}$, independence of $\overline{A}$ and B, and independence of $\overline{A}$ and $\overline{B}$. In Table 6.8, each entry is the product of corresponding row and column probabilities.

A coin, tossed twice, illustrates the idea of independence in another context. The four possible outcomes, which we assume to be equally likely, can be represented as (H, H), (H, T), (T, H), and (T, T); each outcome has a probability of 1/4. The probability of getting a head on any one toss is 1/2 and so is the probability of getting a tail. To check independence of the two tosses, we note that $\Pr[(H, H)] = \Pr(\text{heads on both tosses}) = 1/4$ and $\Pr(\text{head on first toss}) \times \Pr(\text{head on 2nd toss}) = 1/2 \times 1/2 = 1/4$. We can similarly check (H, T), (T, H), and (T, T).

Mutually exclusive events are represented in a diagram by nonoverlapping figures, because then "A and B" is the empty event. Thus, for mutually exclusive events, $\Pr(A \text{ and } B) = \Pr(\text{empty event}) = 0$.

On the other hand, *independent* events are events A and B such that

$$\Pr(A \text{ and } B) = \Pr(A) \times \Pr(B).$$

Independent events are *not* mutually exclusive; in fact, they overlap *just enough*, so that "A and B" has probability equal to the product of the separate probabilities.

6.8 RANDOM SAMPLING; RANDOM NUMBERS

6.8.1 Random Devices

We have already described elsewhere a number of devices designed to achieve randomness. These will be reviewed briefly.

We can obtain a random sample from a deck of 52 playing cards by shuffling the deck adequately and dealing a specified number of cards from the top of the deck. Each of the four hands in bridge is a random sample of 13 from the deck of 52. A random sample from the population of 52 weeks in the year can be obtained by identifying each week with a different card and then dealing a sample of the desired size.

A population of 1000 beads in a jar may be numbered from 1 to 1000 or $001, 002, \ldots, 999, 000$. If the beads are otherwise identical, a random sample of 15 may be obtained by mixing the beads in the jar and, without looking at the numbers, taking out 15 beads. A random sample can be drawn from any arbitrary finite population by numbering (or otherwise identifying) each individual in the population, placing in the jar beads labelled with those identities, and drawing a sample of beads from the jar. The sample consists of those individuals whose identities were drawn.

Flipping a fair coin yields a probability of 1/2 for a tail and for a head. Thus, if a tail denotes 0 and a head 1, coin-tossing provides a mechanism for obtaining the integers 0 and 1 with a probability of 1/2 each. Rolling a balanced die yields a probability of 1/6 for each integer between 1 and 6. There is also a "die" that will provide the integers $0, 1, \ldots, 9$ with equal probabilities. It is a long cylinder, the cross-section of which is a regular 10-sided polygon; that is, the cylinder has 10 faces (in addition to the ends) numbered $0, 1, \ldots, 9$. When it is rolled the probability of each integer coming up is 1/10. A spinner (Figure 6.1) over a circle that is divided into n equal segments will provide n equally likely outcomes.

6.8.2   Random Numbers

Unfortunately, these mechanical devices for drawing at random are subject to biases. A deck of cards may be shuffled inadequately; the jar of beads may not be mixed well; and a die may have imperfect edges or be weighted asymmetrically. To avoid these difficulties special numerical methods involving computers have been developed to create processes that generate events with specified probabilities; these methods have essentially no biases. Such a method is used by an expert to generate a series of random events. Since the expert and a computer may not be on hand when the series of random events is needed, the expert records the series in advance. When the series of events is a sequence of integers, the record is a table of random numbers. Appendix V is an example.

Let us pursue the principle of random numbers and their use in drawing random samples. Given a device which has as equally likely outcomes the *digits* 0, 1, . . . , 9, we have what are called *random digits*. We have

$$\Pr(i) = \frac{1}{10}, \qquad i = 0, 1, \ldots, 9.$$

We can determine two random digits, say i and j, each having one of the values 0, 1, . . . , 9. The two-digit number ij formed from these two random digits can be any one of the numbers 00, 01, . . . , 09, 10, 11, . . . , 99. Since the two digits i and j are (presumably) independent, the probability of the pair ij is the product of the probability of the first digit i and the probability of the second digit j,

$$\Pr(ij) = \Pr(i) \times \Pr(j) = \frac{1}{10} \times \frac{1}{10} = \frac{1}{100}.$$

Thus all pairs are equally likely. To draw one individual randomly from 100, we number the individuals from 00 to 99, obtain the random pair of digits, and select the corresponding individual.

What if the population size is not a power of 10? For example, suppose we wish to draw randomly from among 4 individuals. Let us identify them as 1, 2, 3, and 4. The probability of the event consisting of the integers 1, 2, 3, 4 is

$$\Pr(1, 2, 3, 4) = \frac{4}{10}.$$

The conditional probability of the integer i ($i = 1, 2, 3, 4$) given the event 1, 2, 3, 4 is

$$\Pr(i|1, 2, 3, 4) = \frac{1/10}{4/10} = \frac{1}{4},$$

which is the probability associated with 4 equally likely outcomes. A procedure for drawing one individual at random from a population of 4 is to take random digits until one obtains one of the digits 1, 2, 3, or 4; this digit is a random selection among 4. Any of the digits 5, 6, . . . , 9, 0 is discarded. (A more efficient procedure is to let also 5 denote 1, 6 denote 2, 7 denote 3, and 8 denote 4, and discard only 9 and 0.)

If we have a population consisting of N individuals, we may number them from 1 to N. If N is less than 10, we draw random digits singly until we obtain a digit between 1 and N, inclusive. If N is more than 10, but less than 100, we draw pairs of random digits until we obtain a pair between 01 and N, inclusive.

Appendix V contains "random" digits. This means that the procedure by which they were obtained is supposed to produce random digits. Actually, they are produced on a computer by numerical operations that give equal proportions of the 10 digits.* Sometimes they are called "pseudo-random" digits.

To illustrate how to use Appendix V, suppose we want random pairs of digits. We pick a starting point by some more or less random device, such as flipping a coin to determine the page, and pointing a pencil with eyes closed to determine the starting point on that page. We should decide in advance whether to proceed vertically (down the page) or horizontally (across the page). We use the first pair of digits and then proceed to the next pair. (After using one row, or column, we go to the next one.)

6.8.3   Sampling with Replacement

One procedure for obtaining a sample of n individuals from a population of N individuals is simply to draw n random numbers (between 1 and N) and let these determine the sample. In this procedure, however, there might be some duplications; some individuals might appear in the sample several times. This is called *sampling with replacement* because it corresponds to drawing beads from a jar when one replaces the bead that has been drawn before drawing the next bead.

6.8.4   Sampling without Replacement

Now consider taking a random sample such that all the individuals in the sample are different. If we want a sample of 2 individuals

---

*One method is to take as the "random number" the remainder obtained when dividing the product of two large numbers by a third large number.

from a population of 4, the first is selected randomly as described above by drawing random digits until 1, 2, 3, or 4 appears. Suppose 3 appears. That leaves 1, 2, and 4 in the population. Then one of these three is drawn randomly. We draw random digits until we obtain 1, 2, or 4; the result determines the second member of the sample. Thus, to draw a sample of n from a population of N, we draw the first member of the sample randomly, then the second member of the sample from the population which is left, etc. This is called *sampling without replacement.*

Sampling without replacement is considered in detail in Section 7.3.2 and Section 8.7.

*6.9   BAYES' THEOREM

6.9.1   Bayes' Theorem in a
Simplified Case

In Chapter 5 we have seen that the 4 entries in a 2×2 table can be added to obtain column totals, row totals, and the overall total; from these, percentages can be computed by using the row totals, the column totals, or the overall total as bases. Since we can construct the original table from the total frequency and certain percentages, all the information is available from the total frequency and these percentages.

We have seen how conditional probabilities are obtained from other probabilities in a similar fashion. Bayes' Theorem in the simplified case is used in obtaining conditional probabilities one way from conditional probabilities the other way.

We consider events A, $\overline{A}$, B, $\overline{B}$, and events formed from them. Suppose we are given $\Pr(B)$, $\Pr(A|B)$, and $\Pr(A|\overline{B})$ as indicated in Table 6.10. If we subtract $\Pr(B)$ from 1, we obtain $\Pr(\overline{B})$, as Step 1 of Table 6.11. Then we calculate $\Pr(A \text{ and } B) = \Pr(A|B) \times \Pr(B)$ and $\Pr(A \text{ and } \overline{B}) = \Pr(A|\overline{B}) \times \Pr(\overline{B})$ as Step 2 of Table 6.11. We add these to obtain $\Pr(A)$ as Step 3. Finally we obtain the conditional probability $\Pr(B|A) = \Pr(A \text{ and } B)/\Pr(A)$. The full calculation is Bayes' Theorem.*

BAYES' THEOREM

$$\Pr(B|A) = \frac{\Pr(B) \times \Pr(A|B)}{\Pr(B) \times \Pr(A|B) + \Pr(\overline{B}) \times \Pr(A|\overline{B})}.$$

*Due to the Reverend Thomas Bayes (1763).

TABLE 6.10
Given Probabilities

	B	$\bar{B}$	
A			
$\bar{A}$			
	$\Pr(B)$		1

	B	$\bar{B}$	
A	$\Pr(A\mid B)$	$\Pr(A\mid\bar{B})$	
$\bar{A}$			
	1	1	

TABLE 6.11
Derived Probabilities

STEP 1

	B	$\bar{B}$	
A			
$\bar{A}$			
	$\Pr(B)$	$\Pr(\bar{B})$	1

	B	$\bar{B}$	
A	$\Pr(A\mid B)$	$\Pr(A\mid\bar{B})$	
$\bar{A}$			
	1	1	

STEP 2

	B	$\bar{B}$
A	$\Pr(A \text{ and } B) = \Pr(A\mid B) \times \Pr(B)$	$\Pr(A \text{ and } \bar{B}) = \Pr(A\mid\bar{B}) \times \Pr(\bar{B})$
$\bar{A}$		
	$\Pr(B)$	$\Pr(\bar{B})$

STEP 3

	B	$\bar{B}$	
A	$\Pr(A \text{ and } B)$	$\Pr(A \text{ and } \bar{B})$	$\Pr(A)$
$\bar{A}$			

STEP 4

	B	$\bar{B}$	
A	$\Pr(B\mid A) = \dfrac{\Pr(A \text{ and } B)}{\Pr(A)}$		1
$\bar{A}$			

6.9.2 Examples

Now let us follow an example to see how Bayes' Theorem can be used. Suppose a new simple and economical test for a certain disease has been invented; the test is good, but not perfect. To validate the test, an experiment is performed. A sample of 1000 persons known to have the disease ("Ill") and a sample of 5000 known not to have the disease ("Not Ill") are tested. These sample sizes are decided by practical budgetary and administrative considerations and are not representative of the incidence of the disease in the overall population. In fact, doctors know that at any one time about 1 in 200 persons in the population have the disease. The results of the experiment are shown in Table 6.12. Of those ill 99% are detected, while only 5% of those not ill give a positive response. Thus the test seems reasonably good. (Note that Table 6.12 is a 2 × 2 table in which the column totals were fixed in advance by the experimental plan.)

TABLE 6.12
Summary of Experimental Results

	ILL	NOT ILL
+	990 (99%)	250 (5%)
−	10 (1%)	4750 (95%)
	1000 (100%)	5000 (100%)

(hypothetical data)

In terms of Table 6.10 A and $\overline{A}$ represent positive and negative responses, respectively, and B and $\overline{B}$ represent Ill and Not Ill, respectively. We shall extrapolate to the entire population:

$$\Pr(A|B) = \Pr(+|\text{Ill}) = .99,$$

$$\Pr(A|\overline{B}) = \Pr(+|\text{Not Ill}) = .05.$$

In the entire population, we have the doctors' knowledge that

$$\Pr(B) = \Pr(\text{Ill}) = \frac{1}{200} = .005.$$

Suppose the test is used to detect the disease in members of the general public. What proportion of those having a positive reaction to the test should we expect actually to have the disease? In other words, if a person gives a positive reaction to the test, what is the conditional probability that that person is ill, $\Pr(B|A)$?

To answer this question we imagine the test being given to everyone in the whole population. If the population size were 200,000,000, the figure .005 (1 in 200) provided by doctors for the overall incidence of the disease would tell us that the total number of Ill is about $.005 \times 200,000,000 = 1,000,000$. The number Not Ill is then about 199,000,000 (Step 1). These numbers go into the bottom row of Table 6.13. Now we fill in the rest of the table so that the percentages in each column are as in Table 6.12 (Step 2). After summing across rows we have

Table 6.14 (Step 3), and thereby obtain 9.05% as the expected percentage of persons with a positive reaction who are actually ill (Step 4). This process is equivalent to using the formula

$$\Pr(\text{Ill}|+) = \frac{\Pr(+|\text{Ill}) \times \Pr(\text{Ill})}{\Pr(+|\text{Ill}) \times P(\text{Ill}) + \Pr(+|\text{Not Ill}) \times \Pr(\text{Not Ill})}$$

TABLE 6.13
Calculated Frequencies

	ILL	NOT ILL	
+	990,000 (99%)	9,950,000 (5%)	
−	10,000 (1%)	189,050,000 (95%)	
	1,000,000 (100%)	199,000,000 (100%)	200,000,000

TABLE 6.14
Table of Percentages

	ILL	NOT ILL	TOTAL
+	990,000 (9.05%)	9,950,000 (90.95%)	10,940,000 (100%)
−	10,000 (.005%)	189,050,000 (99.995%)	189,060,000 (100%)
	1,000,000 (0.5%)	199,000,000 (99.5%)	200,000,000 (100%)

with $\Pr(\text{Ill}) = .005$ [so that $\Pr(\text{Not Ill}) = .995$], and $\Pr(+|\text{Ill}) = .99$, $\Pr(+|\text{Not Ill}) = .05$. Information about the overall population [$\Pr(\text{Ill}) = .005$] is combined with information about the diagnostic test [$\Pr(+|\text{Ill}) = .99$, $\Pr(+|\text{Not Ill}) = .05$] to yield new information [$\Pr(\text{Ill}|+) = .0905$].

In spite of the relative low rate of positives among those not ill (5%), less than 1 in 10 of those who react positively to the test actually have the disease. This illustrates a difficulty of mass screening programs [Dunn and Greenhouse (1950)]. To be useful, the test would presumably have to have a smaller rate of positives among those not ill.

One more example will serve to indicate how information may be combined using Bayes' Theorem. Suppose that the voters of Ohio can be divided into two groups with regard to their views on foreign affairs. Call these liberal and conservative. Further, suppose that the division is known to be 40% liberal and 60% conservative, so that the probability that a person picked at random would be a conservative would be .60. Let us suppose that the two groups respond differently to the question, "Do you think that the U.S. should leave the U.N.?", 1/10 of the liberals answering Yes, while 1/2 of the conservatives respond Yes. We would like to know the probability that an individual chosen at random would be a liberal, *given the information that that individual had responded to this question in a certain way.*

We formulate the problem in terms of symbols:

B = The individual is a liberal.
$\overline{B}$ = The individual is a conservative.
A = The individual thinks the U.S. should leave the U.N.
$\overline{A}$ = The individual does not think the U.S. should leave the U.N.

$$\Pr(A|B) = .10, \qquad \Pr(A|\overline{B}) = .50,$$

$$\Pr(B) \quad = .40, \qquad \Pr(\overline{B}) \quad = .60.$$

Given these probabilities, we want to find $\Pr(B|A)$, the probability that a randomly chosen individual is a liberal, given that the response is Yes to the test question. Applying Bayes' Theorem, we find

$$\Pr(B|A) = \frac{\Pr(B) \times \Pr(A|B)}{\Pr(B) \times \Pr(A|B) + \Pr(\overline{B}) \times \Pr(A|\overline{B})} = \frac{.4 \times .1}{(.4 \times .1) + (.6 \times .5)}$$

$$= \frac{.04}{.34} = \frac{2}{17} = .12.$$

Similarly, the probability that the individual is a liberal given that the answer is No is

$$\Pr(B|\overline{A}) = \frac{.4 \times .9}{(.4 \times .9) + (.6 \times .5)} = \frac{.36}{.66} = \frac{6}{11} = .54.$$

Thus, the observed information (the response to the question) allows us to make a better guess of whether the individual is liberal or conservative. The original probabilities, .4 for liberal and .6 for conservative, are called the *prior* (or *a priori*) probabilities, while the altered probabilities, given the information, are called the *posterior* (or *a posteriori*) probabilities. If a person responds A ("get out of U.N."), the prior probability of B (liberal) is converted from .4 to the posterior probability of .12, while if a person responds $\overline{A}$ ("stay in U.N."), the posterior probability of B (liberal) is .54. See Table 6.15.

TABLE 6.15
Posterior Probabilities Resulting from Application of Bayes' Theorem

	B: LIBERAL	$\overline{B}$: CONSERVATIVE	TOTAL
A: YES, WE SHOULD GET OUT OF THE U.N.	.12	.88	1.00
$\overline{A}$: NO, WE SHOULD STAY IN THE U.N.	.54	.46	1.00

6.9.3 Use of Subjective Prior Probabilities in Bayes' Theorem

Although the formula that is called Bayes' Theorem is a perfectly valid mathematical rule, it has been the center of considerable discussion and controversy. The controversy has to do with use of the rule when the *prior probabilities*, $\Pr(B)$ and $\Pr(\bar{B}) = 1 - \Pr(B)$, are subjective. In the medical diagnosis example, the prior probability $\Pr(B)$, the frequency of illness in the population, is really a relative frequency based on the clinic's experience. As such, it is objectively based and readily interpretable. In other examples, the prior probabilities may be only subjective probabilities. Some people feel that multiplying an ordinary probability by a subjective probability is like adding 4 oranges to 5 apples; they just don't mix.

As an example, consider the following:

B = "The phenomenon known as extra sensory perception exists";

A = "A test displaying some ESP is successful."

For a given test of ESP, the probability of the test being a success if there is no ESP, $\Pr(A|\bar{B})$, could perhaps be computed on the basis of randomness; for instance if the test consisted of predicting which face of a fair die will appear on a given toss. We might also be able to agree on what $\Pr(A|B)$ should be.

Skeptics, however, might insist that the probability $\Pr(B)$ is very close to zero, while some others might argue that it should be assumed to be .5. By applying Bayes' Theorem and working out the conditional probabilities, skeptics and their opponents would obtain quite different results. If, for example, $\Pr(A|\bar{B}) = .1$ and $\Pr(A|B) = .2$, then, a person who assigns a prior subjective probability of .01 that ESP exists will, when the test is successful, assign a posterior probability of

$$\Pr(B|A) = \frac{.01 \times .2}{.01 \times .2 + .99 \times .1} = \frac{.002}{.002 + .099} = \frac{2}{101}.$$

However, a person who assigns a prior probability of .5 will compute a posterior probability as

$$\Pr(B|A) = \frac{.5 \times .2}{.5 \times .2 + .5 \times .1} = \frac{.10}{.15} = \frac{2}{3}.$$

Since the test was successful, in each case the *posterior probability*, $\Pr(B|A)$, is higher than the prior probability; but the two posterior probabilities are still very different.

A statement such as "on the basis of the results of experiments the probability that ESP effects exist is .99" must depend on one's *a priori* belief concerning the effect. More generally, statements of the form, "it is 99% probable that the hypothesis is true," must depend upon what prior probability was used, and the prior that suits one person may not suit another. Nevertheless, if the experiment results in evidence favorable to the hypothesis, everybody's posterior probability will be larger than his or her prior probability was (even though some may be large and others small).

SUMMARY

Operations relating to two events A and B, and to the corresponding probabilities, can be understood by reference to the 2×2 frequency table.

The event "A or B" denotes the collection of all outcomes contained either in A *or* in B (or in both). The event "A and B" denotes the collection of all outcomes contained both in A *and* in B.

If A and B are *mutually exclusive* events, that is, have no outcome in common, then

$$\Pr(A \text{ or } B) = \Pr(A) + \Pr(B).$$

If A and B are *any* two events,

$$\Pr(A \text{ or } B) = \Pr(A) + \Pr(B) - \Pr(A \text{ and } B).$$

If a set A is *contained* in a set B, then $\Pr(A) \leq \Pr(B)$.

The *conditional probability* of B given A, defined when $\Pr(A) > 0$, is

$$\Pr(B|A) = \frac{\Pr(A \text{ and } B)}{\Pr(A)}.$$

The events A and B are *independent* if $\Pr(A \text{ and } B) = \Pr(A) \times \Pr(B)$. In this case,

$$\Pr(B|A) = \Pr(B|\bar{A}) = \Pr(B)$$

and

$$\Pr(A|B) = \Pr(A|\bar{B}) = \Pr(A).$$

CLASS EXERCISE

6.1 Flip or shake two distinguishable coins (labelled coins 1 and 2 below) in such a way that you feel confident that the outcome for each coin (head or tail) is random. Record the outcomes for the two coins.

Repeat this process until the pair of coins has been tossed 25 times.

(a) Record the frequency of outcomes in table I below.

(b) Record the proportion of outcomes in table II.

(c) Assuming the coins are unbiased and are tossed independently, indicate in table III the probabilities of various outcomes for the toss of one pair of coins.

Save the results for use in Chapter 7.

I
Observed Frequencies

II
Observed Proportions

	COIN 1	
	Head	Tail
COIN 2 Head		
COIN 2 Tail		

	COIN 1	
	Head	Tail
COIN 2 Head		
COIN 2 Tail		

III
Probabilities

	COIN 1	
	Head	Tail
COIN 2 Head		
COIN 2 Tail		

EXERCISES

6.2 Let A be the event that a person is a college graduate and B the event that a person is wealthy. In each case give in words the event whose probability is represented.

(a) $\Pr(A \text{ or } B)$

(b) $\Pr(A \text{ and } B)$

(c) $\Pr(A \text{ or } B) - \Pr(A \text{ and } B)$

(d) $1 - \Pr(A)$

(e) $1 - \Pr(B)$

(f) $1 - \Pr(A \text{ or } B)$

6.3 Let A be the event that a person is a male and B the event that a person is more than six feet tall. In each case give in words the event whose probability is represented.

(a) $\Pr(A \text{ or } B)$

(b) $\Pr(A \text{ and } B)$

(c) $\Pr(A \text{ or } B) - \Pr(A \text{ and } B)$

(d) $1 - \Pr(A)$

(e) $1 - \Pr(B)$

(f) $1 - \Pr(A \text{ or } B)$

6.4 (Sec. 6.4) Let C be the event that a college student belongs to a fraternity or sorority and D the event that a college student has a scholastic grade-point average of B or better. Express each of the following symbolically.

(a) the probability that a college student belongs to a fraternity or sorority

(b) the probability that a college student has a grade-point average of B or better

(c) the probability that a college student has a grade-point average less than B

(d) the probability that a college student does not belong to a fraternity or sorority and has a grade-point average of B or better.

6.5 (Sec. 6.4) If A and B are mutually exclusive events with $\Pr(A) = .3$ and $\Pr(B) = .4$, find the following.

(a) $\Pr(A \text{ or } B)$ (b) $\Pr(A \text{ and } B)$ (c) $\Pr(\overline{A} \text{ and } \overline{B})$

6.6 (Sec. 6.7) If A and B are independent events with $\Pr(A) = .2$ and $\Pr(B) = .5$, find the following.

(a) $\Pr(A \text{ or } B)$ (b) $\Pr(A \text{ and } B)$ (c) $\Pr(\overline{A} \text{ and } \overline{B})$

6.7 (Sec. 6.4) Which of the following pairs of events A and B are mutually exclusive?

(a) A: being the child of a lawyer
   B: being born in Chicago

(b) A: being under 18 years of age
   B: voting legally in a U.S. election

(c) A: owning a Chevrolet
   B: owning a Ford

6.8 (Sec. 6.7) Which of the following pairs of events A and B are independent?

(a) A: getting a six on a roll of a die
   B: getting a six on the next roll

(b) A: being intoxicated
   B: having an auto accident

(c) A: being on time for class
   B: the weather being good

6.9 By the "odds" against the occurrence of an event A we mean the ratio $\Pr(\overline{A})/\Pr(A)$. Give the odds for each of the following.

(a) against the occurrence of A if $\Pr(A) = 1/4$

(b) against the occurrence of A if $\Pr(A) = 1/3$

(c) against the occurrence of A if $\Pr(A) = 1/2$

(d) against the occurrence of A if $\Pr(A) = 3/4$

(e) for the occurrence of A if $\Pr(A) = 3/4$

6.10 (Sec. 6.6)   Give the probability of getting an ace and a queen, not necessarily in that order, in two draws from a standard deck of 52 playing cards if
   (a) the first card is not replaced before the second card is drawn,
   (b) the first card is replaced before the second card is drawn.

6.11 (Sec. 6.6)   Give the probability of getting three aces in three successive drawings from a deck of cards
   (a) if each card is replaced before the next is drawn,
   (b) if the successive cards are not replaced.

6.12 (Sec. 6.7)   In rolling a balanced die, what is the probability that the first 3 will occur on the fourth try?

6.13 (Sec. 6.6)   In drawing one card at a time without replacement from a deck of cards, what is the probability that the first spade will be the second card drawn?

6.14 (Sec. 6.6)   From a sample of 2000 persons, each tested for color-blindness, Table 6.16 was obtained.

TABLE 6.16
Frequencies of Colorblindness

	MALE	FEMALE	TOTAL
NORMAL EYESIGHT	980	936	1916
COLORBLIND	72	12	84
TOTAL	1052	948	2000

(hypothetical data)

(a) What is the relative frequency of occurrence of colorblindness?
(b) What is the relative frequency of colorblindness among males, that is, the conditional frequency of colorblindness, given that the individuals in question are males?
(c) What is the relative frequency of colorblindness among females?
(d) Assuming these 2000 persons to be a random sample from some given population, relative frequencies obtained from the table are estimates of the corresponding probabilities in the population. For example, the answer to (a) is an estimate of the overall rate of color-blindness in the population. Suppose it is known that the population is half males and half females. (Then our sample is slightly non-representative, in the sense that we obtained 1052/2000, or 52.6% males rather than 50%.) How would you use this information to alter your estimate of the rate of colorblindness in the population?

6.15 (Sec. 6.6)   Suppose we toss a coin twice. The probability of getting two heads is 1/4. What is the conditional probability of getting two heads, given that at least one head is obtained?

6.16 (Sec. 6.6)   Suppose we toss a coin twice. What is the conditional probability of the sequence *HT* given that one *H* and one *T* occurred?

***6.17** (Sec. 6.9)   A test for a rare disease is positive 99% of the time in case of disease and is negative 95% of the time in case of no disease. The rate of occurrence of the disease is .01. Compute the conditional probability of having the disease, given that the test is positive. How do you account for the fact that this probability is so low, in spite of the fact that a high percentage of those with the disease test positively and a high percentage of those without the disease test negatively?

***6.18** (Sec. 6.9)   Suppose that we have carried out a study relating voting behavior to social class, and find that the population can be adequately represented by two social classes, High and Low, having relative sizes 1/3 and 2/3, respectively. Of the people in the High social class, 75% voted Republican, while 36% in the Low social class voted Republican. Making use of Bayes' Theorem, find the conditional probability that a person chosen at random will be from the High social class if he or she votes Republican. What is the conditional probability that a person who votes Republican is from the Low social class?

6.19 (Sec. 6.6)   In the Land of Aridona it rains on one out of ten days; that is, the probability of rain on any given day is .1. The conditional probability of rain on any one day, given that it rained on the previous day, is .6. Compute the probability that it will rain on two given consecutive days.

***6.20** (continuation)   Compute the conditional probability that it will rain on two given consecutive days given that it rained on the day before. Assume

$$\Pr(\text{Rain on Day 3} \mid \text{Rain on Days 1 and 2})$$
$$= \Pr(\text{Rain on Day 3} \mid \text{Rain on Day 2}).$$

***6.21** (Sec. 6.9)   In a certain routine medical examination, a common disorder, *D*, is found in 20% of the cases; that is, $\Pr(D) = .2$. The disorder is detected by a moderately expensive laboratory analysis. The disorder is, however, related to a symptom, *S*, which can be determined without a laboratory analysis.

Among those who have the disorder, 80% are found to have symptom *S*; that is, $\Pr(S \mid D) = .8$. Among those without the disorder, only 5% are found to have the symptom; that is, $\Pr(\overline{S} \mid \overline{D}) = .05$.

(a) Find the probability that presence of the symptom correctly predicts the disorder; that is, find $\Pr(D \mid S)$.

(b) Find the probability that lack of the symptom incorrectly predicts the disorder; that is, find $\Pr(D|\bar{S})$.

(c) Why is it reasonable that $\Pr(D|S)$ exceeds $\Pr(D)$?

6.22 You have one of those strings of Christmas tree lights where the whole string lights up only if each bulb is okay. Your string has four bulbs. If the probability is .9 that any one bulb is okay, what is the probability

(a) that the whole string lights up?

(b) that the string does not light up?

6.23 A family of three go to a photographic studio to have a picture made. A picture is a success if each of the three persons looks good in the picture; otherwise, it is a failure. Suppose that the probability that any one person looks good in the picture is .4. Assume independence.

(a) What is the probability that the picture is a success?

(b) Suppose the photographer makes two pictures. What is the probability that at least one of them is a success?

(c) If the photographer is good, do you think the assumption of independence is warranted?

6.24 (continuation) How many pictures need the photographer make to be 50% sure of getting at least one picture that is a success?

6.25 Eye color is governed by a pair of genes, one of which is contributed by the father, the other by the mother. Let B denote a gene for brown eyes and b a gene for blue eyes. An individual with a BB genotype has brown eyes. And, since B gene dominates the b gene, an individual with a Bb genotype will have brown eyes; only the bb genotype leads to blue eyes. Assume there are only the two genes B and b for eye color and that the father and mother each contribute one of these genes with a probability of $1/2$, independently of one another. What are the probabilities of the various genotypes for the offspring? What are the probabilities of the various eyecolors for the offspring?

*6.26 Under the assumptions of Exercise 6.25, and given that a child has brown eyes, what is the conditional probability that the child's genotype is BB (rather than Bb or bB)?

6.27 Under the assumptions of Exercise 6.25, suppose that a mother has blue eyes and the father has brown eyes. What is the conditional probability that the father has the BB genotype, given that the couple has

(a) a child with brown eyes?

(b) two children with brown eyes?

(c) three children with brown eyes?

6.28 Suppose there were a society in which couples had children until they had a boy. Find the probability that the number of children in a family is 1, 2, 3, . . . , respectively.

6.29 Suppose couples had children until they had at least one child of each sex. Find the probability that the number of children in a family is 2, 3, . . . , respectively.

6.30 Fill in the following 2 × 2 tables.

	B	$\bar{B}$	TOTAL
A	100	200	
$\bar{A}$	80		
TOTAL		300	

	B	$\bar{B}$	TOTAL
A			
$\bar{A}$	30	5	
TOTAL	90		100

	B	$\bar{B}$	TOTAL
A			
$\bar{A}$	.20	.40	
TOTAL	.30		1.00

	B	$\bar{B}$	TOTAL
A	.35	.25	
$\bar{A}$	.15		
TOTAL			1.00

6.31 (Sec. 6.4)   You have a probability of .6 for hitting the dart board each time you throw a dart, and successive throws are independent. What is the probability that you hit the dart board at least once in two throws?

6.32 (Sec. 6.4)   Figure 6.8 represents a system of two components, A and B, *in series,* which means the system functions if and only if both components function. Suppose the probability that A functions is .9, the probability that B functions is .9, and these events are independent. What is the probability that the system functions?

Figure 6.8 *Two components in series*

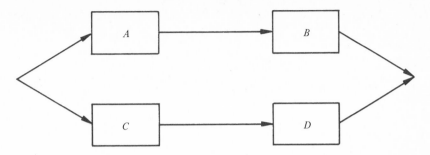

Figure 6.9 *Two parallel subsystems*

6.33 (continuation) Figure 6.9 represents a system with two *parallel* subsystems, *A–B* and *C–D*. If one of the subsystems fails, the other can carry on. Thus the system fails only if both subsystems fail. Assuming that *C* is similar to *A*, *D* to *B*, and that the functioning of the two subsystems is independent, what is the probability that the system functions?

6.34 (continuation) How many parallel subsystems are needed to make the probability that the system functions at least .95?

6.35 (Sec. 6.9) The application of statistical methods to the problem of deciding the authorship of the Federalist Papers was discussed in Section 1.1.6.

Suppose that a work, the author of which is known to be either Madison or Hamilton, contains a certain key phrase. Suppose further that this phrase occurs in 60% of the papers known to have been written by Madison, but in only 20% of those by Hamilton. Finally, suppose a historian gives subjective probability .3 to the event that the author is Madison (and consequently .7 to the complementary event that the author is Hamilton). Compute the historian's posterior probability for the event that the author is Madison.

MATHEMATICAL EXERCISES

6.36 Show that if *A* and *B* are independent, then $\Pr(A \text{ or } B) = \Pr(A) + \Pr(B) - \Pr(A) \times \Pr(B)$.

6.37 This example shows that the conditions $\Pr(A \text{ and } B) = \Pr(A) \times \Pr(B)$, $\Pr(A \text{ and } C) = \Pr(A) \times \Pr(C)$, and $\Pr(B \text{ and } C) = \Pr(B) \times \Pr(C)$ do not imply that $\Pr(A \text{ and } B \text{ and } C) = \Pr(A) \times \Pr(B) \times \Pr(C)$, which is part of the definition of (mutual) independence of *A*, *B*, and *C*; pairwise independence does not imply independence. Suppose that there are only four possible outcomes, denoted by the integers 1, 2, 3, 4, and these are equally likely. Take $A = (1, 2)$, $B = (2, 3)$, $C = (1, 3)$. Show that $\Pr(A \text{ and } B) = 1/4 = 1/2 \times 1/2 = \Pr(A) \times \Pr(B)$, and similarly for $\Pr(A \text{ and } C)$ and $\Pr(B \text{ and } C)$, but $\Pr(A \text{ and } B \text{ and } C) = 0 \neq 1/8 = \Pr(A) \times \Pr(B) \times \Pr(C)$.

6.38   This example shows that $\Pr(A \text{ and } B \text{ and } C) = \Pr(A) \times \Pr(B) \times \Pr(C)$ does not imply $\Pr(A \text{ and } B) = \Pr(A) \times \Pr(B)$. Suppose there are 8 equally likely outcomes, denoted by the integers $1, 2, \ldots, 8$. Take $A = (1, 2, 3, 4)$, $B = (1, 3, 4, 5)$, $C = (1, 5, 6, 7)$. Show that $\Pr(A \text{ and } B \text{ and } C) = \Pr(A) \times \Pr(B) \times \Pr(C)$ but $\Pr(A \text{ and } B) \neq \Pr(A) \times \Pr(B)$.

*6.39   (a) Show that the positive numbers $p_i = \log_{10}(i + 1) - \log_{10}(i)$, $i = 1, 2, \ldots, 9$, satisfy the condition $\Sigma_{i=1}^{9} p_i = 1$, so that these numbers can be a set of probabilities for the space of outcomes $1, 2, \ldots, 9$. (b) This set of probabilities provides a probability distribution for the frequency of left-hand digits. Find a large table, say in an almanac, and make a sample frequency distribution of the left-hand digits. Compare the sample frequencies with the probabilities p_i.

*6.40   Show that if A_1, A_2, A_3, and A_4 constitute a set of mutually exclusive events such that the union of the A's is the whole sample space and B_1, B_2, B_3 constitute a set of mutually exclusive events such that the union of the B's is the whole sample space, then

$\Pr(B_1 | A_1)$

$$= \frac{\Pr(B_1) \times \Pr(A_1 | B_1)}{\Pr(B_1) \times \Pr(A_1 | B_1) + \Pr(B_2) \times \Pr(A_1 | B_2) + \Pr(B_3) \times \Pr(A_1 | B_3)} .$$

[Hint: Use 3×4 tables parallel to the use of 2×2 tables in Section 6.9.1.]

REFERENCES

Bayes, Thomas (1763). "An essay towards solving a problem in the doctrine of chances." *Philosophical Transactions of the Royal Society*, 53: 370–418. [Reprinted in *Biometrika* 45 (1958): 293–315.]

Dunn, John E., Jr., and Samuel W. Greenhouse (1950). *Cancer Diagnostic Tests: Principles and Criteria for Development and Evaluation*. Public Health Service Publication No. 9. Government Printing Office, Washington, D.C.

Rosenblatt, Joan R., and James J. Filliben (1971). "Randomization and the Draft Lottery." *Science* 171: 306–308.

7 SAMPLING
DISTRIBUTIONS

INTRODUCTION

Much of our statistical information comes in the form of samples from populations of interest. In order to develop and evaluate methods for using sample information to obtain knowledge of the population, it is necessary to know how closely a descriptive quantity such as the mean or the median of a sample resembles the corresponding population quantity. In this chapter, the ideas of probability will be used to study the sample-to-sample variability of these descriptive quantities.

7.1 PROBABILITY DISTRIBUTIONS

7.1.1 Random Variables

In Chapter 6 we discussed probabilities of events described *categorically*. Categories define the event that a person is a lawyer and the event that a person is a doctor, the event that a person has a certain disease and the event that a person does not have the disease. We are also interested in events described *numerically*, such as the event that a person regularly reads 3 magazines and the event that a person regularly reads 2 magazines, the event that a person has an annual income of $8,000 and the event that a person has an annual income not greater than $16,000, the event that a tree bears 560 apples and the event that a tree bears 72 apples.

Table 7.1 gives the distribution of families by number of children under 18 years of age for the United States in 1969. This table is a frequency distribution for a population. The relative frequency of families with 3 children is .105; the relative frequency of families with 2 children is .169. If a family is chosen at random, then .105 is the probability that

TABLE 7.1

Distribution of Families by Number of Children Under 18 Years of Age (United States, 1969)

Number of children	0	1	2	3	4 or more
Proportion of families	.442	.181	.169	.105	.103

SOURCE: *Statistical Abstract of the United States: 1970*, p. 38.

this random family will have 3 children. Relative frequencies for a population are probabilities for random sampling. A (relative) frequency distribution for a population is called a *probability distribution*.

Often in describing a probability distribution we focus on the *variable* and say "distribution of *number of children*" rather than "distribution of *families* by number of children." If a family were chosen at random, the probability that the *number of children* would be 3 is .105, the probability that the number of children would be 2 is .169, etc. The variable, "number of children," is a *random variable* in the sense that its value depends on which family is chosen.

Let x represent the number of children. Then the notation $x = 3$ represents the *event* that a family has 3 children, and $\Pr(x = 3)$ stands for the probability that a randomly chosen family has 3 children. Thus, $\Pr(x = 3) = .105$.

Generally, some particular variables are observed for each member of a sample. In the case of a sample of families, one may observe the number of children, the total family size, the family income bracket, the number of autos, and the number of magazines read. Each of these is a random variable.

We want to be able to discuss the probability distribution of a *continuous* variable. As an example, consider the heights of 1293 11-year-old boys; Table 2.11 gives the numbers of boys with heights in one-inch intervals. The distribution of heights of the 1293 boys is approximated by considering all heights within an interval as equal to the midpoint of that interval. The height of a boy picked at random from among the 1293 is a random variable. We can think of Table 2.11 replaced by a frequency table with smaller intervals. If the initial population is very large—and the method of measurement is very accurate—the intervals can be very small. As noted in Chapter 2, the idealization of a histogram with very small intervals is a smooth curve.

Figure 7.1 shows a histogram and a smooth curve. The relative frequency of heights between 54.5 and 56.5 inches is represented by the area of the two bars at 55 inches and 56 inches, that is, by the area of the histogram from 54.5 inches to 56.5 inches, because the area of the entire histogram is taken to be 1. The area under the histogram is about

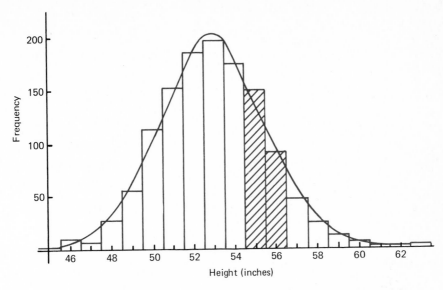

Figure 7.1 *Histogram of heights of 11-year-old boys, with corresponding continuous probability distribution (Table 2.11)*

the same as the area under the smooth curve from 54.5 inches to 56.5 inches. When we sample randomly, the probability of a height between 54.5 and 56.5 is equal to this area. In fact, the probability of obtaining a measurement in any interval is the area over that interval.

7.1.2 Cumulative Probability

Cumulative probabilities are analogous to the cumulative relative frequencies discussed in Chapter 2. Table 7.3 gives the cumulative probabilities corresponding to the probability distribution of Table 7.2. For each number of children, the corresponding cumulative probability is simply the proportion of families having that number of children or fewer. Cumulative probabilities can also be obtained for continuous random variables.

TABLE 7.2
Probability Distribution of Number of Children

Number of children	0	1	2	3	4	5
Proportion of families	.2	.2	.3	.1	.1	.1

(hypothetical data)

TABLE 7.3
Cumulative Distribution Function

Number of children	0	1	2	3	4	5
Proportion of families having this many children or fewer	.2	.4	.7	.8	.9	1.0

SOURCE: Table 7.2

7.1.3 The Mean and Variance of a Probability Distribution

The Mean of a Probability Distribution. In Chapter 3 we noted that the mean of a sample of n individuals can be written as

$$\bar{x} = \frac{f_1 v_1 + f_2 v_2 + \cdots + f_m v_m}{n},$$

where the values $v_1, v_2, \ldots, v_m$ occur in the sample with respective frequencies $f_1, f_2, \ldots, f_m$. Dividing each f_j by n permits this expression to be written as

$$\bar{x} = \frac{f_1}{n} v_1 + \frac{f_2}{n} v_2 + \cdots + \frac{f_m}{n} v_m.$$

This is the same as

$$\bar{x} = r_1 v_1 + r_2 v_2 + \cdots + r_m v_m,$$

where the ratios $r_1 = f_1/n$, $r_2 = f_2/n$, $\ldots$, $r_m = f_m/n$, are the *relative frequencies* of the values $v_1, v_2, \ldots, v_m$. Thus $\bar{x}$ is a *weighted average* of the values $v_1, v_2, \ldots, v_m$, the weights being the relative frequencies r_1, $r_2, \ldots, r_m$.

Applying these same arithmetic operations to the hypothetical data in Table 7.2, we have

$$(.2 \times 0) + (.2 \times 1) + (.3 \times 2) + (.1 \times 3) + (.1 \times 4) + (.1 \times 5)$$
$$= 0 + .2 + .6 + .3 + .4 + .5$$
$$= 2.0 \text{ children per family.}$$

This value, 2.0, is the *mean* of the variable, "number of children," or, equivalently, the mean of the corresponding probability distribution, since probabilities are relative frequencies for a population.

In general, the mean of a probability distribution is

$$p_1 v_1 + p_2 v_2 + \cdots + p_m v_m,$$

where $p_1, p_2, \ldots, p_m$ are the probabilities associated with the values $v_1, v_2, \ldots, v_m$. That is, the mean of a probability distribution is a weighted average of the values $v_1, v_2, \ldots, v_m$, where the weights are the probabilities $p_1, p_2, \ldots, p_m$: it is the *probability-weighted average* of the values $v_1, v_2, \ldots, v_m$.

Finally, we note that the mean of a probability distribution is the same as the population mean, μ, of the variable under consideration. For, just as the sample mean, $\bar{x}$, can be written

$$\bar{x} = \frac{\sum\limits_{i=1}^{n} x_i}{n} = \sum_{j=1}^{m} r_j v_j,$$

so is it true that the mean, μ, for a finite population can be written

$$\mu = \frac{\sum\limits_{i=1}^{N} y_i}{N} = \sum_{j=1}^{m} p_j v_j,$$

where $y_1, y_2, \ldots, y_N$ are the individual values in the population.

The Variance of a Probability Distribution. The mean of a probability distribution is the probability-weighted average of the values $v_1, v_2, \ldots, v_m$. The *variance* of a probability distribution is the probability-weighted average of the squared deviations $(v_1 - \mu)^2$, $(v_2 - \mu)^2$, $\ldots, (v_m - \mu)^2$:

variance of probability distribution

$$= p_1 (v_1 - \mu)^2 + p_2 (v_2 - \mu)^2 + \cdots + p_m (v_m - \mu)^2$$

$$= \sum_{j=1}^{m} p_j (v_j - \mu)^2.$$

For the probability distribution of Table 7.2, we have $\mu = 2.0$, so that the variance is

$$.2 \times (0 - 2.0)^2 + .2 \times (1 - 2.0)^2 + .3 \times (2 - 2.0)^2 + .1 \times (3 - 2.0)^2$$
$$+ .1 \times (4 - 2.0)^2 + .1 \times (5 - 2.0)^2$$

$$= .2 \times (-2.0)^2 + .2 \times (-1.0)^2 + .3 \times (0.0)^2 + .1 \times (1.0)^2 + .1 \times (2.0)^2$$
$$+ .1 \times (3.0)^2$$

$$= .2 \times 4 + .2 \times 1 + .3 \times 0 + .1 \times 1 + .1 \times 4 + .1 \times 9$$

$$= .8 + .2 + 0 + .1 + .4 + .9$$

$$= 2.4.$$

The *standard deviation* of this probability distribution is $\sqrt{2.4} = 1.55$.

Finally, we note that the notion of variance of a probability distribution agrees with the notion of the variance σ^2 of the corresponding population, for

$$\sum_{j=1}^{m} p_j (v_j - \mu)^2 = \frac{\sum_{i=1}^{N} (y_i - \mu)^2}{N} = \sigma^2.$$

The Mean and Variance of a Continuous Random Variable. We want to be able to discuss the mean and variance of a continuous variable. In Section 7.1.1 we considered the heights of 11-year-old boys as given in Table 2.11. From this table we can calculate the mean and the variance of the distribution of heights by using the midpoint of each interval as the height of all boys grouped in that interval. There may be small differences between the mean and variance computed this way and the mean and variance computed on the basis of the 1293 heights if they were measured more accurately than to the nearest inch. We could reduce these differences by using intervals smaller than 1 inch.

If the population is large, the intervals can be very small. The prob-

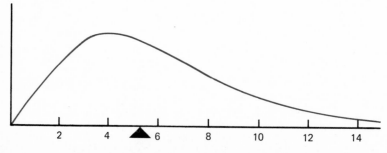

Figure 7.2 *The mean is the center of gravity*

ability distribution of a continuous random variable represented by a smooth curve has a mean and variance which are approximated by the mean and variance of an approximating histogram.

We have defined probability distributions, means, and variances on the basis of the random sampling of one individual from a given finite population. Probability distributions and their means and variances can also be defined on a theoretical basis or from mathematical formulas.

The mean is the center of gravity of the distribution. If a flat metal plate were cut to the shape of the distribution, the plate would balance at the mean (Figure 7.2).

7.1.4 Uniform Distributions

The Discrete Uniform Distributions. One of the simplest probability distributions is that of a digit taken at random. In this case the integers 0, 1, 2, . . . , 9 are to be equally likely. The probabilities are

$$\Pr(\text{integer chosen} = i) = \frac{1}{10}, \qquad i = 0, 1, 2, \ldots, 9.$$

This probability distribution is called the *uniform distribution* on the integers 0, 1, 2, . . . , 9. This distribution provides the *probability model* for "choosing a digit at random."

Similarly we can consider choosing one of the ten values, 0.0, 0.1, 0.2, . . . , 0.9, at random. In this case

$$\Pr\left(\text{value chosen} = \frac{i}{10}\right) = \frac{1}{10}, \qquad i = 0, 1, \ldots, 9.$$

These probabilities are graphed in Figure 7.3.

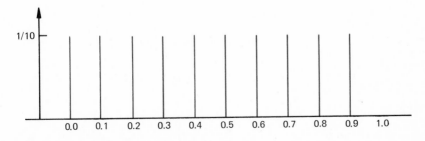

Figure 7.3 *Uniform distribution on tenths*

The use of *pairs* of random digits leads to the uniform distribution on 0.00, 0.01, 0.02, . . . , 0.99, with probabilities

$$\Pr\left(\text{value chosen} = \frac{i}{100}\right) = \frac{1}{100}, \qquad i = 0, 1, 2, \ldots, 99.$$

We can also consider the uniform distribution on the one thousand values 0.000, 0.001, 0.002, . . . , 0.999. The graphs of these probability distributions look like Figure 7.3, but with 100 and 1000 bars, respectively.

The Continuous Uniform Distributions. If we continue this process of considering uniform distributions over the numbers between 0 and 1 with more and more decimal places, in the end the resulting distribution is the *continuous uniform distribution* over the interval from 0 to 1. This probability distribution assigns a probability of 1/2 to the interval 0 to 1/2, to the interval 1/2 to 1, to the interval 1/4 to 3/4; it assigns a probability of 1/4 to any interval of length 1/4. The continuous uniform distribution on the interval 0 to 1 assigns to any subinterval of the interval from 0 to 1 a probability equal to its length.

The continuous uniform distribution provides a model for the experiment of spinning a pointer on a pivot (Figure 6.1). The pointer is spun and the angle it makes with the horizontal line is observed. We consider all angles as equally likely to occur. Let us measure the angle as a fraction of 360°; that is, as a fraction of the entire circle. Then

$$\Pr(a < \text{angle} < b) = b - a,$$

where a and b are each between 0 and 1, and b is larger than a. (If the circle is divided into 10 equal parts, labelled 0, 1, . . . , 9, then the pointer can be used to obtain random digits.)

Figure 7.4 shows a square of length and height 1. The probability that the pointer stops between a and b is the shaded area between a and b.

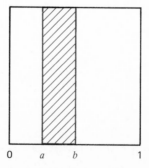

0   a b 1

Figure 7.4 *The uniform distribution on the interval 0 to 1*

7.2 THE FAMILY OF NORMAL DISTRIBUTIONS

7.2.1 The Normal Distributions

Many continuous statistical variables have distributions that look similar. The histograms of such biological measurements as heights and weights, and such psychological measurements as IQ scores have common features. The histogram of Figure 7.1, based on the frequencies of heights to the nearest inch of 1293 Boston schoolboys (Table 2.11), shows these features. The modal class is 52.5–53.5 inches, with a midpoint of 53 inches. In classes on either side of 53 inches the frequencies decrease slowly at first, then rapidly, and then slowly again. The frequency for 54 inches is about the same as that for 52 inches, the frequency for 55 inches is about the same as that for 51 inches, etc.; that is, the histogram is approximately symmetric about 53. If the class intervals had been half as large, the histogram would have been smoother. This is approximated by a smooth curve, as shown in Figure 7.1.

It turns out that in many cases the smooth curve has the shape of the graph of a particular mathematical relationship. This mathematical relationship or formula specifies a theoretical probability distribution, known as the *normal* distribution. [It is also known as the *Gaussian*

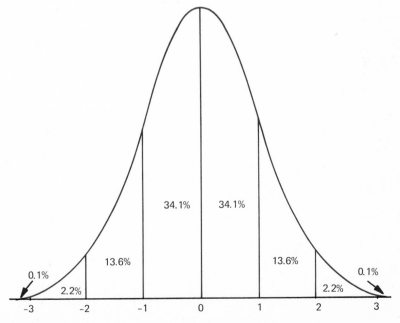

Figure 7.5 *Graph of the standard normal distribution*

distribution, after the famous German mathematician, K. F. Gauss (1777–1855); in France it is called Laplace's distribution, after P. S. Laplace (1749–1827).]

A particular normal distribution is graphed in Figure 7.5. It is symmetric, unimodal, and bell-shaped. Based on the total area under the curve, the percentages in the figure indicate the area under the curve between various vertical lines. For instance, 34.1% of the total area is between 0 and +1 and the same percentage is between −1 and 0. In probability terms this expresses the idea that the probability of a value between 0 and 1 is .341. This corresponds to the interpretation of relative area in a histogram as indicating relative frequency.

> *Mathematical Note.* If z denotes the horizontal axis and y the vertical axis, the curve in Figure 7.5 is the graph of
>
> $$y = \frac{1}{\sqrt{2\pi}}\, e^{-z^2/2}.$$

The area between this curve and the horizontal axis is 1.

There is a whole family of normal distributions. The horizontal scale can be expanded and the peak can be moved to various points. The curve for each distribution is symmetric, unimodal, and bell-shaped. Three different normal distributions are shown in Figures 7.6, 7.7, and 7.8.

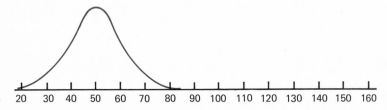

Figure 7.6 *Normal distribution with center at 50*

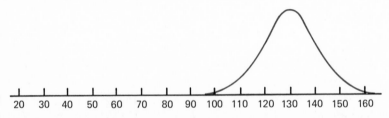

Figure 7.7 *Normal distribution with same spread as that of Figure 7.6, but with center at 130*

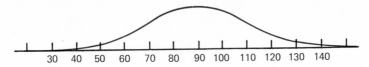

Figure 7.8 *Normal distribution with large spread and center at 90*

7.2.2 Different Normal Distributions

The histogram of heights of 11-year-old boys (Figure 7.1) shows roughly the typical shape of a normal distribution. The histogram of heights of 12-year-old boys (Exercise 7.20) is similar but is centered at 54.7 inches. The histograms of heights of boys in one-year age brackets will have similar shapes but different means and percentiles, and perhaps different variabilities. Heights of cats would give a histogram which would again approximate the normal distribution, but the location and spread would differ from those for boys. In each case the histogram can be approximated by a member of the family of normal distributions.

All the normal distributions have the same form and are defined by a similar mathematical formula. The particular normal distribution is specified by a mean and a standard deviation. The mean μ gives the location along the horizontal axis, and the standard deviation σ tells how spread out the distribution is; the larger σ is, the more spread out is the distribution. The distribution in Figure 7.6 has $\mu = 50$ and $\sigma = 10$; in Figure 7.7, $\mu = 130$ and $\sigma = 10$; and in Figure 7.8, $\mu = 90$ and $\sigma = 20$. The pair of characteristics μ and σ are the *parameters* of the family of normal distributions.

Mathematical Note. These curves are graphs of

$$y = \frac{1}{\sigma\sqrt{2\pi}} \, e^{-(z-\mu)^2/(2\sigma^2)}.$$

The normal distribution with mean 0 and standard deviation 1, graphed in Figure 7.5, is called the *standard* normal distribution.

We saw in Chapter 3 that if we subtracted a number from every measurement the mean of the transformed measurements was the original mean minus this number. In particular, if we subtract the mean μ of the original measurements constituting a population, the mean of the transformed measurements is 0. In Chapter 4 we saw that if we divided every measurement by a positive number, the standard deviation of the transformed measurements was the original standard deviation divided by the number. In particular, if we divided by the standard deviation σ of the original measurements constituting a population, the standard deviation of the transformed measurement is 1. If we subtract the mean μ of the original measurements and then divide by the standard deviation σ of the original measurements, we obtain transformed measurements with mean 0 and standard deviation 1. If the original measurement is x, the transformed measurement is

$$z = \frac{x - \mu}{\sigma} .$$

This particular coding of the random variable x is called the *standardized version* of x and is sometimes denoted by z. Such a score z is called a *standard score*.

If a random variable x has a *normal* distribution with mean μ and standard deviation σ, the coded variable also has a normal distribution, namely the normal distribution with mean 0 and standard deviation 1, the *standard* normal distribution, the formula for which is given in Section 7.2.1.

7.2.3   Cumulative Normal Probabilities

From Table 2.11 or Figure 7.1 we can find the number of boys with heights between 50.5 and 52.5, or between 50.5 and 55.5, or less than 55.5. We can convert these frequencies to relative frequencies. If we sample one boy randomly from this population, the relative frequencies are probabilities.

As we saw in Chapter 2, the cumulative relative frequencies are obtained by adding the relative frequencies over intervals. From Table 2.11, for example, the cumulative relative frequency to 45.5 is .0015, to 46.5 is .0015 + .0062 = .0077, to 47.5 is .0015 + .0062 + .0046 = .0123, etc. Conversely, the relative frequency of any interval can be obtained from two cumulative relative frequencies by subtraction. For instance, the relative frequency of 46.5 to 47.5 is the cumulative relative frequency at 47.5 of .0123 minus the cumulative relative frequency at 46.5 of .0077.

When we cumulate probabilities, we obtain a *cumulative distribution function*. For example, for the normal distribution as graphed in Figure 7.5, we can obtain the cumulative distribution as tabulated in Appendix I. Table 7.4 is a short version of this table. The numbers in these tables are obtained by adding the probabilities in Figure 7.5 to the left of the stated value of z.

We shall denote the cumulative distribution function of the standard normal distribution by $\Phi(z)$, called (capital) *phi* of z. It can be defined

TABLE 7.4
A Short Table of the Standard Normal Cumulative Distribution

z	PROBABILITY TO z
-3	.001
-2	.023
-1	.159
0	.500
1	.841
2	.977
3	.999
∞	1.000

mathematically as the area to the left of z under the standard normal curve, as shown in Figure 7.9. Since the total area is 1, the area to the right of z is $1 - \Phi(z)$.

We remarked that normal curves are symmetric. The standard normal curve is symmetric about the value $z = 0$ (the mean); that is, if we turn the graph of the standard normal distribution around through 180 degrees, using a vertical axis through the value $z = 0$ as a pivot, then the resulting figure will coincide with the original figure. Consequently, the area to the right of z is equal to the area to the left of $-z$; that is, $1 - \Phi(z) = \Phi(-z)$. Values of $\Phi(z)$ can be found in Appendix I. The table

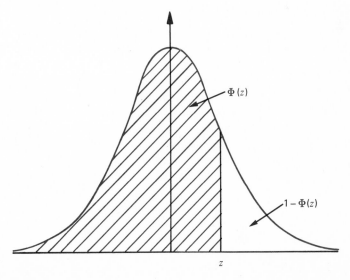

Figure 7.9 *Cumulative distribution function as area to left of a given point*

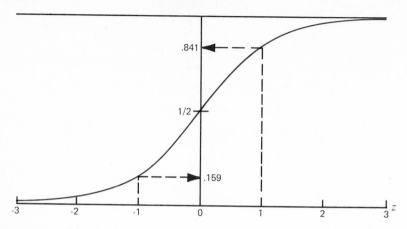

Figure 7.10 *The cumulative distribution function of the standard normal distribution*

gives values of $\Phi(z)$ for positive values of z, which is sufficient. If we want to find, say, $\Phi(-1.5)$, we write

$$\Phi(-1.5) = 1 - \Phi(1.5) = 1 - .9332 = .0668.$$

A graph of the cumulative standard normal distribution is given in Figure 7.10.

7.2.4 Calculating Probabilities from Normal Distributions

The various normal distributions differ with respect to mean (center) and standard deviation (spread). For every normal distribution the probability between the mean and one standard deviation above the mean is .341. Similarly, the probability between the mean and one standard deviation below the mean is .341. Within one standard deviation of the mean lies 68.2% (about two-thirds) of the total population. The probabilities of every normal distribution correspond to Figure 7.5 and Table 7.4 if measurements are given in terms of the number of standard deviations from the mean.

The scoring of one of the usual IQ tests has been set up so that the mean is 100 and the standard deviation is 20, as drawn in Figure 7.11. What proportion of the population has an IQ of 120 or less? To answer this question, we note that 120 is one standard deviation more than the mean. In the lower part of Figure 7.11 we have drawn the axis cor-

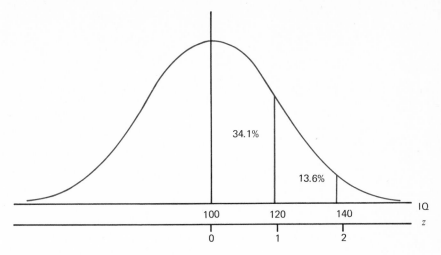

Figure 7.11 *Distribution of IQ scores*

responding to a standard normal variable. To find the probability of a score less than or equal to 120 we add .341 (the probability of a score between 100 and 120) to .500, the probability of a score less than 100; the probability is .841.

As another example, consider the probability that a person sampled at random has an IQ not greater than 110. Since the standard deviation is 20, we have

$$\Pr(IQ \le 110) = \Pr\left(\frac{IQ - 100}{20} \le \frac{110 - 100}{20}\right)$$

$$= \Pr(z \le 1/2) = \Phi(1/2) = .6915.$$

The interpretation is that 69.15% of the population has an IQ less than or equal to 110. The "Pr" symbol for probability can be read as "proportion of the population"; "$\Pr(IQ \le 110)$" then becomes "proportion of the population having an IQ not greater than 110."

Percentage Points of Normal Distributions. What IQ score is exceeded by only 10% of the population; that is, what score is the *90th percentile*? From Appendix I we can see that in σ units this score is between 1.28, which gives .8997, and 1.29, which gives .9015. It is useful to have a table like Appendix II, a table of percentiles* of the standard normal distribution. Such a table is the inverse of Appendix I. From Appendix II we see that the 90th percentile of the standard normal

*Percentage points (percentiles) of distributions will be given with three-decimal accuracy to aid the student by preserving uniformity throughout the text. It is not implied that such accuracy is necessary or even desirable in practice.

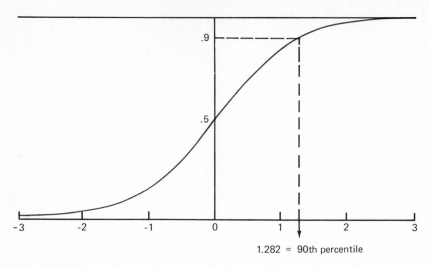

1.282 = 90th percentile

Figure 7.12 *How to find percentiles of the standard normal distribution*

distribution is 1.282. (See Figure 7.12.) To convert this to an IQ score we multiply by σ and add μ to obtain $20 \times 1.282 + 100 = 25.64 + 100 = 125.64$. This is the 90th percentile of the population of IQ scores.

What is the 40th percentile of the standard normal distribution? By the symmetry of the bell-shaped normal curve, the standard score corresponding to a proportion .40 is the negative of the standard score corresponding to a proportion .60. The 60th percentile is 0.253. Thus the 40th percentile is −0.253.

Weight, height, and IQ do not have distributions that are *exactly* normal, but experience has shown that normal distributions often provide good approximations to their distributions.

Grading on the Curve. Sometimes teachers grade their pupils according to how far above or below the average their scores are in standard deviation units. If the mean score on a test is $\mu = 79$, the standard deviation σ is 9, and Bill's score was 88, then Bill's score is one standard deviation above the mean; we say his standard score is +1. If Charles' score was 61, then his standard score is −2, because 61 is two standard deviations below the mean. If x_i is the score of the ith pupil, then that pupil's standard score z_i is

$$z_i = \frac{x_i - \mu}{\sigma}.$$

Standard scores have a mean equal to 0 and a standard deviation equal to 1.

Sometimes teachers treat standard scores as a sample from the standard normal distribution. This would be approximately so if the raw scores $x_1, x_2, \ldots, x_n$ were themselves a sample from some normal distribution.

Using Appendix I, it can be seen that if a teacher used the grading policy outlined in Table 7.5, and if the teacher's pupils' scores were normally distributed, in the long run, he or she would be assigning the percentages of A's, B's, etc. shown in the last column of this table. Ac-

TABLE 7.5
A Grading Policy

GRADE	STANDARD SCORE	PERCENTAGE OF STUDENTS
A	At least 1.5	6.68
B	.5 to 1.5	24.17
C	−.5 to .5	38.30
D	−1.5 to −.5	24.17
F	Less than −1.5	6.68

(hypothetical data)

cording to this grading policy, Bill, whose standard score is +1, would get a B, while Charles, whose standard score is −2, would get an F.

7.3 SAMPLING FROM A POPULATION

7.3.1 Random Samples

In Section 6.8 we saw how a random sample could be drawn from a specified population of a finite number of individuals. A random sample (without replacement) of 13 cards from a deck of 52 cards is obtained by dealing 13 cards from the deck after shuffling; a random sample of 50 beads from a jar of 1000 beads is obtained by mixing the jar of beads and blindly taking out 50 beads; and a random sample of 50 students from a class of 1000 students is obtained by numbering the students and using 50 different random numbers between 1 and 1000. All possible sets of 13 cards are equally likely to be drawn, and all sets of 50 students are equally likely to appear as the random sample.

DEFINITION   *The* size *of a sample is the number of individuals in it.*

In the examples above the size of the bridge hand is 13, and the size of the sample of beads is 50, as is the size of the sample of students.

DEFINITION   *A* random sample *of a given size is a group of members of a population obtained in such a way that all groups of that size are equally likely to appear in the sample.*

Here, the word "random" means that the sample was produced by a *procedure* that is random.

Is a card a spade or is it not a spade? That question can be asked about each of the 13 cards in a bridge hand. Is a bead red or blue? What is the Scholastic Aptitude Test score (SAT) of a student? The same particular *property* or *measurement* is considered for each member of a sample. This property or measurement is a random variable.

The focus of attention may be on the *number* of cards in the bridge hand that are spades, the *number* of beads of each color in the sample, the *mean* of the SAT scores in the sample. Each of these characteristics of the sample as a whole is called a *statistic*.

DEFINITION   *A* statistic *is a characteristic of a sample.*

We know that the number of spades varies from hand to hand. The number of beads that are red will vary in randomly drawn sets of 50 (if there are some red beads in the jar, but not all of them are red). The mean SAT score will vary in samples of students.

7.3.2   Sampling Distributions

Similarly, if we draw random samples of 5 men, "basketball teams," from among students at a college, the individuals composing these teams would vary, and the average heights of these teams would vary over a range of heights. It is unlikely that the mean height for any one of these teams will be equal to the mean of the height of the players on the varsity team. Each basketball team selected is a *sample*. The mean height of a team is a *statistic* for that sample.

Different samples usually lead to different values of the statistic: most of the basketball teams will have different average heights. The *sampling distribution* of the statistic gives the probability of each possible value of the statistic.

We illustrate with an example that is simpler than that of choosing

5 persons from a college population. We shall choose samples of size 2 from a population of size 4. This population consists of the four boys Bill, Ed, Jim, and Randy. Their heights are given in Table 7.6. We shall draw random samples of two (different) boys, measure the heights of the two boys, and compute the mean height of the sample.

If we use random digits to draw the sample, we number the boys— 1: Bill, 2: Ed, 3: Jim, and 4: Randy. The first digit drawn between 1 and 4 determines the first boy: the second digit drawn between 1 and 4 different from the first digit specifies the second boy. (See Table 7.7.)

TABLE 7.6
Heights of Four Boys

NAME	HEIGHT
Bill	56″
Ed	60″
Jim	68″
Randy	72″

(hypothetical data)

TABLE 7.7
Samples of Two Boys

PAIRS OF DIGITS IN ORDER	PAIRS OF BOYS IN ORDER	PAIRS OF BOYS
1,2	Bill, Ed	Bill, Ed
2,1	Ed, Bill	
1,3	Bill, Jim	Bill, Jim
3,1	Jim, Bill	
1,4	Bill, Randy	Bill, Randy
4,1	Randy, Bill	
2,3	Ed, Jim	Ed, Jim
3,2	Jim, Ed	
2,4	Ed, Randy	Ed, Randy
4,2	Randy, Ed	
3,4	Jim, Randy	Jim, Randy
4,3	Randy, Jim	

The probability of Bill being drawn as the first boy is 1/4, and the conditional probability of Ed as the second boy given Bill as the first is 1/3. The probability of the pair being Bill, Ed in that order is $1/4 \times 1/3 = 1/12$. Similarly the probability of Ed, Bill in that order is 1/12. The probability of each possible pair of boys is 1/6.

In Table 7.8 we have given the 6 possible samples of boys, the corresponding 6 possible pairs of heights, and the means of the pairs of heights. The example makes it clear that usually different samples lead to different values of the statistic. Here the six samples gave rise to five different values of the mean; the two samples (Bill, Randy) and (Ed, Jim) both gave values of 64 for the sample mean.

TABLE 7.8

Values of the Sample Mean for Each Sample

Sample	Probability of This Sample	Heights (Inches)	Mean Height
Bill, Ed	1/6	56, 60	58
Bill, Jim	1/6	56, 68	62
Bill, Randy	1/6	56, 72	64
Ed, Jim	1/6	60, 68	64
Ed, Randy	1/6	60, 72	66
Jim, Randy	1/6	68, 72	70

TABLE 7.9

Sampling Distribution of the Sample Mean

Value of Sample Mean	Probability
58	1/6
62	1/6
64	2/6
66	1/6
70	1/6

In Table 7.9 we tabulate the 5 possible values of the sample mean and their probabilities. The probability of 64 is the sum of the probability of (Bill, Randy) and the probability of (Ed, Jim). Table 7.9 gives the *sampling distribution* of the mean. (Another example is given in Section 8.7.2.)

7.3.3 Independence of Random Variables

Two events A and B were said (in Section 6.7) to be independent if

$$\Pr(B|A) = \Pr(B).$$

When we are studying two random variables, the events may be defined by values of the two random variables. As an example, consider the two random variables—number of children in the family and number of magazines read regularly—for a person drawn randomly from the population described by Table 7.10. Here, event A may be defined as the occurrence of a particular value of the random variable x, number of children; and the event B may be defined as the occurrence of a particular value of the random variable y, the number of magazines read regularly. The table displays independence in terms of events. For example, let A be the event $x = 1$ and B the event $y = 2$. We compute

$$\Pr(B|A) = \Pr(y = 2|x = 1) = .3,$$
$$\Pr(B) = \Pr(y = 2) = .3,$$

and thus see that A and B are independent. We verify that, for any integer a from 0 to 4 and any integer b from 0 to 3,

$$\Pr(y = b|x = a) = \Pr(y = b).$$

We say that x and y are *independent random variables.*

Another way of stating independence of events A and B (Section 6.7) is

$$\Pr(A \text{ and } B) = \Pr(A) \times \Pr(B).$$

TABLE 7.10
Independent Random Variables

		NUMBER OF MAGAZINES READ REGULARLY				Total
		0	1	2	3	
	0	40 (10%)	200 (50%)	120 (30%)	40 (10%)	400 (100%)
NUMBER	1	20 (10%)	100 (50%)	60 (30%)	20 (10%)	200 (100%)
OF	2	20 (10%)	100 (50%)	60 (30%)	20 (10%)	200 (100%)
CHILDREN	3	10 (10%)	50 (50%)	30 (30%)	10 (10%)	100 (100%)
	4	10 (10%)	50 (50%)	30 (30%)	10 (10%)	100 (100%)
	Total	100 (10%)	500 (50%)	300 (30%)	100 (10%)	1000 (100%)

(hypothetical data)

If the frequencies in Table 7.10 are divided by 1000 to obtain probabilities for random sampling, then we see that

$$\Pr(x = a \text{ and } y = b) = \Pr(x = a) \times \Pr(x = b)$$

for all possible values of a and b.

The idea described in terms of Table 7.10 applies to any pair of random variables x, y for which probabilities $\Pr(x = a \text{ and } y = b)$ are defined.

DEFINITION   *Two discrete random variables x and y are* independent *if*

$$\Pr(x = a \text{ and } y = b) = \Pr(x = a) \times \Pr(y = b)$$

for all possible values of a and b.

We sometimes need independence of several random variables. For example x, y, and z are independent if

$$\Pr(x = a \text{ and } y = b \text{ and } z = c) = \Pr(x = a) \times \Pr(y = b) \times \Pr(z = c)$$

for all possible values a, b, and c.

7.3.4 Sampling from a
Probability Distribution

The definition of a random sample in Section 7.3.1 is appropriate for sampling from a finite population, that is, a population with a finite number of members or elements. Samples are also drawn from "infinite populations," that is, populations defined by probability distributions and random variables. The number of possible tosses of a coin is unlimited. We describe this "infinite population" by the probability p of a head on a single toss of the coin and the probability $q = 1 - p$ of a tail. We consider the outcomes on two tosses to be independent, for example, the probability of a head on each of two tosses is $p \times p$. A random sample of two consists of the outcomes on two independent tosses. A random sample of n consists of the outcomes on n independent tosses.

A continuous random variable has an infinite number of values. A normal random variable is a random variable with a normal distribution; that is, the probability that the variable takes on a value within a given interval is the area under a normal curve over this interval. Two independent normal variables are normal variables such that outcomes defined by the two variables are independent.

DEFINITION   *A* random sample *of size n of a random variable with a given probability distribution is a set of n independent random variables each with this probability distribution.*

7.4 SAMPLING DISTRIBUTION OF A SUM

It is theoretically possible to determine the sampling distribution of the mean or any other statistic computed from a random sample from any given *parent population*. Sometimes this is difficult, sometimes easy.

We have shown how to obtain the sampling distribution of the mean of a random sample of size 2 from a population of size 4. As another example we consider the distribution of the *sum* of two observations from the uniform distribution over the integers 1, 2, . . . , 6. This sample can be drawn by rolling two balanced dice. The outcomes are 36 pairs (1, 1), (1, 2), . . . , (6, 6), each having probability 1/36. Let the outcome on the first die be x_1 and on the second x_2. We want to find the sampling distribution of the sum $x_1 + x_2$. The values of the sum are graphed in Figure 7.13 at the ends of the diagonal line segments. The probability of a given value of the sum is the sum of the probabilities of the points on that line. For instance,

$$\Pr(x_1 + x_2 = 3) = \Pr(x_1 = 1 \text{ and } x_2 = 2) + \Pr(x_1 = 2 \text{ and } x_2 = 1)$$

$$= \frac{1}{36} + \frac{1}{36} = \frac{2}{36}.$$

The probabilities of the sum are given in Table 7.11 and graphed in Figure 7.14. The sampling distribution is unimodal (with a mode of 7 for the sum) and is symmetric about the mode. In fact, the shape of the distribution is somewhat like the normal.

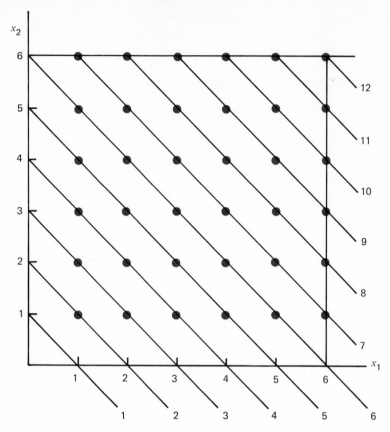

Figure 7.13 *Outcomes of the throw of two dice*

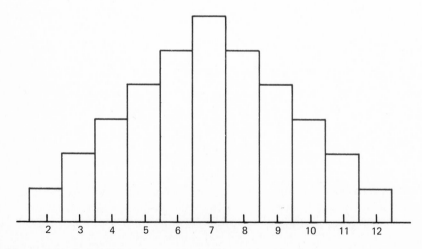

Figure 7.14 *Probabilities of the sum of values on two dice*

TABLE 7.11

Probability Distribution for Sum of Outcomes on Two Dice

Sum	2	3	4	5	6	7	8	9	10	11	12
Mean	1	$1\frac{1}{2}$	2	$2\frac{1}{2}$	3	$3\frac{1}{2}$	4	$4\frac{1}{2}$	5	$5\frac{1}{2}$	6
Probability	$\frac{1}{36}$	$\frac{2}{36}$	$\frac{3}{36}$	$\frac{4}{36}$	$\frac{5}{36}$	$\frac{6}{36}$	$\frac{5}{36}$	$\frac{4}{36}$	$\frac{3}{36}$	$\frac{2}{36}$	$\frac{1}{36}$

Note that the mean of the sampling distribution is 7, which is 2 times the mean of the value on one die, $3\frac{1}{2}$. In general, the mean of the sampling distribution of a sum of two variables is the sum of the means of the two variables; we write

$$\mu_{x_1 + x_2} = \mu_{x_1} + \mu_{x_2}.$$

That the mean of a sum is the sum of the means is true for any number of variables. For three variables, we write

$$\mu_{x_1 + x_2 + x_3} = \mu_{x_1} + \mu_{x_2} + \mu_{x_3};$$

for n variables we write

$$\mu_{x_1 + x_2 + \cdots + x_n} = \mu_{x_1} + \mu_{x_2} + \cdots + \mu_{x_n}.$$

If $x_1, x_2, \ldots, x_n$ represent a random sample from a parent population with mean μ, then each of these variables has mean equal to μ, and

$$\mu_{x_1 + x_2 + \cdots + x_n} = \mu + \mu + \cdots + \mu = n\mu.$$

The variance of the sampling distribution of the sum of the values on the two dice is 35/6, which is 2 times the variance of the values on one die, which is 35/12. In general, the variance of the sampling distribution of a sum of two *independent* variables is the sum of the variances; we write

$$\sigma^2_{x_1 + x_2} = \sigma^2_{x_1} + \sigma^2_{x_2}.$$

The standard deviation is

$$\sigma_{x_1 + x_2} = \sqrt{\sigma^2_{x_1} + \sigma^2_{x_2}}.$$

The analogous rule holds for three, four, or any number of *independent* random variables:*

$$\sigma^2_{x_1 + x_2 + \cdots + x_n} = \sigma^2_{x_1} + \sigma^2_{x_2} + \cdots + \sigma^2_{x_n}.$$

*These rules are proved in Appendix 7A.

The standard deviation is

$$\sigma_{x_1 + x_2 + \cdots + x_n} = \sqrt{\sigma_{x_1}^2 + \sigma_{x_2}^2 + \cdots + \sigma_{x_n}^2}.$$

In particular, for a random sample $x_1, x_2, \ldots, x_n$ from a parent population with variance σ^2, we have

$$\sigma_{x_1 + x_2 + \cdots + x_n}^2 = n\sigma^2,$$

and

$$\sigma_{x_1 + x_2 + \cdots + x_n} = \sqrt{n}\,\sigma.$$

The example of the dice illustrates the general idea of finding the probability distribution of the sum of two random variables each of which takes on a finite number of values. We start with the probabilities of all possible pairs of values; the random variables need not be independent. To find the probability that the sum of the two variables is a given value we add the probabilities of all pairs that give rise to that value of the sum; geometrically, we add the probabilities of the points on a diagonal line as in Figure 7.13. If the random variables are continuous, the procedure is similar, but involves calculus.

To find the probability distribution of the sum of three independent random variables, we find the probability distribution of the sum of the first two of them and then find the probability distribution of the sum of that variable and the third. This procedure can be continued to find the sampling distribution of the sum of any number of independent random variables.

The sampling distribution of the sum is important because the sampling distribution of the sample mean can be found from it. In Table 7.11 the values of the mean of the two variables x_1 and x_2 (the outcomes on the two dice), namely $\frac{1}{2}(x_1 + x_2)$, are given directly below the values of the sum; the respective probabilities apply to the values of the sum and the corresponding values of the mean. In this case, since each value of the mean of the two variables is $\frac{1}{2}$ the corresponding value of the sum of the two variables, the mean of the probability distribution of the mean of the two variables is $\frac{1}{2}$ the mean of the probability distribution of the sum of the two variables. It is $3\frac{1}{2}$, which is the mean of the probability distribution of each component variable, x_1 and x_2. The deviation of the mean of the two variables $\frac{1}{2}(x_1 + x_2)$ from $3\frac{1}{2}$ is $\frac{1}{2}$ of the deviation of the sum $x_1 + x_2$ from 7, and the squared deviation of the mean is $(\frac{1}{2})^2 = \frac{1}{4}$ times the squared deviation of the sum. Since the variance is the average of squared deviations, the variance of the mean (in this case 35/24) is $\frac{1}{4}$ the variance of the sum (in this case 35/6). The variance of the sum is thus $\frac{1}{2}$ the variance of the parent population.

This reasoning leads in general to

$$\mu_{\bar{x}} = \mu,$$

$$\sigma_{\bar{x}}^2 = \frac{\sigma^2}{n}.$$

(See Appendix 7A for a more complete demonstration.)

An important feature of the family of normal distributions (which we cannot demonstrate without the use of calculus) is that the sum of two or more independent random variables with normal distributions has a sampling distribution which is normal. It follows that the sampling distribution of the sample mean is normal if the parent population is normal.

7.5 THE BINOMIAL DISTRIBUTION

7.5.1 Sampling Distribution of the Number of Heads

In many situations we are interested in only two possible outcomes: a person is either male or female, a stamp collector or not; a person will either die or not die during the next year; a roll of two dice will either produce a seven or not.

We can consider each of these situations with only two possible outcomes as formally equivalent to the simple operation of tossing a coin with an arbitrary probability of heads. We associate one of the two possibilities with heads, denoted H; the other, with tails, denoted T. In another context the outcomes are termed "success" and "failure," respectively.

The probability of a head (or success) is labelled p, and the probability of a tail (or failure) is q. Because p and q are probabilities, we know p and q are each between 0 and 1. Since the two events H and T are mutually exclusive and exhaustive, $p + q = 1$. If the coin is fair or unbiased, $p = q = 1/2$. We sometimes call the occurrence of a head or tail with respective probabilities p and q a *Bernoulli trial*.*

We may toss the coin twice; the possible outcomes are two heads, a head on the first toss and a tail on the second, a tail on the first toss and a head on the second, and two tails. If we observe two 80-year-olds

---

*After James Bernoulli (1654–1705), who discussed such trials in his *Ars Conjectandi* (1713).

for a year, both may die (two heads or two successes!), the first may die and the second not, the first may not die and the second may, and both may live. In each case we have two Bernoulli trials. We may consider the probability of a head (or success) on each trial to be p and we may consider the trials to be independent; this means

$$\text{Pr}\,(H \text{ on first toss and } H \text{ on second toss})$$

$$= \text{Pr}\,(H \text{ on first toss}) \times \text{Pr}\,(H \text{ on second toss}),$$

which is $p \times p = p^2$.

We often want to consider an arbitrary number of Bernoulli trials, say n of them. We may toss the coin n times. If we observe 7000 80-year-olds, for a year, then $n = 7000$, and p is the probability that any one of them dies. In a group of 100 persons, each is either a stamp collector or not; here $n = 100$, and p is the probability that an individual is a stamp collector.

The formal definition of a set of n Bernoulli trials is as follows:
   (i)   There are n trials.
   (ii) The same pair of outcomes are possible on all trials.
   (iii) The probability of a specified outcome is the same on all trials.
   (iv) The outcomes on the n trials are independent.

Many problems of statistics fit this model. Usually it is the *total number* of heads that is important; the number of heads in n trials is a *statistic*. We shall now derive the *sampling distribution* of this statistic.

Table 7.12 shows the possible outcomes when a coin is tossed once,

TABLE 7.12
Table of Possible Outcomes When Coin Is Tossed n Times

$n = 1$			T		H	

$n = 2$		T		TH		H
		T		HT		H

$n = 3$	T		TTH		THH		H
	T		THT		HTH		H
	T		HTT		HHT		H

$n = 4$	T		TTTH		TTHTHH		THHH		H
	T		TTHT		THTHTH		HTHH		H
	T		THTT		HTTHHT		HHTH		H
	T		HTTT		HHHTTT		HHHT		H

twice, three times, or four times. We let n be the number of tosses, and we say $n = 1, 2, 3, 4$. The first row of the table represents the two possible outcomes of a single toss; either we get a tail (T) or a head (H). Now we make a second toss. A tail on the first toss can be followed by a tail or by a head on the second toss, resulting in one of the two sequences

$$
\begin{array}{cc}
T & \qquad T \\
T & \qquad H
\end{array}
$$

These two sequences are listed in the left half of the two rows for $n = 2$, that is, under T in the row for $n = 1$, to emphasize that these two sequences can result from T on the first toss. A head on the first toss can also be followed by a tail or a head, resulting in one of the two sequences

$$
\begin{array}{cc}
H & \qquad H \\
T & \qquad H
\end{array}
$$

These two sequences are listed in the two rows for $n = 2$ at the right (under H in the row for $n = 1$). The two (vertical) sequences

$$
\begin{array}{c}
T\,H \\
H\,T
\end{array}
$$

are grouped together because each contains exactly one H.

For $n = 3$, there are eight possible outcomes; each is obtained by appending an H or a T to one of the four outcomes for $n = 2$. These eight possibilities are grouped to show that there is 1 possibility giving 0 heads, 3 possibilities giving exactly 1 head, 3 possibilities giving exactly 2 heads, and 1 possibility giving 3 heads. The 3 ways of getting exactly 2 heads,

$$
\begin{array}{c}
T\,H\,H \\
H\,T\,H \\
H\,H\,T
\end{array}
$$

arose from two sets of possible outcomes for $n = 2$, those which contain exactly 1 head and the one containing exactly 2 heads:

$$
\begin{array}{cc}
T\,H & \qquad H \\
H\,T & \qquad H
\end{array}
$$

Thus the number of possibilities for $n = 3$ containing exactly 2 heads (3) is the number of possibilities for $n = 2$ containing exactly 1 head (2), plus the number of possibilities for $n = 2$ containing exactly 2 heads (1).

Pascal's Triangle. These numerical results are collected into the diagram given in Table 7.13, called *Pascal's triangle.* [Blaise Pascal (1623–1662) was a famous French philosopher and mathematician.]

TABLE 7.13
Pascal's Triangle of Binomial Coefficients to $n = 8$

				1		1				
			1		2		1			
		1		3		3		1		
	1		4		6		4		1	
1		5		10		10		5		1
1	6		15		20		15		6	1
1	7	21		35		35		21	7	1
1	8	28	56		70		56	28	8	1

Note that the 10 in the fifth row is $4 + 6$, the sum of the numbers above it. The 21 in the seventh row is $6 + 15$, the sum of the numbers above it. We denote the number of ways of obtaining exactly r heads in n tosses of a coin by the symbol C_r^n because it is the number of choices ("*combinations*") of r objects out of n; others use $\binom{n}{r}$.

The results of 3 tosses of a coin can be described as a sequence of 3 symbols, each being an H or a T. There are 8 such sequences, listed in the lines for $n = 3$ in Table 7.12.

Probabilities. Now we consider the probabilities of these sequences when the probability of a head is p, the probability of a tail is $q = 1 - p$, and the trials are independent. The three sequences HHT, HTH, and THH each contain 2 heads. (We now write the sequences horizontally for the sake of compactness.) We assert that each of these sequences has probability p^2q. To see this, first consider the sequence HHT. Using the independence of the trials, we see that

$$\text{Pr}(HHT) = \text{Pr}(H) \times \text{Pr}(H) \times \text{Pr}(T) = p \times p \times q = p^2q.$$

Similarly the probability of *any* sequence containing 2 heads and 1 tail in any order will be the product of two p's and one q, which is again p^2q.

What then is the probability of obtaining 2 heads in 3 tosses of a coin with heads probability p? There are 3 sequences containing exactly

2 H's, and each has probability p^2q, so the probability is $3p^2q$ (because we have the probability of an event which is composed of the 3 mutually exclusive events *HHT*, *HTH*, and *THH*).

The probabilities of r heads in n tosses are stated in Table 7.14 for the possible values of r and for $n = 1, 2, 3, 4, 5$. The probabilities can be calculated for $n = 1, 2, 3, 4$ by noting the sequences in Table 7.12 that have exactly r heads. Since it is tiresome to extend Table 7.12 to larger values of n we now show the idea of calculating the probability of r heads in n tosses by deriving the probability of 2 heads in 4 tosses.

We saw in the development of Table 7.12 that 2 heads are obtained in 4 tosses by having 1 head in the first 3 tosses followed by a head on

TABLE 7.14

Binomial Distributions for $n = 1, 2, 3, 4, 5$

NUMBER OF TRIALS, n	NUMBER OF HEADS	PROBABILITY OF NUMBER OF HEADS
1	0	q
	1	p
2	0	q^2
	1	$2pq$
	2	p^2
3	0	q^3
	1	$3pq^2$
	2	$3p^2q$
	3	p^3
4	0	q^4
	1	$4pq^3$
	2	$6p^2q^2$
	3	$4p^3q$
	4	p^4
5	0	q^5
	1	$5pq^4$
	2	$10p^2q^3$
	3	$10p^3q^2$
	4	$5p^4q$
	5	p^5

the fourth toss and by having 2 heads in the first 3 tosses followed by a tail on the fourth toss. Thus

$$\Pr(2 \text{ heads in 4 tosses}) = \Pr(1 \text{ head in 3 tosses}) \times p$$
$$+ \Pr(2 \text{ heads in 3 tosses}) \times q$$
$$= 3pq^2 \times p + 3p^2q \times q$$
$$= (3 + 3)p^2q^2$$
$$= 6p^2q^2.$$

Note that the numerical factor 6 is the number in Pascal's triangle in position 2* of row 4, the exponent of p is 2, the number of heads, and the exponent of q is 2, the number of tails.

The probability of r heads in n tosses is the product of the number of sequences of length n with r heads (the number in Pascal's triangle in position r of row n) and the probability of r heads and $n - r$ tails in a given sequence (for example, $HH \ldots HTT \ldots T$). It is

$$\Pr(\text{exactly } r \text{ heads in } n \text{ tosses}) = C_r^n p^r q^{n-r},$$

for $r = 0, 1, 2, \ldots, n$. This is called the *binomial probability distribution*.

Mathematical Remarks. For those readers familiar with the *factorial function*, defined as $n! = n \times (n - 1) \times \cdots \times 2 \times 1$, we remark that

$$C_r^n = \frac{n!}{r!(n - r)!}.$$

If n is large it is difficult to compute C_r^n by the triangle or by this formula. Appendix IV gives the numerical values of $\Pr(x = r)$, $r = 0, 1, 2, \ldots, n$, for various n and p. To use the table for values of p greater than 1/2, use the rule $\Pr(r \text{ heads in } n \text{ tosses of a coin with heads probability } p) = \Pr(n - r \text{ heads in } n \text{ tosses of a coin with heads probability } q)$. [For extensive tables see National Bureau of Standards (1950).]

The Case $p = 1/2$. If the coin is fair ($p = 1/2$), all of the sequences of n H's and T's have the same probability. For example, when $n = 3$ there are eight sequences and each has probability 1/8. The probabilities are then

$$\Pr(0 \text{ heads}) = \Pr(TTT) = 1/8,$$

$$\Pr(1 \text{ head}) = \Pr(TTH \text{ or } THT \text{ or } HTT)$$
$$= \Pr(TTH) + \Pr(THT) + \Pr(HTT) = 3/8,$$

*There is a position 0 filled with a 1.

$$\text{Pr}\,(2 \text{ heads}) = \text{Pr}\,(THH \text{ or } HTH \text{ or } HHT)$$
$$= \text{Pr}\,(THH) + \text{Pr}\,(HTH) + \text{Pr}\,(HHT) = 3/8,$$

$$\text{Pr}\,(3 \text{ heads}) = \text{Pr}\,(HHH) = 1/8.$$

Table 7.15 gives the probabilities of different numbers of heads for $n = 1, 2, 3, 4, 5$ and $p = 1/2$. In Figure 7.15 the probabilities are graphed. Although the number of heads must be an integer, we have drawn the probability as an area centered at the number of heads.

TABLE 7.15
Binomial Distribution for $n = 1, 2, 3, 4, 5$ and $p = 1/2$

n	0	1	2	3	4	5
1	$\frac{1}{2}$	$\frac{1}{2}$				
2	$\frac{1}{4}$	$\frac{1}{2}$	$\frac{1}{4}$			
3	$\frac{1}{8}$	$\frac{3}{8}$	$\frac{3}{8}$	$\frac{1}{8}$		
4	$\frac{1}{16}$	$\frac{4}{16}$	$\frac{6}{16}$	$\frac{4}{16}$	$\frac{1}{16}$	
5	$\frac{1}{32}$	$\frac{5}{32}$	$\frac{10}{32}$	$\frac{10}{32}$	$\frac{5}{32}$	$\frac{1}{32}$

It will be observed that each histogram is centered at $\frac{1}{2}n$ and is symmetric about this point. The larger n is the more the probabilities are spread out.

The probability that a baby born is a boy is about 1/2. Table 7.15 can be interpreted as the probabilities of various numbers of boys in families of 1 to 5 children. If we observed many families of size 3, for example, we would expect no boy in about 1/8 of them and no girl in about 1/8 of them, and we would expect one boy in about 3/8 of them and one girl in about 3/8 of them.

In many applications the n "coin tosses" correspond to observations of n members of a sample. In order that condition (iv) of the definition of Bernoulli trials hold, that is, in order that the n observations be strictly independent, the sample must be chosen with replacement; but if the population size is large, the observations are nearly independent even if the sampling is without replacement, and the binomial distribution provides a good approximation to the actual distribution, the hypergeometric distribution. [See, for example, Hoel, Port, and Stone (1972).] Suppose we have a jar of $N = 10,000$ beads, of which 5,000 are

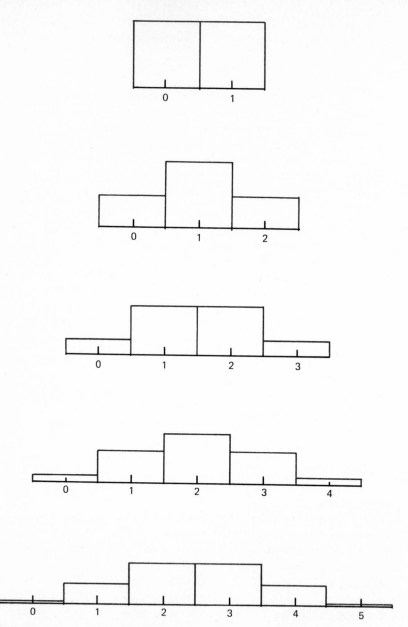

Figure 7.15 *Binomial distributions for n = 1, 2, 3, 4, 5, and p = 1/2*

red, and take a sample of $n = 2$ beads. The probability of a red bead on
the first draw in $5{,}000/10{,}000 = 1/2$, and the conditional probability of
a red bead on the second draw given that the first bead was red is

4,999/9,999 = .49994999, which is almost 1/2. The probability of 2 red beads in drawing 2 beads (without replacement) is

$$\frac{1}{2} \times \frac{4,999}{9,999} = .24997,$$

which is almost 1/4.

7.5.2 Proportion of Heads in Bernoulli Trials

Often we are interested in the *proportion* of heads in the tosses. In n tosses the proportion of heads can be $0/n = 0$, $1/n$, $2/n$, ..., or $n/n = 1$. The probability of a *proportion* being r/n is the same as the probability of the *number* of heads being r. Thus Tables 7.14 and 7.15 apply to proportions as well as numbers of heads.

When it comes to graphing these probabilities, however, there is a difference because the proportions must lie between 0 and 1. In Table

TABLE 7.16
Probability Distributions of Proportions for $n = 1, 2, 3, 4, 5$ and $p = 1/2$

$n = 1$	Proportions	0	1			
	Probabilities	$\frac{1}{2}$	$\frac{1}{2}$			
$n = 2$	Proportions	0	$\frac{1}{2}$	1		
	Probabilities	$\frac{1}{4}$	$\frac{1}{2}$	$\frac{1}{4}$		
$n = 3$	Proportions	0	$\frac{1}{3}$	$\frac{2}{3}$	1	
	Probabilities	$\frac{1}{8}$	$\frac{3}{8}$	$\frac{3}{8}$	$\frac{1}{8}$	
$n = 4$	Proportions	0	$\frac{1}{4}$	$\frac{1}{2}$	$\frac{3}{4}$	1
	Probabilities	$\frac{1}{16}$	$\frac{4}{16}$	$\frac{6}{16}$	$\frac{4}{16}$	$\frac{1}{16}$
$n = 5$	Proportions	0	$\frac{1}{5}$	$\frac{2}{5}$	$\frac{3}{5}$	$\frac{4}{5}$	1
	Probabilities	$\frac{1}{32}$	$\frac{5}{32}$	$\frac{10}{32}$	$\frac{10}{32}$	$\frac{5}{32}$	$\frac{1}{32}$

n = 1

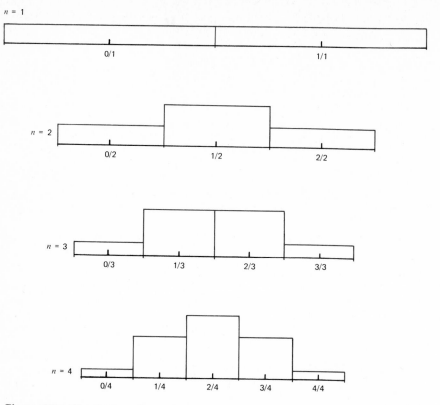

Figure 7.16 *Distribution of sample proportions for n = 1, 2, 3, 4, and p = 1/2*

7.16 we have given the probabilities of proportions of heads in 1, 2, 3, and 4 tosses of an unbiased coin; these are graphed in Figure 7.16. Each probability is represented as an *area*. For instance, the probability of the proportion 1/2 when *n* = 4 is 6/16, or 3/8. The bar is of width 1/4 (from 3/8 to 5/8) and height 3/2 to give area 1/4 × 3/2 = 3/8.* Figures 7.16 and 7.18 show that the probability of a proportion being 1/2 or near 1/2 is greater for larger *n*.

It is sometimes convenient to denote a head by a 1 and a tail by a 0. The sequence *HHT* is then written 110. In *n* trials we let x_i denote the number of heads on the *i*th trial; that is, $x_i = 1$ if there is a head on the *i*th trial and $x_i = 0$ if there is a tail on the *i*th trial. Then the sum of the x_i's,

$$\sum_{i=1}^{n} x_i,$$

*The vertical scale is not the same as the horizontal scale.

is the number of heads in the n trials. For example, in the sequence HHT, $x_1 = 1$, $x_2 = 1$, $x_3 = 0$, and

$$\sum_{i=1}^{3} x_i = x_1 + x_2 + x_3 = 1 + 1 + 0 = 2$$

is the number of heads in this sequence.

The proportion of heads in n trials is the number of heads divided by n, that is,

$$\frac{\sum_{i=1}^{n} x_i}{n},$$

which is the sample mean. Thus, any statements made about sample means apply to the proportion of heads in a number of trials. Conversely, the sampling distribution of the number of heads and proportion of heads are examples of the sampling distributions of sums and means.

7.5.3 The Mean and Variance of the Binomial Probability Distribution

The probability distribution representing a single coin toss is given in Table 7.17. The mean of this probability distribution is

$$\mu = p_1 v_1 + p_2 v_2 = p \times 1 + q \times 0 = p.$$

The variance of this probability distribution is

$$\sigma^2 = p_1(v_1 - \mu)^2 + p_2(v_2 - \mu)^2 = p(1 - p)^2 + q(0 - p)^2$$

$$= pq^2 + q(-p)^2 = pq^2 + p^2 q$$

$$= pq(q + p) = pq \times 1$$

$$= pq.$$

TABLE 7.17
Probability Distribution of a Single Coin Toss

Values	$v_1 = 1$ (= head)	$v_2 = 0$ (= tail)
Probabilities	$p_1 = p$	$p_2 = q$

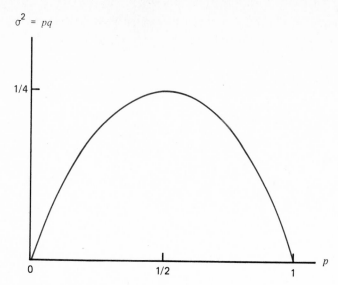

Figure 7.17 *The variance of a Bernoulli variable*

The binomial probability distribution is the distribution of a sum of independent random variables $x_1, x_2, \ldots, x_n$, each having mean p and variance pq. Hence, using the rules for the mean of a sum of variables and for the variance of a sum of independent variables, we have

$$\mu_{x_1 + x_2 + \cdots + x_n} = \mu_{x_1} + \mu_{x_2} + \cdots + \mu_{x_n} = p + p + \cdots + p = np$$

and

$$\sigma^2_{x_1 + x_2 + \cdots + x_n} = \sigma^2_{x_1} + \sigma^2_{x_2} + \cdots + \sigma^2_{x_n} = pq + pq + \cdots + pq = npq.$$

The proportion of heads in n independent Bernoulli trials shall often be denoted by $\hat{p}$. It is the sample mean of $x_1, x_2, \ldots, x_n$, where x_i has the probability distribution of Table 7.17. Thus, for the sampling distribution of $\hat{p}$

$$\mu_{\hat{p}} = p, \ \sigma^2_{\hat{p}} = \frac{pq}{n}.$$

These rules for the means and variances of sums and proportions can be verified numerically for Tables 7.15 and 7.16.

The variance of a distribution is a measure of uncertainty as to what will be the value of a random observation drawn from the distribution. In the case of a Bernoulli variable we are most uncertain about what the outcome will be if $p = 1/2$. We should expect that the variance is largest when $p = 1/2$. To see that this is the case, we write

$$\sigma^2 = pq = p(1 - p) = \left[\frac{1}{2} + \left(p - \frac{1}{2}\right)\right]\left[\frac{1}{2} - \left(p - \frac{1}{2}\right)\right] = \left(\frac{1}{2}\right)^2 - \left(p - \frac{1}{2}\right)^2$$
$$= \frac{1}{4} - \left(p - \frac{1}{2}\right)^2 \leq \frac{1}{4},$$

and note that σ^2 has its maximum value of $1/4$ when $p = 1/2$. [Here we use the rule $(a + b)(a - b) = (a^2 - b^2)$, taking $a = 1/2$ and $b = p - 1/2$.] A graph of $\sigma^2 = pq = p(1 - p)$ is given in Figure 7.17.

7.6 THE LAW OF AVERAGES (LAW OF LARGE NUMBERS)

The idea of the Law of Averages is that in the long run an event will happen about as frequently as it is supposed to, that is, according to its probability. If one tosses an unbiased coin many times, one expects heads in about half the tosses. More precisely, if an event has probability p of occurring then, when a large number of independent trials are made, the probability is great that the proportion of trials on which the event occurs is approximately p.

Table 7.16 gives the probabilities of the possible proportions of heads for $n = 1, 2, 3, 4, 5$ tosses of an unbiased coin. The probabilities that these proportions will deviate from $1/2$ by various amounts are shown in Table 7.18. For example, in 5 tosses of a coin the proportions

TABLE 7.18

Probability That the Proportion of Heads Deviates from 1/2 by Various Amounts for Various Numbers of Tosses

				n			
		1	2	3	4	5	100
Deviation of at most $\dfrac{1}{2}$		1	1	1	1	1	1
Deviation of at most $\dfrac{1}{4}$		0	1	$\dfrac{3}{4}$	$\dfrac{7}{8}$	$\dfrac{5}{8}$	1.00000
Deviation of at most $\dfrac{1}{8}$		0	$\dfrac{1}{2}$	$\dfrac{3}{4}$	$\dfrac{3}{8}$	$\dfrac{5}{8}$	.98796

of 2/5 and 3/5 are within 1/8 of 1/2 and each of these events has probability 10/32; the probability that one of these two events occurs is 5/8, as given in the last row and fifth column of Table 7.18. Also shown in Table 7.18 are the probabilities for 100 tosses, which is a large number of tosses; these are similarly computed from the binomial distribution. The probability that the proportion of heads in 100 tosses will fall between .375 and .625 is .98796; it is almost certain that the proportion will fall between .25 and .75. The calculations depend only on the probability p being 1/2 and independence of events on the n tosses. This concentration of probability near 1/2 can also be seen from Figure 7.16. If n is larger, the concentration around 1/2 is greater.

In Bernoulli trials with a probability p of a head, the probability is high that the sample proportion of heads is close to the probability p if the number of tosses n is sufficiently large. The probability p is the mean of the Bernoulli distribution when a head is given the value 1 and a tail is given the value 0 and the proportion of heads is the sample mean.

The case of independent Bernoulli trials is an example of a more general form of the law of large numbers which states that the probability is high that the sample mean is close to the population mean if the sample size is large. This fact is easily made plausible when the random variable can take on a finite number of values, $v_1, \ldots, v_m$. Then the sample mean $\bar{x}$ can be written as

$$\bar{x} = \sum_{j=1}^{m} r_j v_j,$$

where r_j is the relative frequency of the value v_j in the sample. If the sample size is large, it is likely that each relative frequency r_j will be close to the corresponding population probability p_j. Thus $\sum_{j=1}^{m} r_j v_j$ will be close to $\sum_{j=1}^{m} p_j v_j$; that is, $\bar{x}$ will be close to μ. Thus the probability is large (near 1) that the sample mean $\bar{x}$ differs by only a small amount from the population mean

$$\mu = \sum_{j=1}^{m} p_j v_j.$$

The Law of Averages is also known as the Law of Large Numbers.

LAW OF LARGE NUMBERS   *If the sample size is large, the probability is high that the sample mean is close to the mean of the parent population.*

It was stated in Section 7.4 that the mean of the sampling distribution of the sample mean $\bar{x}$ from a parent population with mean μ is that population mean μ; that is, the sampling distribution of the sample mean is located around the parent population mean. If the standard deviation of the parent population is σ, then the standard deviation of the sampling distribution of the mean of samples of size n is $\sigma/\sqrt{n}$. Thus if n is large, the variability is small and the probability that the sample mean is near μ is large.

> *Mathematical Note.* We can make a rigorous definition of the Law of Large Numbers. Let "sample mean close to the population mean" be defined as
>
> $$|\bar{x} - \mu| < a$$
>
> for an arbitrarily small positive number a (for example, 0.1). Let "the probability is high" be defined as
>
> $$\Pr(|\bar{x} - \mu| < a) > 1 - b,$$
>
> for an arbitrarily small positive number b (for example, 0.05). The Law of Large Numbers states that for given a and b the above probability statement is true if n is at least as large as some suitable number (to be calculated on the basis of a and b).

The Law of Large Numbers is important. It is one justification for using random samples to learn about populations; if the sample is large enough, the sample information is likely to be very accurate.

7.7 THE NORMAL DISTRIBUTION OF SAMPLE MEANS (CENTRAL LIMIT THEOREM)

7.7.1 The Central Limit Theorem

The Law of Large Numbers states that the sampling distribution of the sample mean of almost any parent population concentrates around the mean of that parent distribution. This is a rather crude statement; now we make the more refined assertion that the sampling distribution of a sample mean is approximately a *normal* distribution. While it is an *empirical* fact that the distributions of many measurements such as height and weight are approximately normal, it is a *mathematical* theorem, as well as an empirical fact, that the sampling distribution of a mean is approximately normal. (It has been said that physicists think the mathematicians have shown the ubiquity of the normal distribution and mathematicians think the physicists have found it by experience.)

In Section 7.5 we showed the probability distribution of the number of heads in n tosses of a coin and later we found the probability distribution of the proportion of heads. The number of heads is the sum of n random variables $\sum_{i=1}^{n} x_i$, where $x_i = 1$ if a head appears on the ith trial and $x_i = 0$ if a tail appears; the proportion is the sample mean of these variables. The standard deviation of the sampling distribution of the sum is $\sqrt{npq}$ and the standard deviation of the sampling distribution of the proportion is $\sqrt{pq/n}$.

In Figure 7.18 we draw the probability distributions of the sample proportion for $p = 1/2$ and $n = 2$, 5, and 10, respectively. It will be seen that for $n = 10$ the distribution resembles the normal distribution. For a larger n the resemblance would be still closer.

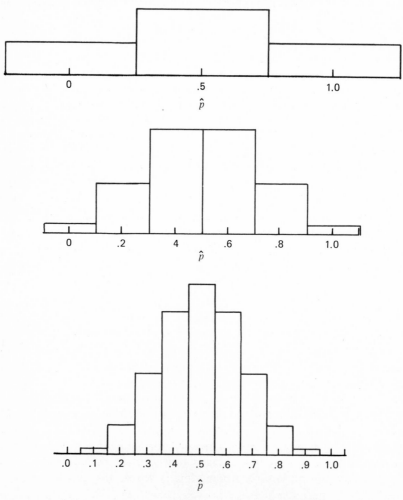

Figure 7.18 *Distributions of sample proportions for $p = 1/2$ and $n = 2, 5, 10$*

The normal distribution that approximates a given sampling distribution is the one with the same population mean and standard deviation. If we have n observations from an infinite population with mean μ and standard deviation σ, the sampling distribution of the sample mean has a mean μ and a standard deviation $\sigma/\sqrt{n}$. The normal distribution with mean μ and standard deviation $\sigma/\sqrt{n}$ approximates this sampling distribution.

Another way of putting the matter is that the *standardized version* of the sample mean,

$$\frac{\bar{x} - \mu_{\bar{x}}}{\sigma_{\bar{x}}} = \frac{\bar{x} - \mu}{\sigma/\sqrt{n}},$$

has a sampling distribution which is approximated by the standard normal distribution; that is,

$$\Pr\left(\frac{\bar{x} - \mu}{\sigma/\sqrt{n}} \leq z\right) \doteq \Phi(z),$$

where we use the sign $\doteq$ to mean "is approximately equal to." The larger n is, the better the approximation.

Figures 7.19 and 7.20 (truncated on the right) show the distributions of $\bar{x}$ for two parent populations for $n = 2, 6,$ and 25.

It should be emphasized that the theorem applies almost* regardless of the nature of the parent population, that is, almost regardless of the distribution from which $x_1, x_2, \ldots, x_n$ are a random sample. Even though the parent population in Figure 7.20 is skewed, the sample means have a distribution that is approximately normal. How large n must be to have a "good" approximation does depend, however, upon the shape of the parent population.

We can now make some general summary statements about the sampling distribution of the sample mean $\bar{x}$.

RULE 1   *The mean of the sampling distribution of $\bar{x}$ is equal to the mean μ of the parent population:*

$$\mu_{\bar{x}} = \mu.$$

RULE 2   *The variance of the sampling distribution of $\bar{x}$ is the variance of the parent population divided by the sample size:*

$$\sigma_{\bar{x}}^2 = \frac{\sigma^2}{n}.$$

---

*A complete mathematical statement of the Central Limit Theorem may be found, for example, in Hoel, Port, and Stone (1972).

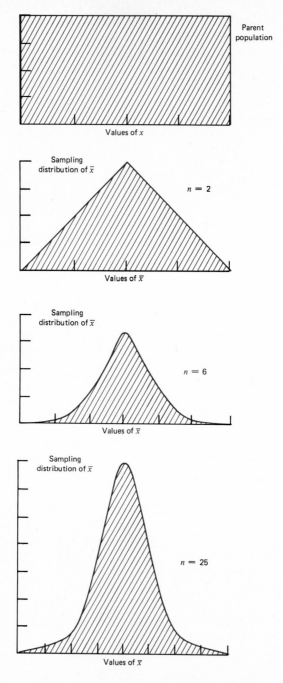

Figure 7.19 *Distributions of sample means from samples of various sizes taken from a uniform parent population*

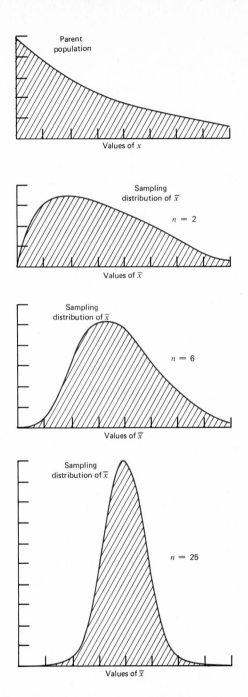

Figure 7.20 *Distributions of sample means from samples of various sizes taken from a skewed parent population*

RULE 3   *The* standard deviation *of the sampling distribution of x̄ is the standard deviation of the parent population divided by the square root of the sample size:*

$$\sigma_{\bar{x}} = \frac{\sigma}{\sqrt{n}}.$$

RULE 4   *The* sampling distribution *of the mean of a random sample from a* normal *distribution is also a normal distribution. The sampling distribution of the mean of a random sample from some other distribution is approximately a normal distribution.*

In Rules 2 and 3 it is assumed that the sampling is with replacement or from an infinite population. Modifications of these rules are given in the next chapter; they are not important if the size of the population is large relative to the size of the sample, which is usually the case.

7.7.2   Normal Approximation to the Binomial Distribution

As we pointed out earlier, the proportion of heads $\hat{p}$ in n trials is the sample mean of Bernoulli variables. Hence the Central Limit Theorem applies to $\hat{p}$ and implies that

$$\frac{\hat{p} - \mu_p}{\sigma_{\hat{p}}} = \frac{\hat{p} - p}{\sqrt{pq/n}}$$

has a distribution which is approximated by that of the standard normal variable:

$$\Pr\left(\frac{\hat{p} - p}{\sqrt{pq/n}} \leq z\right) \doteq \Phi(z).$$

The approximation is good if n is large and p is not too small or large (too near 0 or 1). (As a rule of thumb we can say that the approximation may be used when $np > 5$ and $nq > 5$ if the "correction for continuity" explained in Appendix 7B is used.)

The probability that the proportion of successes is less than or equal to .6 can be approximated as follows:

$$\Pr(\hat{p} \leq .6) = \Pr\left(\frac{\hat{p} - p}{\sqrt{pq/n}} \leq \frac{.6 - p}{\sqrt{pq/n}}\right)$$

$$\doteq \Phi\left(\frac{.6 - p}{\sqrt{pq/n}}\right).$$

What is the probability that the proportion of heads in 100 tosses of a fair coin is less than or equal to .6? We have

$$\Pr(\hat{p} \le .6) \doteq \Phi\left(\frac{.6 - .5}{\sqrt{.5 \times .5/100}}\right) = \Phi\left(\frac{.1}{.5/10}\right) = \Phi\left(\frac{.1}{.05}\right)$$

$$= \Phi(2) = .9772.$$

If we let $x = n\hat{p}$ be the number of heads, then (multiplying both numerator and denominator by n)

$$\frac{\hat{p} - p}{\sqrt{pq/\sqrt{n}}} = \frac{n(\hat{p} - p)}{n\sqrt{pq/n}} = \frac{x - np}{\sqrt{npq}} = \frac{x - \mu_x}{\sigma_x};$$

that is, the standardized version of the *number* of successes is equal to the standardized version of the *proportion* of successes, $\hat{p} = x/n$. To approximate the probability that the number of successes is less than or equal to 60, we write

$$\Pr(x \le 60) = \Pr\left(\frac{x - np}{\sqrt{npq}} \le \frac{60 - np}{\sqrt{npq}}\right)$$

$$\doteq \Phi\left(\frac{60 - np}{\sqrt{npq}}\right).$$

If $n = 100$ and $p = 1/2$, this is

$$\Phi\left(\frac{60 - 100 \times 1/2}{\sqrt{100 \times 1/2 \times 1/2}}\right) = \Phi\left(\frac{60 - 50}{1/2 \times 10}\right) = \Phi\left(\frac{10}{5}\right) = \Phi(2) = .9772.$$

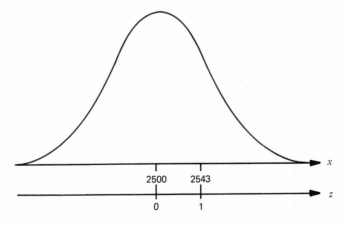

Figure 7.21 *Sampling distribution of the number of Catholics in a random sample of 10,000 from a population in which the proportion p of Catholics is .25*

We obtained the same answer as before because the proportion of successes is .6 if the number of successes in 100 tosses is 60.

EXAMPLE   Suppose the population is 25% Catholic (Figure 2.3). In a sample of 10,000, we expect about 2500 Catholics. What is the probability that the number of Catholics in a sample of $n = 10,000$ will be less than or equal to 2540? Let x be the number of Catholics in the sample. Here $\mu = np = 10,000 \times .25 = 2,500$, $\sigma = \sqrt{npq} = \sqrt{10,000 \times .25 \times .75} = \sqrt{10,000} \sqrt{.25} \sqrt{.75} = 100 \times .500 \times .866 = 100 \times .433 = 43.3$. The value 2540 is about one σ unit greater than 2500; that is, $x = 2543.3$ corresponds to $z = +1$, and $\Pr(x \leq 2540) \doteq \Pr(x \leq 2543.3) \doteq \Phi(1) = .841$.

For a more accurate refinement of this use of the normal approximation see Appendix 7B.

SUMMARY

Probability Distributions.   When an individual is drawn at random from a population, a measurement on that individual is a *random variable*. The set of probabilities of the possible values of the random variable constitutes a *probability distribution*. Random variables and probability distributions can also be defined mathematically.

One family of probability distributions is made up of the *uniform* distributions, which describe, for example, the probabilities involved in "choosing a number at random." One family of theoretical distributions is the family of *normal* distributions, which approximate the empirical distributions of many natural variables. The member of the family is defined by the mean and standard deviation (or, equivalently, the variance).

About two-thirds (68.2%) of a normal population have values within one standard deviation of the mean. The *standard normal distribution* is the distribution having a mean of 0 and a standard deviation of 1.

A *random sample* of size n is a subset of n individuals obtained from the population in such a way that all groups of that size have the same chance of being the actual sample. In particular, this implies that all individuals have the same chance of being included in the actual sample.

Sampling Distributions.   A *statistic* is a characteristic of a sample, such as its mean, median, range, or standard deviation.

A *sampling distribution* is the probability distribution of a statistic. It gives the probability associated with each possible value of the statistic. In this context the distribution from which the sample is drawn is called the *parent distribution*.

The sampling distribution of the number of "heads" in n independent trials of the same two-outcome (heads/tails) experiment is the *binomial distribution*. It gives the probability of obtaining exactly r heads in such an experiment, $r = 0, 1, 2, \ldots, n$.

Sample Means. The *Law of Averages* (Law of Large Numbers) states that in large samples the mean of a random sample has a good chance of being close to the population mean.

The mean of the sampling distribution of the mean is the population mean. The variance of the sampling distribution of the mean is the variance of the parent population divided by the sample size.

Mathematically speaking, the mean of a random sample from a normal parent population has a normal distribution. More generally, almost regardless of the shape of the parent distribution, the mean of a random sample has a sampling distribution which is approximated by a normal distribution. This fact is known as the *Central Limit Theorem*.

In fact, the normal distribution plays a central role in probability and statistics, not only because many natural phenomena have distributions that are well approximated by normal distributions, but also because many statistics (not just the sample mean) have sampling distributions which are well approximated by normal distributions, almost regardless of the shape of the parent distribution.

The proportion of heads in n independent coin-toss trials is an example of the mean of a sample.

APPENDIX 7A   Rules for the Mean and Variance of a Sum and of a Coded Variable

In this appendix we shall prove that the mean of the sampling distribution of the sum of random variables is the sum of the means of those random variables and that the variance of the sum is the sum of the variances if the random variables are independent. We consider only random variables which have a finite number of values. We shall use μ_x to denote the mean of the random variable x and σ_x^2 to denote the variance of the random variable x.

RULE 1

$$\mu_{x_1 + x_2} = \mu_{x_1} + \mu_{x_2}.$$

PROOF: Suppose x_1 takes the values $v_1, v_2, \ldots, v_{m_1}$ and x_2 takes the values $w_1, w_2, \ldots, w_{m_2}$. Then

$$\mu_{x_1 + x_2} = \sum_{j=1}^{m_1} \sum_{k=1}^{m_2} \Pr(x_1 = v_j \text{ and } x_2 = w_k)(v_j + w_k)$$

$$= \sum_{j=1}^{m_1} \sum_{k=1}^{m_2} \Pr(x_1 = v_j \text{ and } x_2 = w_k)v_j$$

$$+ \sum_{k=1}^{m_2} \sum_{j=1}^{m_1} \Pr(x_1 = v_j \text{ and } x_2 = w_k)w_k$$

$$= \sum_{j=1}^{m_1} \Pr(x_1 = v_j)v_j + \sum_{k=1}^{m_2} \Pr(x_2 = w_k)w_k$$

$$= \mu_{x_1} + \mu_{x_2}.$$

RULE 2

$$\mu_{y_1 + y_2 + y_3} = \mu_{y_1} + \mu_{y_2} + \mu_{y_3}.$$

PROOF: In the preceding rule take $x_1 = y_1$ and $x_2 = y_2 + y_3$. Then

$$\mu_{y_1 + y_2 + y_3} = \mu_{y_1 + (y_2 + y_3)}$$

$$= \mu_{y_1} + \mu_{y_2 + y_3}$$

$$= \mu_{y_1} + (\mu_{y_2} + \mu_{y_3})$$

$$= \mu_{y_1} + \mu_{y_2} + \mu_{y_3}.$$

RULE 3

$$\mu_{x_1 + x_2 + \cdots + x_n} = \mu_{x_1} + \mu_{x_2} + \cdots + \mu_{x_n}.$$

PROOF: Just as the case $n = 3$ reduced to the case $n = 2$, the case $n = 4$ reduces to the case $n = 3$, etc., so the rule is true for every value of n.

Note that there is no requirement that the random variables be independent for these three rules to hold.

In what follows we shall need to consider the means of the sampling distributions of statistics such as $(x_1 + x_2)^2$. This mean is written $\mu_{(x_1 + x_2)^2}$. Such notation is rather cumbersome, and instead we shall use $E[(x_1 + x_2)^2]$; the symbol E came to be used in this context because the mean is the "expected value."

RULE 4   *If x_1 and x_2 are independent,*

$$\sigma^2_{x_1 + x_2} = \sigma^2_{x_1} + \sigma^2_{x_2}.$$

PROOF:

$$
\begin{aligned}
\sigma^2_{x_1 + x_2} &= E[(x_1 + x_2) - \mu_{x_1 + x_2}]^2 \\
&= E[x_1 + x_2 - (\mu_{x_1} + \mu_{x_2})]^2 \\
&= E[(x_1 - \mu_{x_1}) + (x_2 - \mu_{x_2})]^2 \\
&= E[(x_1 - \mu_{x_1})^2 + (x_2 - \mu_{x_2})^2 + 2(x_1 - \mu_{x_1})(x_2 - \mu_{x_2})] \\
&= E(x_1 - \mu_{x_1})^2 + E(x_2 - \mu_{x_2})^2 + 2E(x_1 - \mu_{x_1})(x_2 - \mu_{x_2}) \\
&= \sigma^2_{x_1} + \sigma^2_{x_2} + 2E(x_1 - \mu_{x_1})(x_2 - \mu_{x_2}).
\end{aligned}
$$

The rule follows because of the following lemma.*

LEMMA 1   *If x_1 and x_2 are independent,*

$$E(x_1 - \mu_{x_1})(x_2 - \mu_{x_2}) = 0.$$

PROOF: The independence of x_1 and x_2 implies that

$$\Pr(x_1 = v_j \text{ and } x_2 = w_k) = \Pr(x_1 = v_j) \times \Pr(x_2 = w_k).$$

Therefore, the mean of $(x_1 - \mu_{x_1})(x_2 - \mu_{x_2})$ is

$$
\begin{aligned}
E(x_1 - \mu_{x_1})&(x_2 - \mu_{x_2}) \\
&= \sum_{j=1}^{m_1} \sum_{k=1}^{m_2} (v_j - \mu_{x_1})(w_k - \mu_{x_2}) \Pr(x_1 = v_j) \times \Pr(x_2 = w_k) \\
&= \sum_{j=1}^{m_1} (v_j - \mu_{x_1}) \Pr(x_1 = v_j) \times \sum_{k=1}^{m_2} (w_k - \mu_{x_2}) \Pr(x_2 = w_k) \\
&= 0.
\end{aligned}
$$

Remark.   The quantity $E(x_1 - \mu_{x_1})(x_2 - \mu_{x_2})$ is the *covariance* between x_1 and x_2, denoted by $\text{Cov}(x_1, x_2)$.

RULE 5   *If $\text{Cov}(x_1, x_2) = 0$, then*

$$\sigma^2_{x_1 + x_2} = \sigma^2_{x_1} + \sigma^2_{x_2}.$$

*A *lemma* is a rule proved for use in the proof of another rule.

Rule 5 differs from Rule 4 because it is possible that $\text{Cov}(x_1, x_2) = 0$ without x_1 and x_2 being independent.

RULE 6   *If* $\text{Cov}(x_i, x_j) = 0$ *for* $i \neq j$, $i = 1, \ldots, n$, $j = 1, \ldots, n$, *then*

$$\sigma^2_{x_1 + x_2 + \cdots + x_n} = \sigma^2_{x_1} + \sigma^2_{x_2} + \cdots + \sigma^2_{x_n}.$$

PROOF: The proof for the case $n = 3$ follows from the fact that $\text{Cov}(x_1, x_2 + x_3) = \text{Cov}(x_1, x_2) + \text{Cov}(x_1, x_3)$. Then the proof for $n = 4$ reduces to the case $n = 3$, $n = 5$ reduces to $n = 4$, etc.

COROLLARY   *If* $x_1, x_2, \ldots, x_n$ *are independent, then*

$$\sigma^2_{x_1 + x_2 + \cdots + x_n} = \sigma^2_{x_1} + \sigma^2_{x_2} + \cdots + \sigma^2_{x_n}.$$

PROOF: $x_1, x_2, \ldots, x_n$ independent implies $\text{Cov}(x_i, x_j) = 0$ for $i \neq j$. Hence Rule 6 applies.

Mean, Variance, and Standard Deviation of a Coded Random Variable. The mean of cx is $c\mu_x$ and the variance of cx is $c^2\sigma^2_x$. In symbols we write

$$\mu_{cx} = c\mu_x \quad \text{and} \quad \sigma^2_{cx} = c^2\sigma^2_x.$$

When x takes the value v_j, the variable cx takes the value cv_j, so to prove the first equality, for example, we write

$$\mu_{cx} = \sum_{j=1}^{m} p_j(cv_j) = c \sum_{j=1}^{m} p_j v_j = c\mu_x.$$

The standard deviation of cx is $|c|$ times the standard deviation of x:

$$\sigma_{cx} = |c|\sigma_x.$$

Similarly

$$\mu_{y+d} = \mu_y + d, \qquad \sigma^2_{y+d} = \sigma^2_y, \qquad \sigma_{y+d} = \sigma_y.$$

In words, if we add an amount d to a variable, the mean is increased by the amount d; but the variance and standard deviation, being measures of variability and not location, are unchanged.

If we combine these equalities by taking $y = cx$, we obtain

$$\mu_{cx+d} = c\mu_x + d, \qquad \sigma^2_{cx+d} = c^2\sigma^2_x, \qquad \sigma_{cx+d} = |c|\sigma_x.$$

Mean, Variance, and Standard Deviation of the Sample Mean. A particular case of coding occurs when x is the sum of n observations, $c = 1/n$, and $d = 0$; then y is the sample mean.

APPENDIX 7B   The Correction for Continuity

The heights of 11-year-old boys were recorded in Table 2.11 to the nearest inch and were graphed in Figure 2.14 in a histogram in which the bars are of width one inch. In Figure 2.15 (and Figure 7.1) the histogram is repeated together with a continuous probability distribution which approximates it. This continuous distribution is a normal distribution that has the same mean and standard deviation as the histogram. In both cases area represents relative frequency or probability; the entire area in the histogram or under the curve is 1. The most marked difference between the histogram and the normal curve is that the former is discrete. This fact should be taken into account when the normal curve is used to approximate the relative frequencies or probabilities of the histogram.

In the histogram the relative frequency of 54 inches, for example, is .1446 and is represented by the area of the bar over the interval 53.5 to 54.5, the area of the entire histogram being 1. The relative frequency of heights up to and including 54 inches is the area to the left of 54.5, namely .7393; it is the sum of the areas of the histograms to the left of 54.5. The area under the normal curve that approximates this area of the histogram is the area over the same part of the horizontal axis, namely to the left of 54.5. We can calculate this area under the normal curve by referring to Appendix I. The mean of the distributions is 52.9 and the standard deviation is 2.46. Therefore the value 54.5 is $(54.5 - 52.9)/2.46 = 1.6/2.46 = 0.65$ standard deviations from the mean. From Appendix I we find that the area to the left of this point is .7422, which is close to the observed relative frequency, .7393.

If, instead, we calculate the area under the normal curve to the left of 54, we obtain .6700, which is not as close to .7393. The reason for obtaining a poorer approximation is seen in Figure 7.1; the area under the normal curve to the left of 54 approximates the area of the histogram also to the left of 54, thus omitting half of the bar representing a height of 54 inches.

To compute the normal approximation to the probability of a height of 55 or 56 inches, we note that in a histogram these heights are represented by two bars whose bases extend from 54.5 to 56.5. Therefore we find the normal approximation to the probability of a height less than 56.5 and subtract from it the normal approximation to the probability of a height less than 54.5. The height 56.5 is $(56.5 - 52.9)/2.46 = 1.46$ standard deviations above the mean. From Appendix I we find that the probability corresponding to 1.46 is .9279. The height 54.5 is 0.65 standard deviations above the mean, and the corresponding probability is .7422. The difference is $.9279 - .7422 = .1857$. This is close to the actually observed relative frequency (from Table 2.11), which is $.1206 + .0742 = .1948$.

The general idea is that the area under the smooth normal curve should be calculated in the way best to approximate the relevant area of the histogram. The problem arises because of the fact that the histogram represents a discrete variable (in this case, height *to the nearest inch*); the above resolution of this problem is called *correction for continuity*.

This idea applies also to the use of the normal distribution to approximate theoretical probability distributions. In the approximation to the binomial distribution, the continuity correction is made by adding or subtracting 1/2 from the number of successes, as appropriate. For example, the probability that the number of heads is less than or equal to 60 is approximated by the area to the left of 60.5 under the normal curve with mean np and standard deviation $\sqrt{npq}$. If $n = 100$ and $p = 1/2$, we have

$$\Pr(x \le 60) = \Pr(x \le 60.5)$$

$$= \Pr\left(\frac{x - 50}{5} \le \frac{60.5 - 50}{5}\right)$$

$$= \Pr\left(\frac{x - 50}{5} \le 2.1\right)$$

$$\doteq \Phi(2.1)$$

$$= .9821,$$

as compared to the exact probability of .9824. The normal approximation without the continuity correction is .9772, an error of .0048 instead of .0003. Of course, we obtain the same approximation to the probability that the number of heads is less than 61, for we write

$$\Pr(x < 61) = \Pr(x \le 60.5)$$

and continue as before.

If $n = 10$ and $p = 1/2$ (so $np = nq = 5$), $\Pr(x \le 4) = .3770$ and the normal approximation with continuity correction is .3759; and $\Pr(x \le 2) = .0547$ and the normal approximation with the continuity correction is .0579. The normal approximation without the continuity correction gives .2635 and .0289, respectively; these are not satisfactory.

In general, when a continuous distribution is used to approximate the distribution of a discrete variable, the area of the continuous distribution that is closest to the area of the histogram is calculated. Usually this means calculating to a value halfway between the two adjacent discrete values.

CLASS EXERCISES

7.1 Exercise 6.1 called for 25 tosses of two distinguishable coins by each student. Each student should report the number of heads obtained in each of the two sequences of 25 tosses. The results are to be compiled and recorded in a table like the one below. Compute the relative frequencies and the cumulative relative frequencies.

NUMBER OF HEADS OBSERVED IN 25 TOSSES OF A COIN	FREQUENCY	RELATIVE FREQUENCY	CUMULATIVE RELATIVE FREQUENCY	EXPECTED RELATIVE FREQUENCY	EXPECTED CUMULATIVE RELATIVE FREQUENCY
5				.002	.002
6				.005	.007
7				.015	.022
8				.032	.054
9				.061	.115
10				.097	.212
11				.133	.345
12				.155	.500
13				.155	.655
14				.133	.788
15				.097	.885
16				.061	.946
17				.032	.978
18				.015	.993
19				.005	.998
20				.002	1.000
				1.000	

7.2 (continuation) Compile the observed relative frequencies of the deviations from 1/2 as reported by members of the class.

SIZE OF DEVIATION FROM 1/2	OBSERVED RELATIVE FREQUENCY	EXPECTED RELATIVE FREQUENCY
to at most 1/2		
to at most 1/4		
to at most 1/8		
to at most 1/16		

7.3 (continuation)   Let x be the number of heads occurring in 25 tosses, and consider the following events.
   (a) 16 or fewer heads occur, that is, $x \leq 16$.
   (b) 10 or fewer heads occur, that is, $x \leq 10$.
   (c) 14 or more heads occur, that is, $x \geq 14$.
   (d) No more than 16 and no fewer than 14 heads occur, that is, $14 \leq x \leq 16$.
For each of these events, give
      (1) the observed relative frequency of occurrence (from Exercise 7.1);
      (2) the normal approximation to the probability, without the correction for continuity;
      (3) the normal approximation to the probability, with the correction for continuity (Appendix 7B).

   EXERCISES

   7.4 (Sec. 7.3.2)   Compute the sampling distribution of the mean for samples of size 3, drawn without replacement from the population of Table 7.6.

   7.5 (Sec. 7.5)   Compute the probability of getting 3 heads and 3 tails in 6 flips of a fair coin. Check your answer in Appendix IV.

   7.6 (Sec. 7.5)   Compute the probability of getting 4 heads and 4 tails in 8 flips of a fair coin. Check your answer in Appendix IV.

   7.7 (Sec. 7.5)   From Appendix IV, find the probability of getting exactly 2 fives in 6 rolls of a balanced die.

   7.8 (Sec. 7.5)   From Appendix IV, find the probability of getting no more than 2 fives in 6 rolls of a balanced die.

   7.9 (Sec. 7.5)   From Appendix IV, find the probability of getting exactly 2 fives in 5 rolls of a balanced die.

   7.10 (Sec. 7.5)   From Appendix IV, find the probability of getting no more than 2 fives in 5 rolls of a balanced die.

   7.11 (Sec. 7.5)   If in a multiple-choice test there are 8 questions with 4 answers to each question, what is the probability of getting 3 or more correct answers just by chance alone?

   7.12 (Sec. 7.5)   If in a multiple-choice test there are 10 questions with 4 answers to each question, what is the probability of getting 3 or more correct answers just by chance alone?

7.13 (Sec. 7.5)   The probability of recovery from a certain disease is .75. Find the distribution of the number of recoveries among 3 patients and draw the histogram of this binomial distribution.

7.14 (Sec. 7.5)   The probability of a patient recovering from a certain disease is .75. Find the distribution of the number of recoveries among 4 patients and draw the histogram of this binomial distribution.

7.15 (Sec. 7.2)   Using Appendix I, find the areas (probabilities) corresponding to the following intervals.

(a) 0 to 1.65,   (c) −1.30 to 0,
(b) 1.95 to ∞,   (d) −∞ to 1.00.

7.16 (Sec. 7.2)   Using Appendix I, find the areas (probabilities) corresponding to the following intervals.

(a) 0 to 1.25,   (c) −1.50 to 0,
(b) 1.65 to ∞,   (d) −∞ to 2.00.

7.17 (Sec. 7.2)   The College Entrance Examination scores are approximately normally distributed with a mean of 500 and a standard deviation of 100 for a particular subgroup of examinees (those who are high school students taking a college preparatory course). For purposes of this exercise, we shall treat this distribution as if it were continuous and not use the continuity correction, which would give somewhat greater accuracy. Let y represent the score on the College Entrance Examination. Find

(a) the probability of a score less than 650, that is, $\Pr(y < 650)$;
(b) the probability of a score between 550 and 650, that is, $\Pr(550 < y < 650)$;
(c) the probability of a score between 400 and 550, that is, $\Pr(400 < y < 550)$;
(d) the probability of a score less than 304 or greater than 696, that is, $\Pr(y < 304) + \Pr(y > 696)$;
(e) the probability of a score greater than 665, that is, $\Pr(y > 665)$;
(f) the score such that the probability of exceeding it is .0668, that is, the value v of y such that $\Pr(y > v) = .0668$. What percentile is v?

7.18 (Sec. 7.2)   Scores on a certain IQ test are approximately normally distributed with a mean of 100 and a standard deviation of 20 for a certain population of school children. For purposes of this exercise, treat this distribution as continuous, ignoring the continuity correction. Find the proportion of the population having a score

(a) less than 130,
(b) between 110 and 130,
(c) between 80 and 110,
(d) less than 61 or greater than 139,
(e) greater than 133.

Find the score which is the 90th percentile of this IQ distribution.

7.19 (Sec. 7.2)   You are flying from New York to San Francisco and must change planes in Omaha. The New York to Omaha flight is scheduled to arrive at 9:00 PM. You hold a reservation for a flight which leaves promptly at 10:00 PM, and the only other flight out of Omaha that evening leaves promptly at 10:30. While on the flight to Omaha you read that your airline has been cited by the FAA because the arrival times of its New York to Omaha flight have a probability distribution which is normal with mean debarkation time 9:15 and standard deviation of 30 minutes. Assume that you have no baggage and that to change flights takes no time.

(a) What is the probability that you make the 10:00 connection?

(b) What is the probability that you miss the 10:00 connection, but make the 10:30 connection?

(c) What is the probability that you spend the night in Omaha?

TABLE 7.19

Heights of 12-Year-old Boys

HEIGHT	FREQUENCY	RELATIVE FREQUENCY
45.5–46.5	1	.0008
46.5–47.5	3	.0023
47.5–48.5	7	.0055
48.5–49.5	13	.0103
49.5–50.5	31	.0247
50.5–51.5	73	.0582
51.5–52.5	111	.0886
52.5–53.5	176	.1404
53.5–54.5	189	.1508
54.5–55.5	198	.1580
55.5–56.5	162	.1293
56.5–57.5	106	.0846
57.5–58.5	77	.0615
58.5–59.5	49	.0391
59.5–60.5	31	.0247
60.5–61.5	10	.0080
61.5–62.5	9	.0072
62.5–63.5	4	.0032
63.5–64.5	1	.0008
64.5–65.5	1	.0008
65.5–66.5	1	.0008
	1253	

SOURCE: Bowditch (1877).

7.20 The frequencies of heights of 1253 12-year-old boys is given in Table 7.19. Graph the histogram of this distribution.

7.21 (continuation) Compare this histogram with Figure 7.1, which is for 11-year-old boys.

7.22 (Sec. 7.4) Find the probability distribution of the sum of the values on three dice by using the probability distribution of the sum of the values on two dice (given in Table 7.11) and the value on one die. Graph the probabilities. Does the graph look more normal than Figure 7.14? Verify that the mean is $3 \times 3\frac{1}{2}$ and the variance is $3 \times 35/12$.

7.23 (Sec. 7.2.4) Find the range of standard scores corresponding to each letter grade according to grading policy I in Table 7.20.

7.24 (Sec. 7.2.4) Find the range of standard scores corresponding to each letter grade according to grading policy II in Table 7.20.

7.25 (Sec. 7.2.4) Find the range of standard scores corresponding to each letter grade according to grading policy III in Table 7.20.

7.26 (Sec. 7.2.4) Find the range of standard scores corresponding to each letter grade according to grading policy IV in Table 7.20.

TABLE 7.20
Grading Policies

POLICY No.	PERCENTAGE OF STUDENTS RECEIVING GRADE				
	A	B	C	D	F
I	15%	20%	30%	20%	15%
II	10	20	40	20	10
III	10	40	40	0	10
IV	30	30	30	0	10

*7.27 The cards in a deck are dealt to Foster, Phil, Gordon, and you, thirteen each, as in bridge. Phil is your partner, and you and he have 7 of the clubs. That leaves 6 clubs split between Gordon and Foster. What is the probability that Gordon and Foster each have 3 of them?

*7.28 (continuation) What is the probability
(a) that Foster has 2 clubs and Gordon has 4?
(b) that there is a 4–2 split, that is, that Foster has 2 of the 6 clubs or that Gordon has 2 of them?

*7.29 (continuation) What is the probability of a 3–3 split or a 4–2 split? (That is, what is the probability of the event which consists of a 3–3 split *or* a 4–2 split?)

*7.30 The cards in a deck are dealt to Foster, Phil, Gordon, and you, thirteen each, as in bridge. Phil is your partner, and you and he have 9 of the clubs. That leaves 4 clubs split between Gordon and Foster. What is the probability that Gordon and Foster each have 2 clubs?

*7.31 (continuation) What is the probability
  (a) that Foster has 1 of the 4 clubs and Gordon has 3 of them?
  (b) that there is a 3–1 split, that is, that Foster has 3 of them or that Gordon has 3 of them?

*7.32 (continuation) What is the probability of a 2–2 split or a 3–1 split? (That is, what is the probability of the event which consists of a 2–2 split *or* a 3–1 split?)

7.33 From Exercise 7.13 find the probability that the proportion of recoveries differs from .75 by not more than (a) 1/2, (b) 1/4, (c) 5/12, and (d) 3/4.

7.34 From Exercise 7.14 find the probability that the proportion of recoveries differs from .75 by not more than (a) 0, (b) 1/4, (c) 1/2, and (d) 3/4.

7.35 Calculate the mean and standard deviation of the distribution of the number of recoveries in Exercise 7.13 by Section 7.1.3.

7.36 Calculate the mean and standard deviation of the distribution of the number of recoveries in Exercise 7.14 by Section 7.1.3.

MATHEMATICAL EXERCISES

7.37 (Sec. 7.3.1) Four balls in an urn are marked *a, b, c, d*. List all the different unordered samples of size 2, assuming the sampling is done without replacement.

7.38 (Sec. 7.3.1) Five balls in an urn are marked *a, b, c, d, e*. List all the different unordered samples of size 2, assuming the sampling is done without replacement.

7.39 (Sec. 7.3.1) In the example in the text about drawing a sample of 2 among 4 boys, show that the use of random digits does in fact give each of the six possible samples a probability of 1/6.

7.40 (Sec. 7.5.3) Show directly from the definition of the mean that the mean of a binomial distribution is equal to np. [Hint: Note that $rC_r^n = nC_{r-1}^{n-1}$.]

7.41 For a variable taking values $0, 1, 2, \ldots, n$ with probabilities $p_0, p_1, p_2, \ldots, p_n$, define the kth *moment* by $\mu_k = \Sigma_{r=0}^n r^k p_r$ and the kth *factorial moment* by

$$\alpha_k = \sum_{r=0}^n r(r-1) \ldots (r-k+1)p_r.$$

(a) Show that $\alpha_2 = \mu_2 - \mu$.
(b) Using (a), show that $\sigma^2 = \alpha_2 + \mu - \mu^2$.

7.42
(a) Show from its definition that the value of α_2 for the binomial distribution is $n(n-1)p^2$.
(b) Using (a) and 7.41(b) show that for the binomial distribution $\sigma^2 = npq$.

REFERENCES

Bowditch, H. P. (1877). "The growth of children." *Report of the Board of Health of Massachusetts*, VIII.

Computation Laboratory, Harvard University (1955). *Tables of the Cumulative Binomial Probability Distribution.* Harvard University Press, Cambridge.

Hoel, Paul G., Sidney C. Port, and Charles J. Stone (1972). *Introduction to Probability Theory.* Houghton Mifflin, Boston.

National Bureau of Standards (1950). *Tables of the Binomial Probability Distribution.* Applied Mathematics Series 6. U.S. Department of Commerce, Washington.

U.S. Bureau of the Census (1970), *Statistical Abstract of the United States.* U.S. Department of Commerce, Washington.

Zehna, P. W., and D. R. Barr (1970). *Tables of the Common Probability Distributions.* U.S. Naval Postgraduate School, Monterey, California.

PART FOUR
STATISTICAL INFERENCE

8 USING A SAMPLE TO ESTIMATE POPULATION CHARACTERISTICS

INTRODUCTION

One of the main objectives of statistical inference is the *estimation* of population characteristics on the basis of the information in a sample. Responses in political polls are used to estimate the proportions of voters in favor of different candidates. The statistical evaluation of the polio vaccine trial involved the estimation of proportions of polio cases among those inoculated with vaccine and those inoculated with placebo. Thus, a sample proportion is an estimate of the population proportion; the arithmetic mean of a sample is an estimate of the population mean; and the sample variance is an estimate of the population variance. In this chapter we shall discuss how well these estimates, as well as estimates of medians and other percentiles, may correspond to the population quantities. We shall discuss not only estimation procedures in which the statistician states a single best guess as to the value of a parameter, but also procedures in which the statistician states a range of plausible values for the parameter.

8.1 ESTIMATION OF A MEAN BY A SINGLE NUMBER

We introduce the idea of estimating a population characteristic by discussing how a patient's true cholesterol level might be estimated from several cholesterol determinations on a single blood sample. Because there is variability among cholesterol measurements due to the technique and equipment used, the true cholesterol level is considered to be the mean of a hypothetical population of repeated determinations of cholesterol concentration in a single blood sample. Three determinations are treated as a random sample from this population. Suppose the values (milligrams of cholesterol in 100 milliliters of blood) obtained are 215, 222, and 205 units. Then the true cholesterol level of the patient is estimated by the arithmetic sample mean,

$$\frac{215 + 222 + 205}{3} = 214 \text{ units.}$$

Suppose that from previous extensive testing of the procedure used here it is known that a population of many repeated determinations of blood cholesterol levels has a standard deviation of 12.6 units. Then, from Chapter 7 we know that the standard deviation of the mean of three independent determinations is $12.6/\sqrt{3} = 12.6/1.73 = 7.28$. The sample mean is thus a more precise estimate of the true value than is a single observation, which has a standard deviation of 12.6.

Analogy Between Population Mean and Sample Mean. Now we discuss these ideas in a more general way. Consider a variable which in a population of N individuals has the values $y_1, y_2, \ldots, y_N$. The population mean μ of this variable is defined by the arithmetic operations expressed in the formula

$$\mu = \frac{\sum\limits_{k=1}^{N} y_k}{N}.$$

Given a sample of n individuals with observations $x_1, x_2, \ldots, x_n$, it is natural to estimate the population mean μ by applying the same arithmetic operations to the sample; this produces the estimate

$$\bar{x} = \frac{\sum\limits_{i=1}^{n} x_i}{n},$$

the arithmetic mean of the sample. That is, we use the sample mean $\bar{x}$ to estimate the population mean μ because $\bar{x}$ is constructed from the sample values in the same way that μ is constructed from population values.

Sampling Distribution of the Sample Mean. From sample to sample the arithmetic mean will vary; that is, it has a sampling distribution. This sampling distribution has a mean $\mu_{\bar{x}}$ which is equal to μ, the mean of the parent population. Thus the sampling distribution of $\bar{x}$ is centered on the parameter being estimated, μ.

The variance of the sampling distribution of $\bar{x}$ is $\sigma_{\bar{x}}^2$, which is related to the variance of the parent population by

$$\sigma_{\bar{x}}^2 = \frac{\sigma^2}{n} .$$

The standard deviation of the sampling distribution of the mean is thus

$$\sigma_{\bar{x}} = \frac{\sigma}{\sqrt{n}} .$$

The sampling distribution tends to be concentrated near the value of the parameter being estimated because the variability is reduced.

The parent population of cholesterol determinations is known from previous testing to be approximately normally distributed, and the sampling distribution of the sample mean is close to a normal distribution; therefore, since the shape of a *normal* distribution depends only upon its mean and standard deviation, the mean μ and standard deviation $\sigma/\sqrt{n}$ describe the sampling distribution well.

Unbiasedness. Because $\mu_{\bar{x}} = \mu$, we say that $\bar{x}$ is an "unbiased" estimate of μ.

DEFINITION *A statistic is an* unbiased estimate *of a parameter if the mean of the sampling distribution of the statistic is the parameter.*

When an estimate is unbiased, its sampling distribution has the desirable property of being *centered* at the parameter, in the sense that the *mean* of that sampling distribution is equal to the parameter. In this chapter we shall also discuss unbiased estimates of parameters other than a mean, such as a variance.

The Standard Error. The standard deviation of the sampling distribution of $\bar{x}$, namely $\sigma/\sqrt{n}$, is called the *standard error* of $\bar{x}$. The larger the sample size, the smaller the standard error and the more the probability is concentrated around the parameter, μ.

Summary. To sum up, we use $\bar{x}$ to estimate μ because it is formed by applying the same operations to the sample values as are applied to the population values to form μ. Further, we can say that $\bar{x}$ is unbiased and has a standard deviation which is smaller than that of a single observation. These facts imply that $\bar{x}$ has a good chance of being close to μ.

8.2   ESTIMATION OF VARIANCE
AND STANDARD DEVIATION

If the standard deviation is not known, it must be estimated from the sample. If the weights of a random sample of three players from a large list of football players are $x_1 = 225$, $x_2 = 232$, and $x_3 = 215$ pounds, then the sample mean is $\bar{x}_1 = (225 + 232 + 215)/3 = 224$, and the sample variance is

$$
\begin{aligned}
s^2 &= \frac{\sum\limits_{i=1}^{3} (x_i - \bar{x})^2}{3 - 1} \\[2mm]
&= \frac{(225 - 224)^2 + (232 - 224)^2 + (215 - 224)^2}{2} \\[2mm]
&= \frac{1^2 + 8^2 + (-9)^2}{2} \\[2mm]
&= \frac{1 + 64 + 81}{2} \\[2mm]
&= \frac{146}{2} \\[2mm]
&= 73.
\end{aligned}
$$

The sample standard deviation, s, is $\sqrt{73}$, or 8.54. This is an estimate of σ, the population standard deviation.

Analogy Between Population Variance and Sample Variance: Unbiasedness.   Consider again a variable which in a population of N individuals has the values $y_1, y_2, \ldots, y_N$. The population variance σ^2 of this variable is defined by the arithmetic operations expressed in the formula

$$
\sigma^2 = \frac{\sum\limits_{k=1}^{N} (y_k - \mu)^2}{N},
$$

where the population mean μ is as defined in the previous section. Given observations $x_1, x_2, \ldots, x_n$ arising from a random sample of n individuals, it is natural to estimate σ^2 by applying the same arithmetic operations to the sample. If we know the value of the population mean μ we can do this, producing the estimate

$$\frac{\sum_{i=1}^{n} (x_i - \mu)^2}{n}.$$

This is an unbiased estimate of the population variance. Usually, however, we cannot use this estimate because the population mean μ is unknown. Then we substitute for μ the sample mean $\bar{x}$ and use the estimate

$$s^2 = \frac{\sum_{i=1}^{n} (x_i - \bar{x})^2}{n-1}.$$

As mentioned in Chapter 4, we divide by $n-1$ instead of n. The reason is that the resulting estimate is unbiased; that is, the mean of the sampling distribution of s^2 is σ^2:

$$\mu_{s^2} = \sigma^2.$$

We emphasize that s^2 is an unbiased estimate of σ^2, *regardless* of the nature of the parent population. In this sense the sampling distribution of s^2 is centered on σ^2. The proof that s^2 is unbiased is given in Appendix 8A.

The square root of the sample variance is the sample standard deviation s; it is used to estimate the population standard deviation σ. The sampling distribution of s tends to be concentrated around the population standard deviation σ. The larger the sample size n the greater the concentration of s^2 around σ^2 and s around σ.

8.3 AN INTERVAL OF PLAUSIBLE VALUES FOR A MEAN

8.3.1 Confidence Intervals When the Standard Deviation Is Known

If the statistician reports only a *point estimate*, that is, a single value, someone may ask how sure he or she is that the estimate is correct. Usually the point estimate will not be *exactly* correct. Since this is the case, the precision of the estimate must be indicated in some way.

One way to do this is to give the standard deviation of the sampling distribution of the estimate along with the estimate itself, thus permitting others to gauge the concentration of the sampling distribution of the estimate. Another way is to specify an *interval,* which depends on both the point estimate and the standard deviation of its sampling distribution, and which contains relatively plausible values for the parameter.

The upper endpoint of this *interval estimate* is obtained by adding to the sample mean a certain quantity, which is a multiple of the standard deviation of the sampling distribution of the mean. The lower endpoint of the interval is obtained by subtracting this quantity from the sample mean.

In the example above, we supposed that the distribution of repeated determinations of cholesterol concentration for a single sample had a standard deviation σ equal to 12.6 units, and that the distribution around the mean μ is approximately normal. Three determinations of the cholesterol concentration in a single sample are considered to be a random sample from such a normal distribution.

The mean of random samples of size 3 has a sampling distribution which has mean equal to μ, the true concentration, and standard deviation equal to $\sigma/\sqrt{3} = 12.6/1.73 = 7.28$. Furthermore, this sampling distribution is normal. (It is normal because the parent distribution is normal; and, if the sample size were larger, it would be nearly normal even if the parent population were not, by the Central Limit Theorem.) Thus the probability is about two-thirds (.682) that a sample mean $\bar{x}$ is within one standard deviation, that is, 7.28 units, of μ. Similarly, the probability is .95 that $\bar{x}$ is within 1.96 standard deviations of μ. This is because 2.5% of the area under the standard normal curve* is to the right of the point $z = 1.96$, and 2.5% of the area is to the left of the point $z = -1.96$, and so 95% of the area is between -1.96 and $+1.96$. In cholesterol units, 1.96 standard deviations is $1.96 \times 7.28 = 14.3$ units. Thus the probability is .95 that $\bar{x}$ is within 14.3 units of μ, as shown in Figure 8.1.

The probability is .95 that $\bar{x}$ falls within 14.3 units of μ; that is, the probability is .95 that the distance between μ and $\bar{x}$ is less than 14.3. With probability .95 we will obtain a value of $\bar{x}$ such that μ is less than $\bar{x} + 14.3$ and greater than $\bar{x} - 14.3$:

$$\Pr(\bar{x} - 14.3 < \mu < \bar{x} + 14.3) = .95;$$

the probability is .95 of drawing a sample with mean $\bar{x}$ such that the interval† $(\bar{x} - 14.3, \bar{x} + 14.3)$ contains the population mean μ.

---

*Since the 97.5th percentile, which we use often, is 1.960, we drop the 0 even though our usual rule (Section 7.2.4) is to give a percentile to three decimals.

†The notation (a, b) denotes the interval of all numbers between a and b.

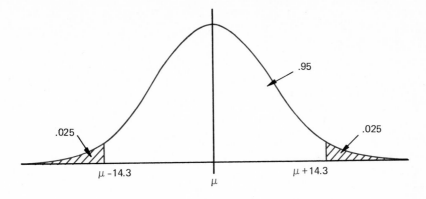

Values of $\bar{x}$

Figure 8.1 *Sampling distribution of $\bar{x}$*

The procedure of interval estimation in this case is to compute the sample mean $\bar{x}$ and give the interval

(8.1) $$(\bar{x} - 14.3, \ \bar{x} + 14.3)$$

as the interval of plausible values for the population mean. In the example of Section 8.1, $\bar{x}$ was 214. The upper endpoint of this confidence interval is $\bar{x} + 14.3 = 214 + 14.3$, or about 228 units; the lower endpoint is $\bar{x} - 14.3 = 214 - 14.3$, or about 200 units.

The probability .95 or 95% associated with the interval (8.1) is called the *level of confidence* or *confidence coefficient*, and the interval estimate (8.1) is called a *confidence interval*.

A review of the steps leading to (8.1) shows that we have subtracted from and added to $\bar{x}$ the multiple 1.96 of the standard deviation of the mean. If the sample of size n is drawn from a population with standard deviation σ, a 95% confidence interval for the population mean μ is

$$\left(\bar{x} - 1.96 \frac{\sigma}{\sqrt{n}}, \ \bar{x} + 1.96 \frac{\sigma}{\sqrt{n}} \right).$$

This is the case because 1.96 is the 97.5 percentile of the standard normal distribution. Because 2.576 is the 99.5 percentile of the standard normal distribution,

$$\left(\bar{x} - 2.576 \frac{\sigma}{\sqrt{n}}, \ \bar{x} + 2.576 \frac{\sigma}{\sqrt{n}} \right)$$

is a 99% confidence interval for μ. Confidence intervals with other confidence levels can be obtained by substituting for 1.96 or 2.576 the

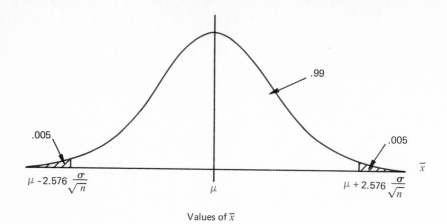

Figure 8.2 *Sampling distribution of x̄*

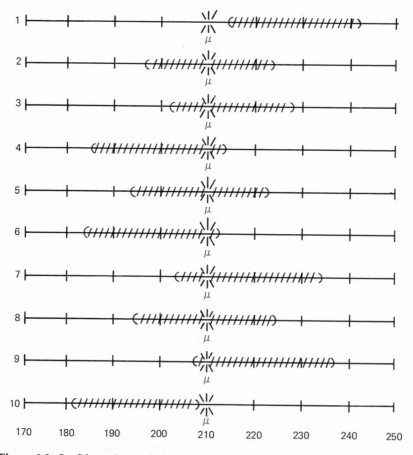

Figure 8.3 *Confidence intervals from 10 samples of size 3 (Table 8.1)*

appropriate number from the tables of percentage points of the normal distribution, such as the table in Appendix II.

The procedure depends upon the fact that $\bar{x}$ has a sampling distribution which is approximated by a normal distribution. By the Central Limit Theorem, we know this is true if the sample size is not small. It is true even when n is fairly small if the parent distribution is unimodal and symmetric, except in unusual cases.

Interpretation. The probability statements leading to (8.1) are based on the sampling distribution of $\bar{x}$. The *probability* refers to the entire procedure of drawing a sample, calculating the sample mean $\bar{x}$, and adding 14.3 to $\bar{x}$ and subtracting 14.3 from $\bar{x}$ to obtain an interval; the probability is .95 that an interval constructed in this way includes μ, the mean of the population sampled.

After the confidence interval statement has been made, we replace the word "probability" by "confidence" because the procedure has already been carried out and the conclusion that μ lies in the stated interval is now either true or false. As an analogy, suppose the instructor flips an unbiased coin but does not show the result to the class members. Would you bet 50–50 that the face up on the hidden coin is a head? It seems reasonable to treat the unknown face as having a probability of 1/2 of being a head even though the act of flipping has been completed and the face up has been determined. We can call 1/2 our "confidence" that the coin is a head, distinguishing this from the probability associated with the process of flipping. ("Random numbers" printed in a table should similarly be distinguished from numbers yet to appear from a random process.)

TABLE 8.1

95% Confidence Intervals from Ten Samples of Size 3 from a Normal Population with a Mean of 210 and a Standard Deviation of 12.6

SAMPLE NUMBER	SAMPLE	MEAN	CONFIDENCE INTERVAL	INTERVAL INCLUDES 210?
1	216, 227, 240	227.7	(213.4, 242.0)	No
2	206, 209, 214	209.7	(195.4, 224.0)	Yes
3	206, 226, 213	215.0	(200.7, 229.3)	Yes
4	208, 211, 178	199.0	(184.7, 213.3)	Yes
5	203, 208, 217	209.3	(195.0, 223.6)	Yes
6	190, 208, 195	197.7	(183.4, 212.0)	Yes
7	210, 217, 229	218.7	(204.4, 233.0)	Yes
8	206, 202, 219	209.0	(194.7, 223.3)	Yes
9	221, 227, 220	222.7	(208.4, 237.0)	Yes
10	198, 200, 187	195.0	(180.7, 209.3)	No

To illustrate the idea that the confidence level is a probability that refers to the entire procedure, we have drawn 10 random samples of size 3 from a normal population with a mean of 210 and a standard deviation of 12.6. (For each sample, 3 entries from a table of random standardized normal deviates were taken, multiplied by 12.6, and added to 210.) The observations are shown in Table 8.1 and the intervals in Figure 8.3. It happens that 8 of the 10 intervals cover the value $\mu = 210$, although if we repeated the procedure many times we would expect the proportion of intervals including μ to be close to .95.

8.3.2   Confidence Intervals When the Variance Is Estimated

It is often of interest to estimate a population mean when the standard deviation of the population is not known. In this case, the interval estimate will still have the sample mean as its center, but the width of the interval will depend on the sample standard deviation.

As an example consider a random sample of $n = 10$ men from a population in which height is approximately normally distributed. Suppose the sample yields a mean height of $\bar{x} = 68.7$ inches and a standard deviation of 2.91 inches. We construct a 95% confidence interval by using the estimate of the standard deviation of the sample mean, which is $s/\sqrt{10} = 2.91/\sqrt{10} = 2.91/3.16 = 0.92$. We shall add to and subtract from the sample mean an appropriate multiple of 0.92. Because of the additional error introduced by having to estimate σ, this multiple will be larger than the 1.96 used to achieve 95% confidence when σ is known.

Looking back on the development of the previous section, we can see that we know the probability that μ is in the interval

$$\left(\bar{x} - 1.96\frac{\sigma}{\sqrt{n}}, \ \bar{x} + 1.96\frac{\sigma}{\sqrt{n}}\right)$$

is .95 because $\bar{x}$ has a sampling distribution which is at least approximately normal with mean μ and standard deviation $\sigma/\sqrt{n}$. A mathematically equivalent way of saying this is to say that the "standardized" variable

$$z = \frac{\bar{x} - \mu_{\bar{x}}}{\sigma_{\bar{x}}} = \frac{\bar{x} - \mu}{\sigma/\sqrt{n}}$$

has the *standard* normal distribution.

Here we shall consider the sampling distribution of the analogue

$$t = \frac{\bar{x} - \mu}{s/\sqrt{n}} .$$

Because σ, a parameter, has been replaced by s, a statistic which only estimates σ, we can expect that t has more sampling variability than z.

Student's t-Distribution. When the sampling is from a parent distribution that is normal, the statistic t has a sampling distribution known as *Student's t-distribution.**

This distribution resembles the standard normal distribution, but it is more spread out and less peaked than the standard normal distribution, and consequently has a bigger variance. Because some of the variability in t is due to s and the amount of variability in s depends on the sample size n, the distribution of t depends on n. More specifically, the distribution depends upon the divisor used in defining s^2, in this case, $n - 1$. This quantity is the number of "degrees of freedom."† We write t_f to denote a t-variable when we want to make it explicit that the number of degrees of freedom is equal to f. Figure 8.4

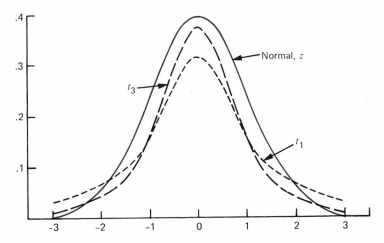

Figure 8.4 *Probability distributions of z, t_1, and t_3*

*W. S. Gossett, a chemist at Guinness' Brewery, Dublin, in the early 1900s, was the first to use this statistic. His company insisted that he publish only under a pen name, and he chose "Student."

†The number of degrees of freedom of a statistic is, loosely speaking, the number of unrestricted variables associated with it. In this case the statistic is s^2 and the number of degrees of freedom is $n - 1$, for the deviations are restricted by one condition, namely, $\sum_{j=1}^{n} (x_i - \bar{x}) = 0$.

shows the t-distributions with 1 and 3 degrees of freedom, together with the normal distribution. The curve marked t_1 is much less peaked and has much bigger tails than the normal; t_3 is more like the normal. When the number of degrees of freedom is large, the corresponding t-distribution is very much like the standard normal.

Some percentage points of Student's t-distribution for various degrees of freedom are given in Appendix III. To include 95% of the area, the interval for z is -1.96 to $+1.96$. For t_9 it is -2.262 to $+2.262$, for t_3 it is -3.182 to $+3.182$, and for t_1 it is -12.706 to $+12.706$.

Confidence Intervals Based on Student's t. For a 95% confidence interval based on a sample of 10 observations, the appropriate percentile is 2.262. Thus, as shown in Figure 8.5, the probability is .95 that t will be between -2.262 and $+2.262$.

In analogy to the case where σ is known, to form a 95% confidence interval for μ based on $n = 10$, $\bar{x} = 68.7$, and $s = 2.91$, we add to and subtract from 68.7 the multiple 2.262 of $s/\sqrt{n} = 0.92$; that is, $2.262 \times 0.92 = 2.1$. The confidence interval has endpoints $68.7 - 2.1 = 66.6$ and $68.7 + 2.1 = 70.8$; it is (66.6, 70.8).

In general, the 95% confidence interval based on a sample of size $n = 10$ is

$$(8.2) \qquad \left(\bar{x} - 2.262 \frac{s}{\sqrt{10}} ,\ \bar{x} + 2.262 \frac{s}{\sqrt{10}} \right).$$

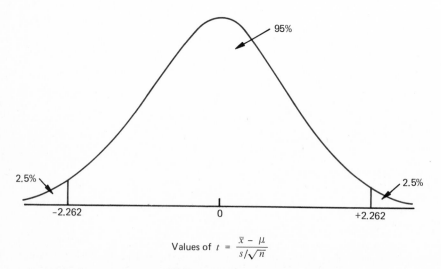

Figure 8.5 *Student's t-distribution for n = 10 (degrees of freedom = 9)*

This is because 2.262 is the upper 2.5% point (that is, 97.5 percentile) of the Student's t-distribution with $10 - 1 = 9$ degrees of freedom. *The 95% confidence interval for μ based on a sample of size n is*

$$\left(\bar{x} - t_{n-1}(.025)\frac{s}{\sqrt{n}} , \ \bar{x} + t_{n-1}(.025)\frac{s}{\sqrt{n}} \right),$$

where $t_{n-1}(.025)$ denotes the upper 2.5% point of the Student's t-distribution with n − 1 degrees of freedom (Figure 8.6). For other levels of confidence, .025 is replaced by the appropriate number, for example, by .005 for a 99% confidence interval.

Mathematical Derivation. Now we shall give the mathematical steps that lead to these confidence intervals. Figure 8.5 indicates that 95% of the area under the curve is between −2.262 and +2.262, that is,

$$\Pr\left(-2.262 < t_9 < 2.262\right) = .95.$$

The definition of t makes this

$$\Pr\left(-2.262 < \frac{\bar{x} - \mu}{s/\sqrt{10}} < 2.262\right) = .95.$$

Multiplication through by $s/\sqrt{10}$ gives

$$\Pr\left(-2.262\,\frac{s}{\sqrt{10}} < \bar{x} - \mu < 2.262\,\frac{s}{\sqrt{10}}\right) = .95.$$

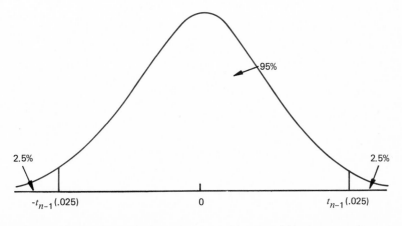

Figure 8.6 *Student's t-distribution with n − 1 degrees of freedom*

TABLE 8.2

95% Confidence Intervals Based on Student's t from Ten Samples of Size 10 from a Normal Population with a Mean of 69 and a Standard Deviation of 3

SAMPLE NUMBER	SAMPLE	MEAN	STANDARD DEVIATION	CONFIDENCE INTERVAL	INTERVAL INCLUDES 69?
1	72.1 67.6 72.8 79.6 70.7 63.4 69.6 72.6 67.5 68.2	70.41	4.33	(67.3, 73.5)	Yes
2	73.2 67.3 69.1 70.0 77.8 74.9 68.2 70.2 70.3 68.9	70.99	3.30	(68.6, 73.4)	Yes
3	71.7 67.5 67.4 70.8 71.6 66.2 73.7 69.5 63.3 70.1	69.18	3.09	(67.0, 71.4)	Yes
4	72.5 65.8 69.0 71.3 71.9 71.1 72.3 67.1 68.2 66.9	69.63	2.50	(67.8, 71.4)	Yes
5	64.5 67.5 68.5 68.6 72.1 69.6 70.3 71.2 67.7 67.7	68.77	2.17	(67.2, 70.3)	Yes
6	66.9 71.3 64.1 68.0 67.5 62.8 67.6 68.3 71.6 67.6	67.57	2.71	(65.7, 69.5)	Yes
7	73.1 69.7 70.1 71.3 69.5 66.8 71.9 64.4 68.2 69.4	69.44	2.52	(67.0, 71.2)	Yes
8	67.6 74.0 68.8 65.3 67.5 71.6 67.5 63.1 60.5 68.3	67.42	3.87	(64.6, 70.2)	Yes
9	73.2 65.5 66.3 72.7 68.4 68.3 72.7 61.3 67.3 69.2	68.49	3.73	(65.8, 71.1)	Yes
10	63.6 68.2 72.7 72.1 67.5 64.1 68.6 67.8 67.1 70.7	68.24	3.01	(66.1, 70.4)	Yes

The pair of inequalities states that $\bar{x}$ differs from μ by less than $2.262s/\sqrt{10}$ in either direction or that the distance between $\bar{x}$ and μ is less than $2.262s/\sqrt{10}$. Another way of indicating the probability is

$$\Pr\left(-2.262\,\frac{s}{\sqrt{10}} < \mu - \bar{x} < 2.262\,\frac{s}{\sqrt{10}}\right) = .95;$$

the addition of $\bar{x}$ throughout yields

$$\Pr\left(\bar{x} - 2.262\,\frac{s}{\sqrt{10}} < \mu < \bar{x} + 2.262\,\frac{s}{\sqrt{10}}\right) = .95.$$

This says that the probability is .95 of drawing a sample with mean $\bar{x}$ and standard deviation s such that the population mean μ lies in the interval (8.2).

We have drawn 10 random samples of size 10 from a normal population with a mean of 69 and a standard deviation of 3. The results are given in Table 8.2. Note that the intervals vary in length, because the length depends upon s, the estimate of the standard deviation. The length of the first interval is $73.5 - 67.3 = 6.2$, the length of the second is $73.4 - 68.6 = 4.8$, etc. It happened that all ten intervals covered μ (Figure 8.7); in the long run, we would expect 95% of the intervals to cover μ.

The t-distribution with various degrees of freedom has been derived mathematically on the basis of sampling from a normal distribution and the tables are computed on that basis. If the number of degrees of free-

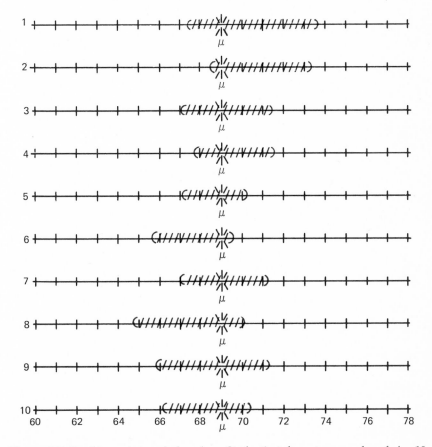

Figure 8.7 *Confidence intervals based on Student's t from ten samples of size 10* (*Table 8.2*)

dom is large, the percentage points of the *t*-table are very close to the corresponding values in the normal table because there is then little sampling error in *s*; that is, with high probability *s* is close to σ. Hence, if *n* is large, the normal tables can be referred to regardless of whether σ is known and used or σ is unknown and *s* is used. Roughly speaking, the normal tables can be used in place of the *t*-table if $n \geq 30$; the reader can verify this statement by comparing values in the *t*-table with those in the normal table. If *n* is large, the Central Limit Theorem gives assurance that $\bar{x}$ is approximately normal and the normal tables can be used, whether or not the parent population is normal. If *n* is small and the parent population is normal or approximately normal, the normal tables can be referred to if σ is known and used, and the *t*-tables can be referred to if σ is unknown and *s* is used.

It should be noted that if σ is used the width of the confidence interval tends to be small if σ is small or if *n* is large. If *s* is used, the width of the interval is itself subject to sampling variability.

Table 8.3 gives a summary of procedures. The number *k* represents the appropriate percentage point of the standard normal distribution or a Student's *t*-distribution.

There is a gap in our kit of procedures, namely, if *n* is small and the parent population is quite different from normal. Then other methods must be used; we shall not study them here.

TABLE 8.3

Procedures for Interval Estimation of a Mean

STANDARD DEVIATION KNOWN.   INTERVAL IS $(\bar{x} - k\sigma/\sqrt{n}, \bar{x} + k\sigma/\sqrt{n})$

Parent Population	*Sample Size*	
	LARGE	SMALL
Considered approximately normal	normal tables	normal tables
Not considered approximately normal	normal tables	

STANDARD DEVIATION NOT KNOWN.   INTERVAL IS $(\bar{x} - ks/\sqrt{n}, \bar{x} + ks/\sqrt{n})$

Parent Population	*Sample Size*	
	LARGE	SMALL
Considered approximately normal	normal tables	*t*-tables
Not considered approximately normal	normal tables	

Number of Digits to Report. The reader may well wonder why we have not yet discussed the matter of how many digits to give when reporting the result of a calculation. This simple question is without a simple answer. A sensible answer to the question of how many digits to report involves consideration of the standard deviation of the result as well as the number of significant figures in the original data and the nature of the calculations involved. Usually one would not want to have to compute a standard deviation merely to decide how many digits to report. However, when we compute a confidence interval, we do compute a standard deviation and have at hand information relevant to how many digits it is sensible to report. There is no sense in reporting a mean as 167.243 when the associated 90% confidence interval is (147, 187). The ten's digit (the 6 in 167.243) is slightly uncertain; surely there is no point in reporting the 2, 4, and 3. To report "167" would suffice. Of course it would be good to give the confidence interval, too, or just say: "167 ± 20 (90% confidence interval)."

8.4 ESTIMATION OF A PROPORTION

8.4.1 Point Estimation of a Proportion

Polling a random sample of 400 registered voters from a large population reveals that 224 intend to vote for Jones. The sample proportion $224/400 = .56$ estimates the proportion p of the population intending to vote for Jones. We denote the estimate of p by $\hat{p}$ and write $\hat{p} = .56$. In general, the caret (^) or "hat" may be read "estimate of"; for example, $\hat{p}$ denotes an "estimate of p."

We can put this estimate in the context of the preceding section by letting x_i equal 1 or 0, according to whether or not the ith individual intends to vote for Jones. Polling the 400 voters leads to observations $x_1, x_2, \ldots, x_{400}$. The estimate $\hat{p}$ is simply the arithmetic mean of these dichotomous observations.

The variance of any of the dichotomous observations x is $\sigma^2 = pq$, where $q = 1 - p$. Since $\hat{p}$ is simply the arithmetic mean $\bar{x}$ of these observations, the variance of $\hat{p}$ is

$$\sigma_{\hat{p}}^2 = \sigma_{\bar{x}}^2 = \frac{\sigma^2}{n} = \frac{pq}{n} .$$

As noted in Chapter 7, $pq = p(1 - p)$ is 0 for $p = 0$; increases with p to $p = 1/2$, where $pq = 1/4$; and then decreases to $p = 1$, where

$pq = 0$. Its maximum is 1/4, and it is approximately at its maximum for values from about $p = 1/4$ to $p = 3/4$. We estimate pq as $\hat{p}\hat{q}$, where $\hat{q} = 1 - \hat{p}$.

8.4.2 Interval Estimation of a Proportion

A confidence interval for the proportion p of the population intending to vote for Jones is based on the idea that for sufficiently large sample size the sample mean is approximately normally distributed; in the procedure discussed in Section 8.3.1, the variance σ^2 is replaced by its estimate $\hat{p}\hat{q}$.

For the numerical example in Section 8.4.1 the procedure is as follows. The estimate of the variance of $\hat{p}$ is $\hat{p}\hat{q}/n = .56 \times .44/400 = .2464/400$; the estimate of the standard deviation is the square root of this, $.496/20$, or $.0248$. For a 95% confidence interval, this is multiplied by 1.96, the appropriate percentage point of the normal distribution. This gives $1.96 \times .0248 = .0486$, which is about .05. The upper and lower limits are, respectively,

$$.56 + .05 = .61 \quad \text{and} \quad .56 - .05 = .51.$$

The justification for this procedure is that the standardized version of $\hat{p}$, which is

(8.3)
$$\frac{\hat{p} - \mu_{\hat{p}}}{\sigma_{\hat{p}}} = \frac{\hat{p} - p}{\sqrt{pq/n}},$$

has approximately the standard normal distribution. (As mentioned in Chapter 7, the normal approximation is good if $np > 5$ and $nq > 5$ and the continuity correction is used; larger n is needed if the correction is not used.) Since n is large here, we replace pq in (8.3) by $\hat{p}\hat{q}$; that is, we use the fact that

$$\frac{\hat{p} - p}{\sqrt{\hat{p}\hat{q}/n}}$$

is approximately the same as (8.3) and hence has approximately the standard normal distribution. A 95% confidence interval in general is

(8.4)
$$\left(\hat{p} - 1.96\sqrt{\frac{\hat{p}\hat{q}}{n}}, \; \hat{p} + 1.96\sqrt{\frac{\hat{p}\hat{q}}{n}} \right).$$

To obtain a confidence interval for any other level of confidence we replace 1.96 by the appropriate number from the normal tables. For example, for 99% we take 2.576 because the probability of a standard normal variable being greater than 2.576 is .005, corresponding to one-half of 1%.

SHORT CUT FOR 95% INTERVAL   The 95% confidence interval can be approximated in a simple way because 1.96 is about 2 and the maximum of $\sqrt{pq}$ is $\sqrt{1/4} = 1/2$. Thus $1.96\sqrt{pq}/\sqrt{n} < 1/\sqrt{n}$. The interval

(8.5)
$$\left(\hat{p} - \frac{1}{\sqrt{n}}, \ \hat{p} + \frac{1}{\sqrt{n}} \right)$$

is slightly wider than (8.4) but not much if $\hat{p}$ is between 1/4 and 3/4; it has confidence level greater than .95 in any case.

Small Samples.   If the sample size is small, the confidence interval should be based on the exact binomial sampling distribution. Charts exist from which the confidence limits can be obtained, given the sample size and the value of the sample proportion. Beyer (1972) gives charts for obtaining 95% and 99% confidence limits when $n = 5, 10, 15, 20, 30, 50, 100, 250,$ or 1000. Pearson and Hartley (1954) give a more extensive set of charts.

The normal approximation can be improved by use of the continuity correction, as discussed in Appendix 8B.

8.5 ESTIMATION OF A MEDIAN OR OTHER PERCENTILE

8.5.1 Point Estimation of a Percentile

The Median.   As an estimate of the population median it is natural to take the median of a random sample from that population. As an example, consider the lifetimes of a sample of 15 light bulbs which were (in hours)

$$499, \quad 26, \quad 614, 231, 719,$$
$$2063, 2723, \quad 466, 709, 904,$$
$$2303, \quad 27, 1374, 981, 654.$$

These numbers, arranged from smallest to largest, are

$$26, \quad 27, \quad 231, \quad 466, \quad 499,$$
$$614, \quad 654, \quad 709, \quad 719, \quad 904,$$
$$981, \quad 1374, \quad 2063, \quad 2303, \quad 2723.$$

The median of these 15 lifetimes is the 8th ranking one, or 709 hours. This is an estimate of the population median. For large random samples the sample median has a good chance of being close to the population median.

From Figure 8.8 one sees that the sample distribution of light-bulb lifetimes is highly skewed to the right. Although 15 is only a small sample size, this is evidence that the population distribution is skewed. For many purposes the median would be a better measure of location than the mean.

For *symmetric* distributions the mean and the median coincide. To estimate this center of the distribution we can use either the median of the sample or the mean of the sample.

For samples from *normal* distributions the sample mean is the better of these two estimates. It can be shown, in fact, that for large random samples from normal distributions the variance of the sample median is about $\pi/2$ (about $1\frac{1}{2}$) times that of the sample mean; that is, the variance of the sample median is about $(\pi/2)(\sigma^2/n)$, or about $1.57\sigma^2/n$. This implies that the *median* of 157 observations is only as precise an estimate of μ as the *mean* of 100 observations.

If there is a possibility that the sample contains "outliers" (erroneous extreme values), then the median would be better than the mean, for the median is not sensitive to the value of these extreme observations.

Percentiles. The 90th percentile of a random sample from a population is an estimate of the 90th percentile of the population.

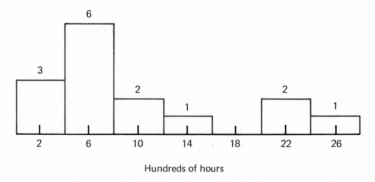

Hundreds of hours

Figure 8.8 *Frequency distribution of lifetimes of light bulbs*

However, when we consider the population distribution to be normal, as we would with the distribution of heights of 5-year-old boys in Table 2.14, we can estimate the 90th percentile as $\bar{x} + 1.282s$, since 1.282 is the 90th percentile of the standard normal distribution (see Appendix II), so that $\mu + 1.282\sigma$ is the 90th percentile of a normal distribution with mean μ and standard deviation σ.

For the heights of the boys, $\bar{x} = 39.3$ inches and $s = 2.5$ inches. This estimate of the 90th percentile would therefore be

$$39.3 + 1.282 \times 2.5 = 39.3 + 3.2 = 42.5 \text{ inches.}$$

The estimate $\bar{x} + 1.282s$ has smaller variance than does the 90th percentile of the sample, provided that the parent population is normal. This estimate is therefore preferable to the sample percentile in such cases. This fact is, of course, similar to that mentioned above for the median and mean, the median being the 50th percentile which would be estimated by $\bar{x} + 0 \times s$, which is simply $\bar{x}$, the sample mean.

8.5.2 Interval Estimation of a Median

We shall give a way of constructing a confidence interval for the median M of any population. That is, we shall give a *distribution-free* confidence interval for the median. By "distribution-free" we mean that the procedure achieves the desired confidence level regardless of the nature of the parent distribution.

Arrange the sample $x_1, x_2, \ldots, x_n$ in ascending order. Denote the *ordered sample* by $x_{(1)}, x_{(2)}, \ldots, x_{(n)}$. That is, $x_{(1)}$ is the value of the smallest observation and $x_{(n)}$ is the value of the largest.

If the sample size is 6, it follows from the binomial distribution with $p = 1/2$ that the probability that the largest value is less than M (that is, all 6 values less than M) is 1/64 and the probability that the smallest value is greater than M is 1/64. Hence,

$$\Pr(x_{(1)} < M < x_{(6)}) = 1 - 2 \times \frac{1}{64} = \frac{62}{64} = .96875,$$

and the interval $(x_{(1)}, x_{(6)})$ is a confidence interval for M with confidence coefficient .96875. (See Exercises 8.30 and 8.31.) The interval $(x_{(2)}, x_{(5)})$ would be a confidence interval for M with a smaller confidence coefficient.

For an arbitrary sample size n we shall use confidence intervals for M of the form $(x_{(r)}, x_{(n-r+1)})$. [For example, if $n = 6$ and $r = 2$, then $(x_{(2)}, x_{(5)})$ is the interval.] Given n, one can compute the value of r which gives an interval with approximate confidence coefficient .95. (See Appendix 8C.) This value is

$$r = \frac{n+1}{2} - 1.96 \frac{\sqrt{n}}{2} = \frac{n+1}{2} - 0.98\sqrt{n},$$

which is very nearly

$$r = \frac{n+1}{2} - \sqrt{n}.$$

When $\frac{1}{2}(n+1) - \sqrt{n}$ is not an integer, it is rounded down to the largest integer less than it. For a 95% confidence interval for the median of the lifetimes of the population of light bulbs from which the 15 are a sample, we have

$$r = \frac{15+1}{2} - \sqrt{15} = 8.0 - 3.9 = 4.1,$$

which rounds down to 4. The confidence interval is $(x_{(4)}, x_{(12)})$ or (466, 1374).

For an interval with confidence coefficient .99, the formula for r is

$$r = \frac{n+1}{2} - 2.576 \frac{\sqrt{n}}{2} = \frac{n+1}{2} - 1.288\sqrt{n}.$$

This formula is derived (in Appendix 8C) using the normal approximation with continuity correction to the binomial distribution with $p = 1/2$. The rule of thumb, $np > 5$ and $nq > 5$, reduces in this case to $n > 10$. If $n \le 10$, the exact binomial distribution with $p = 1/2$ should be used.

8.6   PAIRED MEASUREMENTS

8.6.1   Mean of a Population of Differences

The individuals of a single sample may be measured twice. Such data arise in "before and after" studies and in studies where pairs of measurements are made. We might consider comparing the sizes of left and right feet or the measured strengths of left and right hands for a sample of right-handed people.

Table 8.4 shows reading scores of 30 pupils before and after the second grade. Each pupil's score *after* second grade is compared with his or her score *before* the second grade by computing the gain: the variable under consideration is the *difference,* score after second grade minus score before second grade. We treat the observed differences as a sample from a normally distributed population of differences. The calculation of the differences, mean of the differences, and standard deviation of the differences is summarized at the foot of Table 8.4. The mean difference is $\bar{d} = 0.51$, and the standard deviation is 0.522. A 95% confidence interval for the corresponding population mean μ_d, assuming normality, is

$$\left(\bar{d} - t_{29}(.025) \frac{s_d}{\sqrt{n}} , \ \bar{d} + t_{29}(.025) \frac{s_d}{\sqrt{n}} \right),$$

TABLE 8.4 *Reading Scores of 30 Pupils Before and After Second Grade*

	Scores				Scores		
Pupil	Before	After	Difference	Pupil	Before	After	Difference
i	x_i	y_i	$d_i = y_i - x_i$	i	x_i	y_i	$d_i = y_i - x_i$
1	1.1	1.7	0.6	16	1.5	1.7	0.2
2	1.5	1.7	0.2	17	1.0	1.7	0.7
3	1.5	1.9	0.4	18	2.3	2.9	0.6
4	2.0	2.0	0.0	19	1.3	1.6	0.3
5	1.9	3.5	1.6	20	1.5	1.6	0.1
6	1.4	2.4	1.0	21	1.8	2.5	0.7
7	1.5	1.8	0.3	22	1.4	3.0	1.6
8	1.4	2.0	0.6	23	1.6	1.8	0.2
9	1.8	2.3	0.5	24	1.6	2.6	1.0
10	1.7	1.7	0.0	25	1.1	1.4	0.3
11	1.2	1.2	0.0	26	1.4	1.4	0.0
12	1.5	1.7	0.2	27	1.4	2.0	0.6
13	1.6	1.7	0.1	28	1.5	1.3	−0.2
14	1.7	3.1	1.4	29	1.7	3.1	1.4
15	1.2	1.8	0.6	30	1.6	1.9	0.3

Source: Records of a second-grade class.

$n = 30$

$\sum_{i=1}^{30} d_i = 15.3$

$\bar{d} = 0.51$

$\sum_{i=1}^{30} d_i^2 = 15.61$

$\sum_{i=1}^{30} (d_i - \bar{d})^2 = 15.61 - \frac{(15.3)^2}{30} = 7.91$

$s_d^2 = 0.273$

$s_d = 0.522$

where s_d is the standard deviation of the 30 differences. From tables we find $t_{29}(.025) = 2.045$. We compute

$$t_{29}(.025)\frac{s_d}{\sqrt{n}} = 2.045\frac{0.522}{\sqrt{30}} = 2.045 \times 0.095 = 0.19;$$

so that the confidence interval is

$$(.51 - .19,\ .51 + .19),\qquad \text{or}\qquad (.32, .70).$$

In the example, let
y_i = reading score of ith pupil after second grade,
x_i = reading score of the ith pupil before second grade, and
$d_i = y_i - x_i$ = gain in reading score achieved by ith pupil.
The statistical procedure is to compute the differences

$$d_i = y_i - x_i,\qquad i = 1, \ldots, n.$$

This reduces the data set of $2n$ numbers (x_i, y_i), $i = 1, \ldots, n$, to a set of n numbers d_i, $i = 1, \ldots, n$.

The example fits the following model, where we have a random sample of *pairs:*

$$(x_1, y_1),\ (x_2, y_2),\ \ldots,\ (x_n, y_n).$$

Here "random sample" means that the n individuals are a random sample of individuals. Thus there is statistical independence among pairs.

The sample mean of the differences, $\bar{d}$, is an estimate of the population mean, μ_d. The standard deviation s_d is computed from the differences $d_1, \ldots, d_n$.

It should be noted that $\bar{d} = \bar{y} - \bar{x}$; that is, the average gain in the class is identical to the average score after second grade less the average score before second grade. Similarly, the population mean difference μ_d is the difference of the population means $\mu_y - \mu_x$.

8.6.2 Matched Samples

The method of "matched samples" can be used to make a comparison between two treatments, such as ways of teaching reading or effectiveness of drugs. As an example, suppose two brands of feed for baby pigs, Brand X and Brand Y, are to be compared. It is known that piglets within a litter are more similar than piglets from different litters. To avoid confusing differences between litters with differences between brands, an investigation is based on feeding Brand X to one piglet in a litter and Brand Y to another. Thus, from each of n litters two piglets are chosen randomly; one of each pair is selected randomly to be fed Brand X and the other Brand Y. The pigs are weighed at the end of six weeks. Let

y_i = weight of the pig from the ith litter fed Brand Y,

x_i = weight of the pig from the ith litter fed Brand X, and

$d_i = y_i - x_i$ = difference between weights of the two pigs from the ith litter,

for $i = 1, \ldots, n$. Then $d_1, \ldots, d_n$ may be considered as a sample of n from a (hypothetical) population of differences of the effects of Brand Y and Brand X. The analysis described in Section 8.6.1 can be used to estimate μ_d, the population mean difference, and to give a confidence interval for μ_d.

The weights of the n pigs fed Brand X can be considered as a sample from a general population of pigs being fed Brand X; call the population mean μ_x. Similarly, the weights of the other n pigs constitute a sample from a general population, with mean μ_y, being fed Brand Y.

The pairs of weights $(x_1, y_1), \ldots, (x_n, y_n)$ are considered as a sample from a population of pairs. The population mean of differences μ_d is the difference of the respective means, $\mu_y - \mu_x$. Such a sample of pairs is often called a pair of "matched samples."

In this example, taking a pair of pigs from a litter "matches" the pigs in each pair. In a psychological study the investigator might pair individuals according to IQ or temperament and choose randomly one person from each pair to be given a treatment (say, put in a situation of tension) while the other person in the pair is considered as a control.

8.7 IMPORTANCE OF SIZE OF POPULATION RELATIVE TO SAMPLE SIZE

8.7.1 The Finite Population Correction

In many instances the population of interest consists of a finite number of individuals or objects, such as all the light bulbs produced in a certain month, the blood cells in 0.10 ml of blood, the counties in the United States, or members of the national labor force. As was seen in Section 6.8.4, after one unit has been drawn from a finite population the remaining population is different from the original one because of the omission of that unit. After the second unit is drawn, the remaining population is different from the original because of omission of the two units. If the size of the original population is large relative to the size of the sample this effect is not very large. For instance, consider a jar of 10,000 beads of which 5,000 are white and 5,000 are black. The original proportion of white beads is 1/2, which is the probability of drawing a white bead at random from the original population. After drawing a few beads this probability has not been changed much. After 20 beads have been drawn, the probability that the 21st bead is white is at least 4980/9980 = .4991 (which is the case if the first 20 beads drawn are white) and at most 5000/9980 = .5009 (which is the case if the first 20 beads drawn are black).

If more white beads than black have been drawn, the probability of a white bead is less than 1/2; and if more black beads have been drawn, the probability is greater than 1/2; the next bead drawn is more likely to bring the ratio of white beads in the sample back towards 1/2 than away from it.

If the sampling is done with replacement, there is greater variability than when the sampling is done without replacement. In particular, the variability of the sampling distribution of the *sample mean* is greater when sampling with replacement than when sampling without replacement.

This fact manifests itself mathematically as follows. The formula given in Chapter 7 for the variance of the sampling distribution of the mean,

$$\sigma_{\bar{x}}^2 = \frac{\sigma^2}{n} ,$$

where σ^2 is the variance of the parent population, is exactly true for sampling *with* replacement or for sampling from an infinite population. For sampling *without* replacement from a population of size N the variance of the sampling distribution of the mean is

$$\sigma_{\bar{x}}^2 = \frac{N-n}{N-1} \frac{\sigma^2}{n} .$$

The standard deviation of the sampling distribution of the mean is thus

$$\sigma_{\bar{x}} = \sqrt{\frac{N-n}{N-1}} \frac{\sigma}{\sqrt{n}} .$$

The factors $(N-n)/(N-1)$ and $\sqrt{(N-n)/(N-1)}$, which appear here, are called *finite population correction factors*. The values of the latter factor for various n and N are given in Table 8.5. Multiplying this factor by $\sigma/\sqrt{n}$ gives $\sigma_{\bar{x}}$. (The row labelled ∞ corresponds to sampling with replacement.) It will be seen that the factor is about 1 for all values of N from 10,000 to 1,000,000 except for $n = 3,600$ and $N = 10,000$, where the sample size is a relatively large fraction of the population size; that is, $n/N = .36$.

TABLE 8.5

Values of $\sqrt{(N-n)/(N-1)}$ for Various N and n

POPULATION SIZE	SAMPLE SIZE n		
N	100	900	3600
10,000	.995	.955	.80
100,000	.9995	.9955	.982
1,000,000	.99995	.99955	.9982
∞	1.00000	1.00000	1.00000

For samples as large as $n = 100$ or larger, the sampling distribution of the sample mean is nearly normal (except for extremely skewed parent distributions). Figures 8.9 and 8.10 show the sampling distribution of the sample mean for a parent population with mean 150 and standard deviation 20 and for population sizes of $N = 10,000$ and $N = 1,000,000$. When the sample size is 900, as in Figure 8.9, there is little difference between the two sampling distributions. When the sample size is 3600 (four times as large), the sampling distribution for $N = 10,000$ is more concentrated than for $N = 1,000,000$. These examples emphasize that the variability depends primarily on the sample size and only slightly on what proportion of the population is sampled unless that proportion is fairly large.

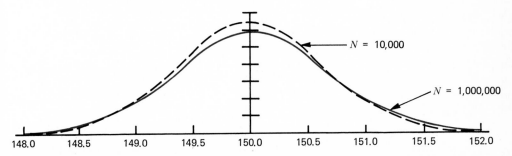

Figure 8.9 *Sampling distributions of the mean of samples of size $n = 900$ when $\mu = 150$ and $\sigma = 20$*

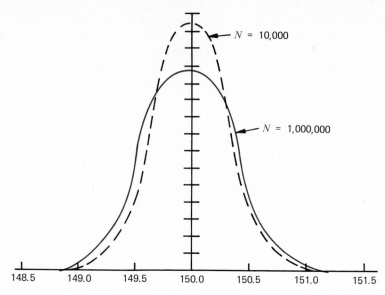

Figure 8.10 *Sampling distributions of the mean of samples of size* $n = 3600$ *when* $\mu = 150$ *and* $\sigma = 20$

The size of $\sigma_{\bar{x}}$ is important because it affects the width of confidence intervals (as well as the power of tests of hypotheses, as explained in Chapter 9). The values of $\sigma_{\bar{x}}$ corresponding to the four curves in Figures 8.9 and 8.10 are listed in Table 8.6.

TABLE 8.6

Standard Deviation of the Sampling Distribution of the Mean of a Random Sample from a Parent Population with $\sigma = 20$ *when* $n = 900$, *3600, and* $N = 10{,}000$, *1,000,000, and Infinity*

n	N	$\sigma_{\bar{x}}$
900	∞	.667
900	1,000,000	.666
900	10,000	.637
3,600	∞	.333
3,600	1,000,000	.333
3,600	10,000	.273

Table 8.7 gives mathematical expressions for the variances and standard deviations of sums and means, for sampling with replacement and sampling without replacement.

TABLE 8.7
Means, Variances, and Standard Deviations of Sums and Means

		Mean	Variance	Standard Deviation
Parent population	In general	μ	σ^2	σ
	For Bernoulli variables	p	pq	$\sqrt{pq}$
Sampling distribution of sample *sums* with replacement	In general	$n\mu$	$n\sigma^2$	$\sigma\sqrt{n}$
	For Bernoulli variables	np	npq	$\sqrt{npq}$
Sampling distribution of sample *sums* without replacement	In general	$n\mu$	$\dfrac{N-n}{N-1}n\sigma^2$	$\sigma\sqrt{n\dfrac{N-n}{N-1}}$
	For Bernoulli variables	np	$\dfrac{N-n}{N-1}npq$	$\sqrt{\dfrac{N-n}{N-1}npq}$
Sampling distribution of sample *means* with replacement	In general	μ	$\dfrac{\sigma^2}{n}$	$\dfrac{\sigma}{\sqrt{n}}$
	For Bernoulli variables	p	$\dfrac{pq}{n}$	$\sqrt{\dfrac{pq}{n}}$
Sampling distribution of sample *means* without replacement	In general	μ	$\dfrac{N-n}{N-1}\dfrac{\sigma^2}{n}$	$\dfrac{\sigma}{\sqrt{n}}\sqrt{\dfrac{N-n}{N-1}}$
	For Bernoulli variables	p	$\dfrac{N-n}{N-1}\dfrac{pq}{n}$	$\sqrt{\dfrac{N-n}{N-1}\dfrac{pq}{n}}$

If we draw a random sample without replacement from a large population the result is almost the same as sampling *with* replacement. This is because, even if we replaced the individuals drawn, there is only a very slight chance that the same individual would be included in the sample more than once since the population is large. Consequently, if the sample size is small relative to the population size, sampling without replacement will give about the same results as sampling with replacement, and we need make no adjustment.

As a rule of thumb, we can say that if n/N is less than 1/10, the finite population correction may be ignored. When the correction is ignored, the variability of the sample mean will be overestimated, and the resulting confidence interval will be conservative in the sense that it is widened and the confidence level is underestimated.

When we do need to make the correction, we can use the approximation

$$\sqrt{\frac{N-n}{N-1}} \doteq \left(1 - \frac{1}{2}\frac{n}{N}\right);$$

this finite population correction factor is approximately one minus half the proportion sampled. The reader can verify this approximation numerically. (See also Mathematical Exercise 8.26.)

EXAMPLE   The officials of a town of population 5000 wish to make a survey to determine whether a majority of its citizens favor a certain proposition. Of a random sample of 1200 persons, 960 are in favor of the proposition; thus $\hat{p} = .80$. A 95% confidence interval is given by

$$\left(.80 - 1.96\sqrt{\frac{N-n}{N-1}\frac{\hat{p}\hat{q}}{n}}, \ .80 + 1.96\sqrt{\frac{N-n}{N-1}\frac{\hat{p}\hat{q}}{n}}\right).$$

The estimated standard deviation of the sample mean is

$$\sqrt{\frac{N-n}{N-1}\frac{\hat{p}\hat{q}}{n}} = \sqrt{\frac{5000-1200}{5000-1} \times \frac{.80 \times .20}{1200}} = .01006.$$

(The finite population correction is .8718.) Since $1.96 \times .01006 = .020$, the interval is (.780, .820). Thus we would be confident in saying that somewhere between 78% and 82% of the people are in favor of the proposition.

8.7.2   An Example Comparing Sampling Distributions With and Without Replacement

The effects of sampling without and with replacement can be clarified by going through a simple experiment. Consider a population of $N = 3$ objects: 1 red ball, 1 blue ball, and 1 yellow ball. A sample of $n = 2$ balls is to be drawn to estimate the proportion of red balls in this population, which is $p = 1/3$. A red ball is a "success" and is given the value 1. The sample proportion of red balls $\hat{p}$ is used to estimate p.

Sampling Without Replacement.   The six possible pairs of drawings for this case, in terms of colors and numerical values, are given in Table 8.8. The possibilities are equally likely to occur; hence, each has probability 1/6. The values that $\hat{p}$ can take on are $v_1 = 0$ and $v_2 = 1/2$ with respective probabilities

$$\Pr(\hat{p} = 0) = \frac{1}{6} + \frac{1}{6} = \frac{2}{6} = \frac{1}{3},$$

$$\Pr\left(\hat{p} = \frac{1}{2}\right) = \frac{1}{6} + \frac{1}{6} + \frac{1}{6} + \frac{1}{6} = \frac{4}{6} = \frac{2}{3},$$

which are tabulated in Table 8.9.

TABLE 8.8
Sampling Without Replacement

POSSIBLE PAIRS IN TERMS OF COLOR		POSSIBLE PAIRS IN TERMS OF 1 AND 0		PROBABILITY	PROPORTION $\hat{p}$
1st Draw	*2nd Draw*	*1st Draw*	*2nd Draw*		
R	B	1	0	$\frac{1}{6}$	$\frac{1}{2}$
R	Y	1	0	$\frac{1}{6}$	$\frac{1}{2}$
B	R	0	1	$\frac{1}{6}$	$\frac{1}{2}$
B	Y	0	0	$\frac{1}{6}$	0
Y	R	0	1	$\frac{1}{6}$	$\frac{1}{2}$
Y	B	0	0	$\frac{1}{6}$	0

TABLE 8.9
Sampling Distribution of $\hat{p}$ Based on Sampling Without Replacement

v_j	$\Pr(\hat{p} = v_j)$
0	$\frac{1}{3}$
$\frac{1}{2}$	$\frac{2}{3}$

Sampling with Replacement. There are now $3^2 = 9$ possible samples, which are recorded in Table 8.10. The values that $\hat{p}$ can take on are $v_1 = 0$, $v_2 = 1/2$, and $v_3 = 1$ with respective probabilities

$$\Pr(\hat{p} = 0) = \frac{1}{9} + \frac{1}{9} + \frac{1}{9} + \frac{1}{9} = \frac{4}{9},$$

$$\Pr\left(\hat{p} = \frac{1}{2}\right) = \frac{1}{9} + \frac{1}{9} + \frac{1}{9} + \frac{1}{9} = \frac{4}{9},$$

$$\Pr(\hat{p} = 1) = \frac{1}{9},$$

which are tabulated in Table 8.11.

TABLE 8.10

Sampling with Replacement

| Possible Pairs in Terms of Color | | Possible Pairs in Terms of 1 and 0 | | Probability | Proportion |
1st Draw	2nd Draw	1st Draw	2nd Draw		$\hat{p}$
R	R	1	1	$\frac{1}{9}$	1
R	B	1	0	$\frac{1}{9}$	$\frac{1}{2}$
R	Y	1	0	$\frac{1}{9}$	$\frac{1}{2}$
B	R	0	1	$\frac{1}{9}$	$\frac{1}{2}$
B	B	0	0	$\frac{1}{9}$	0
B	Y	0	0	$\frac{1}{9}$	0
Y	R	0	1	$\frac{1}{9}$	$\frac{1}{2}$
Y	B	0	0	$\frac{1}{9}$	0
Y	Y	0	0	$\frac{1}{9}$	0

TABLE 8.11

Sampling Distribution of $\hat{p}$ Based on Sampling with Replacement

v_j	$\Pr(\hat{p} = v_j)$
0	$\dfrac{4}{9}$
$\dfrac{1}{2}$	$\dfrac{4}{9}$
1	$\dfrac{1}{9}$

Comparison of Sampling With and Without Replacement. Histograms of the two sampling distributions of $\hat{p}$ are given in Figure 8.11. We see that sampling without replacement more often gives values of $\hat{p}$ that are closer to p. Of the three values 0, 1/2, and 1, the value 1/2 is closest to the true population proportion $p = 1/3$. The value 0 is next best, and the value 1 is furthest away from p. The probability of obtaining a value of 1/2 is 2/3 under sampling without replacement and 4/9 under sampling with replacement, a difference of 2/9. In comparing sampling without and with replacement, we see that half of this difference of 2/9 goes with the value 0, the other half with the value 1. Hence the "best" value is less probable under sampling with replacement; the two other values, more probable.

Means of the Sampling Distributions. The mean of the sampling distribution without replacement is

$$0 \times \frac{1}{3} + \frac{1}{2} \times \frac{2}{3} = 0 + \frac{1}{3} = \frac{1}{3} = p,$$

and the mean of the sampling distribution with replacement is

$$0 \times \frac{4}{9} + \frac{1}{2} \times \frac{4}{9} + 1 \times \frac{1}{9} = 0 + \frac{2}{9} + \frac{1}{9} = \frac{3}{9} = \frac{1}{3} = p.$$

In either case the mean is p; $\hat{p}$ is an unbiased estimate. Under either sampling method, the sampling distribution is centered at the "true" value, p. Hence, on the basis of the mean of the sampling distribution, there is no reason to prefer one method to the other.

Variance of the Sampling Distribution. We can use the formula

$$\sigma_{\hat{p}}^2 = \sum_j v_j^2 \Pr(\hat{p} = v_j) - \mu^2$$

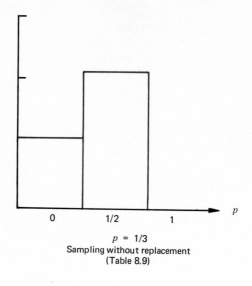

p = 1/3
Sampling without replacement
(Table 8.9)

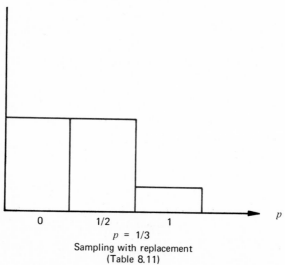

p = 1/3
Sampling with replacement
(Table 8.11)

Figure 8.11 *Sampling distributions of $\hat{p}$ (Tables 8.9 and 8.11)*

to find the variance of $\hat{p}$. If the sampling is done without replacement, this is

$$\sigma_{\hat{p}}^2 = 0^2 \times \frac{1}{3} + \left(\frac{1}{2}\right)^2 \times \frac{2}{3} - \left(\frac{1}{3}\right)^2$$

$$= 0 + \frac{1}{4} \times \frac{2}{3} - \frac{1}{9}$$

$$= 0 + \frac{1}{6} - \frac{1}{9}$$

$$= \frac{1}{18}.$$

If the sampling is done with replacement, we obtain

$$\sigma_{\hat{p}}^2 = 0^2 \times \frac{4}{9} + \left(\frac{1}{2}\right)^2 \times \frac{4}{9} + 1^2 \times \frac{1}{9} - \left(\frac{1}{3}\right)^2$$

$$= 0 + \frac{1}{4} \times \frac{4}{9} + 1 \times \frac{1}{9} - \frac{1}{9}$$

$$= 0 + \frac{1}{9} + \frac{1}{9} - \frac{1}{9}$$

$$= \frac{1}{9}.$$

We note that the variance of a single trial is

$$\sigma^2 = pq = \frac{1}{3} \times \frac{2}{3} = \frac{2}{9}.$$

For sampling without replacement the finite population correction factor agrees with our computation:

$$\sigma_{\hat{p}}^2 = \frac{N-n}{N-1} \times \frac{\sigma^2}{n} = \frac{N-n}{N-1} \times \frac{pq}{n} = \frac{3-2}{3-1} \times \frac{2/9}{2} = \frac{1}{2} \times \frac{1}{9} = \frac{1}{18}.$$

For sampling with replacement we have

$$\sigma_{\hat{p}}^2 = \frac{\sigma^2}{n} = \frac{pq}{n} = \frac{2/9}{2} = \frac{1}{9}.$$

Comparing the variances, we see that the method of sampling without replacement is subject to less variability than sampling with replacement. Hence, sampling without replacement leads to more precise estimates. Another example of sampling without replacement was discussed in Section 7.3.2.

*8.8   GENERAL CONCEPTS OF ESTIMATION

8.8.1   General Nature of an Estimation Problem

We consider the general problem of estimating a population characteristic or parameter θ on the basis of a sample drawn from that population. We may have a choice of statistics to use. For instance, since the mean of a symmetric distribution (such as a normal distribution) is also the median of that distribution, we could use either the sample mean or sample median as an estimate. To estimate a population variance when the mean is unknown, we might use either s^2, which is $\Sigma_{i=1}^{n} (x_i - \bar{x})^2/(n-1)$, or alternatively, $\Sigma_{i=1}^{n} (x_i - \bar{x})^2/n$, the arithmetic average of the squared deviations from the sample mean. How do we choose between various competing estimates?

In order to think about answers to this question, let us review the setting. (i) An unknown parameter, θ, is to be estimated. (ii) The sample from the parent distribution having parameter θ contains information about θ. We use a function of the sample (a statistic) as an estimate; that is, we compute a number (the estimate) from the sample. (iii) The estimate varies from sample to sample; that is, the estimate has a sampling distribution. The decision as to which of several estimates is "best" must depend upon comparison of salient characteristics of their sampling distributions.

8.8.2   Bias

If an estimate is to be good, it is essential that the center of its sampling distribution be equal to, or nearly equal to, the true value of the parameter. When we use the *mean* to indicate the center of the sampling distribution, we are led to the concept of an "unbiased estimate." A statistic $\hat{\theta}$ is said to be an *unbiased estimate* of a parameter θ if the mean of its sampling distribution is θ, that is, if $\mu_{\hat{\theta}} = \theta$. For example, the mean of the sampling distribution of the sample mean is the mean of the parent population, so the sample mean is an unbiased estimate of the population mean. One reason for using s^2 instead of $\Sigma_{i=1}^{n} (x_i - \bar{x})^2/n$ is that s^2 is an unbiased estimate of σ^2 whereas the latter on the average underestimates σ^2. The fact that the mean of the sampling distribution of an unbiased estimate is equal to the parameter insures that the sampling distribution of the estimate is centered (in terms of the mean) at the right value.

The *bias* of an estimate $\hat{\theta}$ is the mean of its sampling distribution minus the parameter value θ, namely $\mu_{\hat{\theta}} - \theta$. (The bias of an unbiased estimate is 0.) If the bias is small, even though different from 0, we may find this small bias acceptable. (See Section 8.8.4.)

8.8.3   Variability

It is important that the sampling distribution of an estimate be concentrated around the parameter value. The standard deviation, or equivalently the variance, of the sampling distribution of an estimate $\hat{\theta}$ is a measure of its concentration about its mean, $\mu_{\hat{\theta}}$. When an estimate is unbiased, $\mu_{\hat{\theta}} = \theta$, the variance is a measure of concentration of the sampling distribution about θ itself. Hence, we want the variance of the sampling distribution to be small. When we have a choice between two unbiased estimates, we usually choose the one with smaller variance. For instance, we have discussed the fact that to estimate the mean of a normal distribution the sample mean is preferred to the sample median because the variance of the sample mean is σ^2/n whereas the variance of the sample median is approximately $1.57\sigma^2/n$.

For large samples, the sampling distributions of many useful statistics are well approximated by normal distributions, so that the mean and the standard deviation of the sampling distribution characterize it. Hence, we emphasize the desirability of a small bias and a small standard deviation.

8.8.4   Mean Squared Error

Instead of considering a measure of location and a measure of variability to characterize the sampling distribution of an estimate, we

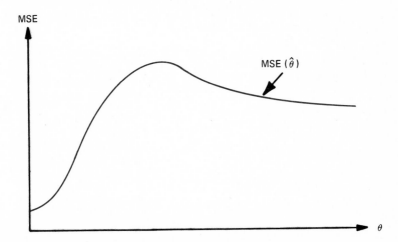

Figure 8.12 *Mean squared error of an estimate $\hat{\theta}$*

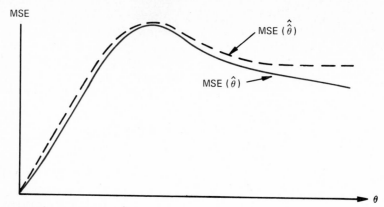

Figure 8.13 $\hat{\theta}$ is better than $\hat{\hat{\theta}}$ (as in Fig. 8.14)

may want to use a single measure. A possible measure is the *mean squared error*. The mean squared error of an estimate $\hat{\theta}$, MSE $(\hat{\theta})$, is defined as the average value of $(\hat{\theta} - \theta)^2$.

The interpretation of MSE $(\hat{\theta})$ is as follows: The quantity $(\hat{\theta} - \theta)^2$ is the square of the distance between the parameter and its estimate; that is, it is the squared error. The mean squared error is the average of the squared distance between the estimate and the parameter. The size of the mean squared error may depend upon the value of the parameter θ; Figure 8.12 shows how the mean squared error may depend on the value of the parameter.

If one estimate $\hat{\theta}$ has a MSE-curve that is always below that of another estimate $\hat{\hat{\theta}}$, as in Figure 8.13, we say that $\hat{\theta}$ is better than $\hat{\hat{\theta}}$.

Often the MSE curves of two estimates cross, as in Figure 8.14. Then neither estimate is universally better than the other. (If we knew which

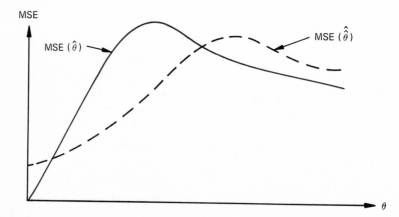

Figure 8.14 *Neither estimate is better*

region the true value of θ fell in, we might be able to make a choice of one estimate over the other.)

The criterion of small mean squared error has the nice property of combining in itself the two criteria of small bias and small variance, for it can be shown that

(8.6) $$\text{MSE}\,(\hat{\theta}) = \text{Var}\,(\hat{\theta}) + [\text{Bias}\,(\hat{\theta})]^2.$$

8.9 GENERAL REMARKS ON CONFIDENCE INTERVALS

8.9.1 Confidence Intervals Based on Approximately Normal Estimates

We have considered problems of statistical inference about means and proportions in one population, in two populations, and in one bivariate population. All the problems we discussed fit the following model. We have a point estimate $\hat{\theta}$ of θ. The estimate $\hat{\theta}$ is (at least approximately) normally distributed. The standard deviation of the estimate, $\sigma_{\hat{\theta}}$ (that is, its standard error), is known or else we can construct an estimate of it $(\hat{\sigma}_{\hat{\theta}})$ using the data at hand. The 95% confidence interval for θ when the standard deviation of the estimate is known is then of the form

$$(\hat{\theta} - 1.96\sigma_{\hat{\theta}},\ \hat{\theta} + 1.96\sigma_{\hat{\theta}}).$$

A good, rough-and-ready, approximate 95% confidence interval is obtained by adding to and subtracting from $\hat{\theta}$ the quantity $2\sigma_{\hat{\theta}}$. When $\sigma_{\hat{\theta}}$ is unknown we substitute the estimate $\hat{\sigma}_{\hat{\theta}}$.

8.9.2 Effect of Sample Size

The length of the confidence interval depends upon the standard deviation of the estimate used. The standard deviation is usually inversely proportional to the square root of the sample size. Thus to cut the length of the confidence interval in half, we have to increase the sample size by a factor of 4, from n to $4n$.

8.9.3 Effect of Confidence Coefficient

For a random sample of size $n = 1600$ from a normal distribution with $\sigma = 100$, a 95% confidence interval for μ is, since $\sigma_{\bar{x}} = \sigma/\sqrt{n} = 2.5$,

$$(\bar{x} - 1.96 \times 2.5,\ \bar{x} + 1.96 \times 2.5) \quad \text{or} \quad (\bar{x} - 4.90,\ \bar{x} + 4.90).$$

The width of this interval is

$$2 \times 4.90 = 9.80.$$

A 99% interval is

$$(\bar{x} - 2.576 \times 2.5, \ \bar{x} + 2.576 \times 2.5) \qquad \text{or} \qquad (\bar{x} - 6.44, \ \bar{x} + 6.44).$$

The width of this interval is

$$2 \times 6.44 = 12.88,$$

which is larger than 9.80, the width of the 95% interval. The greater the confidence, the wider the interval must be.

*8.9.4   Asymmetrical Confidence Intervals

One-Sided Interval.   If we are interested in how long light bulbs burn, we may be especially interested in providing a plausible *lower* limit for the mean lifetime.

For random samples of size n from a normal distribution with mean μ and standard deviation σ, the probability is .95 that the value of the sample mean $\bar{x}$ will be less than $\mu + 1.645\sigma/\sqrt{n}$:

$$\Pr\left(\bar{x} < \mu + 1.645\,\frac{\sigma}{\sqrt{n}}\right) = .95.$$

If we subtract $1.645\sigma/\sqrt{n}$ from both sides of the inequality, we obtain

$$\Pr\left(\bar{x} - 1.645\,\frac{\sigma}{\sqrt{n}} < \mu\right) = .95,$$

or, equivalently,

$$\Pr\left(\mu > \bar{x} - 1.645\,\frac{\sigma}{\sqrt{n}}\right) = .95.$$

Thus the interval

$$\left(\bar{x} - 1.645\,\frac{\sigma}{\sqrt{n}}, \ \infty\right)$$

is a 95% confidence interval for μ. It is called a *one-sided* interval because the upper limit is infinity.

If we are testing a new drug for persons with high blood pressure, we may be interested in an upper limit for the mean blood pressure of persons treated with the drug. Such an interval takes the form

$$\left(-\infty, \ \bar{x} + 1.645 \ \frac{\sigma}{\sqrt{n}}\right)$$

if the confidence coefficient is to be .95. If the confidence coefficient is to be .99, the interval is

$$\left(-\infty, \ \bar{x} + 2.326 \ \frac{\sigma}{\sqrt{n}}\right).$$

Intervals with Unequal Tails. A metals company produces gold ingots which are supposed to weigh 28 ounces. To check this, a government inspector weighs a sample of ingots, computes $\bar{x}$, and forms a confidence interval for μ. The inspector is more interested in protecting the public from short weight than in protecting the metals company. Hence, it is a more serious mistake to include erroneously large values of μ in the interval than it is to include erroneously small values. He or she could use as the 95% confidence interval the interval described by

$$\left(\bar{x} - 2.326 \ \frac{\sigma}{\sqrt{n}}, \ \bar{x} + 1.751 \ \frac{\sigma}{\sqrt{n}}\right).$$

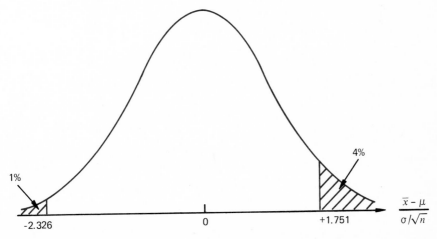

Figure 8.15 *Sampling distribution of* $\dfrac{\bar{x} - \mu}{\sigma/\sqrt{n}}$

This is

$$\left(\bar{x} - z(.01) \frac{\sigma}{\sqrt{n}} \, , \; \bar{x} + z(.04) \frac{\sigma}{\sqrt{n}} \right).$$

This interval is derived from the fact that

$$\Pr\left(-2.326 < \frac{\bar{x} - \mu}{\sigma/\sqrt{n}} < +1.751 \right) = .95.$$

(See Figure 8.15.) The probabilities in the *tails* are made equal to 1% and 4% instead of both being equal to $2\frac{1}{2}$%.

A 95% confidence interval with both tails having probability $2\frac{1}{2}$% is called an *equal-tails* interval. It can be shown that when the sampling distribution of the estimate is symmetric and unimodal, as is the case with $\bar{x}$ based on a random sample from a normal population, then among all intervals with confidence coefficient $1 - \alpha$, the equal-tails interval (that is, the one with both tails having probability $\alpha/2$) is shortest. (See Exercise 8.27.)

SUMMARY

An estimate of a parameter is *unbiased* if the mean of its sampling distribution is equal to that parameter, no matter what the value of the parameter is. The sample mean is an unbiased estimate of the population mean. The sample variance is an unbiased estimate of the population variance.

A *confidence interval* with *level of confidence* equal to $1 - \alpha$ is an interval produced by a procedure which gives intervals that include the true parameter value with probability $1 - \alpha$.

Student's t-distribution takes into account the extra variability introduced by having to estimate the population standard deviation.

A short-cut, approximate 95% interval for a proportion is obtained by adding to and subtracting from the sample proportion the quantity $1/\sqrt{n}$, where n is the sample size.

A short-cut, approximate 95% interval for any parameter the estimate of which is approximately normally distributed is obtained by adding to and subtracting from the estimate twice its standard deviation, or estimated standard deviation.

The sample median is an estimate of the population median.

The sample size is much more important than the population size in determining the precision of the sample mean. This is how polls based on thousands of individuals can be relevant in predicting the behavior of a country of millions of people.

APPENDIX 8A   Unbiasedness of s^2

We shall now show that s^2 is an unbiased estimate of σ^2 when sampling from an infinite population; that is, that the mean of the sampling distribution of s^2 is equal to σ^2:

(8.7) $$\mu_{s^2} = \sigma^2.$$

Since

$$s^2 = \frac{\displaystyle\sum_{i=1}^{n} (x_i - \bar{x})^2}{n-1},$$

$$\mu_{s^2} = \frac{E\left[\displaystyle\sum_{i=1}^{n} (x_i - \bar{x})^2\right]}{n-1},$$

where E, for "mathematical expectation," means averaging with respect to the probability distribution of $x_1, \ldots, x_n$. Thus to prove (8.7) we have to show that

$$\frac{E\left[\displaystyle\sum_{i=1}^{n} (x_i - \bar{x})^2\right]}{n-1} = \sigma^2,$$

or

$$E\left[\sum_{i=1}^{n} (x - \bar{x})^2\right] = (n-1)\sigma^2.$$

Recall from Chapter 4 that

$$\sum_{i=1}^{n} (x_i - \mu)^2 = \sum_{i=1}^{n} (x_i - \bar{x})^2 + n(\bar{x} - \mu)^2.$$

This is the same as

$$\sum_{i=1}^{n} (x_i - \bar{x})^2 = \sum_{i=1}^{n} (x_i - \mu)^2 - n(\bar{x} - \mu)^2.$$

The mathematical expectation is

$$E\left[\sum_{i=1}^{n} (x_i - \bar{x})^2\right] = E\left[\sum_{i=1}^{n} (x_i - \mu)^2\right] - nE\left[(\bar{x} - \mu)^2\right].$$

The first term on the right-hand side is

$$E \sum_{i=1}^{n} (x_i - \mu)^2 = \sum_{i=1}^{n} E (x_i - \mu)^2$$

$$= \sum_{i=1}^{n} \sigma^2$$

$$= n\sigma^2.$$

The second term contains

$$E\left[(\bar{x} - \mu)^2\right] = E\left[(\bar{x} - \mu_{\bar{x}})^2\right]$$

$$= \sigma_{\bar{x}}^2$$

$$= \frac{\sigma^2}{n}.$$

Thus, collecting terms, we have

$$E\left[\sum_{i=1}^{n} (x_i - \bar{x})^2\right] = n\sigma^2 - n\frac{\sigma^2}{n}$$

$$= n\sigma^2 - \sigma^2$$

$$= (n-1)\sigma^2,$$

as was to be shown.

Estimation of the Standard Deviation.   We use s, the square root of s^2, to estimate σ, which is the square root of σ^2. The sampling distribution of s tends to be concentrated around σ because the sampling distribution of s^2 tends to be concentrated around σ^2.

This follows from the fact that s^2 is an unbiased estimate of σ^2, no matter what the parent population is.

Note, however, that even though s is not an unbiased estimate of σ, it is usually used as the estimate of σ. What constitutes an unbiased estimate of σ depends upon the particular parent population. The problem of constructing an unbiased estimate of σ, even in the case of a normal parent distribution, is beyond the scope of this book.

APPENDIX 8B   The Continuity Adjustment

Proportions.   The number of heads in n tosses of a coin can be any of the values $0, 1, \ldots, n$. In a histogram of the binomial distribution, the probability of exactly r heads is represented by the area of the rectangle whose base extends from $r - 1/2$ to $r + 1/2$. This probability is approximated by the area under the appropriate normal curve from $r - 1/2$ to $r + 1/2$. The adding and subtracting of 1/2 is called the continuity adjustment, since we need to make this adjustment when approximating a discrete distribution by a continuous one.

As was seen in Appendix 7B, if a and b are two integers, then

$$\Pr(a < \# \text{ heads } < b)$$

is approximated by

$$\Phi\left(\frac{b - 1/2 - np}{\sqrt{npq}}\right) - \Phi\left(\frac{a + 1/2 - np}{\sqrt{npq}}\right).$$

To make this probability approximately .95 we write

$$\frac{b - 1/2 - np}{\sqrt{npq}} = 1.96 \qquad \text{and} \qquad \frac{a + 1/2 - np}{\sqrt{npq}} = -1.96.$$

This gives

$$b = np + 1/2 + 1.96\sqrt{npq} \qquad \text{and} \qquad a = np - 1/2 - 1.96\sqrt{npq}.$$

Thus

$$.95 = \Pr\left(np - \frac{1}{2} - 1.96\sqrt{npq} < \# \text{ heads } < np + \frac{1}{2} + 1.96\sqrt{npq}\right),$$

or dividing through by n, since $\hat{p} = \# \text{ heads}/n$,

$$.95 = \Pr\left(p - \frac{1}{2n} - 1.96\frac{\sqrt{pq}}{\sqrt{n}} < \hat{p} < p + \frac{1}{2n} + 1.96\frac{\sqrt{pq}}{\sqrt{n}}\right),$$

which is equivalent to

$$.95 = \Pr\left[-\left(1.96\frac{\sqrt{pq}}{\sqrt{n}} + \frac{1}{2n}\right) < \hat{p} - p < \left(1.96\frac{\sqrt{pq}}{\sqrt{n}} + \frac{1}{2n}\right)\right].$$

Thus, with probability .95, $\hat{p}$ is within a distance

$$1.96\,\frac{\sqrt{pq}}{\sqrt{n}} + \frac{1}{2n}$$

of p; that is, p is within that distance of $\hat{p}$ with probability .95:

$$.95 = \Pr\left(\hat{p} - \frac{1}{2n} - 1.96\,\frac{\sqrt{pq}}{\sqrt{n}} < p < \hat{p} + \frac{1}{2n} + 1.96\,\frac{\sqrt{pq}}{\sqrt{n}}\right).$$

When we substitute $\hat{p}\hat{q}$ for pq we obtain an approximate 95% confidence interval of

$$\left(\hat{p} - \frac{1}{2n} - 1.96\,\frac{\sqrt{\hat{p}\hat{q}}}{\sqrt{n}}\,,\ \hat{p} + \frac{1}{2n} + 1.96\,\frac{\sqrt{\hat{p}\hat{q}}}{\sqrt{n}}\right).$$

The effect of the continuity correction is to *increase* the width of the confidence interval.

APPENDIX 8C   Derivation of the Distribution-Free Confidence Interval for the Median

Now we give the derivation of a distribution-free confidence interval for the population median M discussed in Section 8.5.2. We illustrate a 95% confidence interval.

The interval $(x_{(r)}, x_{(n-r+1)})$ will be a 95% confidence interval if r is the largest integer for which

$$\Pr(x_{(r)} < M < x_{(n-r+1)}) \geq .95,$$

or

$$\Pr(x_{(r)} \geq M) + \Pr(x_{(n-r+1)} \leq M) \leq .05.$$

This gives

$$.025 \geq \Pr(x_{(r)} \geq M)$$

$$= \Pr(\text{at most } r - 1 \text{ observations} \leq M).$$

If the variable y is the number of observations less than M, then y has a binomial distribution with parameters n and $p = 1/2$ because the probability of an observation being less than or equal to the median is $1/2$. Its mean is $n/2$, its variance $n/4$, and its standard deviation $\sqrt{n}/2$.

The above requirement is

$$.025 \geq \Pr(y \leq r - 1).$$

Making the continuity adjustment, we have

$$.025 \geq \Pr(y \leq r - 1/2).$$

This is the same as

$$.025 \geq \Pr\left(\frac{y - n/2}{\sqrt{n}/2} \leq \frac{r - 1/2 - n/2}{\sqrt{n}/2}\right)$$

$$\doteq \Pr\left(z \leq \frac{r - 1/2 - n/2}{\sqrt{n}/2}\right),$$

where z has the standard normal distribution. Since

$$.025 = \Pr(z \leq -1.96),$$

this gives

$$\frac{r - 1/2 - n/2}{\sqrt{n}/2} = -1.96,$$

or

$$r = \frac{n}{2} + \frac{1}{2} - 1.96 \frac{\sqrt{n}}{2}$$

$$= \frac{n + 1}{2} - 0.98\sqrt{n}.$$

(Thus, approximately, $r = (n + 1)/2 - \sqrt{n}$.)

For an interval with confidence coefficient $1 - \alpha$, the formula for r is

$$r = \frac{n + 1}{2} - \frac{1}{2}z\left(\frac{\alpha}{2}\right)\sqrt{n}.$$

Usually r will not be an integer. A confidence interval that is conservative in the sense of giving confidence *at least* $1 - \alpha$ is obtained by taking r to be the integer below the value given by the formula.

EXERCISES

8.1 (Sec. 8.3.1) A population distribution of incomes has a standard deviation of $1,000. If the mean income for a random sample of 100 persons was $10,300, estimate the population mean by a 95% confidence interval.

8.2 (continuation) Repeat Exercise 8.1, but make a 99% confidence interval.

8.3 (continuation) Repeat Exercise 8.1, but make a 90% confidence interval.

8.4 (Sec. 8.3.1) A population distribution of weights has a standard deviation of 20 pounds. If the mean weight for a random sample of 400 persons was 161 pounds, estimate the population mean by a 95% confidence interval.

8.5 (continuation) Repeat Exercise 8.4, but make a 99% interval.

8.6 (continuation) Repeat Exercise 8.4, but make a 90% interval.

8.7 (Sec. 8.3.1) The mean diastolic blood pressure for a random sample of 25 people was 90 millimeters of mercury. If the standard deviation of individual blood pressure readings is known to be 10 millimeters of mercury, estimate the population mean with a 95% confidence interval.

8.8 (Sec. 8.3.1) Suppose it is known that the standard deviation of the lifetimes of electric light bulbs is 480 hours, and we obtain a mean lifetime of 493 hours for a sample of 64 bulbs.
 (a) Give a 95% confidence interval for the true mean lifetime.
 (b) The electric company claims the true mean lifetime is 500 hours. Is their claim consistent with the data? [Hint: Does the confidence interval include the value of 500 hours?]

8.9 (Sec. 8.3.2) The average cholesterol level for a random sample of 25 people of age 30 was $\bar{x} = 214$ units (milligrams of cholesterol per 100 milliliters of blood). The sample variance was $s^2 = 456$. (Note that the variability in the observations is due to two sources: differences between the subjects' true cholesterol levels and random errors made in the biochemical determination of the individual cholesterol levels.)
 (a) Give an 80% two-sided confidence interval for the population mean cholesterol level. Note that the relatively small sample size here suggests use of Student's t.
 (b) What property of the parent distribution will justify using this procedure for the confidence interval?

8.10 (Sec. 8.3.2) Make a 90% two-sided confidence interval for the population mean height of 5-year-old boys, using the data of Exercise 2.8 (Table 2.14) as a sample.

8.11 (Sec. 8.3.2) Make a 90% two-sided confidence interval for the population mean height of 11-year-old boys, treating the 1293 boys represented in Table 2.11 as a sample.

8.12 (continuation) These data were published in 1877. Do you think 11-year-old boys today have the same height distribution? Document your opinion, if possible, by reference to a book or article or to an expert such as a pediatrician or teacher of physical education.

8.13 Three dozen people were invited to a party and asked to "RSVP." Table 8.12 is a cross classification according to RSVP and attendance.

TABLE 8.12
Cross-Tabulation of 36 Persons, by RSVP and Attendance

		ATTENDED		
		No	*Yes*	*Total*
RESULT OF RSVP	*No response*	7	5	12
	No	4	1	5
	Yes	1	18	19
	Total	12	24	36

Suppose you invite to a party 55 persons whose behavior could be expected to be similar to these three dozen, and 38 say they will attend, 5 say they will not, and the other 12 do not respond. How would you predict the attendance at your party?

8.14 In early 1960, a pollster reported the following data based on interviews with registered voters. Each respondent was asked, "If John Kennedy were the Democratic candidate for president and Richard Nixon were the Republican candidate, which one would you like to see win?" The percentages are tabulated below. (The respondents who

	OCTOBER 1959	JANUARY 1960	MARCH 1960
Nixon	52.3%	51.1%	50.2%
Kennedy	47.7%	48.9%	49.8%

were undecided were not included in the calculation of the percentages.) The write-up included a statement that public opinion samplings of this size are subject to a margin of deviation of approximately 4% when the findings are at or very near the 50% mark.

(a) Compute 95% and 99% confidence intervals for the true proportion in favor of Nixon at each of the three time periods. You may take $n = 1000$.

(b) Discuss what further information you would need in order to make probability statements about the differences between poll results obtained at different times. Are the proportions at the three times independent?

(c) What do you think was meant by a "margin of deviation of approximately 4% . . . ?"

8.15   Twins are of two types: identical and fraternal. We want to estimate the number of identical twins in the population.

Fraternal twins are as often of different sex as of the same sex. Identical twins are, of course, of the same sex. We know the total number of twins in the population is 1,000,000. We count the number of twins in pairs of different sex and find 300,000. Give an estimate of the number of identical twins in the population. Show work to indicate your reasoning.

8.16 (Sec. 8.5.1)   The lifetimes of a random sample of 9 light bulbs were 1066, 1776, 1492, 753, 70, 353, 1984, 1945, 1914. Estimate the 90th percentile of the population.

8.17 (Sec. 8.5.1)   The lifetimes of a random sample of 15 light bulbs are given in Section 8.5. Estimate the 75th percentile of the population.

8.18 (Sec. 8.6)   In order to make a standardized assessment of their instructional project in terms of child development, the Ravenswood Children's Center of East Palo Alto, California, employed the Bettye Caldwell Preschool Inventory (revised) as an investigative tool to attain a measurement of individual growth. This test has questions ranging

TABLE 8.13
Summary of Scores on Preschool Inventory

		RIGHT	WRONG (OR DON'T KNOW)
Pretest (administered	Poorest Score	10	54
between 11/30/70	Best Score	63	1
and 12/4/70)	Mean Score	41	23
Post-test (administered	Poorest Score	10	54
between 5/9/71	Best Score	64	0
and 5/14/71)	Mean Score	47	17

SOURCE: Ravenswood City (Elementary) School District, E. Palo Alto, California.

from the simple identification of body parts to the manipulation of color- and shape-related objects.

The test was administered to 112 children during the pretesting period. There were 84 of that original group still in attendance at the time of the post-testing period, five months later. Only the test scores of these 84 children were analyzed and tabulated.

The range of achievement from low to high is reflected by the statistics in Table 8.13.

(a) What is the mean gain in number right?
(b) What additional information would you need to make a confidence interval for the population mean gain?

8.19 (Sec. 8.6)   Verify for the data in Table 8.4 that $\bar{d} = \bar{y} - \bar{x}$.

8.20 (Sec. 8.7.1)   In a national study of physicians the standard deviation of hours worked per week was 7 hours. A random sample of 49 physicians was taken from the staff of Mayo Clinic and a mean work-week of 50 hours was obtained. Give a 95% confidence interval for the mean work-week of all physicians in the Mayo Clinic. Assume there are 401 physicians altogether.

8.21   In order to estimate the number N of fish in a lake the following plan is to be used: (i) Catch 500 fish, tag them all, and release them. (ii) Then catch 600 fish and observe how many of them are tagged; call this number x. We shall assume that all fish—tagged or not—have the same probability of being caught. Since

$$\frac{500}{N} \doteq \frac{x}{600},$$

so that

$$N \doteq \frac{300{,}000}{x},$$

it is intended to use

$$\hat{N} = \frac{300{,}000}{x}$$

as an estimate of the total number of fish in the lake. In order to decide whether it is worthwhile to carry out this experiment, the sampling distribution of $\hat{N}$ is computed. Explain what is meant by the "sampling distribution" and how it can be useful in deciding whether the experiment should be carried out.

8.22   The distribution of number of children in a random sample of 10 families is given in Table 8.14. A statistician described the sample by saying that the average number of children is 2. What statistic was used as an estimate? Do you think this was a good choice?

TABLE 8.14
Distribution of Number of Children in Ten Families

Number of children per family	0	1	2	3
Number of families	1	1	7	1

8.23   What statistic is being used as an estimate in the following military problem from Thucydides [Warner (1954), p. 172]?

> "[The problem for the Athenians was] . . . to force their way over the enemy's surrounding wall. . . . Their methods were as follows: they constructed ladders to reach the top of the enemy's wall, and they did this by calculating the height of the wall from the number of layers of bricks at a point which was facing in their direction and had not been plastered. The layers were counted by a lot of people at the same time, and though some were likely to get the figure wrong, the majority would get it right, especially as they counted the layers frequently and were not so far away from the wall that they could not see it well enough for their purpose. Thus, guessing what the thickness of a single brick was, they calculated how long their ladders would have to be. . . ."

8.24   To estimate the mean length of logs floating down the river, we sit on the bank for 10 minutes and measure the lengths of the logs that pass during those 10 minutes and take the mean of those measurements.
(a) Is the resulting estimate unbiased? Why?
(b) Suppose we choose n logs at random from among those in the river and measure their lengths. Is the resulting estimate unbiased?

MATHEMATICAL EXERCISES

8.25 (Sec. 8.6)   Show algebraically that $\bar{d} = \bar{y} - \bar{x}$.

8.26 (Sec. 8.7.1)   Show that

$$\sqrt{\frac{N-n}{N-1}} \doteq \left(1 - \frac{n/N}{2}\right).$$

[Hint: Argue that $(N - n)/(N - 1) \doteq 1 - n/N$; then use the fact that $(1 - x)^2 \doteq 1 - 2x$ if x is small, and take $x = n/(2N)$.]

*8.27 (Sec. 8.9.4)   A random sample of size n is drawn from a normal distribution with known standard deviation. Show that, among all confidence intervals for the mean with confidence coefficient $1 - \alpha$, the equal-tails interval is the shortest.

8.28 (continuation)   Show that the result of Exercise 8.27 is true even if the standard deviation is estimated. [Hint: The reasoning is the same as for Exercise 8.27, but the relevant sampling distribution is a Student's t-distribution instead of a normal distribution.]

*8.29 (Sec. 8.8.4)   Prove that

$$\text{MSE } (\hat{\theta}) = \text{Var } (\hat{\theta}) + [\text{Bias } (\hat{\theta})]^2.$$

[See formula (8.6).]

8.30 (Sec. 8.5.2)   A random sample of four observations x_1, x_2, x_3, x_4 is to be made on a continuous variable.
(a) What is the probability that all four observations will be less than the population median?
(b) What is the probability that all four observations will be greater than the population median?
(c) Let $x_{(1)}, x_{(2)}, x_{(3)}, x_{(4)}$ denote the ordered sample ($x_{(1)} \leq x_{(2)} \leq x_{(3)} \leq x_{(4)}$). What is the probability that the interval $(x_{(1)}, x_{(4)})$ will fail to contain the population median?
(d) If the interval $(x_{(1)}, x_{(4)})$ is used as a confidence interval for the population median, what is the level of confidence?

8.31 (Sec. 8.5.2)   A random sample of 6 observations $x_1, x_2, \ldots, x_6$ is to be made on a continuous variable.
(a) What is the probability that all 6 observations will be less than the population median?
(b) What is the probability that all 6 observations will be greater than the population median?
(c) Let $x_{(1)}, x_{(2)}, \ldots, x_{(6)}$ denote the ordered sample ($x_{(1)} \leq x_{(2)} \leq \ldots \leq x_{(6)}$). What is the probability that the interval $(x_{(1)}, x_{(6)})$ will fail to contain the population median?
(d) If the interval $(x_{(1)}, x_{(6)})$ is used as a confidence interval for the population median, what is the level of confidence?

8.32 (Sec. 8.5.2)   Let $x_1, x_2, \ldots, x_n$ be a random sample of observations of a continuous random variable. Let $x_{(1)}, x_{(2)}, \ldots, x_{(n)}$ be the ordered sample. If $(x_{(1)}, x_{(n)})$ is used as a confidence interval for the population median, what is the level of confidence?

8.33   Suppose $\hat{\theta}$ is an unbiased estimate for θ. Give an argument to show that when the sampling distribution of $\hat{\theta}$ is symmetric, the equal-tails confidence intervals for θ are also symmetric, in the sense that the same number is added to and subtracted from $\hat{\theta}$ to form the confidence interval.

*8.34   (Sec. 8.8.4)   If it is known that a proportion p is not near zero or one, the estimate $\tilde{p} = (x + 1)/(n + 2)$, where x is the number of successes, might be more sensible than the usual estimate $\hat{p} = x/n$. Why?

8.35   Mass-produced items are marked with consecutive serial numbers as they are produced, starting with the number 1. It is desired to estimate the total number N of items which have been produced. The serial numbers of n such items are observed. Consider these n serial numbers as a random sample from the population of N serial numbers. This population is, then, the first N integers. The mean of the population is $(N + 1)/2$. In the case of each of the following estimates, state whether or not you would consider it to be a reasonable estimate of N, and why. Let $x_{(1)}, x_{(2)}, \ldots, x_{(n)}$ denote the ordered sample, $x_{(1)}$ being the minimum and $x_{(n)}$ the maximum.

(a) $\hat{N} = x_{(n)}$

(b) $\hat{N} = \dfrac{n + 1}{n} x_{(n)}$

(c) $\hat{N} = x_{(n)} + (x_{(1)} - 1)$

(d) $\hat{N} = 2\bar{x}$

(e) $\hat{N} = 2\bar{x} - 1$.

REFERENCES

Beyer, William H., ed. (1972). *Handbook of Tables for Probability and Statistics.* 2nd ed. Chemical Rubber Co., Cleveland, Ohio.

Pearson, E. S., and H. O. Hartley, eds. (1954). *Biometrika Tables for Statisticians.* Vol. 1. Cambridge University Press, Cambridge, England.

Warner, Rex, trans. (1954). *Thucydides, History of the Peloponnesian War.* Penguin Books, Baltimore, Maryland.

Zehna, P. W., and D. R. Barr (1970). *Tables of Common Probability Distributions.* U.S. Naval Postgraduate School, Monterey, California.

9 ANSWERING QUESTIONS ABOUT POPULATION CHARACTERISTICS

INTRODUCTION

It is often the purpose of a statistical investigation to answer a Yes or No question about some characteristic of a population. An election candidate, for example, may employ a pollster to determine whether the proportion of voters intending to vote for him does or does not exceed 1/2. The polio vaccine trial was designed so that medical researchers could decide whether the incidence rate of polio is or is not smaller in a population of persons inoculated with the vaccine than in a population of persons not inoculated with the vaccine. Industrial quality control involves determination as to whether the average strength, lifetime, or concentration of the products in each manufacturing batch fall within acceptable limits.

Chapter 8 dealt with the use of samples for the *estimation* of population characteristics. (How many? How tall? etc.) This chapter will discuss methods of using these estimates to provide Yes or No answers to questions about the measured population characteristics. Is the proportion of voters intending to vote for Jones greater than 1/2? Is the incidence rate of polio smaller among those inoculated with the vaccine? Statistical answers to such questions are not infallible, for a sample never provides complete or perfect information. A sample can, on occasion, lead to an incorrect conclusion. That is, the conclusions drawn from statistical inference are subject to errors due to *sampling variability*. This chapter will show how to take account of sampling variability and how to evaluate the probabilities of such errors.

Although the focus here will be on questions concerning means and proportions, the principles discussed apply to questions concerning other population parameters as well.

9.1 DECIDING WHETHER A POPULATION MEAN IS LARGE

9.1.1 The Procedure of Testing a Hypothesis

It is often desirable to know whether some characteristic of a population is larger than a specified value, or whether the obtained value of a given parameter is less than a value hypothesized for purposes of comparison. We may want to know, for instance, if the average IQ of children of scientists exceeds the average IQ of all children. Thus, if the mean IQ of all children is known to be 100, we want to know whether the mean IQ in the population consisting of scientists' children exceeds 100. We may want to know whether or not the average reading speed of second-grade children for material of a certain level of difficulty is less than 120 words per minute, or whether the average concentration of carbon monoxide in downtown Chicago is or is not less than some specified noxious concentration.

Since the value of the population characteristic is unknown, the information provided by a sample from the population is used to answer the question of whether or not the population quantity is larger than the specified or hypothesized value. In the present discussion the sample is assumed to be *random*. We know that when we estimate the value of a parameter there is some sampling error; an estimate hardly ever coincides exactly with the parameter. Similarly, when a sample is used to answer a question about the population, sampling error may sometimes lead us astray and hence must be taken into account.

A question that might be investigated by drawing a sample from a population is whether students now are studying more than they used to. In the spring of 1970 there was a great deal of political activity on campus, and interest in academic matters appeared to be at an ebb. Some have asserted that since that time students have taken more interest in learning and scholarship, while others claim that there has been no change. Is the amount of studying about the same as before, or is it greater?

Suppose that during the spring of 1970 all undergraduates at a certain university reported the number of hours spent on their studies (attending lectures, reading in the library, etc.) during a certain week; the average was 40 hours; the standard deviation, 10 hours. This year a sample of 25 students was taken by assigning consecutive numbers to all registered students and including in the sample those 25 students whose numbers corresponded to 25 different numbers from a table of random numbers. Each student in the sample was interviewed about current interests and activities. Among other things, the student reported how many hours were spent on studies during the week which corresponded to the week of the previous investigation. We ask the question, "In the population of *all* students at the university, is the average time spent on studies 40 hours per week, or is it greater?" We use the information in the sample to arrive at the answer.

It seems reasonable that we should calculate the mean time spent on studies by the 25 students in the sample and compare that with the 40 hours established as the (population) average earlier. If the sample mean is much larger than 40 hours, we will be led to believe that the population average this year is greater than before. However, if the sample mean is less than 40 or only slightly more than 40, we would not conclude that the average time of study has increased. The investigator needs a way of determining whether the sample mean is large enough to warrant the conclusion that the population mean is large. We shall explain such a statistical procedure.

The possibility that the population mean is still 40 is called the *null hypothesis,* denoted H. The possibility that it is greater than 40 is called the *alternative hypothesis.* If the null hypothesis is true, there has been no change (that is, the change is "null"), and further investigation may be uninteresting. If the investigator decides the alternative is true, he or she will want to see how the students' attitudes and habits have changed and why, and will make a follow-up study. (At this point we do not admit the possibility that the population mean is less than 40.)

For random samples of 25, we know that the mean of the sampling distribution of the sample mean, $\mu_{\bar{x}}$, is equal to the mean of the parent population, μ, and that the standard deviation of this sampling distribution, $\sigma_{\bar{x}}$, is the standard deviation of the parent population, σ, divided by 5, the square root of the sample size. If the parent population is about the same as in the earlier year, then its standard deviation, σ, is 10, and the standard deviation of the sampling distribution is $\sigma/5 = 10/5 = 2$. (The population of all students in the university is large enough that the finite population correction can be ignored.)

The sampling distribution of the mean is approximately normal in this case with a standard deviation of 2. The probability is .05 that the sample mean exceeds the population mean by more than 1.645 standard deviations. If the mean of the population is 40, the probability is .05 that the sample mean is greater than $40 + 1.645 \times 2 = 43.29$. Suppose we use the rule that if the sample mean exceeds 43.29 we shall reject the null hypothesis; otherwise we accept the null hypothesis. The procedure is diagrammed in Figure 9.1.

If we use this rule, then when the null hypothesis is true we have a

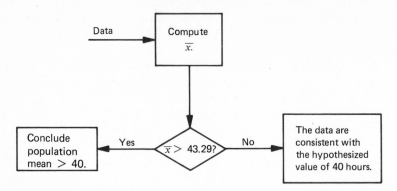

Figure 9.1 *Schematic diagram of a hypothesis test concerning the average number of hours studied*

probability of .05 of making the mistake of rejecting it. This probability is called the *significance level* of the test. We choose the significance level to be a small probability so that the risk of wasting time and money in a follow-up study when in fact the average number of hours studied has not changed is small.

The value 43.29 is called the *significance point*. Values greater than 43.29 are *significantly different* from 40, at the 5% level.

If the mean for our sample of students is 46.2, the above rule would lead us to reject the null hypothesis and conclude that the population mean is greater than 40; that is, we would conclude that students spend more time on the average on their studies now than previously. We know that a sample mean this large or larger is unlikely (has probability of less than .05) if the null hypothesis is true, and hence we are loath to believe the null hypothesis is true.

This type of procedure permits us to control the probability of rejecting the null hypothesis when in fact it is true. If we want to be more cautious about such a false rejection, we could take a smaller probability as the significance level. If we used .01, then the cut-off point would be $40 + 2.326 \times 2 = 44.652$. The sample mean of 46.2 would lead to rejection of the null hypothesis with this significance level, too. On the other hand, a sample mean of 44.2 would lead us to reject the null hypothesis at the 5% significance level, but not at the 1% level.

In the study of group pressure described in Section 1.1.5, the null hypothesis was that the probability of making errors in the experimental situation, under group pressure, was the same as when not under pressure. The numbers of errors made by the two sets of subjects are so dramatically different that the null hypothesis must be rejected. The consequence of this was intensive study of various aspects of group pressure; if the experiment had not led to rejection of the null hypothesis, that experimental set-up would not have been used in future experiments, and the concept of group pressure would have been studied in a different way.

In the case of testing of polio vaccine, the consequence of rejection of the null hypothesis was mass inoculation of children all over the country. Although the scientific question was whether the vaccine was or was not effective, the practical result was such that the experiment led to a decision whether to use the vaccine or not. Sometimes, as in this case, a test is considered a *decision procedure*; that is, some action will be taken or not taken according to the outcome of the test. The result of the test is not simply to reject or accept the null hypothesis; it is to take some action or not.

The group-pressure and polio vaccine examples will be discussed further in Chapter 10, which deals with comparison of two samples.

As another example, we refer to the path-breaking study of Otto Klineberg (1935) on the influence of the northern environment on the intelligence test scores of Blacks. A large number of Black 12-year olds were given IQ tests. The mean score for boys born in New York City was 86.9 with a standard deviation of about 29. The 56 boys who had lived in New York City for one or two years were considered as a sample from the population of newly arrived migrants from the South. A question is whether the mean IQ score of the "recent-migrant population" is equal to that of the northern-born residents. We test the null hypothesis that the population mean is 86.9 against the alternative hypothesis that it is lower than 86.9. The standard deviation of the sampling distribution of means of random samples of size 56 is $29/\sqrt{56} = 29/7.48 = 3.89$. At the 1% level of significance the null hypothesis is to be rejected if the sample mean is less than the hypothesized mean by more than 2.326 standard deviations; that is, if it is less than $86.9 - 2.326 \times 3.89 = 86.9 - 9.1 = 77.8$. The observed sample mean of 64.2 leads to rejection of the hypothesis.* (The study showed that the average IQ's were higher for boys living in New York City longer, and similar results held for girls.)

9.1.2 Types of Error: Effectiveness of Procedures

Types of Error.   In the hours-of-study example the significance level is the probability of concluding that students are now spending more time on their studies than formerly, if indeed, on the average, they are still spending exactly the same amount of time (40 hours per

---

*Table 8 of Klineberg (1935).

week). By choosing the significance point appropriately, the test can be constructed so as to make this probability small. However, there is the other possibility of error in the procedure, namely, that of concluding that students are *not* studying more if they actually *are*. There are *two* kinds of errors. If the students average 40 hours of study, we may decide incorrectly that they are studying more; this is called a *Type I error*.

DEFINITION   *A* Type I error *is the error of rejecting the null hypothesis when it is true.*

If the students average more than 40 hours of study, we may erroneously conclude they do not; this is a *Type II error*.

DEFINITION   *A* Type II error *is the error of accepting the null hypothesis when it is false.*

TABLE 9.1
Possible Errors in Investigation of Hours of Study

		CONCLUSION ABOUT CURRENT POPULATION MEAN	
		40 Hours	*> 40 Hours*
TRUE VALUE OF CURRENT POPULATION MEAN	*40 Hours*	Correct conclusion	Type I error
	> 40 Hours	Type II error	Correct conclusion

TABLE 9.2
Possible Errors in Drawing Statistical Conclusions

		CONCLUSION	
		Accept Null Hypothesis	*Reject Null Hypothesis*
TRUE "STATE OF NATURE"	*Null Hypothesis*	Correct conclusion	Type I error
	Alternative Hypothesis	Type II error	Correct conclusion

These possibilities are set out in Table 9.1. The kind of error that can be made depends on the actual state of affairs (which, of course, is unknown to the investigator). The investigator does not know which type of error corresponds to actuality and so would like to keep the probabilities of both types of errors small.

These two possible types of error exist in any hypothesis-testing problem. They are displayed in Table 9.2 in general terms.

The probability of a Type II error, that is, the probability of accepting the null hypothesis when it is in fact false, depends on "how false" it is. If the population mean number of hours of study is very great, say 50 hours, it is very unlikely that we would draw a random sample with a mean so small that we would accept the null hypothesis. On the other hand, if the population mean is only a little more than 40, that fact will be hard to discover; we are likely to accept the null hypothesis.

Figure 9.2 shows three sampling distributions of the mean for a sample size 25 drawn from normal distributions with a standard deviation of 10 and means of 40, 42, and 44, respectively. When the population mean is 40, the probability that the sample mean exceeds 43.29 is .05, which is the significance level of the test which rejects the null hypothesis of a mean of 40 when the sample mean exceeds 43.29. The probability of the sample mean exceeding 43.29 is .259 when the population mean is 42, and it is .639 when the population mean is 44. The probability of rejecting the null hypothesis that the population mean is 40 is high when the true value of the population mean is large.

Power. Table 9.3 gives the probability that the sample mean exceeds 43.29, for various values of μ, the mean of the population sampled. The probabilities are computed as indicated in Figure 9.2. The graph of these values is given in Figure 9.3. The probability of rejection as dependent upon the mean μ of the parent population is called the *power function* of the test. (We say that the probability of rejection is a *function* of the mean of the parent population; that is, given any value of the mean, there is a corresponding value of the rejection probability.) The probability of acceptance, which is 1 minus the probability of rejection, is called the *operating characteristic* of the test (commonly referred to as the "O.C. curve").

The power increases as μ increases, equals the level .05 when $\mu = 40$, and increases to 1 as μ gets larger. For values of μ less than 40, the probability of rejection is less than .05. Thus, the procedure distinguishes a large value of μ from a value less than or equal to 40. Consequently, we extend the null hypothesis to be H: $\mu \leq 40$. If the data are

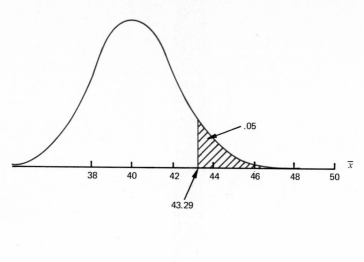

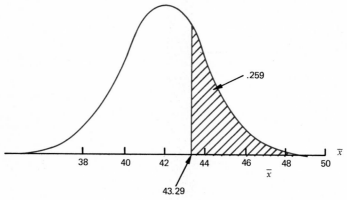

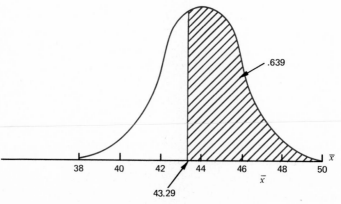

Figure 9.2 *Probability of $\bar{x} > 43.29$ for $\mu = 40, 42, 44$*

TABLE 9.3
Power Function of 5%-Level Test of H: $\mu \leq 40$ Against Alternative: $\mu > 40$
When $\sigma = 10$ and $n = 25$

VALUE OF μ	POWER
38	.004
39	.016
40	.050
41	.126
42	.259
43	.442
44	.639
45	.804
46	.912
47	.968
48	.991

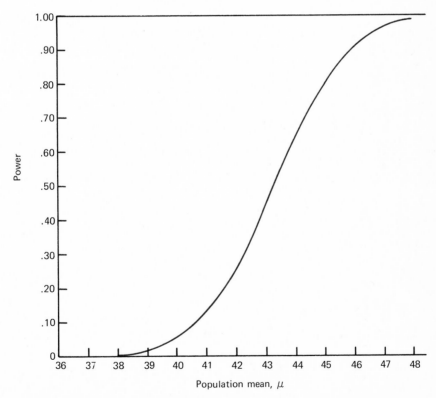

Figure 9.3 *Power function of 5%-level test of H: $\mu \leq 40$ against Alternative: $\mu > 40$ when $\sigma = 10$ and $n = 25$ (Table 9.3)*

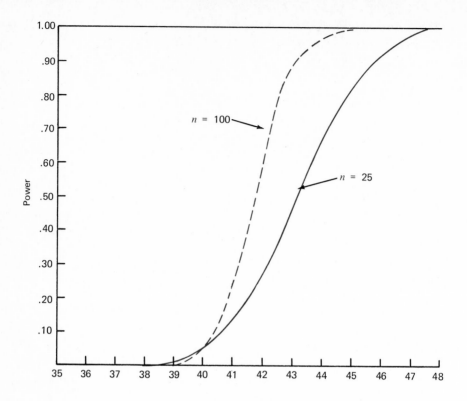

Figure 9.4 *Power functions of 5%-level test of $\mu \le 40$ when $\sigma = 10$ for $n = 25$ and $n = 100$*

such as to rule out a value of 40, they also rule out values smaller than 40. In our example, when the investigator asks whether the average number of hours of study is greater than before, the answer would be "No" if he or she knew that the average is *less* than before.*

For large values of μ, the investigator is almost certain to reject the null hypothesis, but for values of μ only a little larger than the hypothesized value of $\mu = 40$ the probability of rejection is only a little larger than the significance level of .05.

Effect of Sample Size. Of course, the performance of a test procedure is better if the sample size is larger. The power function for $n = 100$ reflects this fact. As seen in Figure 9.4, this curve is above the

---

*Writing the null hypothesis as an inequality as well as an equality is not followed in all texts.

one for $n = 25$ when $\mu > 40$; the power is greater and the probability of Type II error is less. For instance, if $\mu = 42$ the probability of rejecting the null hypothesis with a sample of 25 is .259, while the probability with a sample of 100 is .639.

Effect of Significance Level. The choice of significance level has an effect on the power function. The test of the null hypothesis $H: \mu \leq 40$ against the alternative $\mu > 40$ at significance level 1% calls for rejection of the null hypothesis if the sample mean is greater than 44.66. The probability of a Type I error, rejecting the null hypothesis when it is true, is reduced from .05 to .01, but the probability of rejecting the null hypothesis is also less for *any* value of μ. Table 9.4 and Figure 9.5 give the power function of the test; that is, the probability of rejecting the null hypothesis for various values of μ. For example, for the 1%-level tests, if $\mu = 42$, the probability of rejecting H is only .092, while it is .259 for the 5%-level test. The probability of a Type II error is increased, for any value of μ. Reducing the probability of Type I error increases the probabilities of Type II error.

Choice of Significance Level. In choosing the significance level, it is necessary to balance the undesirabilities of the two types of error. We want to discover a difference from the null hypothesis if it exists, but we do not want to claim there is a difference if it does not

TABLE 9.4
Power Function of 1%-Level Test of H: $\mu \leq 40$ Against Alternative: $\mu > 40$ When $\sigma = 10$ and $n = 25$

VALUE OF μ	POWER
38	.001
39	.002
40	.010
41	.034
42	.092
43	.203
44	.371
45	.568
46	.749
47	.879
48	.953
49	.985
50	.996

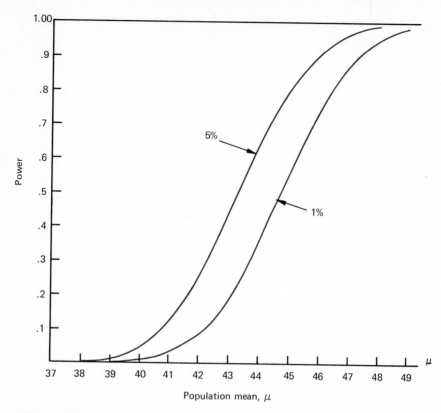

Figure 9.5 *Power functions of 1%- and 5%-level tests of $\mu \le 40$ against $\mu > 40$ when $\sigma = 10$ and $n = 25$ (Tables 9.3 and 9.4)*

exist. Although 5% and 1% are conventional levels of significance, there is no reason to restrict the choice to these. We can choose a significance level of 0.5% if we want to be careful in rejecting the null hypothesis or choose 10% if we want a relatively small chance of overlooking a difference from the null hypothesis.*

Rational choice of significance level involves simultaneous consideration of the power. If a 1%-level test gives reasonable power against reasonable alternatives, use it. Otherwise, use a 5%- or even a 10%-level test.

Summary. We conclude this section by stating the procedure in general terms. To test the null hypothesis $H: \mu \le \mu_0$, where μ_0 is a

---

*There is a technical reason for caution in using extremely small significance levels: the normal approximation to the sampling distribution of the mean is sometimes not very good for very extreme values.

specified number, against the alternative $\mu > \mu_0$ at significance level α, reject the null hypothesis if

$$\bar{x} > \mu_0 + z(\alpha) \frac{\sigma}{\sqrt{n}},$$

where $\bar{x}$ is the sample mean, n is the sample size, σ is the known standard deviation of the parent population, and $z(\alpha)$ is the number such that the area of the standard normal distribution to the right of this value is α. Such an *upper-tail* test was made in the hours-of-study example.

The test of $H: \mu \geq \mu_0$ against $\mu < \mu_0$ at the level α is to reject if

$$\bar{x} < \mu_0 - z(\alpha) \frac{\sigma}{\sqrt{n}}.$$

This is called a *lower-tail* test. A lower-tail test was made in the Klineberg example.

Note that the test of $H: \mu \leq \mu_0$ against $\mu > \mu_0$ (the upper-tail test) is equivalent to rejecting H if

$$\frac{\bar{x} - \mu_0}{\sigma/\sqrt{n}} > z(\alpha).$$

For example, if we choose the 5% level, then we reject H if $\bar{x}$ exceeds μ_0 by more than 1.645 standard deviations; that is, we reject H if

$$\frac{\bar{x} - \mu_0}{\sigma/\sqrt{n}} > 1.645.$$

The variable

$$\frac{\bar{x} - \mu_0}{\sigma/\sqrt{n}}$$

is

$$\frac{\bar{x} - \mu_0}{\sigma_{\bar{x}}},$$

the standardized value of $\bar{x}$ when $\mu = \mu_0$. The number 1.645 is the 95th percentile, or upper 5 percentage point, of the standard normal distribution. Similarly, the 5% lower-tail test consists in rejecting H if

$$\frac{\bar{x} - \mu_0}{\sigma/\sqrt{n}} < -1.645.$$

9.2 DECIDING WHETHER A POPULATION MEAN DIFFERS FROM A GIVEN VALUE

9.2.1 Procedures and Power

In many cases we want to know whether some characteristic of a population is equal to a given value or is different from it—either by being larger *or* smaller. We may want to know whether students are about as active in their studying now as previously, and we would be interested in a difference in either direction from the earlier average. The null hypothesis is that the population mean study time is 40 hours and the alternative is that the mean time is *different* from 40; it can be either less than 40 or greater than 40. Whereas previously we wanted to know whether studying had or had not *increased*, we are now interested in discovering whether studying has *changed* (either increased *or* decreased).

The procedure is to reject the null hypothesis if the observed sample mean is *either* very large *or* very small. The values that are "very large" and "very small" have a small probability when the null hypothesis is true. We know that the probability that a standard normal variable is more than 1.96 standard deviations away from the population mean is .05. If we take a sample of 25 from a normal population with mean 40 and standard deviation 10, the probability is .05 that the sample mean is outside the interval with endpoints $40 - 1.96 \times 2 = 36.08$ and $40 + 1.96 \times 2 = 43.92$. The test procedure is to reject the null hypothesis if the sample mean is less than 36.08 or greater than 43.92. The signif-

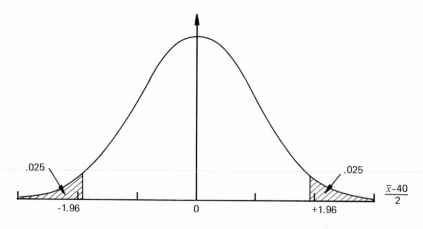

Figure 9.6 *The sum of the probabilities of the two tails shown is* .025 + .025 = .05

icance level is 5%. The test is called a *two-tailed* test because the probability .05 is the sum of the probabilities in the two tails of the normal distribution. (See Figure 9.6.) Another way of stating the rule for rejecting H is to say that H is rejected if

$$\frac{\bar{x} - 40}{2} > 1.96 \quad \text{or} \quad \frac{\bar{x} - 40}{2} < -1.96.$$

The number 1.96 is the 97.5 percentile, or upper $2\frac{1}{2}$ percentage point, of the standard normal distribution. The number -1.96 is the 2.5 percentile, or lower $2\frac{1}{2}$ percentage point.

The 1%-level test of the hypothesis H: $\mu = 40$ against the alternative $\mu \neq 40$ is to reject the null hypothesis if the sample mean is farther than 2.576 standard deviations away from 40; that is, if the sample mean is less than $40 - 2.576 \times 2 = 34.85$ or greater than $40 + 2.576 \times 2 = 45.15$.

TABLE 9.5
Power Function of 5%-Level Two-Tailed Test of H: $\mu = 40$ Against Alternative: $\mu \neq 40$ When $\sigma = 10$ and $n = 25$

Value of μ	Power
30	.999
31	.994
32	.979
33	.938
34	.851
35	.705
36	.516
37	.323
38	.168
39	.079
40	.050
41	.079
42	.168
43	.323
44	.516
45	.705
46	.851
47	.938
48	.979
49	.994
50	.999

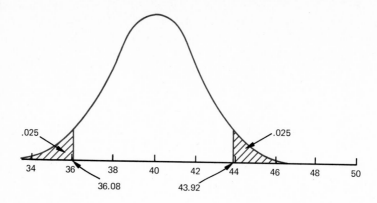

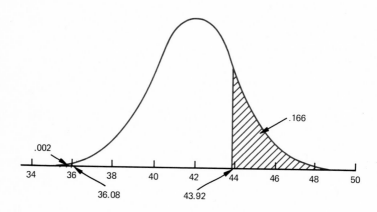

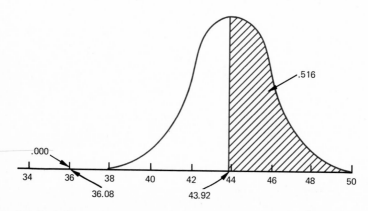

Figure 9.7 *Probability of x̄ < 36.08 or x̄ > 43.92 for μ = 40, 42, 44 (the sum of the two probabilities is the power)*

The effectiveness of a procedure in detecting differences from the null hypothesis is described by the *power function,* the probability of rejecting the null hypothesis for various values of the true population mean μ. The null hypothesis is rejected at the 5% level if the sample mean is greater than 43.92 or less than 36.08. The probabilities of these events are diagrammed in Figure 9.7 for $\mu = 40, 42,$ and 44. These probabilities are the values of the power function for $\mu = 40, 42,$ and 44. The power function for this test is tabulated in Table 9.5 and graphed in Figure 9.8.

It will be observed that the probability of rejecting the null hypothesis when μ is close to the hypothetical value of 40 is about .05, the significance level. The probability increases as μ moves away from 40. Rejection of the null hypothesis is almost certain if the population mean is as small as 30 or as large as 50. The difference between the power function of a one-tailed test, as in Figure 9.3, and that of a two-tailed test is that the two-tailed test has power to reject the null hypothesis if the parameter is *either* smaller *or* larger than the hypothesized value,

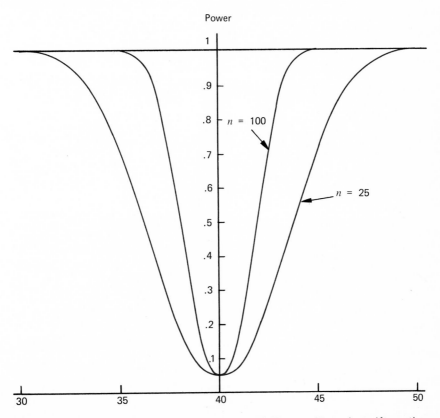

Figure 9.8 *Power functions of 5%-level tests of H: $\mu = 40$ against Alternative: $\mu \neq 40$ when $\sigma = 10$ for $n = 25$ and $n = 100$ (Tables 9.5 and 9.6)*

while the one-tailed test has power against a difference in only one direction. There is a price for this advantage of a two-tailed test; the power against any given alternative value larger than the null hypothesis value is less than for the one-tailed test at the same significance level.

If a larger sample size is used, the power function goes up more steeply as μ moves away from 40; in either direction there is a greater probability of rejecting the null hypothesis when it is not true. Table 9.6 and Figure 9.8 give the power function of the 5%-level test for $n = 100$.

TABLE 9.6

Power Function of 5%-Level Two-Tailed Test of H: $\mu = 40$ Against Alternative: $\mu \neq 40$ When $\sigma = 10$ and $n = 100$

VALUE OF μ	POWER
35	.999
36	.979
37	.851
38	.516
39	.168
40	.050
41	.168
42	.516
43	.851
44	.979
45	.999

If the significance level is 1% instead of 5%, the power function is .01 at 40 and is at every value of μ less than the power function of the 5%-level test in Table 9.6. The advantage of the 1%-level test is that the probability of Type I error is smaller; the disadvantage is that the Type II error probabilities are increased. The power function of the 1%-level test of H: $\mu = 40$ against the alternative $\mu \neq 40$ when $\sigma = 10$ and $n = 25$ is tabulated in Table 9.7 and graphed in Figure 9.9.

Usually the investigator wants merely to distinguish between a parameter value close to a hypothesized value and one far from that value. In our example, it is not of much interest to discover that the mean study time is 40.5 hours instead of 40 hours. The difference of $\frac{1}{2}$ hour is small compared to errors of measurement and compared to other more important differences between student behavior now and earlier.

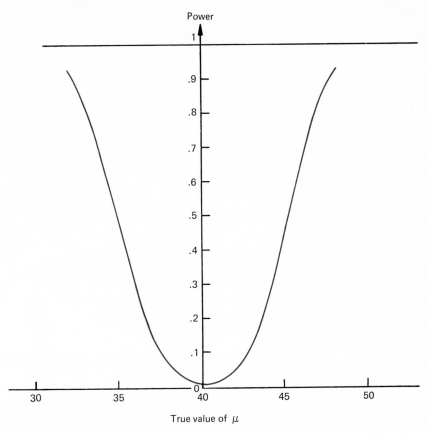

Figure 9.9 *Power function of the 1%-level test of H: $\mu = 40$ against Alternative: $\mu \neq 40$ when $\sigma = 10$ and $n = 25$ (Table 9.7)*

    An investigator might even consider a difference of 1 hour as un-interesting, considering a mean study time of anywhere between 39 and 41 as roughly the same as one of 40 hours. From Table 9.5 it is seen that the power is only .079 when μ is 39 or 41. If one wants to extend the null hypothesis to include values from 39 to 41, one is, in effect, using a significance level of 7.9%, in the sense that the probability of rejecting the null hypothesis H that μ is between 39 and 41 is at most .079 when μ is in this range.*

---

    *If the sample size is large, however, the probability of rejection may be greater than desired for values of the parameter fairly close to the hypothesized value. For example, with $n = 100$, the power when $\mu = 41$ is .168. If this is felt to be too high, the investigator could use a smaller nominal significance level so that the probability of rejection near the hypothesized value is acceptably small.

TABLE 9.7
Power Function of 1%-Level Two-Tailed Test of H: $\mu = 40$
Against Alternative: $\mu \neq 40$ When $\sigma = 10$ and $n = 25$

VALUE OF μ	POWER
32	.923
33	.822
34	.664
35	.470
36	.282
37	.141
38	.058
39	.020
40	.010
41	.020
42	.058
43	.141
44	.282
45	.470
46	.664
47	.822
48	.923

9.2.2 Relation of Two-Tailed Tests to Confidence Intervals

To test *H: $\mu = \mu_0$* against the alternative hypothesis $\mu \neq \mu_0$ at the 5% level, we accept H if $\mu_0 - 1.96\sigma/\sqrt{n} < \bar{x} < \mu_0 + 1.96\sigma/\sqrt{n}$. This pair of inequalities is equivalent to $\bar{x} - 1.96\sigma/\sqrt{n} < \mu_0 < \bar{x} + 1.96\sigma/\sqrt{n}$. But the set of all values of μ_0 that satisfy this pair of inequalities is just the 95% confidence interval for μ. Thus, the confidence interval consists of all "acceptable" values of μ, that is, all those values μ_0 for which the null hypothesis *H: $\mu = \mu_0$* would be accepted. The value μ_0 is in or out of the 95% confidence interval according as the corresponding hypothesis *H: $\mu = \mu_0$* is accepted or rejected in a two-sided test at the 5% level. Stated the other way, *the null hypothesis $\mu = \mu_0$ for any number μ_0 is accepted or rejected according to whether the number is inside or outside of the confidence interval.*

The confidence interval provides the information that a hypothesis test does, but also more. We consider four cases which may arise when the null hypothesis is a specified number μ_0. These are classified in Table 9.8 and are sketched in Figure 9.10.

TABLE 9.8
Length of Confidence Interval and Inclusion of Hypothetical Value

		Does Confidence Interval Include μ_0?	
		Yes	No
Length of Confidence Interval	Short	(a)	(b)
	Long	(c)	(d)

In case (a) the interval is short and contains the hypothesized value. The null hypothesis is accepted with some conclusiveness since the investigator has confidence that the true value of the population mean is near μ_0; the interval may include only parameter values which have the same meaning for the investigator as μ_0, that is, parameter values which are so close to μ_0 as to be equivalent to μ_0 in practical terms.

In case (b) the interval is short and does not contain the hypothesized value. The null hypothesis is rejected and the investigator is guided by some knowledge of what the true value of the parameter is; he or she may have confidence that it is quite different from the hypothetical value.

In case (c) the interval is long and contains the hypothetical mean. The null hypothesis is accepted, but other values of the population mean are acceptable; some of these are quite far from the hypothetical value.

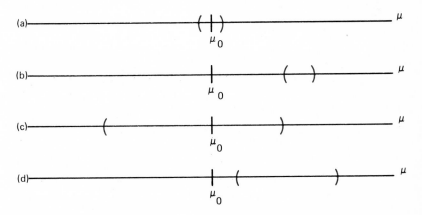

Figure 9.10 *Length of confidence interval and inclusion of hypothetical value (Table 9.8)*

In case (d) the interval is long and does not contain the hypothesized value. The null hypothesis is rejected, but the acceptable values of the population mean range widely; the investigator may not get much guidance.

9.3 TESTING HYPOTHESES ABOUT A MEAN WHEN THE STANDARD DEVIATION IS UNKNOWN

When an investigator asks a question about a population mean, frequently the standard deviation of the population is unknown as well. The procedures described in Sections 9.1 and 9.2 were based upon knowledge of σ, the population standard deviation. When σ is unknown, the sample standard deviation is used to estimate it. This increases the sampling error in the procedure; we take account of the added statistical error by basing significance points on the t-distribution instead of the normal distribution.

When the standard deviation is known, the test of the null hypothesis $H: \mu = \mu_0$ or $H: \mu \leq \mu_0$ is based on the fact that

$$\frac{\bar{x} - \mu_0}{\sigma/\sqrt{n}}$$

has the standard normal distribution if the sample is obtained from a normal population with mean μ_0 and standard deviation σ. When the standard deviation is unknown, the test is based on the fact that

$$\frac{\bar{x} - \mu_0}{s/\sqrt{n}}$$

has the t-distribution with $n - 1$ degrees of freedom if the null hypothesis that $\mu = \mu_0$ is true; then

$$\Pr\left(\frac{\bar{x} - \mu_0}{s/\sqrt{n}} > t_{n-1}(\alpha)\right) = \alpha.$$

(See Figure 9.11.)

EXAMPLE   A way of testing the "level of aspiration" of individuals is as follows. Before taking a test of general knowledge of literature, each individual is given two sample questions and then asked to estimate his or her score on a test made up of 50 similar questions. The

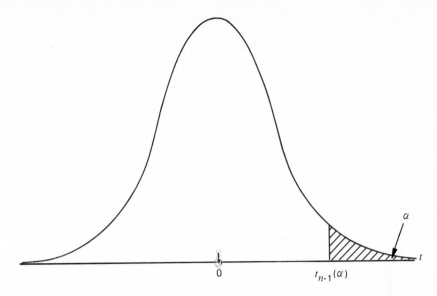

Figure 9.11 *Tail probability of Student's t-distribution with n − 1 degrees of freedom*

guess is taken as the individual's level of aspiration. It is known from a long series of such experiments that the average guess (aspiration level) is 27. Each of 22 psychology students was told that "the average score for psychology students is 37.2" and asked to estimate his or her score. For this group of 22 persons the average aspiration level was found to be $\bar{x} = 31.09$, and the sample standard deviation was $s = 8.95$.

The null hypothesis is that the mean level of aspiration of such students is not affected by the additional "information." The alternative hypothesis is that such students are influenced by the information in such a way as to *raise* their mean aspiration levels. To test

$$H: \mu \leq 27 \text{ against } Alternative: \mu > 27$$

the quantity

$$\frac{\bar{x} - \mu_0}{s/\sqrt{n}} = \frac{31.09 - 27}{8.95/\sqrt{22}} = \frac{4.09}{8.95/4.68} = \frac{4.09}{1.94} = 2.10$$

is used. For significance level $\alpha = .05$, the significance point of the t-distribution with $n - 1 = 21$ degrees of freedom is $t_{21}(.05) = 1.721$. Since $2.10 > 1.721$, the null hypothesis $\mu_0 \leq 27$ is rejected. [If a significance level of 1% had been used the null hypothesis would have been accepted since $t_{21}(.01) = 2.518$.]

To test the null hypothesis $H: \mu = \mu_0$ against the alternative $\mu \neq \mu_0$, one rejects the null hypothesis if $(\bar{x} - \mu_0)/(s/\sqrt{n})$ exceeds $t_{n-1}(\alpha/2)$ or falls short of $-t_{n-1}(\alpha/2)$.

The power functions of the t-tests are similar to those of the tests based on the normal distribution as described in Sections 9.1 and 9.2. For any test (one-tailed or two-tailed) at a given significance level α based on a sample of given size n, the test based on the normal distribution has higher power than the test based on the t-distribution: the cost of not knowing the standard deviation is loss of power.

Confidence Intervals.   As indicated in the previous section, a test of a hypothesis concerning the mean may be performed by constructing the corresponding confidence interval, and the confidence interval in fact provides more information than the hypothesis test.

Nonnormality.   By the Central Limit Theorem the sample mean is approximately normally distributed, almost regardless of the shape of the parent population. The approximation is good if the sample size is not small. Hence, the procedures of this section apply even when the parent population is not normal, provided the sample size is not too small. (See Table 8.3.)

It is a good idea to examine the histogram of the sample if the sample size is large enough to provide a meaningful histogram. In order to make use of approximate normality, a larger sample size is required if the histogram is extremely skewed than if the histogram is roughly symmetric. One way of dealing with the case of extreme skewness is to use the sign test for the median (Section 9.5).

Discussion.   The t-test is based upon the sample mean and standard deviation. In the example on level of aspiration the null hypothesis concerns the *mean* score for persons receiving the "information." Other questions may be of interest. For example, do *all* people respond to the "information," or are only some affected? Inspection of the histogram of the scores would help decide this.

Figures 9.12 and 9.13 might be histograms of scores made by two different groups of 25 students on a standardized reading test after they had received training in special reading techniques. The mean of this test for all students is known to be 50. The two sample distributions have the same mean (53) and almost exactly the same standard deviation (2.71 for Figure 9.12 and 2.72 for Figure 9.13). Thus, the value of Student's t would be almost exactly the same in the two cases. But the histogram in Figure 9.13 is bimodal, with one mode near 50 and another near 55, suggesting that there are two clusters of people, the lower

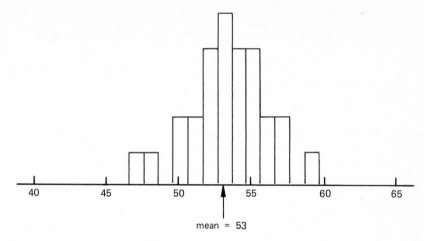

Figure 9.12 *Scores of 25 students: Mean = 53, standard deviation = 2.71*

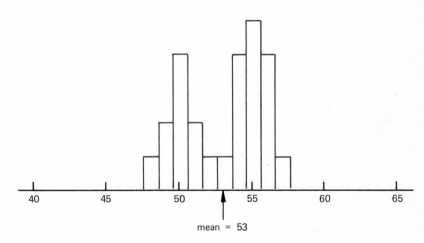

Figure 9.13 *A bimodal distribution with the same mean and standard deviation as Figure 9.12*

cluster consisting of people who were not helped by the training. Examination of Figure 9.13 would suggest that the training is not effective for everyone, and this in turn would suggest further research questions. What characterizes pupils for whom the training is not effective? Do they have low IQ's? Is their first language other than English? How can we train *these* pupils to read with more comprehension?

9.4   TESTING HYPOTHESES ABOUT A PROPORTION

9.4.1   Testing Hypotheses About the Probability of a Success Against One-Sided Alternatives

Information about the probability of an event is given by the proportion of trials in which that event occurs. The sample proportion is used to answer questions about the probability, in particular, the question of whether the probability is equal to some specified number. If the sample proportion is very different from the hypothesized probability, the null hypothesis will be rejected. As in the other cases involving means, the procedure depends on whether the alternative hypothesis is that the parameter is greater than, less than, or either greater or less than the hypothesized value.

The procedure for testing hypotheses about the mean of a normal distribution can be applied to testing hypotheses concerning the probability of a success because the normal distribution provides an approximation to the binomial distribution. More precisely, let x be the number of successes in n Bernoulli trials with probability p of a success. The proportion $\hat{p} = x/n$ is approximately normally distributed with mean $\mu_{\hat{p}} = p$ and variance $\sigma_{\hat{p}}^2 = pq/n$; this fact provides the basis for testing the null hypothesis that the probability p is a specified value.

Suppose psychotherapists have claimed success in treating a certain mental disorder. A study is to be made of their claims by treating 200 patients having the disorder and then observing the recovery rate. Previous studies have shown that 2/3 of all victims of this disorder recover "spontaneously," that is, without treatment.

Suppose there are 150 recoveries, that is, a sample proportion $150/200 = .75$ of the 200 patients recover. We know that if one samples randomly from a dichotomous population with probability $p = 2/3$ of success, then the sampling distribution of the sample proportion $\hat{p}$ has approximately a normal distribution with mean $p = 2/3$ and standard deviation $\sqrt{pq}/\sqrt{n} = \sqrt{(2/3)(1/3)}/\sqrt{200} = 1/30$. Then

$$\Pr\left(\frac{\hat{p} - 2/3}{1/30} > 1.645\right) \doteq .05.$$

The inequality is equivalent to

$$\hat{p} > \frac{2}{3} + 1.645 \times \frac{1}{30} = .722;$$

that is, the probability of obtaining a sample proportion greater than

.722 is about .05 when sampling from a population with $p = 2/3$. Since the observed proportion is .75, which is greater than .722, the null hypothesis is rejected at the 5% level of significance.*

This is an example of the following general problem. On the basis of a sample of size n one wishes to test at level α the null hypothesis that the probability of a success is a specified number p_0 against the alternative that the probability is larger than p_0. If the null hypothesis is true, then the standardized variable

$$\frac{\hat{p} - p_0}{\sqrt{p_0 q_0}/\sqrt{n}}$$

has a sampling distribution that is approximately normal. (Here $q_0 = 1 - p_0$.) The probability is α that the standardized variable is greater than $z(\alpha)$. Thus, one rejects the null hypothesis if

$$\frac{\hat{p} - p_0}{\sqrt{p_0 q_0}/\sqrt{n}} > z(\alpha),$$

that is, if

$$\hat{p} > p_0 + z(\alpha)\frac{\sqrt{p_0 q_0}}{\sqrt{n}}.$$

It should be noted that the null hypothesis specifies not only the mean p_0 but also the standard deviation $\sqrt{p_0 q_0}$ of the parent population.

9.4.2 Effectiveness of the One-Tailed Tests

We shall discuss the effectiveness of the one-tailed test in terms of the example of a one-tailed test of $H: p \leq 2/3$ against the alternative $p > 2/3$ based on a sample of size $n = 200$. For a 5%-level test the significance point for $\hat{p}$ is .722. If p is the probability of a success, then

$$\frac{\hat{p} - p}{\sqrt{pq}/\sqrt{200}}$$

has a sampling distribution which is approximated by the standard normal distribution. The probability that $\hat{p} > .722$ is the probability that

$$\frac{\hat{p} - p}{\sqrt{pq}/\sqrt{200}} > \frac{.722 - p}{\sqrt{pq}/\sqrt{200}},$$

---

*Here the continuity correction has not been used though it would give greater accuracy. See Appendix 9A.

TABLE 9.9

Power of Tests of Null Hypothesis H: $p \leq 2/3$ at 5% and 1% Levels with Sample Size of 200 Against Alternative: $p > 2/3$

p	$\sqrt{pq}/\sqrt{200}$	5%-LEVEL TEST $\dfrac{.722 - p}{\sqrt{pq}/\sqrt{200}}$	Power	1%-LEVEL TEST $\dfrac{.744 - p}{\sqrt{pq}/\sqrt{200}}$	Power
.60	.0346	3.517	.000	4.170	.000
.65	.0337	2.128	.017	2.798	.003
.67	.0332	1.557	.059	2.238	.012
.69	.0327	0.969	.166	1.661	.048
.71	.0321	0.364	.358	1.069	.142
.73	.0314	−0.264	.604	0.455	.325
.75	.0306	−0.925	.822	−0.186	.574
.77	.0297	−1.626	.948	−0.865	.806
.79	.0288	−2.372	.991	−1.587	.944
.81	.0277	−3.188	.999	−2.373	.991
.83	.0266	−4.071	1.000	−3.222	1.000

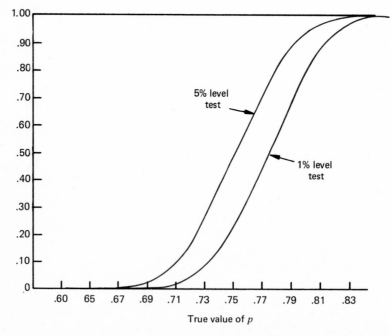

Figure 9.14 *Power functions of 5%- and 1%-level tests of H: $p \leq 2/3$ against Alternative: $p > 2/3$ (Table 9.9)*

and the probability of this event is approximately

$$\Pr[z > (.722 - p)/(\sqrt{pq}/\sqrt{200})].$$

Table 9.9 gives the power of this test and also the power of the 1%-level test, which has the significance point .744, for various values of p. Figure 9.14 gives graphs of the two power functions.

9.4.3 Two-Sided Alternative

When the alternative includes values of the parameter larger and smaller than the hypothetical value, the procedure is to reject the null hypothesis if the sample proportion is sufficiently larger or smaller than that value. To test the null hypothesis $H: p = p_0$ against the alternative $p \neq p_0$ at level α we reject the null hypothesis if

$$p > p_0 + z(\tfrac{1}{2}\alpha) \frac{\sqrt{p_0 q_0}}{\sqrt{n}}$$

or if

$$p < p_0 - z(\tfrac{1}{2}\alpha) \frac{\sqrt{p_0 q_0}}{\sqrt{n}} .$$

EXAMPLE In a study of racial awareness and prejudice, Clark and Clark (1958) investigated the question of whether Black children distinguish the "race of dolls, under the assumption that the behavior of distinguishing indicates racial awareness and possibly prejudice (preference)."

In this study, 253 Black children (ages 3–7) were presented four dolls, two white and two nonwhite. Each child was told: "Give me the doll that you would like to play with." The results were as follows:

$$
\begin{array}{ll}
& \text{83 children chose a nonwhite doll} \\
& \text{169 children chose a white doll} \\
& \underline{\text{1 child chose no doll}} \\
\text{Total} \quad & \text{253 children}
\end{array}
$$

We shall consider our sample to consist of the 252 children who chose one of the dolls; that is, $n = 252$.

In this study randomness arises from two sources: (i) selection of children—the 253 children are a sample from a large population (actually, 134 from Arkansas and 119 from Massachusetts) and (ii) choice of doll—the phenomenon under observation is nondeterministic (random) in the sense that a given child who has a preference for one type of doll may not *always* choose that color of doll. Each child's behavior was observed on only one occasion; on another occasion the child may have behaved differently.

Here the success probability p represents the probability of selecting a child who will choose a nonwhite doll, and $q = 1 - p$ then represents the probability of selecting a child who will choose a white doll. Our null hypothesis is that $p = 1/2$ ($= q$); that is, that a child selected is just as likely to choose one type of doll as the other. What kind of result would cause us to decide that a preference exists? Even if the null hypothesis is true, we know that we cannot expect *exactly* half of the children to choose white dolls and half to pick nonwhite dolls. The sample proportion will not be exactly 1/2. There will be some sampling variability. But if the sample proportion is *quite different* from 1/2 we will find the null hypothesis untenable.

On which side of 1/2 would we expect the proportion to be if there is racial awareness or preference? Toward white dolls (p less than 1/2)? Toward nonwhite dolls (p greater than 1/2)? Or either way (p not equal to 1/2)? In the absence of strong prior information we take our alternative hypothesis to be $p \neq 1/2$. In general, we are making fewer prior assumptions when we use a two-sided alternative hypothesis.

If the null hypothesis is true, the standard deviation of the sample proportion is

$$\sqrt{\frac{1}{2} \times \frac{1}{2}} \Big/ \sqrt{252} = .0315.$$

We reject the null hypothesis at the 5% level of significance if the observed sample proportion is greater than $.5 + 1.96 \times .0315 = .562$ or less than $.5 - 1.96 \times .0315 = .438$.

The proportion who chose a nonwhite doll was $\hat{p} = 83/252 = .329 < .438$; so we reject the null hypothesis that $p = 1/2$. That is, at least at the 5% level we have ruled out chance as a plausible explanation of preference for one type of doll over the other. We take this as evidence that the hypothesis of no racial awareness or preference is untenable.

Other more detailed questions, for example, the effect of age and state of residence upon the proportion choosing nonwhite dolls, were investigated in the study.

The children were given other instructions too; for example, "Give me the doll that is a nice doll." Altogether, the results were a rather striking demonstration of racial awareness in young children.

9.4.4 Effectiveness of Two-Tailed Tests

The Power Function.   In this example the probability of rejecting the null hypothesis when p is the true proportion is the probability that either

$$\frac{\hat{p} - p}{\sqrt{pq}/\sqrt{252}} < \frac{.438 - p}{\sqrt{pq}/\sqrt{252}} \quad \text{or} \quad \frac{\hat{p} - p}{\sqrt{pq}/\sqrt{252}} > \frac{.562 - p}{\sqrt{pq}/\sqrt{252}},$$

which is approximately

$$\Pr\left(z < \frac{.438 - p}{\sqrt{pq}/\sqrt{252}}\right) + \Pr\left(z > \frac{.562 - p}{\sqrt{pq}/\sqrt{252}}\right).$$

Values of this function for various values of p are given in Table 9.10 and graphed in Figure 9.15. The corresponding curve is symmetric about the point $p = 1/2$. (If the null hypothesis were not $p = 1/2$, the power curve would not be symmetric.)

TABLE 9.10
Power Function of the 5%-Level Test of Null Hypothesis
H: $p = 1/2$ Against Alternative: $p \neq 1/2$ for $n = 252$

p	$\dfrac{.438 - p}{\sqrt{pq}/\sqrt{252}}$	$\Pr\left(z < \dfrac{.438 - p}{\sqrt{pq}/\sqrt{252}}\right)$	$\dfrac{.562 - p}{\sqrt{pq}/\sqrt{252}}$	$\Pr\left(z > \dfrac{.562 - p}{\sqrt{pq}/\sqrt{252}}\right)$	POWER
.35	2.93	.998	6.73	.000	.998
.40	1.23	.891	5.15	.000	.891
.45	−0.38	.351	3.56	.000	.351
.50	−1.96	.025	1.96	.025	.050
.55	−3.56	.000	0.38	.351	.351
.60	−5.15	.000	−1.23	.891	.891
.65	−6.73	.000	−2.93	.998	.998

Note that the power is small near $p = 1/2$; for $p = .45$ or $.55$ it is only .351. This is not undesirable in the sense that we really want to accept the null hypothesis if p is nearly $1/2$; usually we do not mean our null hypothesis to be the single value $p = 1/2$. We would consider the effect not appreciable even if $p = .49$, say; such a small effect would be unimportant. Then we must ask what we consider to be an "important" value of p. Perhaps we would feel that $p = .4$ or $p = .6$ would represent

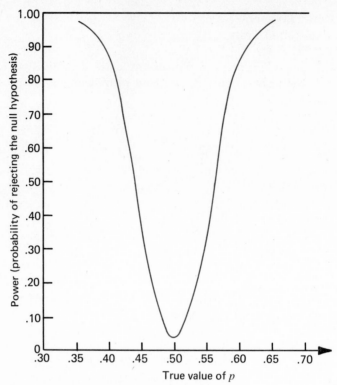

Figure 9.15 *Power function of the 5%-level test of H: p = 1/2 against Alternative: p ≠ 1/2 for n = 252 (Table 9.10)*

TABLE 9.11
Power Function of the 5%-Level Test of Null Hypothesis
H: p = 1/2 Against Alternative: p ≠ 1/2 for n = 126

p	$\dfrac{.413-p}{\sqrt{pq}/\sqrt{126}}$	$\Pr\!\left(z<\dfrac{.413-p}{\sqrt{pq}/\sqrt{126}}\right)$	$\dfrac{.587-p}{\sqrt{pq}/\sqrt{126}}$	$\Pr\!\left(z>\dfrac{.587-p}{\sqrt{pq}/\sqrt{126}}\right)$	POWER
.35	1.47	.929	5.78	.000	.929
.40	0.30	.618	4.28	.000	.618
.45	−0.85	.198	3.11	.001	.199
.50	−1.96	.025	1.96	.025	.050
.55	−3.11	.001	.085	.198	.199
.60	−4.28	.000	−0.30	.618	.618
.65	−5.78	.000	−1.47	.929	.929

an important deviation from $p = 1/2$. We want the power to be large for such values of p. We see from the table that with a sample of size $n = 252$ the power is .891, if $p = .4$ or .6—there is about a nine-in-ten chance of rejecting $H: p = 1/2$.

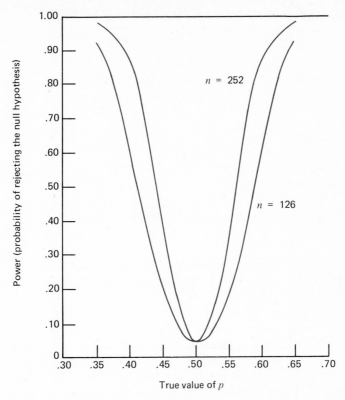

Figure 9.16 *Power functions when* $n = 126$ *and* $n = 252$ *of tests of H:* $p = 1/2$
against Alternative: $p \neq 1/2$ *at the 5% level (Table 9.12)*

TABLE 9.12
Power Functions of 5%-Level Tests Based on Samples of Size 252 and 126

	POWER WHEN	
p	$n = 252$	$n = 126$
.35	.998	.929
.40	.891	.618
.45	.351	.199
.50	.050	.050
.55	.351	.199
.60	.891	.618
.65	.998	.929

Effect of Sample Size. Suppose the sample size were only half
as large; that is, suppose $n = 252/2 = 126$. The 5%-level two-tailed test

is to reject the null hypothesis if $\hat{p} < .413$ or $\hat{p} > .587$ since

$$1.96 \ \sqrt{\tfrac{1}{2} \times \tfrac{1}{2}} / \sqrt{126} = .087.$$

The power function of the test is given in Table 9.11 and graphed in Figure 9.16. Table 9.12 and Figure 9.16 permit easy comparison of the power functions of the tests based on samples of size 252 and 126.

9.5 TESTING HYPOTHESES ABOUT A MEDIAN: THE SIGN TEST

The median is a measure of location that does not depend on the extreme values in a population and can be defined even for an ordinal scale. The *sign test* for a hypothesized value of the median similarly does not depend on extreme values and is suitable for data on an ordinal scale (as well as on an interval or ratio scale). Moreover, the sign test is very easy to carry out. (At the 5% significance level with a large sample the two-tailed test can be done without using pencil and paper.)

The median divides the population in half in the sense that when drawing at random the probability is 1/2 of obtaining a value less than the median and 1/2 of obtaining a value greater than the median. (We assume that the probability of drawing an observation equal to the median is 0.) To test the hypothesis that a specified number, say M_0, is the median of a population, one can test the equivalent hypothesis that the probability of drawing an observation less than M_0 is 1/2. If the proportion of observations in a sample less than M_0 is sufficiently different from 1/2, the hypothesis is rejected.

Let us now put this in formal terms. Let M be the median* of the population sampled and suppose that the probability of M is 0; that is,

$$\Pr(x < M) = \frac{1}{2} = \Pr(x > M),$$

where x is the variable. The null hypothesis is

$$H: M = M_0,$$

where M_0 is a specified number. Let the alternative hypothesis be $M \neq M_0$. An equivalent null hypothesis is $\Pr(x < M_0) = 1/2$, and an equivalent alternative hypothesis is $\Pr(x < M_0) \neq 1/2$. Suppose the test is to be made at the 5% level of significance.

In the sample of size n count the number of observations less than M_0 and denote this number by y. Under random sampling, y has a binomial

---

*We assume that the median is uniquely defined.

distribution with a success probability of $\Pr(x < M_0)$. Under the null hypothesis this probability is $1/2$; the mean and standard deviation of the sample ratio y/n are $1/2$ and $\sqrt{(1/2)(1/2)(1/n)} = 1/(2\sqrt{n})$. In large random samples

(9.1)

$$\frac{\dfrac{y}{n} - \dfrac{1}{2}}{\dfrac{1}{2} \times \dfrac{1}{\sqrt{n}}}$$

has a sampling distribution which is well approximated by the standard normal distribution. The null hypothesis is rejected if (9.1) is either less than -1.96 or greater than 1.96. (If 1.96 is replaced by 2, the procedure is to reject if

$$\frac{\left|\dfrac{y}{n} - \dfrac{1}{2}\right|}{\dfrac{1}{2} \times \dfrac{1}{\sqrt{n}}} > 2,$$

which is equivalent to

$$\left|\frac{y}{n} - \frac{1}{2}\right| > \frac{1}{\sqrt{n}}.$$

This test is easily performed.)

For other significance levels the value of (9.1) is referred to the normal tables. (When the continuity correction is used, the numerator of (9.1) is modified by $1/(2n)$ to make it numerically smaller.) If the alternative hypothesis is $M < M_0$, an equivalent alternative hypothesis is $\Pr(x < M_0) > 1/2$; then the hypothesis is rejected at the 5% level if (9.1) is greater than 1.645.

EXAMPLE   Jones claims the median lifetime of light bulbs is 500 hours. This claim is incorrect if the median lifetime is either less than or greater than 500 hours. In a sample of 100 light bulbs 54 burned out before 500 hours had passed. Then (9.1) is

$$\frac{\dfrac{54}{100} - \dfrac{1}{2}}{\dfrac{1}{2} \times \dfrac{1}{\sqrt{100}}} = \frac{.54 - .5}{.5 \times .1} = \frac{.04}{.05} = .8,$$

which does not exceed the 5% significance point of 1.96 for a two-tailed test.

This procedure is called the "sign test" because it is based on counting the number of minus signs among the differences between the observations and M_0. If the parent distribution is symmetric the mean and median are identical and the sign test is appropriate to test a hypothesized value of the mean.

9.6   PAIRED MEASUREMENTS

9.6.1   Testing Hypotheses About the Mean of a Population of Differences

Pairs.   Paired measurements arise when two measurements are made on one unit of observation. When the *difference* of the two measurements is the variable of interest, a test of the hypothesis that the mean difference is 0 is based on the differences of pairs of measurements in the sample of observed units.

Education was one factor under consideration in Whelpton and Kiser's *Social and Psychological Factors Affecting Fertility*. The sample consisted of $n = 153$ married couples (p. 109). Each husband was asked to state the "highest school grade" completed by his wife. Each wife was also asked the highest grade she had completed. The information obtained for each husband is compared to that obtained from his wife to see whether husbands tend to exaggerate their wives' educational accomplishments. The difference between the husband's and wife's reports is the variable of interest. We want to test the null hypothesis that the population mean difference is 0 against the alternative that it is positive; the alternative reflects the idea that a respondent may tend to exaggerate a prestige item such as education.

Let y_i be the husband's statement of the highest grade completed by the wife in the ith couple and x_i be the wife's statement of her highest grade. The differences $d_i = y_i - x_i$, $i = 1, \ldots, n$, were calculated. The average difference was $\bar{d} = 0.32$ years; the standard deviation of the differences was 1.07 years. The value of the test statistic is

$$t = \frac{0.32}{1.07/\sqrt{153}} = 3.70.$$

A one-tailed test is used here because the alternative is a positive population mean difference. Student's t-distribution with 152 degrees of freedom is almost the same as the standard normal distribution. For a one-tailed test at the 1% level, the null hypothesis is rejected if $z > 2.326$. Since $3.70 > 2.326$, the null hypothesis is rejected. The data have ruled out chance as a plausible explanation of the observed mean difference. The observed difference may be due to the fact that husbands tend to exaggerate their wives' education while wives tend to be accurate; or it may be due to the fact that the wives exaggerate, but the husbands exaggerate more; or it may even be due to some extraneous factor.

Matched Samples. As we mentioned in Section 8.6, some studies are carried out on the basis of pairs of individuals who have been matched on some variable. In an experiment to compare workbooks for reading instruction in elementary school, pupils could be matched into pairs according to IQ, one pupil of each pair being chosen to receive instruction using Workbook A, the other to receive instruction using Workbook B. The pairs are formed by listing the pupils in order of IQ and pairing the top two, the next two, etc. A table of random numbers can be used to assign a digit to the first member of each pair, who is assigned to A if the number is odd, to B if it is even. (See Table

TABLE 9.13

Use of Random Numbers to Assign Pair Members to A or B

PUPIL	IQ	RANDOM DIGITS FOR FIRST MEMBER OF PAIR	WORKBOOK
Tom	131	9	A
Bill	128		B
John	114	6	B
Albert	109		A
Harry	103	7	A
Steve	102		B
Randy	99	3	A
Bart	97		B
etc.			

(hypothetical data)

9.13.) Then comparison of the workbooks cannot be biased by systematically assigning the brighter children to Workbook B.

Let y_i be the score of the pupil in the ith pair using Workbook B and let x_i be the score of the pupil in the ith pair using Workbook A. Then $d_i = y_i - x_i$ is an observation from a population of differences of scores of pairs of pupils, the first using Workbook B and the second

Workbook A. A *t*-test can be carried out to test the null hypothesis that the mean population difference μ_d is 0.

If μ_y is the mean score of the (hypothetical) population of pupils using Workbook B and μ_x is the population mean for Workbook A, then $\mu_d = \mu_y - \mu_x$. The test can be considered a test of the null hypothesis $\mu_y = \mu_x$.

An alternative method of carrying out this experiment is to select randomly one half the pupils in the class for work with Workbook A and assign Workbook B to the remaining half. This experimental procedure would also be unbiased, but it could be expected to be less efficient. Performance depends on the level of intelligence as well as on which workbook is used, and some differences between pupils' scores on the two workbooks might be due to differences in intelligence.

9.6.2   Testing the Hypothesis of Equality of Proportions

The measurements in a pair can also be dichotomous. In the toss of a pair of coins each can show a head or a tail (Exercise 6.1). The husband in a couple can vote Republican or Democratic, and the wife can vote Republican or Democratic; one may ask whether more married women than married men tend to vote Republican.

Turnover Tables.   In Section 5.1 we remarked that one of the ways a 2 × 2 table arises is when we ask the same persons the same question at two different times. The resulting data are termed "change-in-time data," and the resulting 2 × 2 table a "turnover table."

Jones, a political candidate, had polls taken in August and October; she wished to compare the proportions of voters favoring her at the two different times. If the same persons were polled both times, the results would appear as in Table 9.14. The results can be summarized in a "turnover table" like Table 9.15. The number of persons changing to Jones was 224, more than the 176 who changed away from her, causing the proportion of the sample in favor of Jones to grow from 37% in August to 40% in October. Does this reflect a real change in the population from which these 1600 persons are a sample? The population is represented in Table 9.16. The proportions of the population in favor of Jones in August and October are $p_1 = (A + B)/N$ and $p_2 = (A + C)/N$, respectively. We wish to test the hypothesis of equality of these two proportions. Note that p_2 has C where p_1 has B; they are equal if and only if $B = C$; that is, if and only if the number changing to Jones is the same as the number changing away from Jones. Moreover, the equality $B = C$ is equivalent to $C/(B + C) = 1/2$. Note that $B + C$ is the total number of persons who *changed* their minds, and C is the number who switched to Jones. The number $p = C/(B + C)$ is the proportion, among those who changed, who changed to Jones. We wish to test the hypothesis that $p = 1/2$.

TABLE 9.14
Table of Results of Poll

Identification of Individual	Did Individual Favor Jones? August	October
1	Yes	Yes
2	Yes	No
.	.	.
.	.	.
.	.	.
n	No	Yes

TABLE 9.15
Turnover Table for Poll

		October For Jones	For Other Candidates	Total
	For Jones	416	176	592 (37%)
August	For Other Candidates	224	784	1008 (63%)
	Total	640 (40%)	960 (60%)	1600 (100%)

(hypothetical data)

TABLE 9.16
Turnover Table for the Population

		October Jones	Others	Total
	Jones	A	B	$A + B$
August	Others	C	D	$C + D$
	Total	$A + C$	$B + D$	$N = A + B + C + D$

The persons in the sample who changed their minds were the 224 who switched to Jones and the 176 who switched away from Jones. These $224 + 176 = 400$ persons are considered a sample from the population of $C + B$ persons who changed their minds. ($C + B$ is assumed large enough that the finite population correction is not needed.) The observed sample proportion who changed to Jones is

$$\hat{p} = \frac{224}{400} = .56.$$

Under the null hypothesis the population mean of a random proportion is $1/2$ and the standard deviation is $\sqrt{(1/2)(1/2)(1/400)} = 1/40 = .025$. The standardized version of $\hat{p}$ is

$$\frac{.56 - .50}{.025} = 2.4.$$

This exceeds 1.96, the significance point for a two-tailed 5%-level test. The null hypothesis is rejected.

We can describe a general problem of this type as follows. Each of n individuals is asked the same question at two different times. The sample joint frequency distribution takes the form of Table 9.17. The estimate of p, the proportion of changers in the population who change from No to Yes, is the sample proportion

$$\hat{p} = \frac{c}{b + c}.$$

As a test statistic for testing $C/(B + C) = 1/2$, the standardized version of $\hat{p}$ can be used; it is

$$\frac{\hat{p} - \dfrac{1}{2}}{\sqrt{\dfrac{1}{2} \times \dfrac{1}{2} \Big/ (b + c)}}.$$

TABLE 9.17
2 × 2 Table: Change-in-Time Data

		TIME 2 Yes	No	Total
TIME 1	Yes	a	b	$a + b$
	No	c	d	$c + d$
	Total	$a + c$	$b + d$	n

The value of this statistic is compared with a value from the normal distribution to decide whether the sample result is significant.

This problem can also be considered as based on differences. In the example, there are $n = 1600$ individuals; let $x_i = 1$ or 0 according as in August the ith individual favored Jones or not, and let $y_i = 1$ or 0 according as in October he or she favored Jones or not. Using Table 9.18, it can be shown (Exercise 9.29) that the sample mean difference $\bar{d}$ is $(c - b)/n$. Hence, approaching the problem by considering $\bar{d}$ again leads to comparing b and c.

TABLE 9.18
Turnover Table

		OCTOBER: y		
		For Jones	For Another Candidate	Total
AUGUST: x	For Jones	$a\,(0)$	$b\,(-1)$	$a + b$
	For Another Candidate	$c\,(+1)$	$d\,(0)$	$c + d$
	Total	$a + c$	$b + d$	n

NOTE: Values in parentheses are the values of differences.

Matched Samples. Convicts due to be released in one year's time were grouped into matched pairs according to age, type of crime, and IQ. One member of each pair was selected at random for a special rehabilitation program. For each pair it was noted whether the convict in the program returned to prison within one year of release and whether the convict not in the program returned. The frequency table for pairs is Table 9.19. The numbers which pertain to whether the program reduced returns are c and b.

TABLE 9.19
Return of Convicts

		CONVICT NOT IN PROGRAM		
		Returned	Not Returned	Total
CONVICT IN PROGRAM	Returned	a	b	$a + b$
	Not Returned	c	d	$c + d$
	Total	$a + c$	$b + d$	n

9.6.3   The Sign Test for the Median of Differences

The sign test can be used for the hypothesis that the median of a difference variable d is zero. The procedure is that of Section 9.5.

Hyman, Wright, and Hopkins (1962) made a study of 96 participants in a summer Encampment for Citizenship. Each participant was given a Civil Liberties test at the beginning and end of the summer, and the score at the beginning was subtracted from the score at the end to give a difference-score for each participant. A lower score (a negative difference) indicates improvement in the direction of Civil Liberties. The data are shown in Table 9.20. (See Exercises 5.33 to 5.37.)

TABLE 9.20

Individual Changes in Campers' Scores on Civil Liberties During the Summer

		SCORE AT END OF SUMMER											
		0	1	2	3	4	5	6	7	8	9	10	Total
	0	<u>10</u>	2	1	1	2							16
	1	5	<u>2</u>	1									8
	2	3	2	<u>13</u>	1	1							20
SCORE AT	3	2		2	<u>2</u>								6
BEGINNING	4	5	4	4	1	<u>6</u>		1					21
OF SUMMER	5		1		3	1	<u>0</u>			1			6
	6	1		1		4	2	<u>2</u>	2				12
	7			1		1	1		<u>0</u>				3
	8	1								<u>0</u>			1
	9			1						1	<u>0</u>		2
	10										1	<u>0</u>	1
	Total	27	11	23	9	15	3	3	3	2	0	0	96

SOURCE: Hyman, Wright, and Hopkins (1962) Table D-6, p. 392. Copyright © 1962 by The Regents of the University of California; reprinted by permission of the University of California Press.
NOTE: A lower score indicates improvement.

We want to test the hypothesis that in the (hypothetical) population of potential and actual participants in such encampments the median difference is 0 (no change). In performing the test we discard ties (the cases underlined in Table 9.20). There are $10 + 2 + 13 + 2 + 6 + 2 = 35$ of these. Of the remaining $96 - 35 = 61$ individuals, those represented below and to the left of the ties improved (their scores at the end of the summer encampment were lower than their scores at the beginning). There are 48 such individuals. As a check we count to make sure that $61 - 48 = 13$ individuals showed an increase in score. To make the significance test the ratio $\hat{p} = 48/61 = .787$ is compared with 1/2.

The standardized value of $\hat{p}$ is

$$\frac{\hat{p} - \dfrac{1}{2}}{\sqrt{\dfrac{1}{2} \times \dfrac{1}{2}} \Big/ \sqrt{n}} = \frac{\hat{p} - \dfrac{1}{2}}{\dfrac{1}{2} \dfrac{1}{\sqrt{n}}} = \frac{.787 - .5}{\dfrac{1}{2} \times \dfrac{1}{\sqrt{61}}} = \frac{.287}{.0640} = 4.5.$$

This value exceeds the one-tailed, 1%-level significance point of 2.326; the result is highly significant, even according to this simple sign test, which does not even take into account the magnitude of the differences.

Note that to say the median of d is zero is to say $\Pr(d > 0) = 1/2$. Since $d = y - x$, this is $\Pr(y - x > 0) = 1/2$, or $\Pr(y > x) = 1/2$.

In using the sign test we only check to see whether each d_i is greater than zero, that is, whether $y_i > x_i$. And for this, the variables x and y need only be defined on an ordinal scale. On the other hand, valid use of tests involving sample means requires that the data be defined on at least an interval scale—an assumption researchers often wish to avoid making when dealing with test scores and with data from experiments in which subjects compare or rank items. Furthermore, the sign test is valid even when the d_i's have different variances (which tends to invalidate the t-test).

9.7 GENERAL COMMENTS ON TESTING HYPOTHESES

9.7.1 Achieved Level of Significance

In this chapter we have considered the use of a sample to answer a Yes-or-No question about a population. The question has been phrased as a null hypothesis, tested against an alternative hypothesis, and answered when we accept or reject the null hypothesis. We have presented each test procedure as carried out at a predetermined level of significance to control the probability of a Type I error. The choice of significance level in turn determines the probability of Type II error for each possible alternative. Rational choice actually must involve simultaneous consideration of both types of error.

Several investigators may evaluate the seriousness of Type I and Type II errors differently and hence may be led to different choices of significance level. How can various investigators communicate their statistical results? One possibility is that a researcher state the value of the statistic proposed for the significance test and let each reader complete the test procedure by choosing an appropriate significance level and finding the significance point(s) in the appropriate table. This suggestion is obviously impractical because it does not carry the inference procedure to conclusion.

Another possibility is to report the range of significance levels at which the hypothesis would be rejected. Since rejection at one significance level implies rejection at any larger level, one only needs to report the smallest significance level at which one would reject the null hypothesis. This smallest level is called the "achieved level of significance."

In the example in Section 9.3 the value of the t-statistic was 2.10. It was referred to the t-table with 21 degrees of freedom with rejection of the null hypothesis if 2.10 exceeded the significance point. It was noted that the 5% significance point was 1.721 and the 1% significance point was 2.518. There is rejection at the 5% level, but not at the 1% level. Further reference to the t-table shows that the 2.5% significance point is 2.080. The observed value of 2.10 is barely significant at the 2.5% level of significance. We say that the achieved level of significance (of a one-tailed test) is 2.5% or .025.

The achieved level of significance is often denoted by P and is sometimes referred to as the "P-value." In this example we would write "$P = .025$."

The reader knows that if he or she feels the appropriate significance level is a larger number, for instance, .04, then the null hypothesis would be rejected.

The achieved level of significance can be interpreted as a measure of the implausibility of the null hypothesis. In the above example, if we took another sample of 22, the probability of obtaining a t-statistic exceeding the previously observed value of 2.10 is $P = .025$ if the null hypothesis is true. If from the original data, a larger value of t had been observed, say about 2.52, the P value would be smaller, about 1%, and one would consider the null hypothesis more implausible. In this sense one can compare different kinds of evidence; a sample of another size would also lead to a t-statistic and the P-value could be stated.

If a null hypothesis is tested against a two-sided alternative hypothesis, the suitable test is two-tailed. The P-value in such a case is again the smallest significance level at which the null hypothesis is rejected.

9.7.2 Sample Size Needed to Achieve a Given Power at a Fixed Level

In this chapter we have taken the sample size n as given in advance. Often the investigator may select this quantity, for instance, by deciding how many respondents to interview in a poll. An investigator who is particularly interested in a test of significance may predetermine the sample size in order to achieve a specified significance level and a specified power at some alternative of interest.

As an example, we consider a problem of testing the null hypothesis that the mean μ of a population is a specified value, μ_0; that is, we consider testing $H: \mu = \mu_0$, using the mean $\bar{x}$ of a sample of size n. Suppose we consider a fixed alternative value $\mu_1 > \mu_0$, and ask how large n must be in order that the probability of rejecting the null hypothesis be .90 when the true value of μ is a specified value, μ_1, and the level α is fixed at .05. The test is to reject the null hypothesis if

$$\bar{x} > \mu_0 + 1.645 \frac{\sigma}{\sqrt{n}} .$$

We want the probability of this event when $\mu = \mu_1$ to be .90. This event is

$$\frac{\bar{x} - \mu_1}{\sigma/\sqrt{n}} > \frac{\mu_0 - \mu_1}{\sigma/\sqrt{n}} + 1.645.$$

The probability of this event is

(9.2)
$$\Pr\left(z > \frac{\mu_0 - \mu_1}{\sigma/\sqrt{n}} + 1.645\right),$$

since $(\bar{x} - \mu_1)/(\sigma/\sqrt{n})$ is distributed according to the standard normal distribution. We want this probability to be as large as .90. We know from tables that

(9.3)
$$\Pr(z > -1.282) = .90.$$

Thus, from the two equations (9.2) and (9.3) we find

$$\frac{\mu_0 - \mu_1}{\sigma/\sqrt{n}} + 1.645 = -1.282.$$

This gives

(9.4)
$$n = \frac{(1.645 + 1.282)^2}{[(\mu_1 - \mu_0)/\sigma]^2} = \frac{8.567}{[(\mu_1 - \mu_0)/\sigma]^2} .$$

The value n will not be an integer; usually we would take the sample size to be the next larger integer.

Suppose that $\mu_0 = 100$, $\mu_1 = 110$, and $\sigma = 20$. Then $(\mu_1 - \mu_0)/\sigma = 0.5$, and (9.4) gives $n = 8.567/(0.5)^2 = 34.268$. Thus, if we took $n = 35$, our 5%-level test would have power greater than .90 when $\mu = 110$.

More generally, if we want to have power equal to $1 - \beta$ when $\mu = \mu_1$ and we are making a one-tailed test of $H: \mu = \mu_0$ at level α, we need to have

$$n = \frac{[z(\alpha) + z(\beta)]^2}{[(\mu_1 - \mu_0)/\sigma]^2} .$$

Approximate results for two-tailed tests are obtained by replacing α by $\alpha/2$.

When the variance is unknown, it can sometimes be estimated from previous data or from a small preliminary study (a "pilot" study). In dealing with an estimated variance, Student's t-distribution replaces the normal, so the result derived here gives only an approximation, which is a good approximation if the required sample size is large.

9.7.3 The Two-Tailed Significance Test as a Three-Decision Procedure

We have considered testing the null hypothesis that a population mean μ is some given value μ_0 against the two-sided alternative $\mu \neq \mu_0$. In many situations, however, the conclusion to be drawn if $\mu < \mu_0$ is quite different from the conclusion to be drawn if $\mu > \mu_0$. If the two-tailed significance test leads to rejection of the null hypothesis, we would conclude $\mu < \mu_0$ if $\bar{x}$ were less than the lower significance point, and we would conclude $\mu > \mu_0$ if $\bar{x}$ were greater than the upper significance point. In Section 9.2 we were asking whether the average number of hours of study was 40. If the average was less than 40, we would think that students have become more frivolous. Subsequent action or investigation would depend on whether we conclude $\mu < 40$ or $\mu > 40$.

For many purposes a two-tailed test is a three-decision procedure. To analyze the effectiveness of a procedure in these terms, the power function should be replaced by two functions, the probability of reaching each of the two conclusions alternative to the null hypothesis.

If Type II errors in one direction are more serious than those in the other, it is appropriate to use unequal tail probabilities, for example, 1% and 4% instead of $2\frac{1}{2}$% in both tails. (This is analogous to the asymmetrical confidence intervals discussed in Section 8.9.4.)

9.7.4 The Finite Population Correction

In this chapter we have studied tests of significance in terms of sampling from infinite populations. In most instances we have based the tests on the fact that the sample mean is normally distributed, or approximately so, and have calculated the standard deviation on the grounds that the population was infinite. In many cases the sampling is from a finite population and then the standard deviation of the sample mean should be calculated using the finite population correction factor (unless n/N is less than .10).

SUMMARY

A hypothesis tested is called a *null hypothesis*. It is tested against an *alternative hypothesis*.

A *Type I error* consists of rejecting the null hypothesis when it is true. We control the probability of such an error to be equal to some small prescribed probability; this prescribed probability is called the *significance level*.

A *Type II error* consists of accepting the null hypothesis when it is false. The probability of rejecting the null hypothesis when it is false is called the *power*. This probability depends upon the true value of the parameter, the sample size, and the significance level. A rule of thumb for the choice of significance level is to use 1% if it gives reasonable power against reasonable alternative parameter values; otherwise, use 5% or even 10%.

When the test statistic is approximately normally distributed and the standard deviation of the estimate is not a function of the parameter, the hypothesis test may be carried out by forming the appropriate confidence interval and noting whether or not the hypothesized parameter value falls in the interval.

The *achieved level of significance* is the probability of obtaining a more extreme value of the test statistic than the value actually obtained, given that the null hypothesis is true. The null hypothesis is rejected if the achieved level of significance is sufficiently small.

APPENDIX 9A   The Continuity Correction

Consider a *discrete* random variable which can take on successive integer values. Then successive possible values of the mean of a sample of size 25, for example, differ by 1/25, or .04. The amount of the continuity correction is one-half of this, or .02. For a sample of size n, the continuity correction is $1/(2n)$. The continuity adjustment consists in adding or subtracting this constant as appropriate. For example, in testing $H: \mu \leq \mu_0$ against the alternative: $\mu > \mu_0$ at the 5% level, we would

$$\text{Reject } H \text{ if} \quad \bar{x} > \mu_0 + \frac{1}{2n} + 1.645 \frac{\sigma}{\sqrt{n}}.$$

For a two-tailed test we would

$$\text{Reject } H \text{ if} \quad \bar{x} > \mu_0 + \frac{1}{2n} + 1.96 \frac{\sigma}{\sqrt{n}}$$

$$\text{or if} \quad \bar{x} < \mu_0 - \frac{1}{2n} - 1.96 \frac{\sigma}{\sqrt{n}}.$$

(A way to remember whether to add or subtract the $1/(2n)$ is to recall that its effect is to make it "harder" to reject the null hypothesis.) If n is large, the continuity correction is small and may be ignored.

Binomial variables and ranks are of course discrete. The continuity correction for the number of successes is $\pm 1/2$, and for the proportion of successes is $\pm 1/(2n)$. The continuity correction for R, the sum of ranks in the signed rank test, is $\pm 1/2$.

APPENDIX 9B The Signed Rank Test

The Wilcoxon *signed rank test* is a test for a hypothesized value of a median or mean which uses more information from the sample than the sign test and less information than the sample mean (and sample standard deviation). It is appropriate when the investigator thinks that the parent distribution is symmetric (or approximately symmetric) about its center and does not want extreme values to have much effect on the procedure.

Each observation is ranked according to how far away it is from the hypothesized value of the center (mean or median), regardless of whether the difference between the observation and the hypothesized value is positive or negative. Then each rank is given the sign of the difference (+ or −) and the analysis is carried out on the signed ranks.

EXAMPLE A Civil Liberties Scale is constructed to be symmetrically distributed over the population of all students with center at zero. (The items which make up the scale are scored positively and negatively.) Scores were obtained for a sample of student leaders in private high schools in New York. It is desired to test whether the median score for the population of such student leaders is the same as the *overall* median score (zero). A 5% level of significance will be used. The scores of 10 student leaders are given in the first column of Table 9.21 (and graphed in Figure 9.17). In the second column is the absolute value of the score, which is the distance of each point from the origin. The score −1 has absolute value 1, which is the smallest of the absolute values and hence gives the rank of 1; the score 2 with absolute value 2 has rank 2; the score −5 with absolute value 5 has rank 3, etc. The ranks of the absolute values are given in the third column. In the fourth column the ranks are given with minus signs attached to the observations which were negative.

We now proceed to analyze the signed ranks. Their sum is 33 and their mean is 3.3. When the null hypothesis is true, the sample mean of signed ranks has a sampling distribution with mean 0. We calculate a sample standard deviation on this basis. The sum of squares of the signed ranks is the same as the sum of squares of the ranks themselves, which is

$$1^2 + 2^2 + \cdots + 10^2 = 385.$$

TABLE 9.21
Computations for the Signed Rank Test

Score	Absolute Value	Rank of Absolute Value	Signed Rank
−10	10	7	−7
−5	5	3	−3
−1	1	1	−1
2	2	2	2
7	7	4	4
8	8	5	5
9	9	6	6
11	11	8	8
15	15	9	9
16	16	10	10
			33

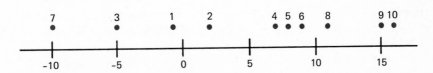

Figure 9.17 *Scores of 10 student leaders on the Civil Liberties Scale (Table 9.21)*

We calculate the standard deviation (of the signed ranks), subtracting 0 (since the mean is 0) and dividing, in this case, by the number of observations to obtain $\sqrt{385/10}$. The "estimated" standard error of the mean (of the signed ranks) is

$$\sqrt{\frac{385/10}{10}} = \frac{\sqrt{385}}{10} = 1.962.$$

Then the standardized sample mean rank

$$\frac{3.3 - 0}{1.962} = 1.68$$

is referred to the standard normal distribution. Since this falls between -1.96 and 1.96, the null hypothesis is accepted.

In general the hypothesis being tested is $H: M = M_0$; in the example, $M_0 = 0$. The procedure is to compute differences $x_i - M_0$, $i = 1, \ldots, n$, rank their absolute values, and then give each the sign of the differences. Let R be the mean of the signed ranks. Assuming a symmetrical distribution about M_0, we can say that if H is true, that is, if M_0 is the true value of the median, then the data points would be scattered as much to the left as to the right of M_0. That is, any one of the ranks is equally likely to be $+$ or $-$. Hence, the sum of all the signed ranks of the observations should be about 0. The sum of squares of the ranks from 1 to n is $n(n + 1)(2n + 1)/6$, and the standard error of the mean signed rank is $\sqrt{(n + 1)(2n + 1)/(6n)}$. The standardized version of R is

$$\frac{R}{\sqrt{(n + 1)(2n + 1)/(6n)}}.$$

The distribution of this statistic is well approximated by that of the standard normal distribution. The signed rank test is more powerful than the sign test since it involves the *magnitudes* of the deviations from M_0 as well as their signs. But an extra assumption, that of symmetry about M_0, is required for its applicability; and of course the simple sign test is easier to carry out. (For more accurate use of the normal approximation, the continuity correction can be used.)

The Signed Rank Test for Differences.   The signed rank test can be applied to differences. The procedure is the same as above except that it is differences which are ranked. This requires a decision as to whether $y_i - x_i$ exceeds $y_j - x_j$. Hence, x and y must be defined on an interval scale.

EXERCISES

9.1 (Sec. 9.1)   Suppose that the mean of the scores of a random sample of 225 eighth-grade pupils on an IQ test was 102.6, and suppose the standard deviation of the test is known to be 12. At the 5% level, would you reject the hypothesis that the population mean is less than or equal to 100 against the alternative that the mean is greater than 100?

9.2 (Sec. 9.1)   Suppose that the mean of the scores of a random sample of 144 eighth-grade pupils on an IQ test was 97.2 and the standard deviation of the test is known to be 20. At the 5% significance level, would you reject the hypothesis that the mean of the distribution is equal to or greater than 100 against the alternative that the mean is less than 100?

9.3 (Sec. 9.6.3)   In the Hyman, Wright, and Hopkins study (Table 9.20) it is reasonable to assume approximate normality of the sample mean difference $\bar{d}$ since the sample size of 96 is large. The sample variance of the differences is 4.89. Test at the 5% level the hypothesis that the population mean difference is zero against the one-sided alternative that the population mean difference is negative (indicating improvement).

9.4 (Sec. 9.3)   Ten specimens are sampled from a lot of 1000 metal plates. The tensile strengths are given in Table 9.22. Assuming normality, test at the 10% level the hypothesis that the mean tensile strength of the lot is at most 60,000 psi against the alternative that it exceeds 60,000 psi. Use the fact that $s = 24,680$.

TABLE 9.22
Tensile Strengths (Pounds per Square Inch)

65,770	81,100
70,100	66,000
64,600	20,000
96,900	84,200
58,400	22,800

(hypothetical data)

9.5 (Sec. 9.4)   In a random sample of 1600 of the state's registered voters, 850 favor Double Talk for governor. Let p represent the proportion of all the state's registered voters in favor of Double Talk. Test $H: p \leq 1/2$ against the alternative $p > 1/2$ at the 5% level.

9.6 (Sec. 9.4)   In a random sample of 3600 of the country's registered voters, 1850 favor Mr. Big for president. Let p represent the proportion of all the country's voters in favor of Mr. Big. Test $H: p \leq 1/2$ against the alternative $p > 1/2$ at the 5% level.

9.7 (Sec. 9.3)   The weights of 25 packages of hamburger meat in a freezer at a grocery store are given in Table 9.23. Are these weights consistent with a population mean of 3 pounds? (Make a two-tailed, 5%-level test of the hypothesis that the population mean weight is equal to 3 pounds against the alternative that it is not equal to 3 pounds.)

TABLE 9.23
Net Weights of 25 Packages of Hamburger Meat (Pounds)

3.00	3.02	2.93	3.05	3.03
2.95	2.95	2.95	2.96	2.94
3.00	2.93	2.95	2.94	2.93
2.94	2.96	2.95	2.95	2.96
3.00	3.06	3.05	3.07	3.05

SOURCE: a freezer in a grocery store.

9.8 (Sec. 9.4)   The Ad Manager says 25% of the cars sold with four-speed manual transmissions are four-door sedans. Super Salesperson, believing that four-door sedans constitute a smaller percentage of total sales, examines the data for a random sample of 400 cars from among the thousands sold with four-speed manual transmissions, finds only 40 four-door cars among them, and tells the Ad Manager that the 25% is wrong. Do you think the Ad Manager's figure of 25% is wrong?

*9.9 (Sec. 9.4)   Four independent tosses of a coin with probability p of a head are made.
   (a) For testing $H: p = 1/2$, what is the significance level of the test, "reject H if the number of heads is 3 or 4"?
   (b) Give the power of this test when $p = .9$. (The alternative hypothesis is $p > 1/2$.)
[Hint: It is necessary to use the binomial distribution. The normal distribution cannot be used because the sample size $n = 4$ is so small. Use the table of binomial probabilities, Appendix IV.]

9.10 (Sec. 9.3)   The average number of matches per box produced by a certain company is supposed to be at least 36. A sample of 9 boxes is taken, and the average number of matches is found to be 34.56.
   (a) On the basis that the number of matches in a box has a distribution which is approximately normal with a standard deviation of 2, test at the 10% level of significance the hypothesis that the true mean is at least 36.
   (b) (Errors of measurement model) Errors are made when the matches are counted. Suppose each count has an approximate normal

distribution with mean equal to the corresponding true count and a standard deviation of 1. Suppose further that the error is statistically independent of the true count. That is, assume that each count has the form $y = x + u$ where x is the true count (approximately normally distributed with $\sigma_x = 2$), the error u is approximately normal with mean zero and unit standard deviation, and x and u are statistically independent. Test the hypothesis using this model. [Hint: $\sigma_y^2 = \sigma_x^2 + \sigma_u^2$.]

9.11 (Sec. 9.4)   Jerome Carlin (1962) presents information about the activities of solo practitioners in Chicago, and the social system within which these lawyers work. Many of these lawyers are largely limited by the availability of cases to those which demand little legal competence. If they are to remain solvent their work must at times be done in ways which are not condoned by the Bar Association's Canon of Ethics. The Chicago Bar Association might provide an arena in which the ethical dilemmas of the solo lawyers could be worked out. However, this probably would not be done if solo lawyers were not active in that association. Carlin presents a table to compare the frequency of solo practitioners holding office in the Chicago Bar Association with the proportion of solo lawyers in practice in Chicago. (See Table 9.24.)

TABLE 9.24 *Chicago Lawyers Holding Official Positions in the Chicago Bar Association, by Status in Practice, 1956*

STATUS IN PRACTICE	PERCENT OF ALL CHICAGO LAWYERS (1956)	OFFICERS AND MEMBERS OF BOARD OF MGRS. (1955–1957)	COMMITTEE CHAIRMEN AND VICE CHAIRMEN (1956)
Individual Practitioners	54%	10	18
Lawyers in firms of 2 to 9	23%	23	29
Lawyers in firms of 10 or more	7%	28	38
Not in private practice	16%	2	10
Total	100%	63	95

Number of lawyers: 12,000

SOURCE: From p. 203 of *Lawyers on Their Own: A Study of Individual Practitioners in Chicago* by Jerome E. Carlin. Copyright © 1962 by Rutgers, The State University. Reprinted by permission of the Rutgers University Press.

Is the leadership of the Chicago Bar Association representative of the solo practitioners or is it biased in favor of other lawyers? Test the null hypothesis that the 95 committee chairmen and vice chairmen have, in effect, been drawn randomly from a population of lawyers in which the proportion of individual practitioners is greater than or equal to .54. Use significance level 1%. The alternative hypothesis is that the proportion in the population drawn from is less than .54.

9.12 In each of the situations below a hypothesis test is required. In each case state the null hypothesis and the alternative hypothesis. If you use letters like p or μ, state what they mean. Do not do any calculations.

(a) It is well known that the lengths of the tails of newborn Louisiana swamp rats have a mean of 4.0 cm and a standard deviation of 0.8 cm. An experimenter is wondering whether or not adverse conditions on the rat mother shorten the tails of the young. In his lab the rat mothers were kept on a rotating turntable. The 131 young rats born from these mothers had a mean tail length of 3.8 cm.

(b) The Gorillaburger Drive-In chain requires its individual franchise owners to use 50 grams as a minimum mean weight for hamburgers; it does this in order to protect the reputation of the chain. The management of the chain suspects that the downtown Sunnyvale franchise has been cheating on weight, and they would like to find conclusive evidence of this. A spy sent by the management purchased quantities of hamburgers on several different occasions; of the 144 hamburgers purchased, the mean was found to be 48.7 grams, and the standard deviation was found to be 3.0 grams.

(c) A roulette wheel has 38 slots; there are 18 red slots, 18 black slots, and 2 green slots. Since the two green slots (0 and 00) are favorable to the casino, a wheel weighted by the casino owners would probably have a higher probability for the green slots. You suspect that the wheel at the Sodom and Gomorrah Casino has been weighted. Along with some friends you watch the wheel for 1000 successive spins. During these 1000 turns the green slots came up 27 times.

9.13 (Sec. 9.7)   The achieved level of significance (P-value) of a two-tailed test is

(a) equal to that of the corresponding one-tailed test,
(b) half that of the corresponding one-tailed test,
(c) twice that of the corresponding one-tailed test.
Explain your answer.

9.14 (Sec. 9.4)   Super-Duper Soda Company has decided to market ginger ale. They have two recipes, a standard recipe and an expensive recipe. They would prefer to use the standard recipe unless there is substantial evidence that the expensive recipe makes a better drink. In a blindfold taste-preference test, 480 of 900 tasters preferred the expensive recipe. If p represents the probability of preference for the expensive recipe, test the null hypothesis $H: p \leq 1/2$ against the alternative $p > 1/2$; use the 5% level of significance.

9.15 (Sec. 9.4)   A doctor reported that the normal death rate for patients with extensive burns (more than 30% of skin area) has been cut by 25% through the use of silver nitrate compresses. Of 46 burned adults treated, 21 could have been expected to die. The number who died was 16. The compress treatment seemed especially beneficial to children. Of 27 children under 13, two died, while the expected mortality rate was six or seven.

The report implies that the expected mortality rate for adult burn patients before the new treatment was about $21/46 = .46$ and that the rate for children was about $6/27 = .22$.

(a) Test the hypothesis that the mortality rate for treated adults is .46 against the alternative that it is less than .46 at the 5% level.

(b) Test the hypothesis that the mortality rate for treated children is .22 against the alternative that it is less than .22 at the 5% level. (The sample size for the children is a bit small for use of the normal approximation, but use it anyway.)

9.16 (Sec. 9.6.1)   An insurance company made a change in the rules for its health insurance plan. Effective January 1, 1968, the insurance company would no longer pay 100% of hospital bills; the patient would have to pay 25% of them. The joint distribution of number of hospital stays in 1967 and 1968 for a random sample of individuals is given in Table 9.25. Was there a significant decrease in usage between 1967 and 1968? (Test the significance of the sample mean difference. Justify the assumption that this statistic is approximately normally distributed.)

TABLE 9.25
Distribution of Hospital Stays in 1967 and 1968

		NUMBER OF HOSPITAL STAYS IN 1968			
		0	1	2	*Total*
NUMBER OF HOSPITAL STAYS IN 1967	0	2219	40	8	2267
	1	65	105	51	221
	2	54	8	50	112
	Total	2338	153	109	2600

(hypothetical data)

9.17 (Sec. 9.6.2)   In a study of attitudes in Elmira, New York, in 1948, Paul Lazarsfeld and associates found the following changes in attitudes toward the likelihood of war in the next ten years among the same persons interviewed in June and again in October. Treat these 597 people as a random sample from a large population, and test the hypothesis that the proportion of the population expecting war was the same at the two times.

TABLE 9.26
2 × 2 Table: Change-in-Time Data

| | | OCTOBER RESPONSE | | |
		Expects War	*Does Not Expect War*	*Total*
JUNE	*Expects War*	194	45	239
RESPONSE	*Does Not Expect War*	147	211	358
	Total	341	256	597

SOURCE: Lazarsfeld, Berelson, and Gaudet (1968).

9.18 (Sec. 9.5.1)   Jones claims the median daily precipitation during spring is one-half inch per day. On 49 of the 90 spring days the rainfall exceeded one-half inch. Using the sign test, test at the 5% level the hypothesis that the true median daily rainfall is one-half inch per day.

9.19 (continuation)   The sign test is based on the assumption that the observations are independent. Do you think this assumption is justified? Is it reasonable to assume, for example, that the conditional probability that rainfall exceeds 1/2 inch tomorrow, given that it exceeded 1/2 inch today, is equal to the probability that rainfall exceeds 1/2 inch tomorrow, given that it did not exceed 1/2 inch today?

9.20 (Sec. 9.6.1)   Use the data of Table 3.28 to test the null hypothesis that the average increase in hours of sleep due to the two drugs tested are the same. Use a two-tailed test at the 5% significance level.

9.21   Let $\bar{x}$ be the mean of a random sample from a normal distribution with unknown mean μ. In testing $H: \mu \leq \mu_0$ against the alternative $\mu > \mu_0$, what is the significance level of the rule:

$$\text{Reject } H \text{ if } \quad \bar{x} > \mu_0?$$

9.22 (continuation)   When might such a rule be appropriate?

9.23 (Sec. 9.3)   On the basis of the data in Appendix 9B, test the null hypothesis that the mean is 0 against a two-sided alternative at the 5% level of significance using a t-test. Compare the result with the signed rank test used in Appendix 9B.

MATHEMATICAL EXERCISES

9.24 (Sec. 9.6)   Show that if $d_i = y_i - x_i$, $i = 1, \ldots, n$, then s_d^2, which is $\Sigma_{i=1}^n (d_i - \bar{d})^2/(n-1)$, is equal to $s_x^2 + s_y^2 - 2s_{xy}$, where the number $s_{xy} = \Sigma_{j=1}^n (x_i - \bar{x})(y_i - \bar{y})/(n-1)$ is the *sample covariance* between x and y.

9.25 (Sec. 9.3.2)   A state university is required to admit all students who apply except for those who score in the lowest 5% on the Verbal Scholastic Aptitude Test (VSAT). Scores from previous years have shown that the scores are very nearly normally distributed with mean 500 and standard deviation 100. That value of the standard normal variable z which is exceeded with probability .95 is -1.645. The VSAT scores x can be standardized by subtracting their mean and dividing by their standard deviation:

$$z = \frac{x - 500}{100}.$$

(Since the VSAT score is an integer, there should really be a continuity adjustment, but it is omitted for simplicity.) Solving the equation for x we have

$$x = 500 + 100z.$$

Hence, that VSAT score which is the minimum acceptable is

$$500 + 100 \times (-1.645) = 335.5;$$

336 is acceptable and 335 is not.

The university uses the previous test results (normal distribution, mean 500, standard deviation 100) in finding the quantitative criterion, because this year's figures will not be available until after the college makes its admissions decisions. So the university rejects those students whose score is 335 or less. One question that must be considered regarding the validity of this procedure is whether this year's VSAT scores will conform to those of previous years. Suppose that the distribution of this year's scores is again approximately normal with standard deviation 100, but that the population mean μ is 550, not 500. The proportion rejected will then be

$$\Pr\left(z < \frac{335.5 - 550}{100}\right) = \Pr(z < -2.145) = .0160 < .0500.$$

Hence, the college will be rejecting not the lower 5%, but rather only the lower 1.6%. In the table are given the results of this calculation for several different values of the true mean μ. These results show the characteristics of the procedure, "Reject those students whose VSAT score is 335 or lower." (The third column, "Pr," in Table 9.27 gives, then,

$$\Pr\left(z < \frac{335.5 - \mu}{100}\right),$$

where μ is the population mean VSAT score for *this* year, given in the first column.)

TABLE 9.27
Probabilities and Power Functions

POPULATION MEAN-VSAT SCORE	POPULATION MEAN IN STANDARD UNITS	PR	POPULATION MEAN OF STUDENT'S TRUE MEASURE	POPULATION MEAN IN STANDARD UNITS	POPULATION MEAN OF DIFFERENCE IN YEARS	POPULATION MEAN OF FAMILY INCOME
625	+1.25	.0019		−1.25		
600	+1.00	.0041		−1.00		
575	+ .75			− .75		
550	+ .50			− .50		
525	+ .25	.0290		− .25		
500	.00	.0500	80	.00		$5500
475	− .25	.0815		+ .25		
450	− .50	.1260		+ .50		
425	− .75			+ .75		
400	−1.00	.2595		+1.00		$5804
375	−1.25	.3465		+1.25		

(hypothetical data)

(a) Fill in the three missing values in the Pr column.
(b) Make a graph of "Pr" versus "Population Mean," a power function, by plotting the 11 points given in the table and connecting them with a smooth curve. The horizontal axis is the axis for the population mean. Use standard units along this axis.

9.26 (Sec. 9.3.2)   Consider a test of simple knowledge of facts (such as a spelling test), so that the true measure of a student's knowledge is simply the percentage of facts known. Assume that if the same student took a large number of similar tests, then, due to variability in the selection of items to be included in the test and in the grading procedure, the test scores would be very nearly normally distributed with mean equal to the student's true measure and standard deviation 6.08. The teacher wants to be 95% sure of passing a student whose true measure is 80.

(a) What is the minimum passing score?

(b) Complete the fourth column of Table 9.27, and interpret the power function which you plotted in Exercise 9.25 in terms of this example. (For example, approximately how great is the probability that a student whose true measure is 72 will fail the test?)

9.27 (Sec. 9.3.2)   Consider the Whelpton-Kiser study (Section 9.6.1), in which husbands were asked the highest school grade completed by their wives. The average difference, husband's report minus wife's report, for the 153 couples in the sample was used to test the null hypothesis that the population mean difference was zero. The standard deviation of the difference in the population was estimated as 1.07 years. We take the sample mean difference as normally distributed (for samples drawn randomly from the given population). The test is to reject the null hypothesis if the observed sample mean difference is greater than

$$\frac{1.07}{\sqrt{153}} \times 1.645 = .142.$$

We want to know the probability of rejecting the null hypothesis for various possible values of the true mean differences in the population.

(a) Fill in the sixth column of Table 9.27 to correspond to the fifth column.

(b) Interpret the Pr column.

9.28 (Sec. 9.3.2)   A sociologist is repeating in a new community a study done previously in another community. As a control, it is important to check that the mean family income in the new community is not appreciably greater than that in the old, where it was $5500. Suppose the standard deviation of family income is $2128 in both towns, and that a preliminary sample of 49 families is taken in the second community. Their incomes are averaged to give $\bar{x}$, the sample mean family income in the new community. Setting up the problem formally in terms of testing a hypothesis, we have

$$H: \mu \leq \$5500, \quad Alternative: \mu > \$5500,$$

where μ is the true mean level of family income in the new community. For a 5%-level test, we reject $H: \mu \leq \$5500$ if

$$\bar{x} > \$5500 + 1.645 \frac{\$2128}{\sqrt{49}}$$

$$= \$5500 + 1.645(\$304)$$

$$= \$6000.$$

We reject at the 5% level the null hypothesis that the mean level of family income is the same in both communities if the sample mean of 49 families in the second community exceeds $6000.

(a) Fill in the last column of Table 9.27.
(b) Interpret the Pr column.

9.29 (Sec. 9.6.2) Show that $\bar{d} = (c - b)/n$. (Refer to Table 9.18.)

9.30 (continuation)   Show that $\Sigma_{i=1}^{n} (d_i - \bar{d})^2 = (b + c) - (c - b)^2/n$. [Remark. The test based on $\bar{d}$ is not exactly the same as the test based on $\hat{p}$, but in large samples the standardized versions of these statistics are very nearly equal.]

REFERENCES

Carlin, Jerome E. (1962). *Lawyers on Their Own: A Study of Individual Practitioners in Chicago*. Rutgers University Press, New Brunswick, New Jersey.

Clark, Kenneth B., and Mamie P. Clark (1958). "Racial identification and preference in Negro children." In *Readings in Social Psychology*, Eleanor E. Maccoby, Theodore M. Newcomb, and Eugene L. Hartley, eds. 3d ed. Holt, Rinehart, and Winston, Inc., New York, pp. 602–611.

Hyman, Herbert H., Charles R. Wright, and Terence K. Hopkins (1962). *Applications of Methods of Evaluation: Four Studies of the Encampment for Citizenship*. University of California Press, Berkeley and Los Angeles.

Klineberg, Otto (1935). *Negro Intelligence and Selection Migration*. Columbia University Press, New York, N.Y. (Reprinted in part in *The Language of Social Research*. Paul F. Lazarsfeld and Morris Rosenberg, eds. The Free Press, Glencoe, Illinois, 1955.)

Lazarsfeld, Paul F., Bernard Berelson, and Hazel Gaudet (1968). *The People's Choice.* 3rd ed. Columbia University Press, New York.

Whelpton, P. K., and Clyde V. Kiser (1950). *Social and Psychological Factors Affecting Fertility. Vol. One. The Household Survey in Indianapolis.* Milbank Memorial Fund, New York.

10 DIFFERENCES BETWEEN
TWO POPULATIONS

INTRODUCTION

Frequently an investigator wishes to compare or contrast two populations—sets of individuals or objects. This may be done on the basis of a sample from each of the two populations, as when average incomes in two groups, average IQ's in two parts of the country, or average levels of achievement of children in two school districts are compared. The polio vaccine trial compared the incidence rate of polio in the hypothetical population of children who might be inoculated with the vaccine and the rate in the hypothetical population of those who might not be inoculated; the two groups of children observed were considered as samples from these respective (hypothetical) populations. This example illustrates an experiment in which *a group receiving an experimental treatment is compared with a "control" group*. Ideally, the control group is similar to the experimental group in every way except that its members are not given the treatment.

Other experiments involve *the comparison of two methods of accomplishing an objective*. A teacher may compare the reading achievement of children taught in two different ways. A metallurgist may compare two methods of baking metal plates to improve their tensile strengths. A physician may compare the effectiveness of two dosages of a medicine used to reduce high blood pressure.

The steps involved in the experimental comparison of two treatments or methods are diagramed in Figure 10.1. A random sample is drawn from the basic population of all individuals. Members of this sample are randomly assigned to Treatment 1 or Treatment 2. There are thus two hypothetical populations of the values of the variable x—one which would result if all individuals in the basic population received Treatment 1 and one which would result if they all received Treatment 2.* The

---

*Note that the word "population" is used to refer not only to a set of *individuals* but also to the set of *values* of x corresponding to those individuals.

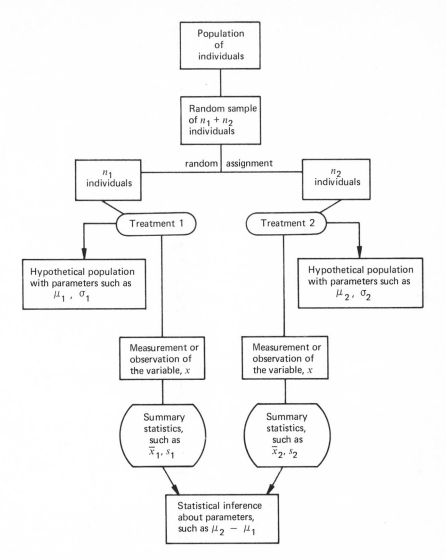

Figure 10.1 *Outline of steps in an experiment for comparing two treatments*

inference refers to parameters of these two hypothetical populations. The individuals given a particular treatment are considered to be a sample from the hypothetical population of individuals given that treatment.

In this chapter we treat problems of making inferences about two populations on the basis of two independent random samples, one from each population.

10.1   COMPARISON OF TWO INDEPENDENT SAMPLE MEANS WHEN THE POPULATION STANDARD DEVIATIONS ARE KNOWN

10.1.1   One-Tailed Tests

In each case treated in this section, the question is whether the means of two populations are equal or whether the mean of the first population is greater than the mean of the second. In formal terms, the null hypothesis is that the first population mean is equal to (or less than) the second, and the alternative hypothesis is that the first mean is greater than the other.

For example, a teacher who thinks that tests of a certain kind give no advantage to cultural group A over cultural group B, but wants some check on the fairness of such tests, administers such a test to a second-grade class. The pupils in the class of group A are considered as a sample from that group and the pupils of group B as a sample from group B. The null hypothesis that the average performance of all members of group A is less than or equal to the average performance for B is tested against the alternative that the average for A is greater. The data are given in Table 10.1 and plotted in Figure 10.2.

TABLE 10.1

Scores on a Standardized Test

A $n_1 = 22$			B $n_2 = 7$
97	83	81	89
95	85	90	104
86	110	121	107
102	119	105	85
119	117	104	70
99	108	96	91
87	74		96
101	93		

SOURCE: A second-grade teacher.

$$\bar{x}_1 = 98.7 \qquad \bar{x}_2 = 91.7$$

The average test score of the A's is greater than the average for the B's, but there is considerable overlap between the two sets of measurements. We ask if the difference between means, $98.7 - 91.7 = 7.0$, is large enough to cause us to believe there is a difference in populations or whether a sample difference of this size can plausibly be ex-

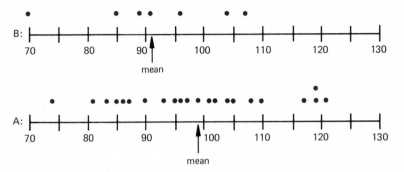

Figure 10.2 *Scores of two groups of children (Table 10.1)*

plained by chance alone, under the hypothesis that the mean scores in the two populations are in fact the same. The null hypothesis is tested at the 10% level of significance in order to achieve adequate power.*

To reach a decision, the difference of 7.0 points is converted to standard deviation units. The standard deviation of scores on this standardized test is known to be 12 points.‡ The variance is $\sigma^2 = 12^2 = 144$. The variance of the difference between two independent sample means is the sum of their variances. (See below.) The mean of a sample of size $n_1 = 22$ has a variance of $\sigma^2/n_1 = 144/22 = 6.55$; the mean of a sample of size $n_2 = 7$ has a variance of $\sigma^2/n_2 = 144/7 = 20.6$. The variance of the difference of the two means, $\sigma^2_{\bar{x}_1 - \bar{x}_2}$, is thus $6.55 + 20.6 = 27.2$. The standard deviation $\sigma_{\bar{x}_1 - \bar{x}_2}$, is the square root of this, or 5.21. The difference in means, 7.0, is equal to $7.0/5.21 = 1.34$ standard deviations.

When we treat the two sets of scores as independent random samples, the two independent sample means are then approximately normally distributed, and their difference is approximately normally distributed. Hence the value 1.34 is to be compared with the standard normal distribution. For a one-sided test at the 10% level, the appropriate percentage point is 1.282. The result obtained, 1.34, is greater than this, so it is significant, at the 10% level, and the null hypothesis is rejected.

*Suppose we would like to be 90% sure of detecting a difference in population means that is as large as 10 points. As is to be explained in Section 10.5, the power of a 5%-level, one-tailed test when $\mu_1 - \mu_2 = 10$, $\sigma = 12$, $n_1 = 22$, and $n_2 = 7$ is

$$\Pr\left(z > 1.645 - \frac{10}{12\sqrt{1/22 + 1/7}}\right) = \Pr(z > -0.275) = .61.$$

[See expression (10.5) in Section 10.5.2.] This is less than the desired power of .90. Since the 5%-level test does not give adequate power against reasonable alternatives, the 10% level is used to achieve larger power.

‡The variability is supposed to be constant in all subpopulations (although at times we may wish to test that assumption).

The samples here are necessarily small. If tests of this kind are un-
fair, the teacher wants to have a good chance of learning the fact in
spite of the paucity of information. The use of a significance level of
10% permits this. The conclusion is not very firm, but is good enough
for the teacher.

We shall now describe the test procedure in general terms when the
alternative is that the first population mean is larger. Two populations
are considered with means μ_1 and μ_2 and variance σ_1^2 and σ_2^2, respectively. The null hypothesis is

$$H: \mu_1 \leq \mu_2$$

or, equivalently,

$$H: \mu_1 - \mu_2 \leq 0.$$

This null hypothesis states that the difference between population
means is zero or less than zero. The alternative hypothesis is $\mu_1 > \mu_2$,
or, equivalently, $\mu_1 - \mu_2 > 0$. The null hypothesis and alternative reflect
the idea of determining whether or not μ_1 is greater than μ_2.

A random sample

$$x_{11}, x_{12}, \ldots, x_{1n_1}$$

is drawn from the first population and another

$$x_{21}, x_{22}, \ldots, x_{2n_2}$$

from the second. The symbol x_{1i} denotes the value for the ith individual
in the first sample, and x_{2j} denotes that of the jth individual in the
second sample. The sample means are

$$\bar{x}_1 = \frac{\sum_{i=1}^{n_1} x_{1i}}{n_1} \quad \text{and} \quad \bar{x}_2 = \frac{\sum_{j=1}^{n_2} x_{2j}}{n_2}.$$

In random sampling the sampling distribution of the sample mean
from the first population is approximately normal with mean μ_1 and
variance σ_1^2/n_1; the sampling distribution of the sample mean for the
second is approximately normal with mean μ_2 and variance σ_2^2/n_2, and
the two sample means are statistically independent. Then the difference between the sample means has approximately a normal distribution with mean

$$\mu_{\bar{x}_1 - \bar{x}_2} = \mu_{\bar{x}_1} - \mu_{\bar{x}_2} = \mu_1 - \mu_2$$

and variance*

$$\sigma^2_{\bar{x}_1 - \bar{x}_2} = \sigma^2_{\bar{x}_1} + \sigma^2_{\bar{x}_2} = \frac{\sigma^2_1}{n_1} + \frac{\sigma^2_2}{n_2}.$$

By the Central Limit Theorem, $\bar{x}_1 - \bar{x}_2$ will have a sampling distribution that is approximately normal, the approximation being good when n_1 and n_2 are not small. Thus, hypothesis tests are constructed by using the fact that the statistic

$$\frac{\bar{x}_1 - \bar{x}_2}{\sigma_{\bar{x}_1 - \bar{x}_2}}$$

has a sampling distribution which is at least approximately a standard normal distribution when $\mu_1 = \mu_2$. (If $\mu_1 < \mu_2$, the sampling distribution has a negative mean.)

For testing the null hypothesis H against the alternative $\mu_1 > \mu_2$ we reject the hypothesis H at the 5% level if

$$\frac{\bar{x}_1 - \bar{x}_2}{\sigma_{\bar{x}_1 - \bar{x}_2}} > 1.645;$$

this is equivalent to

$$\bar{x}_1 - \bar{x}_2 > 1.645\, \sigma_{\bar{x}_1 - \bar{x}_2}.$$

A 1%-level test consists in rejecting H if

$$\frac{\bar{x}_1 - \bar{x}_2}{\sigma_{\bar{x}_1 - \bar{x}_2}} > 2.326;$$

this is equivalent to

$$\bar{x}_1 - \bar{x}_2 > 2.326\, \sigma_{\bar{x}_1 - \bar{x}_2}.$$

Alternatively, we can compute the P-value. In the example, the observed difference in means was 1.34 standard deviations, so the achieved level of significance, P, is .0901 (the area to the right of 1.34). Since the achieved level of significance is less than .10, we reject H at the 10% level. We would not reject H at the 5% level on the basis of these data.

---

*Use the rule $\mu_{x+y} = \mu_x + \mu_y$, taking $x = \bar{x}_1$ and $y = -\bar{x}_2$, so that $x + y = \bar{x}_1 + (-\bar{x}_2) = \bar{x}_1 - \bar{x}_2$. Also, when x and y are independent, $\sigma^2_{x+y} = \sigma^2_x + \sigma^2_y$, and $\sigma^2_{-\bar{x}_2} = \sigma^2_{\bar{x}_2}$.

10.1.2   Two-Tailed Tests

A teacher wants to see if there is a difference between the IQ scores of urban and rural high school students. The teacher is interested in a difference in either direction. Random samples of 100 urban and 25 rural students are obtained, the test is administered, and the resulting data are summarized in Table 10.2. Scores on this IQ test are known from past experience to have a standard deviation of 12 points (that is, a variance of 144).

TABLE 10.2

Scores for Two Groups of Students on an IQ Test Having a Standard Deviation of 12

	GROUP 1 (URBAN)	GROUP 2 (RURAL)
Number of students	$n_1 = 100$	$n_2 = 25$
Mean	$\bar{x}_1 = 103.92$	$\bar{x}_2 = 101.50$
Variance	$\sigma_1^2 = 144$	$\sigma_2^2 = 144$
Variance of mean	$\sigma_{\bar{x}_1}^2 = \dfrac{144}{100} = 1.44$	$\sigma_{\bar{x}_2}^2 = \dfrac{144}{25} = 5.76$

(hypothetical data)

The variance of the difference in means is

$$\sigma_{\bar{x}_1 - \bar{x}_2}^2 = \sigma_{\bar{x}_1}^2 + \sigma_{\bar{x}_2}^2 = 1.44 + 5.76 = 7.20,$$

so that the standard deviation is

$$\sigma_{\bar{x}_1 - \bar{x}_2} = \sqrt{7.20} = 2.68.$$

The difference in means is

$$103.92 - 101.50 = 2.42,$$

or

$$\frac{2.42}{2.68} = 0.90 \text{ standard deviations.}$$

The hypothesis $H: \mu_1 = \mu_2$ is to be tested against the alternative $\mu_1 \neq \mu_2$. At the 1% level, one rejects H if the observed difference exceeds 2.576 in either direction, positive or negative. At the 5% level, one rejects H if the observed difference is larger than 1.96 standard deviations in either direction, positive or negative. Here the null hypothesis is not rejected. In fact, the achieved level of significance is .3682 (because the area under the normal curve below $-.90$ and above $+.90$ is $2 \times .1841$).

10.1.3 Confidence Intervals

A psychologist who is using two groups of students in an experiment involving learning wishes to check that the average IQ is about the same in the two groups. Scores on this IQ test are known to have a standard deviation of 12 points (that is, a variance of 144). The data are summarized in Table 10.3.

The variance of the difference in means is

$$\sigma^2_{\bar{x}_1 - \bar{x}_2} = \sigma^2_{\bar{x}_1} + \sigma^2_{\bar{x}_2} = 5.76 + 3.60 = 9.36,$$

so that

$$\sigma_{\bar{x}_1 - \bar{x}_2} = \sqrt{9.36} = 3.06.$$

The difference in means is

$$103.71 - 102.50 = 1.21.$$

TABLE 10.3
Mean Scores for Two Groups of Students on an IQ Test Having a Standard Deviation of 12

	GROUP 1	GROUP 2
Number of students	$n_1 = 25$	$n_2 = 40$
Mean IQ	$\bar{x}_1 = 103.71$	$\bar{x}_2 = 102.50$
Variance of mean	$\sigma^2_{\bar{x}_1} = \dfrac{144}{25} = 5.76$	$\sigma^2_{\bar{x}_2} = \dfrac{144}{40} = 3.60$

(hypothetical data)

A 95% confidence interval for $\mu_1 - \mu_2$ is obtained by adding to and subtracting from $\bar{x}_1 - \bar{x}_2 = 1.21$ the quantity $1.96\, \sigma_{\bar{x}_1 - \bar{x}_2} = 1.96 \times 3.06 = 6.00$. The interval is $(-4.79, 7.21)$.

If, as in this example, one is more interested in verifying that μ_1 and μ_2 are approximately the same than in providing evidence that they differ, the confidence interval is especially helpful. A short confidence interval, containing the value 0, such as the one obtained, makes it plausible that μ_1 and μ_2 are nearly equal.

The basis for this confidence-interval procedure is that under random sampling

$$\frac{(\bar{x}_1 - \mu_1) - (\bar{x}_2 - \mu_2)}{\sigma_{\bar{x}_1 - \bar{x}_2}} = \frac{(\bar{x}_1 - \bar{x}_2) - (\mu_1 - \mu_2)}{\sqrt{\sigma_1^2/n_1 + \sigma_2^2/n_2}}$$

has a standard normal distribution or (by the Central Limit Theorem) approximately a standard normal distribution. Inequalities from numbers obtained from tables of the normal distribution are manipulated as in Section 8.3.1 to obtain as a 95% confidence interval

$$(\bar{x}_1 - \bar{x}_2 - 1.96 \sqrt{\sigma_1^2/n_1 + \sigma_2^2/n_2}, \qquad \bar{x}_1 - \bar{x}_2 + 1.96 \sqrt{\sigma_1^2/n_1 + \sigma_2^2/n_2}).$$

10.2 COMPARISON OF TWO INDEPENDENT SAMPLE MEANS WHEN THE POPULATION STANDARD DEVIATIONS ARE UNKNOWN BUT TREATED AS EQUAL

If the standard deviations of the two parent populations are unknown, the standard deviation of the sampling distribution of the difference of the sample means has to be estimated. How this estimation is done depends upon whether or not the standard deviations of the parent populations are considered equal when the null hypothesis is true.

We begin with an example. In a certain statistics course the undergraduates claim they should be graded separately from the graduate students on the grounds that graduate students tend to do better. The graduate students claim they should be graded separately from the undergraduates on the grounds that undergraduates tend to do better! We ask the question whether or not on the average undergraduates do about as well as graduate students in statistics courses of this type.

Scores of the 39 undergraduates and the 29 graduates who took the course one year are given in Table 10.4. We shall treat these two groups as samples from the (hypothetical) population of graduate students taking such courses and the (hypothetical) population of undergraduate students taking such courses. The null hypothesis is that the population means are equal. Since we are interested in both positive and negative differences in means, we shall make a two-tailed test.

Because we are not dealing with scores on a standardized test, such as the IQ test of the preceding section, the standard deviations in the two populations are unknown and must be estimated. We suppose that the standard deviations in the two populations are the same.*

TABLE 10.4 *Final Grades in a Statistics Course*

UNDERGRADUATES: $n_1 = 39$						GRADUATES: $n_2 = 29$			
Student	Score	Student	Score	Student	Score	Student	Score	Student	Score
1	153	14	101	27	113	40	145	55	139
2	109	15	109	28	130	41	117	56	107
3	157	16	159	29	139	42	110	57	137
4	145	17	99	30	129	43	160	58	138
5	131	18	143	31	152	44	109	59	132
6	161	19	158	32	161	45	135	60	136
7	124	20	143	33	108	46	141	61	160
8	158	21	153	34	126	47	132	62	94
9	131	22	149	35	147	48	124	63	144
10	120	23	99	36	164	49	153	64	144*
11	153	24	162	37	119	50	126	65	85
12	113	25	122	38	137	51	133	66	109
13	134	26	165	39	145	52	157	67	105
						53	93	68	135
						54	83		

SOURCE: A professor who requests anonymity.
*When two successive values in a list are the same, there is some reason to suspect an error of transcription in typing. In this case, the original data sheet was checked, and the two 144's are in fact correct.

$$\bar{x}_1 = 136.44 \qquad\qquad \bar{x}_2 = 127.00$$

$$\sum_{i=1}^{n_1} (x_{1i} - \bar{x}_1)^2 = 15{,}551 \qquad \sum_{i=1}^{n_2} (x_{2i} - \bar{x}_2)^2 = 13{,}388$$

$$s_1^2 = 409.2 \qquad\qquad s_2^2 = 478.1$$

$$s_1 = 20.2 \qquad\qquad s_2 = 21.9$$

The common value of the standard deviation or of the variance is to be estimated on the basis of the deviations of the values in each sample from their respective sample means. We average the squared deviations over the two samples to estimate the variance by adding together ("pooling") the sums of squared deviations in the two groups,

$$15{,}551 + 13{,}388 = 28{,}939,$$

*The ratio of sample variances is $s_2^2/s_1^2 = 1.17$. That is, s_2^2 is only about 20% larger than s_1^2. It can be shown that for samples of this size this difference is well within the range of variability expected when $\sigma_1^2 = \sigma_2^2$.

and dividing this total by the appropriate number of degrees of freedom. The number of degrees of freedom associated with 15,551 is $n_1 - 1 = 39 - 1 = 38$; the number associated with 13,388 is $n_2 - 1 = 29 - 1 = 28$. The number associated with their sum is $38 + 28 = 66$. The estimate of variance is the pooled sum of squared deviations divided by the number of degrees of freedom; that is,

$$s_p^2 = \frac{28,939}{66} = 438.5.$$

The estimated variance of the sampling distribution of the difference between means (namely, $\sigma_1^2/n_1 + \sigma_2^2/n_2$) is

$$\frac{438.5}{39} + \frac{438.5}{29} = 26.36.$$

The corresponding standard deviation is $\sqrt{26.36} = 5.13$. The difference between means measured in terms of the estimated standard deviation is

$$t = \frac{136.44 - 127.00}{5.13} = \frac{9.44}{5.13} = 1.83.$$

The sampling distribution of the difference of sample means divided by its estimated standard deviation is a Student's t-distribution when the parent population means are equal. The number of degrees of freedom is 2 less than the total number of observations in the two samples.

The number of degrees of freedom, 66, in this case is large enough by our rule of thumb (at least 30 degrees of freedom) to treat t as normal. The 5% significance point is 1.96 and we do not reject the null hypothesis at the 5% level. (The achieved significance level is $P = .0672$.)

Now we explain this procedure in general terms. The null hypothesis is

$$H: \mu_1 = \mu_2.$$

The variances of the parent populations are denoted by σ_1^2 and σ_2^2. In some problems it is reasonable to consider these variances equal when the null hypothesis is true. It is convenient to call this common variance σ^2. Then the variance of the difference of sample means,

$$\sigma_{\bar{x}_1 - \bar{x}_2}^2 = \frac{\sigma_1^2}{n_1} + \frac{\sigma_2^2}{n_2},$$

can be written

$$\sigma_{\bar{x}_1 - \bar{x}_2}^2 = \frac{\sigma^2}{n_1} + \frac{\sigma^2}{n_2}$$

or, factoring out σ^2,

$$\sigma^2_{\bar{x}_1 - \bar{x}_2} = \sigma^2 \left(\frac{1}{n_1} + \frac{1}{n_2} \right).$$

An estimate of $\sigma^2_{\bar{x}_1 - \bar{x}_2}$ is thus

$$s_p^2 \left(\frac{1}{n_1} + \frac{1}{n_1} \right),$$

where s_p^2 is the "pooled" estimate of the common variance σ^2, namely,

$$s_p^2 = \frac{\sum\limits_{i=1}^{n_1} (x_{1i} - \bar{x}_1)^2 + \sum\limits_{j=1}^{n_2} (x_{2j} - \bar{x}_2)^2}{n_1 + n_2 - 2}.$$

The divisor of s_p^2 is the number of degrees of freedom associated with the two sums of squared deviations together, $(n_1 - 1) + (n_2 - 1) = n_1 + n_2 - 2$. The statistic s_p^2 is an unbiased estimate* of σ^2.

If σ^2 is known, the statistic used to make the test is

$$z = \frac{\bar{x}_1 - \bar{x}_2}{\sqrt{\sigma^2(1/n_1 + 1/n_2)}},$$

which has a sampling distribution that is approximately normal. When σ^2 is estimated by s_p^2 the statistic to use is

$$t = \frac{\bar{x}_1 - \bar{x}_2}{\sqrt{s_p^2(1/n_1 + 1/n_2)}} = \frac{\bar{x}_1 - \bar{x}_2}{s_p \sqrt{1/n_1 + 1/n_2}}.$$

The sampling distribution of this statistic in random samples is the Student's t-distribution with $n_1 + n_2 - 2$ degrees of freedom; hence we denote the statistic by t.

The 5%-level test of the null hypothesis $H: \mu_1 = \mu_2$ (or $H: \mu_1 \leq \mu_2$) against the alternative $\mu_1 > \mu_2$ (that is, that the first population mean is larger than the second) is to reject if

$$t > t_{n_1 + n_2 - 2} (.05).$$

This is equivalent to rejecting H if

$$\bar{x}_1 - \bar{x}_2 > t_{n_1 + n_2 - 2} (.05) s_p \sqrt{\frac{1}{n_1} + \frac{1}{n_2}};$$

*The expected values of the sums of squared deviations are $(n_1 - 1)\sigma^2$ and $(n_2 - 1)\sigma^2$, so the expected value of their sum is $(n_1 + n_2 - 2)\sigma^2$.

that is, if the difference in sample means is sufficiently large in terms of the estimated standard error. To carry out a test at another significance level, $t_{n_1+n_2-2}$ (.05) is replaced by the appropriate number from the t-table.

To test the null hypothesis H: $\mu_1 = \mu_2$ against the two-sided alternative $\mu_1 \neq \mu_2$, a two-tailed test is used. At the 5% level, reject H if

$$t > t_{n_1+n_2-2}(.025) \qquad \text{or} \qquad t < -t_{n_1+n_2-2}(.025).$$

For another level of significance, $t_{n_1+n_2-2}$ (.025) is replaced by another number, for example, at 1% by $t_{n_1+n_2-2}(.005)$.

The confidence-interval procedures for the difference $\mu_1 - \mu_2$ as presented in Section 10.1.3 can be appropriately modified; the standard error $\sigma_{\bar{x}_1 - \bar{x}_2}$ is replaced by its estimate $s_p \sqrt{1/n_1 + 1/n_2}$ and the number from the normal table is replaced by the number from the t-table. For example, a 95% confidence interval for $\mu_1 - \mu_2$ is

$$\bar{x}_1 - \bar{x}_2 - t_{n_1+n_2-2}(.025)\, s_p\, \sqrt{\frac{1}{n_1} + \frac{1}{n_2}} < \mu_1 - \mu_2$$

$$< \bar{x}_1 - \bar{x}_2 + t_{n_1+n_2-2}(.025)\, s_p\, \sqrt{\frac{1}{n_1} + \frac{1}{n_2}}\,.$$

10.3 COMPARISON OF TWO INDEPENDENT SAMPLE MEANS WHEN THE POPULATION STANDARD DEVIATIONS ARE UNKNOWN AND NOT TREATED AS EQUAL

As we have seen, the standard deviation of the sampling distribution of the differences of the two sample means is

$$\sigma_{\bar{x}_1 - \bar{x}_2} = \sqrt{\frac{\sigma_1^2}{n_1} + \frac{\sigma_2^2}{n_2}}\,.$$

To test the null hypothesis H: $\mu_1 = \mu_2$ when the population standard deviations are known, one refers

(10.1)
$$\frac{\bar{x}_1 - \bar{x}_2}{\sqrt{\dfrac{\sigma_1^2}{n_1} + \dfrac{\sigma_2^2}{n_2}}}$$

to tables of the standard normal distribution as explained in Section 10.1. When the standard deviations are unknown and there is no reason to believe $\sigma_1 = \sigma_2$ (when H is true) the population standard deviation σ_1 and σ_2 are estimated by the sample standard deviations s_1 and s_2, respectively; the standard deviation of the sampling distribution of the differences in sample means is estimated by

$$\sqrt{\frac{s_1^2}{n_1} + \frac{s_2^2}{n_2}} \, ;$$

and the test is carried out by using

(10.2)
$$\frac{\bar{x}_1 - \bar{x}_2}{\sqrt{\dfrac{s_1^2}{n_1} + \dfrac{s_2^2}{n_2}}}$$

instead of (10.1). If n_1 and n_2 are both reasonably large (each greater than 20 as a rule of thumb), then (10.2) can be referred to the normal distribution since the Central Limit Theorem insures that the sampling distribution of the differences in sample means is at least approximately a normal distribution and there is little sampling variablity in the estimate of $\sigma_{\bar{x}_1 - \bar{x}_2}$. For samples of moderate size the statistic (10.2) can be referred to the t-table with $n_1 + n_2 - 2$ degrees of freedom, unless either of the ratios n_1/n_2 or s_1/s_2 is very different from 1. However, if n_1 or n_2 is small or if σ_1 is believed to be very different from σ_2 (possibly indicated by s_1 being very different from s_2) neither the normal table nor t-table should be used for a test at a specified significance level.

Construction of confidence intervals when the population standard deviations are not considered to be equal is done on the same basis as construction of tests.

10.4 COMPARISON OF TWO INDEPENDENT SAMPLE PROPORTIONS

10.4.1 Hypothesis Tests

A frequently asked question is whether the proportions of individuals or objects in two populations having some characteristic are equal. Is the proportion of voters intending to vote for the Republican candidate in August the same as the proportion in October? Is the probability of contracting polio the same in the population of children inoculated with vaccine and the population of children not inoculated? To answer such a question a sample is taken from each population and the relevant proportions in the two samples are compared.

As an example, we return to the Asch experiment on the effect of group pressure on perception (Section 1.1.5) in which the individuals in the test group were under pressure to give wrong responses. Instead of counting the number of errors made by each subject, account will be taken only of whether a subject made at least one error or made no error. This condensation of Table 1.7 results in Table 10.5.

TABLE 10.5

Summary of Results of One of Asch's Early Experiments

	TEST GROUP	CONTROL GROUP	BOTH GROUPS
ONE OR MORE ERRORS	37 (74.0%)	2 (5.4%)	39 (44.8%)
NO ERRORS	13 (26.0%)	35 (94.6%)	48 (55.2%)
TOTAL	$n_1 = 50$ (100%)	$n_2 = 37$ (100%)	87 (100%)

SOURCE: Table 1.7.

The $n_1 = 50$ cases of individuals subjected to group pressure are considered to constitute a sample from the (hypothetical) population of individuals who might be subjected to pressure. In this population the probability of one or more errors is p_1. Similarly the $n_2 = 37$ cases in the "control group" are treated as a sample from a population not under pressure, with probability p_2 of at least one error. The null hypothesis is $H: p_1 = p_2$. If the probabilities are not equal, it is expected that the probability under pressure is greater. The alternative hypothesis is $p_1 > p_2$. (The null hypothesis could be extended to $p_1 \leq p_2$.) The null hypothesis is to be rejected if the sample proportion $\hat{p}_1$ is much greater than the sample proportion $\hat{p}_2$; that is, if the difference $\hat{p}_1 - \hat{p}_2$ is large. In this case $\hat{p}_1 = .740$ and $\hat{p}_2 = .054$. Is the difference $\hat{p}_1 - \hat{p}_2 = .740 - .054 = .686$ large enough?

In random sampling a sample proportion has approximately a normal distribution, and the difference of two proportions has approximately a normal distribution. When $p_1 = p_2$, the standard deviations of the parent populations are equal, that is, $\sqrt{p_1 q_1} = \sqrt{p_2 q_2}$. Let p be the common value of the probability $p_1 = p_2$ in this case. The estimate of p is the proportion of subjects in both groups making one or more errors, $\hat{p} = .448$. The estimate of the standard deviation of the sampling distribution of the difference between the two independent proportions is

$$\sqrt{\hat{p}(1 - \hat{p}) \left(\frac{1}{n_1} + \frac{1}{n_2} \right)} = \sqrt{.448(1 - .448) \left(\frac{1}{50} + \frac{1}{37} \right)}$$
$$= \sqrt{.247(.0200 + .0270)}$$

$$= \sqrt{.247 \times .0470}$$

$$= \sqrt{.0116}$$

$$= .108.$$

The difference in proportions divided by the estimated standard deviation is

$$\frac{.740 - .054}{.108} = \frac{.686}{.108} = 6.36.$$

Thus even the gross summary of the data provided by the simple 2×2 table gives overwhelming evidence against the hypothesis that group pressure did not affect perception.

Now we explain this procedure in general terms. The observations are Bernoulli variables, taking the values 0 or 1. Each observation in the first sample is from a population with mean $\mu_1 = p_1$ and variance $\sigma_1^2 = p_1 q_1$; similarly, each observation in the second sample is from a population with mean $\mu_2 = p_2$ and variance $\sigma_2^2 = p_2 q_2$. The null hypothesis is

$$H: p_1 = p_2$$

(which may be extended to $p_1 \leq p_2$ if the alternative is $p_1 > p_2$ and may be extended to $p_1 \geq p_2$ if the alternative is $p_1 < p_2$).

Here the means $\bar{x}_1$ and $\bar{x}_2$ are the sample proportions, $\hat{p}_1$ and $\hat{p}_2$. The difference $\bar{x}_1 - \bar{x}_2$ is $\hat{p}_1 - \hat{p}_2$. The standard deviation of the sampling distribution of the difference in sample proportions is

$$\sigma_{\hat{p}_1 - \hat{p}_2}^2 = \frac{p_1 q_1}{n_1} + \frac{p_2 q_2}{n_2}.$$

When $p_1 = p_2 (= p)$,

$$\sigma_{\hat{p}_1 - \hat{p}_2}^2 = \frac{p_1 q_1}{n_1} + \frac{p_2 q_2}{n_2} = pq \left(\frac{1}{n_1} + \frac{1}{n_2} \right).$$

The standardized version of $\hat{p}_1 - \hat{p}_2$ is

$$\frac{\hat{p}_1 - \hat{p}_2}{\sigma_{\hat{p}_1 - \hat{p}_2}} = \frac{(\hat{p}_1 - \hat{p}_2)}{\sqrt{pq(1/n_1 + 1/n_2)}}.$$

The estimate $\hat{p}$ is the proportion in both samples combined: $\hat{p} = (a + b)/n$. (See Table 10.6.)

Thus we use the statistic

$$\frac{\hat{p}_1 - \hat{p}_2}{\sqrt{\hat{p}\hat{q}(1/n_1 + 1/n_2)}},$$

where $\hat{q} = 1 - \hat{p}$. When $p_1 = p_2$, the sampling distribution of this statistic is approximately the same as the standard normal distribution when n_1 and n_2 are not small. (For greater accuracy the continuity correction should be used; see Appendix 10A.)

TABLE 10.6
Classification of Two Samples According to Presence or Absence of Characteristic A

	Sample 1	Sample 2	Both Samples
A	a	b	$a + b$
$\overline{A}$	c	d	$c + d$
	$n_1 = a + c$	$n_2 = b + d$	$n = n_1 + n_2 = a + b + c + d$

10.4.2   Confidence Intervals

Confidence intervals for $p_1 - p_2$ are obtained by using the fact that

$$\frac{(\hat{p}_1 - \hat{p}_2) - (p_1 - p_2)}{\sqrt{\hat{p}_1\hat{q}_1/n_1 + \hat{p}_2\hat{q}_2/n_2}}$$

has a sampling distribution which is approximately the same as the standard normal. A 95% confidence interval for $p_1 - p_2$ is

$$\hat{p}_1 - \hat{p}_2 - 1.96 \sqrt{\frac{\hat{p}_1\hat{q}_1}{n_1} + \frac{\hat{p}_2\hat{q}_2}{n_2}} < p_1 - p_2 < \hat{p}_1 - \hat{p}_2 + 1.96 \sqrt{\frac{\hat{p}_1\hat{q}_1}{n_1} + \frac{\hat{p}_2\hat{q}_2}{n_2}}.$$

Note that this confidence interval does not correspond to the test of the hypothesis $H: p_1 = p_2$. In testing we use $\sqrt{\hat{p}\hat{q}(1/n_1 + 1/n_2)}$ as an estimate of the standard deviation of $\hat{p}_1 - \hat{p}_2$ because the significance level is based on a probability when the null hypothesis is true, whereas in forming an interval estimate we use $\sqrt{\hat{p}_1\hat{q}_1/n_1 + \hat{p}_2\hat{q}_2/n_2}$ because all values of $p_1 - p_2$ are to be included, not just $p_1 - p_2 = 0$.

*10.5 EFFECTIVENESS OF TWO-SAMPLE TESTS

10.5.1 Power

The effectiveness of a test of a hypothesis is described by its power function, that is, by the probabilities of rejection of the null hypothesis at various alternatives. Consider the 5%-level test of the null hypothesis $H: \mu_1 \leq \mu_2$ against the alternative $\mu_1 > \mu_2$ when σ_1 and σ_2 are known. This test is to reject the null hypothesis if one observes sample means $\bar{x}_1$ and $\bar{x}_2$ such that

$$\frac{\bar{x}_1 - \bar{x}_2}{\sigma_{\bar{x}_1 - \bar{x}_2}} > 1.645.$$

The power of this test is

Pr (Reject H when μ_1, μ_2, $\sigma_{\bar{x}_1 - \bar{x}_2}$ are the true parameter values),

for means $\bar{x}_1$ and $\bar{x}_2$ from random samples. This equals

(10.3)
$$\Pr\left(\frac{\bar{x}_1 - \bar{x}_2}{\sigma_{\bar{x}_1 - \bar{x}_2}} > 1.645 \text{ when } \mu_1, \mu_2, \sigma_{\bar{x}_1 - \bar{x}_2} \text{ are the true parameter values}\right).$$

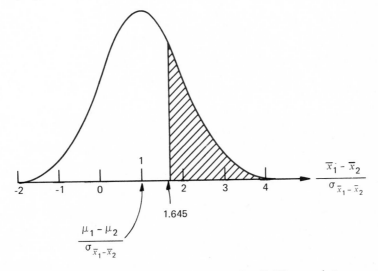

Figure 10.3 *Sampling distribution of test statistic: Small difference between means; small power*

That is, to compute the power we consider the sampling distribution of the test statistic $(\bar{x}_1 - \bar{x}_2)/\sigma_{\bar{x}_1 - \bar{x}_2}$. The power is the area to the right of 1.645 under the curve in Figure 10.3 or Figure 10.4. The sampling distribution is simply a normal distribution with variance 1 and mean $(\mu_1 - \mu_2)/\sigma_{\bar{x}_1 - \bar{x}_2}$. The larger this mean, the more area there is to the right of 1.645, and the greater the power. This mean is larger as the difference between μ_1 and μ_2 is larger, or as the standard deviation $\sigma_{\bar{x}_1 - \bar{x}_2}$ is smaller (that is, as the standard deviation of the parent population is smaller, or the sample sizes larger). Figure 10.4 shows a case in which the power is greater than in Figure 10.3. The areas shown represent the probabilities

(10.4)
$$\Pr\left(z > 1.645 - \frac{\mu_1 - \mu_2}{\sigma_{\bar{x}_1 - \bar{x}_2}}\right)$$

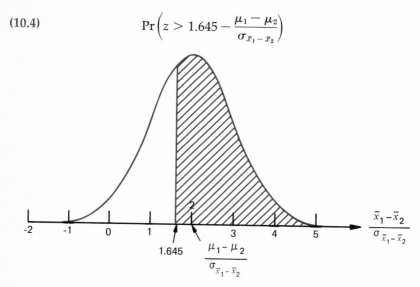

Figure 10.4 *Sampling distribution of test statistic: Large difference between means; large power*

Thus (10.4), considered as a function of $(\mu_1 - \mu_2)/\sigma_{\bar{x}_1 - \bar{x}_2}$, is the power function of the one-tailed 5%-level test. This function is graphed in Figure 10.5.

Figure 10.6 is a graph of the power function of the two-tailed 5%-level test. The probability of rejection is about .05 for values of $(\mu_1 - \mu_2)/\sigma_{\bar{x}_1 - \bar{x}_2}$ near 0 and is large for large absolute values of $(\mu_1 - \mu_2)/\sigma_{\bar{x}_1 - \bar{x}_2}$.

The power of the t-test is similar to that of a normal test, but estimation of the standard deviation adds some sampling variability. This fact is reflected in a decrease in power. However, if n_1 and n_2 are moderately large the power of the t-test can be approximated by that of the normal test.

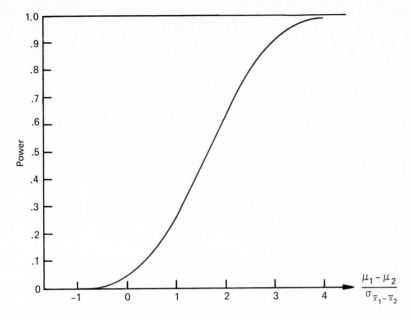

Figure 10.5 *Power function of the one-tailed 5%-level test of H:* $\mu_1 = \mu_2$

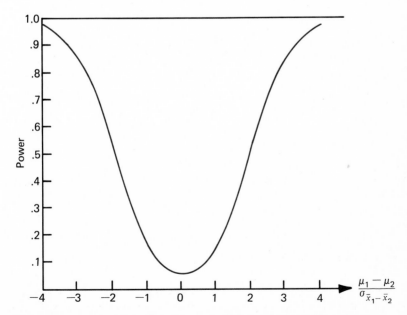

Figure 10.6 *Power function of the two-tailed 5%-level test of H:* $\mu_1 = \mu_2$

10.5.2   Sample Sizes Needed

Suppose that a researcher considers a difference $\mu_1 - \mu_2 = 10$ to be of importance. Then he or she wants to be relatively sure of detecting a difference that large—wants the *power* to be large when $\mu_1 - \mu_2$ is as large as 10. Say the researcher wants to make a one-tailed test at the 5% level and wants the power to be .70 when $\mu_1 - \mu_2 = 10$. Making use of (10.4) and the fact that

$$\Pr(z > .524) = .30, \qquad \Pr(z < -.524) = .30, \qquad \Pr(z > -.524) = .70,$$

we have

$$1.645 - \frac{10}{\sqrt{\sigma_1^2/n_1 + \sigma_2^2/n_2}} \le -.524;$$

that is,

$$\frac{\sigma_1^2}{n_1} + \frac{\sigma_2^2}{n_2} \le \frac{10^2}{(1.645 + 0.524)^2}.$$

In general, if one wishes to achieve power $1 - \beta$ for a specified difference $\mu_1 - \mu_2$, and a one-tailed test at level α is to be made, the sample sizes must be large enough so that

(10.5)
$$\frac{\sigma_1^2}{n_1} + \frac{\sigma_2^2}{n_2} \le \frac{(\mu_1 - \mu_2)^2}{[z(\alpha) + z(\beta)]^2}.$$

For example, suppose $\sigma_1 = \sigma_2 = 12$ and we have an experimental group of size $n_1 = 7$. How large must the size n_2 of the control group be so that, using a one-tailed 5%-level test, we have a 70% chance of detecting a difference $\mu_1 - \mu_2$ as large as 10? Substituting, we obtain

$$\frac{144}{7} + \frac{144}{n_2} \le \frac{10^2}{(1.645 + 0.524)^2},$$

which gives

$$n_2 > 210.$$

Use of these formulas requires knowledge of σ_1 and σ_2. If they are unknown, a pilot study may be run to get an idea of their magnitude or sometimes a reasonable guess may be made. When a *t*-test is made (σ_1 and σ_2 unknown), the *t*-distribution comes into play, and these formulas provide only approximations, which are good if n_1 and n_2 are large.

If we want to have two groups of equal sizes, say $n_1 = n_2 = n$, then we need to have

$$n \geq \frac{(\sigma_1^2 + \sigma_2^2)[z(\alpha) + z(\beta)]^2}{(\mu_1 - \mu_2)^2}.$$

Approximate results for two-tailed tests are obtained by replacing α by $\alpha/2$.

In planning the 1954 polio vaccine trial, the question of sample size was of obvious importance. It was clear that the sample would have to include a large number of children; otherwise it would be likely that there would be no cases of polio in either the control or the experimental group, and it would not be possible to reach a decision as to whether the vaccine was effective. Suppose the rate of incidence of paralytic polio was 30 per 100,000 (that is, 0.0003). If the vaccine was 50% effective, that is, if it reduced the rate to 0.00015, it would be good to have a high probability, say .90, of finding a significant difference. We shall show that this means that the sample size, n, in each group (placebo and vaccine) would have to be about 200,000, as in the actual study.

For proportions, the preceding formula becomes

$$n \geq \frac{(p_1 q_1 + p_2 q_2)[z(\alpha) + z(\beta)]^2}{(p_2 - p_1)^2}.$$

For the polio vaccine trial, assuming $p_1 = .00015$ and $p_2 = .0003$, $\alpha = .05$, $\beta = .10$, this is

$$n \geq \frac{[.00015(1 - .00015) + .0003(1 - .0003)](1.645 + 1.282)^2}{(.0003 - .00015)^2}$$

Now, $.00015(1 - .00015)$ is very nearly $.00015$ and $.0003(1 - .0003)$ is very nearly $.0003$, so we have

$$n \geq \frac{(.00015 + .0003)(2.927)^2}{(.00015)^2} = \frac{.00045 \times 8.57}{(.00015)^2}$$

$$= \frac{8.57}{.00005}$$

$$= 171,400.$$

10.6   TEST FOR LOCATION BASED ON RANKS OF THE OBSERVATIONS

10.6.1   Sign Test for Comparing Locations

In this section, as in Sections 9.5 and 9.6, we discuss some tests which are useful when the sample sizes are too small to assume approximate normality of sample means or when the data are defined on only an ordinal scale. Sometimes, because they are simple, these tests are used even when a *t*-test would be valid.

The so-called "sign test" of the null hypothesis of no difference in locations of two populations consists in comparing the observations in the two samples with the median of the entire set of observations. The null hypothesis is that the distribution of x_1 is the same as the distribution of x_2. The alternative is that the distributions differ with respect to location, that is, that observations from one distribution tend to be larger than those from the other. Two samples of sizes $n_1 = 7$ and $n_2 = 9$ are shown in Figure 10.7. The median, m, of the combined sample is $\frac{1}{2}(18 + 20) = 19$. There is only one value in the first sample which exceeds the median of the combined sample, while there are 7 such values in the second sample. The numbers of observations above and below the median in the two samples are recorded in Table 10.7. The fact that most of the observations in the first sample are less than the median and most in the second greater suggests that the null hypothesis may be implausible.

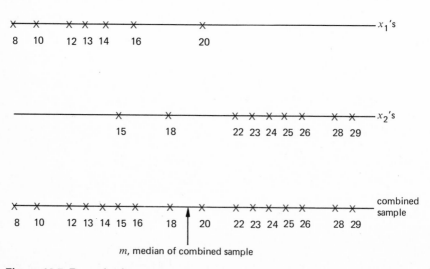

Figure 10.7 *Data plot for two-sample sign test*

TABLE 10.7

Summary Data for Two-Sample Sign Test

	FIRST SAMPLE	SECOND SAMPLE	COMBINED SAMPLES
GREATER THAN COMMON MEDIAN	1 (14.3%)	7 (77.8%)	8 (50.0%)
LESS THAN COMMON MEDIAN	6 (85.7%)	2 (22.2%)	8 (50.0%)
TOTAL	7 (100.0%)	9 (100.0%)	16 (100.0%)

SOURCE: Figure 10.7.

Under the null hypothesis that the two distributions are the same, all of the observations may be considered as a single sample from the common distribution, and the probability of any one observation being greater than the median is 1/2. If the two distributions are different in such a way that the x_2's tend to be larger than the x_1's, the probability that an x_2 observation is greater than the median of the sample exceeds 1/2. The null hypothesis corresponds to the probability of an x_1 observation exceeding the common median being equal to the probability of an x_2 observation exceeding the common median.

The procedure of Section 10.4 for testing equality of probabilities of Bernoulli variables can be used. Let $\hat{p}_1 = 1/7$ be the proportion greater than the common median in the first sample and $\hat{p}_2 = 7/9$ in the second. In the pooled sample $\hat{p}$ must be 1/2 by definition of the common median m.* The test statistic is the standardized version of $\hat{p}_1 - \hat{p}_2$, which is

$$z = \frac{\hat{p}_1 - \hat{p}_2}{\sqrt{\hat{p}\hat{q}\left(\frac{1}{n_1} + \frac{1}{n_2}\right)}} = \frac{\hat{p}_1 - \hat{p}_2}{\sqrt{\frac{1}{2} \times \frac{1}{2} \times \left(\frac{1}{n_1} + \frac{1}{n_2}\right)}} = 2\sqrt{\frac{n_1 n_2}{n_1 + n_2}} (\hat{p}_1 - \hat{p}_2).$$

The procedure is called the "sign test" because the frequencies in Table 10.7 are obtained by counting the number of positive and negative $(x_{1i} - m)$'s and $(x_{2i} - m)$'s. In practice it is necessary to compute only one of the frequencies in the 2×2 table because all the marginal totals are known.

As another numerical example, we take the data "religious preference and worldly success," given in Table 10.8. The median of the combined sample is 94 hundred dollars. When $n_1 + n_2$ is odd, the median is equal to one of the observations. Since this observation cannot be classified

---

*The justification for using the normal distribution here is different from that in Section 10.4 because here the definition of "success" depends on the pooled sample.

as above or below the median, we do not include it in counting observations above and below the median. The frequencies of Protestants and Catholics above and below the median are given in Table 10.9. The test statistic is

$$2\sqrt{\frac{13 \times 9}{13 + 9}}\left(\frac{8}{13} - \frac{3}{9}\right) = 2\sqrt{\frac{117}{22}}\frac{72 - 39}{117} = 2\frac{33}{\sqrt{2574}} = \frac{66}{50.7} = 1.30.$$

At the 5% level of significance the null hypothesis is accepted. (The P-value for a two-tailed test is about .2.)

TABLE 10.8
Incomes (100's of Dollars) of 13 Protestants and 10 Catholics

Protestants:	41, 53, 64, 67, 84, 100, 115, 127, 131, 145, 200, 280, 500
Catholics:	43, 47, 54, 61, 69, 78, 94, 121, 174, 260

(hypothetical data)

TABLE 10.9
Summary Information for Sign Test of Incomes of Catholics and Protestants

		RELIGIOUS PREFERENCE		
		Protestant	*Catholic*	*Total*
INCOME	Greater than 94	8 (61.5%)	3 (33.3%)	11 (50.0%)
	Less than 94	5 (38.5%)	6 (66.7%)	11 (50.0%)
	Total	13 (100.0%)	9 (100.0%)	22 (100.0%)

SOURCE: Table 10.8.

10.6.2   Rank Sum Test for Location of Two Populations

To test the hypothesis of equality of distributions against the alternative that the two distributions differ in location, the sign test is based only on information as to whether each observation is greater than the common median. The *rank sum test* uses more information, namely, the ranks. In this procedure the numerical values of all the observations in two samples are replaced by their ranks (in the combined sample). Then the means of the ranks in the two samples are compared.

As an example we return to the data in Table 10.8. The observations in the two samples are ordered in Table 10.10, with Protestants' incomes underlined, and the ranks are recorded in Table 10.11.

TABLE 10.10
Ranked Incomes of Protestants and Catholics

INCOME	RANK	INCOME	RANK	INCOME	RANK	INCOME	RANK
41	1	64	7	100	13	174	19
43	2	67	8	115	14	200	20
47	3	69	9	121	15	260	21
53	4	78	10	127	16	280	22
54	5	84	11	131	17	500	23
61	6	94	12	145	18		

SOURCE: Table 10.8.

TABLE 10.11
Incomes of 13 Protestants and 10 Catholics: Ranks in the Combined Sample

Protestants:	1,	4,	7,	8,	11,	13,	14,	16,	17,	18,	20,	22,	23
Catholics:	2,	3,	5,	6,	9,	10,	12,	15,	19,	21			

SOURCE: Table 10.8.

The mean rank for Protestants is $174/13 = 13.4$ and the mean rank for Catholics is $102/10 = 10.2$. Is the difference between the means significant? Since the sum of the first 23 integers is $23 \times 24/2 = 276$, we must have $174 + 102 = 276$. The mean or the sum of the ranks for Catholics is determined by the mean or the sum of the ranks for Protestants. We can compare the sum of the ranks for Protestants with its expected value, which is the fraction 13/23 of 276, namely, 156, because 13 out of 23 are Protestants. The variance of the sum under random sampling is

$$n_1 \times n_2 \times (n_1 + n_2 + 1)/12 = 10 \times 13 \times 24/12 = 260;$$

the standard deviation is $\sqrt{260} = 16.25$. The standardized value of the test statistic is

$$\frac{174 - 156}{16.25} = \frac{18}{16.25} = 1.11;$$

the null hypothesis is not rejected at the 5% level of significance. There is not strong evidence of a difference in locations.

(The rank sum test in this case gave a "less significant" result than did the less powerful sign test. That the rank sum test is more powerful means that, *on the average,* it has a higher probability of detecting any real difference.)

In terms of two samples of n_1 and n_2 observations the sum of the ranks is $(n_1 + n_2)(n_1 + n_2 + 1)/2$. The expected value of the sum of ranks for the first sample when the null hypothesis is true is

$$\frac{(n_1 + n_2)(n_1 + n_2 + 1)}{2} \times \frac{n_1}{n_1 + n_2} = \frac{n_1(n_1 + n_2 + 1)}{2}.$$

The variance of the sum when the null hypothesis is true is

$$\frac{n_1 n_2(n_1 + n_2 + 1)}{12}.$$

(See Exercise 10.28.) The test statistic is the difference between the observed sum and its expected value divided by the standard deviation.

Since we need only to be able to rank the data, they need only be defined on an ordinal scale for this test to be valid.

Ties. To deal with ties, the average of the ranks tied for is assigned each member of the tie. The variance of the sum of ranks in the first sample is decreased by the amount

$$\frac{n_1 n_2}{12(n_1 + n_2 - 1)(n_1 + n_2)} \sum_i (i - 1)i(i + 1)t_i,$$

where t_i is the number of ties involving i observations.

TABLE 10.12
Data with Tied Scores

TEST GROUP		CONTROL GROUP	
Score	*Rank*	*Score*	*Rank*
6.6	9.5	4.5	4
5.9	7	3.1	3
7.2	12	5.7	5.5
6.1	8	1.1	2
6.8	11	6.6	9.5
8.3	13	5.7	5.5
	60.5	0.6	1
			30.5

(hypothetical data)

Data from a hypothetical experiment involving a control group and a test group are shown in Table 10.12. A score of 6.6 appears twice, occupying ranks 9 and 10; hence each 6.6 is assigned a rank of 9.5. The score 5.7 occurs twice, occupying ranks 5 and 6; hence each 5.7 is assigned a rank of 5.5. The rank sums are 60.5 for the test group (first sample) and 30.5 for the control group (second sample); we verify that the sum of all the ranks is indeed $13 \times 14/2 = 91$. The expected value of the sum of ranks in the first sample (unchanged by the presence of ties) is

$$\frac{1}{2} n_1 (n_1 + n_2 + 1) = \frac{1}{2} \times 6 \times (6 + 7 + 1) = \frac{1}{2} \times 6 \times 14 = 42.$$

Here there are two ties, each containing two members, so the correction to the variance is

$$\frac{6 \times 7}{12 \times 12 \times 13} (2 - 1)2(2 + 1)2 = \frac{7}{26} = .27;$$

the variance is thus

$$\frac{6 \times 7 \times 14}{12} - .27 = 49 - .27 = 48.73.$$

The standardized value of the test statistic is

$$\frac{60.5 - 42}{\sqrt{48.73}} = \frac{18.5}{6.98} = 2.65$$

(one-tailed $P = .004$).

10.7 GENERAL REMARKS ON COMPARING TWO DISTRIBUTIONS

10.7.1 Effectiveness of Means for Comparing Distributions

Uses and limitations of means for comparing sample distributions were discussed in Sections 3.5 and 4.4.3. Now, having discussed probability, we can add some remarks. If the mean of Population B exceeds that of Population A, we may expect that many members of B will have values for the variable x higher than will many members of A; but some members of B will have lower values than will some members of A. In order to study this notion, let $\Pr(B > A)$ denote the probability that a randomly selected member of Population B will have a value exceeding that of a randomly selected member of Population A, where these members are selected independently. Then a more precise statement is that we expect $\Pr(B > A)$ to be high and $\Pr(A > B)$ [$= 1 - \Pr(B > A)$] to be low. We take $\Pr(A > B)$ as a measure of "reversal." We may expect this probability to be small when $\mu_B - \mu_A$ is large. We now ask to what extent this is true.

We study this question for the case of two normal distributions. It can be shown (see the Mathematical Exercises) that the reversal is the area contained under the standard normal curve to the left of the point $(\mu_A - \mu_B)/\sqrt{\sigma_A^2 + \sigma_B^2}$:

$$\Pr(A > B) = \Phi \left(\frac{\mu_A - \mu_B}{\sqrt{\sigma_A^2 + \sigma_B^2}} \right),$$

where, as discussed in Chapter 7, $\Phi(z)$ is the value of the cumulative standard normal distribution, that is, the area under the standard normal curve to the left of z. This shows exactly how the overlap depends upon the means. It also shows that the reversal depends not only upon the means but also on the standard deviations. Table 10.13 gives the reversal for a few examples. The smaller the reversal probability, the more reliable the information provided by the means. The table indicates that means do more to locate populations when the population variances are small than when they are large. Of course, if the distributions are identical the reversal is 1/2.*

TABLE 10.13
Reversal Probabilities, $\Pr(A > B)$, for Two Normal Populations

		(σ_A, σ_B)	
$\mu_B - \mu_A$	(10, 10)	(10, 20)	(20, 20)
20	.08	.18	.24
10	.24	.33	.36

10.7.2 Summary Statistics

Often the original data set is too voluminous for inclusion in the report of a study. It is essential, however, that the report include

---

*In fact, the reversal is 1/2 if $\mu_A = \mu_B$, even if $\sigma_A \neq \sigma_B$.

enough summary statistics for the sufficiently knowledgeable reader to reproduce the relevant calculations. It is not enough simply to write, "$t_{21} = 1.87$ (one-tailed $P < .05$)"; the values of the sample means and standard deviations must be given. Table 10.14 gives examples of the minimum information that should be included in the report of a statistical study.*

TABLE 10.14
Summary Statistics for Various Statistical Problems

Inference About	Sample Size(s)	Summary Statistic(s)
a mean, μ	n	$\bar{x}, s^2, s$
two means, μ_1, μ_2	n_1, n_2	$\bar{x}_1, \bar{x}_2, s_1^2, s_2^2, s_1, s_2$
a proportion, p	n	$\hat{p}$
two proportions, p_1, p_2	n_1, n_2	$\hat{p}_1, \hat{p}_2$

10.7.3 Control Groups and Randomization

When an experimental comparison of two samples is made, one sample is often an experimental group while the other is a control group (as in the polio vaccine trial in Section 1.1.2).

Ideally, the control group would be the same as the experimental group in every way, except that it would not receive the experimental treatment. One way to try to achieve this ideal, as indicated in Figure 10.1, is to choose both groups from a common pool of individuals, randomly assigning individuals to the experimental group. Random numbers can be used for this purpose. To divide 20 individuals into two groups of size 10, one method is to number the individuals in any order (say, alphabetical) from 1 to 20, then proceed through a table of random numbers, recording the first 10 2-digit numbers between 01 and 20. The individuals with these numbers are assigned to the experimental group.

10.7.4 Finite Population Correction

The tests and confidence intervals in this chapter have been based on random samples of n_1 and n_2 individuals drawn from two infinite populations (or probability distributions). When the populations are finite, the finite population correction is appropriate. If the

---

*A standard deviation is simply the square root of the corresponding variance, so it is redundant to provide both, but this is a helpful redundance.

first sample of n_1 is drawn from a population of N_1 and the second sample of n_2 from a population of N_2, the standard deviation of the difference of sample means is

$$\sqrt{\frac{N_1 - n_1}{N_1 - 1} \frac{\sigma_1^2}{n_1} + \frac{N_2 - n_2}{N_2 - 1} \frac{\sigma_2^2}{n_2}} .$$

The ratio $(N_i - n_i)/(N_i - 1)$ can be neglected if it is sufficiently close to 1 for $i = 1$ or 2 or both. If the standard deviations are estimated, σ_1 and σ_2 are replaced by s_1 and s_2, respectively.

SUMMARY

Many studies involve the comparison of two groups of individuals. The sets of observed individuals are regarded as samples from two populations, and it is to be decided whether these populations differ. Often the comparison focuses on the means.

Comparison of sample means involves consideration of the variance of their difference, which, in the case of two independent samples, is the sum of the variances of the two means.

The comparison of sample proportions is similar to comparison of means, except that equality of two proportions implies equality of the corresponding variances.

The power of the test of equality of two means depends upon the size of the difference between the means relative to the standard deviation of the sampling distribution of the difference between the two sample means.

APPENDIX 10A   The Continuity Adjustment

Proportions.   The continuity adjustment for a difference of two proportions consists in adding to or subtracting from $\hat{p}_1 - \hat{p}_2$ the continuity correction

$$\frac{n_1 + n_2}{2n_1 n_2} .$$

The effect of the adjustment is always to make test statistics less significant and confidence intervals wider. For example, suppose a confidence interval is being computed for $p_1 - p_2$, and $\hat{p}_1$ is larger than $\hat{p}_2$.

Then, in forming the upper limit, $\hat{p}_1 - \hat{p}_2$ is replaced by

$$\hat{p}_1 - \hat{p}_2 + \frac{n_1 + n_2}{2n_1n_2}$$

and, in forming the lower limit, by

$$\hat{p}_1 - \hat{p}_2 - \frac{n_1 + n_2}{2n_1n_2}.$$

Tests Based on Ranks. The quantity $(n_1 + n_2)/2n_1n_2$ is also the continuity correction for the two-sample sign test.

The continuity adjustment for the rank sum tests consists in altering the value of the sum of ranks in the first sample by adding or subtracting 1/2, as appropriate.

APPENDIX 10B   Randomization (Permutation) Tests

Suppose we have independent samples of size three from two populations. If all three observations from Population 1 are larger than all three from Population 2, is that conclusive evidence that the distributions are not the same?

It is easy to show by counting directly that there are 20 possible arrangements of the 6 observations into two groups with 3 observations per group. If the two populations are identical, all of these 20 arrangements are equally likely, so each has probability .05. For testing the hypothesis that the populations are the same against the alternative that the location of Population 1 exceeds that of Population 2, the arrangement actually obtained provides the strongest evidence (in terms of ranks of the observations) against the hypothesis. The associated P-value is .05.

The logic of this procedure is as follows. Suppose we take as a test statistic the difference between the sample means. (Other statistics could be used.) We want to answer the question, "Is the observed value of $\bar{x}_1 - \bar{x}_2$ large?" The answer involves consideration of an appropriate sampling distribution. To make a permutation test, we generate this sampling distribution from the data themselves.

We can suppose that we threw the $n_1 + n_2$ observations into a bowl and drew out n_1 at random, leaving n_2 in. This gives us a fictitious sample of n_1 and another of n_2. There are usually many such fictitious samples. For each we compute the value of $\bar{x}_1 - \bar{x}_2$. This generates a sampling distribution of $\bar{x}_1 - \bar{x}_2$. If the observed difference $\bar{x}_1 - \bar{x}_2$ is

larger than the differences between means for at least 95% of the fictitious samples, at the 5% level we reject the null hypothesis $H: \mu_1 \leq \mu_2$ in
favor of the alternative $\mu_1 > \mu_2$.

Permutation tests can often be used in problems where it is hard to
compute even an approximate distribution of the test statistic. The
reason for calling such a test a "permutation test" is that the possible
drawings of $n_1 + n_2$ observations are the *permutations* of $n_1 + n_2$ objects.

EXERCISES

10.1 (Sec. 10.2)   For the data in Exercise 2.20 (Table 2.23) relating to
students in a course in statistics, compare the mean final grade of students who had taken at least one course on probability and statistics
with the mean final grade of students who had no previous course on
probability and statistics, by making a 95% confidence interval for the
difference between means.

10.2 (Sec. 10.2)   Of the students for whom data are given in Table
10.4, the students numbered 28, 39, 40, 42, 44, 48, 49, 52, 53, 54, 56, 59,
62, 66, and 67 had a native language other than English. Test the null
hypothesis that the mean score of statistics students in the United States
whose native language is English is the same as that of those whose
native language is not English. Make a one-tailed test at the 5% level of
significance and proceed on the basis of equality of variances.

10.3 (continuation)   Arrange the scores in Table 10.4 into a two-way
table like Table 10.15.

TABLE 10.15
Scores of Students in a Statistics Course, by Student Status and Native Language

		STUDENT STATUS	
		Undergraduates	*Graduates*
NATIVE	*English*		
LANGUAGE	*Other*		

(a) Compute the sum of squared deviations for each of the four cells
of the table.
(b) Add these four sums of squared deviations to obtain a pooled
sum of squared deviations.

(c) What is the number of degrees of freedom associated with this pooled sum of squared deviations? [Hint: Each cell contributes one less than the sample size for that cell.]

(d) Compute a pooled estimate of variance by dividing the number of degrees of freedom into the pooled sum of squared deviations. Use this in the denominator of t-statistics in the rest of the exercise.

(e) Test the hypothesis of no difference between undergraduate and graduate means (i) for those whose native language is English and (ii) for the others. Compare the result with that obtained in Section 10.2; interpret any discrepancy.

(f) Test the hypothesis of no difference between means of students whose native language is English and the other students (i) for the undergraduates and (ii) for the graduates. Compare the result with that obtained in Exercise 10.2; interpret any discrepancy.

(g) Does the average effect of language depend upon student status? [Hint: Compute the difference between language groups for under-graduates and for graduates. Then compute the difference between these differences. The estimated variance of this combination of four means has the form

$$s_p^2 \left(\frac{1}{n_1} + \frac{1}{n_2} + \frac{1}{n_3} + \frac{1}{n_4} \right),$$

where s_p^2 is the pooled estimate of variance.]

10.4 (continuation) Write a paragraph summarizing this analysis of the data in Table 10.15.

10.5 (Sec. 10.3) Weiss, Whitten, and Leddy (1972) reported the figures in Table 10.16, which provide a comparison between the lead

TABLE 10.16

Lead in Hair (micrograms of lead per gram of hair)

| | | POPULATIONS | |
		Antique	*Contemporary*
AGE GROUP	Children	$\bar{x} = 164.24$ $s = 124$ $n = 36$ $s^2/n = 428$	$\bar{x} = 16.23$ $s = 10.6$ $n = 119$ $s^2/n = 0.941$
	Adults	$\bar{x} = 93.36$ $s = 72.9$ $n = 20$ $s^2/n = 266$	$\bar{x} = 6.55$ $s = 6.19$ $n = 28$ $s^2/n = 1.37$

SOURCE: Weiss, Whitten, and Leddy (1972). Copyright 1972 by the American Association for the Advancement of Science.

content of human hair removed from persons between 1871 and 1923 (the "antique" population) with that from present day populations. (A microgram is a millionth of a gram.) Make the following comparisons by doing two-tailed *t*-tests at the 1% level, using the separate variance estimates:*

(a) Is there a significant difference between the antique and contemporary groups (i) among children? (ii) among adults?

(b) Is there a significant difference between the children and the adults (i) in the antique group? (ii) in the contemporary group?

10.6 (Sec. 10.4) Two types of needles, A and B, were used for injection of medical patients with a certain substance. The patients were allocated at random to two groups, one to receive the injection from needle A, the other to receive the injection from needle B. Table 10.17 shows the number of patients showing reactions to the injection. Test at the 1% level the null hypothesis of no difference between the true proportions of patients giving reactions to needles A and B. State clearly the alternative hypothesis.

TABLE 10.17
Results of Injections with Two Types of Needles

NUMBER OF PATIENTS

		With Reactions	*Without Reactions*	*Total*
TYPE OF	*A*	44	6	50
NEEDLE	*B*	20	30	50
	Total	64	36	100

(hypothetical data)

10.7 (Sec. 10.4) Suppose that in the context of the previous exercise the figures had been as in Table 10.18. At the 1% level, would you accept the hypothesis of no difference between proportions?

---

*The four group standard deviations exhibit a rather large range of variability and are in almost the same order as the group means, suggesting that a more valid analysis might be obtained by first making a mathematical transformation of the data, such as replacing the raw data by their logarithms. The article itself gives standard errors of the four means, not the standard deviations themselves.

TABLE 10.18

Results of Injections with Two Types of Needles

NUMBER OF PATIENTS

		With *Reactions*	*Without* *Reactions*	*Total*
TYPE OF	*A*	22	3	25
NEEDLE	*B*	10	15	25
	Total	32	18	50

(hypothetical data)

10.8 (Sec. 10.4)   Table 10.19 shows the results of asking a number of persons whether they would favor a law which would require a person to obtain a police permit before he or she could buy a gun. Make the 2 × 2 table that results from omitting those with no opinion. Compute the corresponding percentages. At the 5% level make a two-tailed test of the hypothesis of equality of proportions of men and women favoring the legislation.

TABLE 10.19

Results of Opinion Poll Regarding Gun Permit Legislation

	MEN	WOMEN	BOTH SEXES
FAVOR	396 (66%)	450 (75%)	846 (70.5%)
OPPOSE	174 (29%)	114 (19%)	288 (24.0%)
NO OPINION	30 (5%)	36 (6%)	66 (5.5%)
TOTAL	600 (100%)	600 (100%)	1200 (100%)

(hypothetical data)

10.9 (Sec. 10.1)   Describe the sampling distribution of $\bar{x}_1 - \bar{x}_2$ in terms of repeated independent random samples of sizes n_1 and n_2 from two populations.

10.10 (Sec. 10.6.1)   Carry out the sign test on the two samples shown in Figure 10.7. Use a two-tailed test at the 5% level of significance.

10.11 (Sec. 10.2)   A number of students in a ninth-grade General Mathematics course were divided into two groups. One group (the experimental group) received instruction in a new probability-and-statistics unit; the other group (the control group) did not. Such in-

structional changes are not instituted without the possibility of conse-
quent decline in the learning of traditional skills; accordingly, all the
students were given a test of computational skill before and after the
semester. The score analyzed was the *gain* in skill. The results are
summarized in Table 10.20.

TABLE 10.20   *Summary of Students' Scores on Test of Computational Skill*

Experimental Group	Control Group
$n_1 = 281$	$n_2 = 311$
$\bar{x}_1 = -0.11$	$\bar{x}_2 = 3.54$
$s_1^2 = 74.0$	$s_2^2 = 80.0$
$s_1 = 8.60$	$s_2 = 8.95$

SOURCE: Shulte (1970). Reprinted from *The Mathematics Teacher*, January
1970, Vol. 63, pp. 56–64 (© 1970 by the National Council of Teachers of
Mathematics). Used by permission.
Experimental group: Course included a unit on probability and statistics.
Control group: Course did not include instruction in probability and
statistics.

(a) The pooled sum of squared deviations is 45,583.96. Test the null
hypothesis of no difference between the mean gains in the two
populations. Treat the population standard deviations as equal.
(b) Test the null hypothesis that the mean gain in the experimental
population is zero. Make a two-tailed test.
(c) Test the null hypothesis that the mean gain in the control population is zero.
(d) Write a paragraph interpreting the results of (a), (b), and (c).
(A sentence for each will suffice.)

10.12 (Sec. 10.4)   For the data from the 1954 polio vaccine trial (Sec.
1.1.2, Table 1.4), test the hypothesis of no difference between incidence
rates in the control and experimental populations against the alternative
that the rate in the experimental population is less.

10.13 (Sec. 10.4)   Suppose samples of size 20,000 (instead of 200,000)
had been used for the control and experimental groups in the 1954
polio vaccine trial, and there were 12 cases of paralytic polio in the
control group and 3 in the experimental group. (See Table 10.21. These
rates are about the same as in the actual study.) Is the incidence rate
in the experimental group significantly smaller than that in the control
group?

TABLE 10.21

Results of a Hypothetical Vaccine Trial

	SAMPLE SIZE	NUMBER OF CASES
Control group	20,000	12
Experimental group	20,000	3

(hypothetical data)

*10.14 (Sec. 10.5)   In the polio vaccine trial, what sample sizes would be required to be 90% sure of deciding in favor of the vaccine if it were $33\frac{1}{3}$% effective and the test were to be made at the 5% level? Assume as in the text that the incidence rate for unprotected children was 0.0003. (Thus $33\frac{1}{3}$% effectiveness of the vaccine means that the rate for children inoculated with the vaccine is $\frac{2}{3} \times .0003 = .0002$.)

*10.15 (Sec. 10.5)   In the polio vaccine trial, what sample sizes would be required in order to be 90% sure of deciding in favor of the vaccine if it were $33\frac{1}{3}$% effective and the test were to be made *at the 1% level*? Assume, as in the text, that the incidence rate for unprotected children was 0.0003.

10.16 (Sec. 10.2)   Instead of assuming known standard deviations of 12, use Student's *t*-test to compare the mean IQ scores of the two groups in Table 10.2. In other words:
(a) pretend the σ's are s's, both equal to 12.0, and compute the corresponding sums of squared deviations;
(b) add these sums of squared deviations to obtain a pooled estimate of the assumed common variance;
(c) make the *t*-test. Do you come to the same conclusion as when the standard deviations were assumed known?

*10.17   A university is trimming its budget. The mathematics library has been asked to discontinue its subscription to some of the journals duplicated in other libraries. The librarian sends a questionnaire to twelve mathematics professors, asking them to rate four such journals on a scale from 1 to 10, 10 denoting journals that should be retained and 1 those which could be discontinued. The data are given in Table 10.22. Do you think the data for Professor #10 should be included in the librarian's analysis? Why, or why not?

TABLE 10.22

Ratings of Four Journals by Twelve Professors

PROFESSOR	JOURNAL			
	A	B	C	D
1	8	9	3	2
2	8	10	3	2
3	9	10	3	2
4	8	9	3	3
5	8	8	4	3
6	7	10	4	1
7	7	9	3	2
8	6	8	4	1
9	7	9	2	1
10	2	1	7	9
11	7	9	2	2
12	8	9	2	1

(hypothetical data)

*10.18 (continuation)   Make a two-tailed test at the 5% level that professors have no preference between journals A and B.

MATHEMATICAL EXERCISES

*10.19 (Sec. 10.7)   Compute the values of $\Pr(A > B)$ in Table 10.13. Use the fact that $A - B$ has a normal distribution. (What is the mean of this distribution? What is its standard deviation?)

10.20 (Sec. 10.6.2)   By assigning a 1 to the smallest observation in the combined sample, a 2 to the largest, a 3 to the second largest, a 4 to the second smallest, etc., and computing the sums of the ranks, the rank sum test can be used to test the hypothesis that two populations have the same variance (given that their means are the same).* Carry out this test for the data of the example, "religious preference and worldly success."

10.21 (Sec. 10.6.2)   The rank sum test can be performed by computing the statistic U, which is the total number of times that values from the first sample precede values from the second sample when the combined sample is arranged in order. For example, if the arrangement is

2 1 2 1 2 2 2

*This procedure is due to Siegel and Tukey (1960).

then $U = 7$ because the first 1 precedes four 2's and the second 1 precedes three 2's. Show that

$$U = n_1 n_2 + \frac{n_1(n_1 + 1)}{2} - T,$$

T being the sum of the ranks in the first sample.

*10.22 (Sec. 10.5)   Figure 10.8 shows the power functions of the 5%- and 1%-level, two-tailed tests of $H: \mu_1 - \mu_2 = 0$, when $\sigma_{\bar{x}_1 - \bar{x}_2} = 0.424$. Table 10.23 is a table of these functions.

(a) In the case of the 5%-level test, how large must the difference between the population means be in order that the probability of rejecting the null hypothesis be at least .95? What if the 1%-level test is used? How large must the difference be in order to obtain a power of .50 using the 5%-level test? Using the 1%-level test?

(b) Which rule, the 5% level test or the 1% level test, is more appropriate? Why?

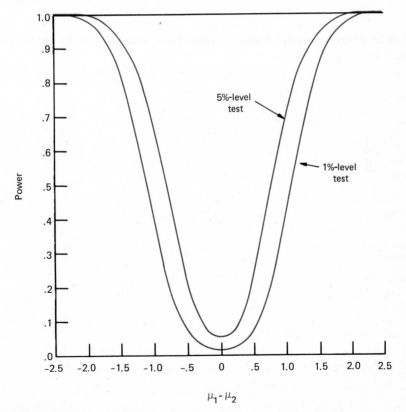

Figure 10.8 *Power functions of 1%- and 5%-level tests (Table 10.23)*

*(c) The power is

$$1 - \Pr\left(-z\left(\frac{\alpha}{2}\right) - \frac{\mu_1 - \mu_2}{0.424} < z < z\left(\frac{\alpha}{2}\right) - \frac{\mu_1 - \mu_2}{0.424}\right).$$

For example, for $\mu_1 - \mu_2 = .5$ and $\alpha = .05$ this is

$$\Pr\left(-1.96 - \frac{0.5}{0.424} < z < 1.96 - \frac{0.5}{0.424}\right)$$

$$= \Pr(-3.14 < z < .78)$$

$$= \Pr(z < .78) - \Pr(z > 3.14)$$

$$= .7823 - .0008$$

$$= .7815.$$

Compute the power of the 5% and 1% tests when $\mu_1 - \mu_2 = .848$.

TABLE 10.23
*Power Functions of 5%-Level and 1%-Level Tests of H: $\mu_1 - \mu_2 = 0$,
When $\sigma_{\bar{x}_1 - \bar{x}_2} = 0.424$*

	POWER	
$\mu_1 - \mu_2$	*5%-Level Test*	*1%-Level Test*
2.5	1.00	1.00
2.0	1.00	.98
1.5	.94	.83
1.0	.65	.41
0.5	.22	.08
0.0	.05	.01
−0.5	.22	.08
−1.0	.65	.41
−1.5	.94	.83
−2.0	1.00	.98
−2.5	1.00	1.00

*10.23 (Sec. 10.2)   When $\mu_1 = \mu_2$, the mean of the combined sample,

$$\bar{x} = \frac{x_{11} + x_{12} + \cdots + x_{1n_1} + x_{21} + x_{22} + \cdots + x_{2n_2}}{n_1 + n_2},$$

is an estimate of the common value of μ_1 and μ_2 and so

$$s'^2 = \frac{\sum\limits_{i=1}^{n_1} (x_{1i} - \bar{x})^2 + \sum\limits_{j=1}^{n_2} (x_{2j} - \bar{x})^2}{n_1 + n_2 - 1}$$

is an estimate of the variance σ^2. Hence one might consider using

$$t' = \frac{\bar{x}_1 - \bar{x}_2}{s'\sqrt{1/n_1 + 1/n_2}}$$

to test $H: \mu_1 = \mu_2$. Show that use of t' is equivalent to use of t. More precisely, show that rejecting H when $|t|$ is large, say $t^2 > c$, is equivalent to rejecting H when $t'^2 > c'$, where

$$c' = \frac{n_1 + n_2 - 1}{1 + \dfrac{n_1 + n_2 - 2}{c}}.$$

*10.24 (Sec. 10.5)   Show that (10.4) is the same as (10.3).

10.25   (Sec. 10.6.3)   List the 20 possible arrangements of 6 observations into two groups, where each group contains 3 observations.

*10.26 (Sec. 10.5)   Derive (10.5).

10.27 (Sec. 10.4)   Show that in terms of Table 10.6

$$\frac{\hat{p}_1 - \hat{p}_2}{\sqrt{\hat{p}\hat{q}\left(\dfrac{1}{n_1} + \dfrac{1}{n_2}\right)}} = \sqrt{n}\,\frac{ad - bc}{\sqrt{(a+b)(c+d)(a+c)(b+d)}}.$$

Note that the right-hand side is $\sqrt{n}$ times the index ϕ given in Section 5.1.3.

10.28 (Sec. 10.6.2)
(a) Show that the variance of the population consisting of the values $1, 2, \ldots, n$ is $(n+1)(n-1)/12$. Use the facts that

$$\sum_{i=1}^{n} i = \frac{n(n+1)}{2} \qquad \text{and} \qquad \sum_{i=1}^{n} i^2 = \frac{n(n+1)(2n+1)}{6}.$$

(b) Show that the variance of the sampling distribution of the mean of a sample of size n_1 ($1 \le n_1 \le n$) from this population is

$$\frac{(n - n_1)(n + 1)}{12 n_1}.$$

Use the finite population correction.

REFERENCES

Shulte, Albert P. (1970). "The effects of a unit in probability and statistics on students and teachers of ninth grade General Mathematics." *The Mathematics Teacher* 63: 56–64.

Siegel, Sidney, and John W. Tukey (1960). "A nonparametric sum of ranks procedure for relative spread in unpaired samples." *Journal of the American Statistical Association* 31: 429–444.

Weiss, D., B. Whitten, and D. Leddy (1972). "Lead content of human hair (1871–1971)." *Science* 178: 69–70.

11 VARIABILITY IN
ONE POPULATION
AND IN TWO POPULATIONS

INTRODUCTION

The old saw, "A chain is only as strong as its weakest link," may be construed as an admonition to consider the *variability* of the links as well as their average strength. In comparing distributions, averages alone are not always adequate. Figures 3.8 and 3.9 show two telephone waiting-time distributions with equal means (1.1 seconds) but very different shapes. In Section 4.4.3 we pointed out that their standard deviations, 0.41 second and 0.69 second, were quite different. If the pupils in each of two school classes have mean IQ's of 100, but Class A has a standard deviation of 10, while Class B has a standard deviation of 20, teaching the relatively homogeneous Class A may be very different from teaching the relatively heterogeneous Class B.

In this chapter we consider methods of answering questions about population variances. We begin with a review of what we learned in Chapters 4 and 8 about the sampling distribution of the variance. This discussion is extended by consideration of a family of probability distributions called the chi-square distributions. We then consider interval estimates and hypothesis tests for a population variance. Finally, we consider statistical inference concerning two population variances. This necessitates the introduction of another family of probability distributions, the F-distributions.

11.1 VARIABILITY IN ONE POPULATION

11.1.1 The Sampling Distribution of the Sum of Squared Deviations

In Section 4.3.1 we defined the variance of a finite population and the variance of a sample. In Section 8.2, in the discussion of the sample variance as a point estimate of the population variance, we noted that the sample variance is an unbiased estimate of the population variance; that is, the mean of the sampling distribution of the sample variance is the population variance, if the population is infinite.

Table 8.2 shows ten random samples of size 10 from a normal distribution with a mean of 69 and a standard deviation of 3, that is, a variance of 9. The values of the sample variance, given in Table 11.1, vary from sample to sample. The mean of these ten sample variances is 10.18, fairly close to the true value of 9. The sample standard deviation is 5.27; they range from 4.7 to 18.8. These facts indicate considerable variability of s^2 in samples of size $n = 10$.

For random samples of $n = 10$ observations from a normal distribution with variance σ^2, there is a theoretical sampling distribution of s^2. (The principle of deducing such a distribution was described in Chapter 7.) The relative frequencies are tabulated in Table 11.2 and graphed in Figure 11.1. Roughly speaking, a number in Table 11.2 indicates the probability that s^2 is in an interval of length 1. For example, the probability that s^2 for $n = 10$ is within 1/2 of 7 is about .104. The entire area under the curve in Figure 11.1 is 1. The probability that s^2 is greater than 14, for instance, is the area below the curve and to the right of 14. A location parameter of the distribution, the mean, is $\sigma^2 = 9$.

Table 11.2 also gives the relative frequencies of s^2 for a sample of size $n = 28$. The curve of relative frequencies, also graphed in Figure 11.1, shows that the probability for this larger sample size is more concentrated about the value $\sigma^2 = 9$. Comparison of the two curves suggests

TABLE 11.1

Values of the Variance from Ten Samples of Size 10 from a Normal Distribution with a Variance of 9 and 80% Confidence Intervals

SAMPLE NUMBER	SUM OF SQUARED DEVIATIONS	SAMPLE VARIANCE	CONFIDENCE INTERVAL	INTERVAL INCLUDES 9?
1	168.9	18.8	(11.5, 40.5)	No
2	98.0	10.9	(6.7, 23.5)	Yes
3	85.7	9.5	(5.8, 20.5)	Yes
4	56.4	6.3	(3.8, 13.5)	Yes
5	42.3	4.7	(2.9, 10.1)	Yes
6	66.1	7.3	(4.5, 15.9)	Yes
7	57.3	6.4	(3.9, 13.7)	Yes
8	134.5	14.9	(9.2, 32.3)	No
9	125.0	13.9	(8.5, 30.0)	Yes
10	81.7	9.1	(5.6, 19.6)	Yes

SOURCE: Table 8.2.

TABLE 11.2
Sampling Distributions of s^2 for $\sigma^2 = 9$ and $n = 10$ and 28

VALUE	HEIGHT OF PROBABILITY CURVE	
OF s^2	$n = 10$	$n = 28$
0	.000	.000
1	.002	.000
2	.016	.000
3	.040	.001
4	.066	.012
5	.087	.042
6	.100	.092
7	.104	.141
8	.101	.166
9	.092	.162
10	.081	.135
11	.069	.099
12	.056	.066
13	.045	.040
14	.036	.022
15	.027	.012
16	.021	.006
17	.016	.003
18	.012	.001
19	.009	.001
20	.006	.000
21	.004	.000
22	.003	.000

that the greater the sample size n, the greater the concentration around the population variance, and the less the variability of s^2.

As explained in Section 4.3.2, the sample variance, $s^2 = \sum_{i=1}^{n}(x_i - \bar{x})^2/(n - 1)$, has $n - 1$ degrees of freedom. (See also Section 8.3.2.) The sampling distribution of s^2/σ^2 does not depend on σ^2 when $x_1, x_2, \ldots, x_n$ are a random sample from a normal distribution with variance σ^2. The distribution of $(n - 1)s^2/\sigma^2$ is known as *the chi-square distribution with $n - 1$ degrees of freedom.* (Chi is the Greek letter χ, corresponding to x.) The curve of relative frequencies is graphed in Figure 11.2 for 6 degrees of freedom. The curves for other numbers of degrees of freedom are similar. If f, the number of degrees of freedom, is greater than 6, the curve is to the right of the curve in Figure 11.2. The mean of the distribution is f, the standard deviation is $\sqrt{2f}$, and the mode (the value of χ^2 at which the maximum of the curve occurs) is

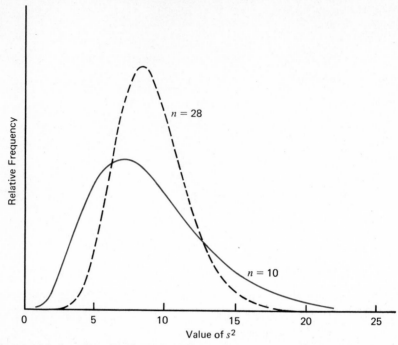

Figure 11.1 *Sampling distributions of s^2 for $\sigma^2 = 9$ and $n = 10$ and 28*

$f - 2$. Thus, the distribution of χ_f^2 is centered at f with a spread proportional to $\sqrt{f}$.

Figure 11.2 indicates $\chi_6^2(.10)$. This is the value of chi-square that is exceeded with probability .10; that is, the 90th percentile of the distribution. A table of values of $\chi_f^2(\alpha)$ is given in Appendix IX. The number of degrees of freedom associated with s^2 is $f = n - 1$. For example, if $n = 10$ ($n - 1 = 9$), we find from the table that $\chi_9^2(.975) = 2.700$ and $\chi_9^2(.025) = 19.022$. Thus, $\Pr(2.700 < \chi_9^2 < 19.022) = .95$. For $n = 20$ ($n - 1 = 19$), we find $\chi_{19}^2(.05) = 30.143$; thus, $\Pr(\chi_{19}^2 < 30.143) = .95$. The family of chi-square distributions is needed for statistical inference for the sample variance because s^2 is distributed as $\sigma^2 \chi_{n-1}^2/(n - 1)$. The curve of relative frequency of s^2 is similar to that of χ_{n-1}^2. Its mean is $\sigma^2(n - 1)/(n - 1) = \sigma^2$, and its standard deviation is $\sigma^2\sqrt{2(n - 1)}/(n - 1) = \sigma^2\sqrt{2}/\sqrt{n - 1}$. The larger n is, the smaller the standard deviation of the sampling distribution of the sample variance. (See Appendix 7A for an explanation.)

11.1.2 Testing the Hypothesis that the Variance Equals a Given Number

We continue our discussion of inference based on the sample variance with a discussion of hypothesis tests.

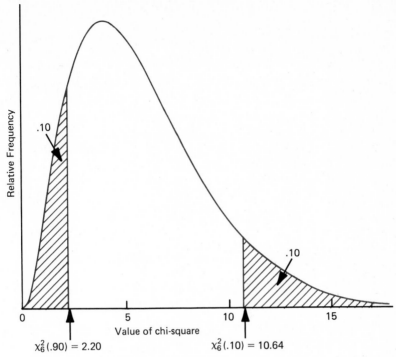

Figure 11.2 *Relative frequency curve of the chi-square distribution for 6 degrees of freedom, with some percentiles indicated*

A One-Tailed Test. All gold ingots produced by a certain company are supposed to have very nearly the same weight; it is important that the weight of the ingots not vary much in either direction from the mean. The standard deviation is an average distance from the mean (the root mean square, defined on page 118). Suppose that a standard deviation of 2 grams or less is tolerable. A sample of 20 ingots yields a standard deviation of 2.1 grams. Is this consistent with a population standard deviation of 2 grams or less? In formal terms, the null hypothesis is H: $\sigma \leq 2$ grams, and the alternative hypothesis is $\sigma > 2$ grams. Suppose that the 5% level is used.

The 20 observations are treated as a sample from a normal distribution. In random sampling, the variable $w = (n - 1)s^2/\sigma^2$ has a chi-square distribution with 19 degrees of freedom. If H is true, then $\sigma^2 = 4$, and the variable $w = 19s^2/4 = 4.75s^2$ is to be referred to the chi-square distribution with 19 degrees of freedom. From the table in Appendix IX we find $\chi^2_{19}(.05) = 30.143$; we shall reject H if the observed w exceeds 30.143. The sample variance of $s^2 = 2.1^2 = 4.41$ gives $w = 4.75 \times 4.41 = 20.9$, which does not exceed 30.143. We therefore accept

H; the conclusion is that weights of the gold ingots do not vary too much.

A Two-Tailed Test.   A standard IQ test has a standard deviation of 4 units in the relevant national population. A sample of 10 pupils is taken from a certain population; the question is raised whether the variability in this population is the same as in the national population. Let σ be the standard deviation of the IQ score in this population. We want to test the hypothesis $H: \sigma = 4$ against the alternative: $\sigma \neq 4$. We shall reject H if the observed sample variance is too small or too large. In random samples of 10 from the normal distribution with variance 16, the variable $w = (n - 1)s^2/\sigma^2 = 9s^2/16$ has a chi-square distribution with $n - 1 = 9$ degrees of freedom. Since $\chi_9^2(.975) = 2.700$ and $\chi_9^2(.025) = 19.022$ (from Appendix IX), we know that in random samples

(11.1) $.95 = \Pr(2.700 \leq w \leq 19.022)$

$= \Pr(2.700 \leq 9s^2/16 \leq 19.022)$

$= \Pr(2.700 \times 16/9 \leq s^2 \leq 19.022 \times 16/9)$

$= \Pr(4.800 \leq s^2 \leq 33.817).$

(The third equality in (11.1) follows from the fact that $a \leq b$ and $c > 0$ imply $ac \leq bc$.) That is, the probability that the variance of a sample of 10 from a normal distribution with $\sigma^2 = 16$ falls between 4.800 and 33.817 is .95. Thus, if an observed sample variance falls in this interval, H is accepted; if the observed variance is either less than 4.800 (too small) or greater than 33.817 (too large), H is rejected. The significance level of this test procedure is .05 because the probability of acceptance when the null hypothesis is true is .95.

This test procedure can be carried out equivalently by comparing $9s^2/16$ directly with $\chi_9^2(.975)$ and $\chi_9^2(.025)$. The power function of this test is $\Pr(s^2 \leq 4.800|\sigma^2) + \Pr(s^2 \geq 33.817|\sigma^2)$ as a function of σ^2. If σ^2 is much less than 16 or much greater, the power (the probability of rejecting H) is close to 1.

11.1.3   Confidence Intervals for the Variance

We continue with the same example to illustrate the construction of a confidence interval for the population variance. Whatever the value of σ^2, the random variable $w = (n - 1)s^2/\sigma^2$ is distributed according to the chi-square distribution with 9 degrees of freedom. Hence,

(11.2) $.95 = \Pr[\chi^2_9(.975) \leq w \leq \chi^2_9(.025)]$

$= \Pr(2.700 \leq w \leq 19.022)$

$= \Pr(2.700 \leq 9s^2/\sigma^2 \leq 19.022)$

$= \Pr(1/2.700 \geq \sigma^2/(9s^2) \geq 1/19.022)$

$= \Pr(9s^2/2.700 \geq \sigma^2 \geq 9s^2/19.022)$

$= \Pr(3.33s^2 \geq \sigma^2 \geq .473s^2).$

(The fourth inequality follows from the fact that if $a > 0$ and $b > 0$, then $a \leq b$ implies $1/a \geq 1/b$.) Hence, a 95% confidence interval for the population variance σ^2 is $(.473s^2, 3.33s^2)$, where s^2 is the observed sample variance. If the observed variance is 7.96, for example, the interval is $(.473 \times 7.96, 3.33 \times 7.96)$, which is $(3.77, 26.5)$. Since the population standard deviation is simply the square root of the population variance, the corresponding interval for the population standard deviation σ is $(\sqrt{3.77}, \sqrt{26.5})$, which is $(1.94, 5.15)$. If we observe that $s^2 = 7.96$, we assert that $1.94 \leq \sigma \leq 5.15$ with confidence .95.

In general, an interval for the population variance σ^2 with confidence coefficient $1 - \alpha$ is

$$[(n - 1)s^2/\chi^2_{n-1}(\alpha/2), (n - 1)s^2/\chi^2_{n-1}(1 - \alpha/2)].$$

The square roots of the limits are confidence limits for the population standard deviation σ.

The confidence interval consists of those values of σ^2 (or σ) that are "acceptable" as hypothesized values in hypothesis tests. In the example with $s^2 = 7.96$, at the 5% level one would accept the hypothesis that the variance is 16 but would reject the hypothesis that it is 35.

The tests and confidence intervals for σ^2, based on chi-square distributions, depend on the sampled population being normal or approximately normal. This feature is to be contrasted with the inference for μ, based on the sample mean (Chapter 9). The Central Limit Theorem implies that in random samples $\bar{x}$ is approximately normally distributed with variance σ^2/n if n is reasonably large, even if the sampled population is different from normal. If the observations are not drawn from a normal population, $(n - 1)s^2/\sigma^2$ may have a sampling distribution quite different from chi-square with $n - 1$ degrees of freedom. Hence, the test and confidence interval procedures just described may be quite inaccurate if the sampled population is different from normal; that is, significance levels and confidence levels may not be the asserted levels.

The technique should be used when there is reason to believe that the normal distribution provides a good approximation to the distribution sampled.

A Sampling Experiment. The 80% confidence intervals for each of the ten samples in Table 8.2 are given in Table 11.1. These intervals were formed by the method just presented, which consists in this case of dividing the sample sum of squared deviations by the 10th and 90th percentiles of the chi-square distributions with 9 degrees of freedom to obtain the upper and lower limits. These percentiles are 4.168 and 14.684. This method of construction of the intervals guarantees that 80% of all samples will give rise to a confidence interval that includes the true value. It happened that, in this sampling of only ten samples, exactly eight included the true value.

11.2   VARIABILITY IN TWO POPULATIONS

11.2.1   The Sampling Distribution of the Ratio of Two Sample Variances

When two populations are being studied, it is often of interest to compare their variabilities. (In Chapter 10 the difference between location parameters, particularly the means, was studied.) Since we usually measure variability by the standard deviation or variance, comparison of variability will be made on the basis of the standard deviations σ_1 and σ_2 of the two populations or on the basis of the variances σ_1^2 and σ_2^2. The ratio σ_1/σ_2 or the ratio σ_1^2/σ_2^2 is a desirable measure of the relationship of variabilities because such a ratio does not depend on the units of measurement. For example, if $\sigma_1 = 2$ feet and $\sigma_2 = 1$ foot, the ratio is $\sigma_1/\sigma_2 = 2$; if the unit of measurement were inches, the ratio would be $24/12 = 2$, the same as in feet. (Note that if $\sigma_1 - \sigma_2$ were used, the comparison would depend on the units.) We would say that the first population is twice as spread as the second. If the variabilities of the two populations are the same, $\sigma_1/\sigma_2 = 1$ and $\sigma_1^2/\sigma_2^2 = 1$.

When we have a sample from each of two populations, we use the sample variances s_1^2 and s_2^2 to estimate the population variances σ_1^2 and σ_2^2. If we want to use the sample variances to make inferences about the comparison of population variances, it is appropriate to use the ratio of variances s_1^2/s_2^2. It turns out that the ratio of variances is more useful than the ratio of standard deviations because the distribution of a variance ratio is simpler and its interpretation is easier in other problems. (See Chapter 13.)

When the samples are drawn from normal distributions with $\sigma_1^2 = \sigma_2^2$, the sampling distribution of s_1^2/s_2^2 is known as the F-distribution with $m = n_1 - 1$ and $n = n_2 - 1$ degrees of freedom; here n_1 and n_2 are the sizes of the two samples. [Note: $(n_j - 1)s_j^2/\sigma_j^2$ has a chi-square distribution with $n_j - 1$ degrees of freedom, $j = 1, 2$.] Since both s_1^2 and s_2^2 are positive, $F = s_1^2/s_2^2$ is positive. The sampling distribution of F for 10 and 20 degrees of freedom is tabulated in Table 11.3 and graphed in Figure 11.3. Since the distribution of s_j^2 is rather concentrated around σ_j^2, the distribution of F is rather concentrated around 1 when $\sigma_1^2 = \sigma_2^2$. The value exceeded by a probability α is denoted by $F_{m,n}(\alpha)$. These values are tabulated in Appendix X for various values of m, n, and α. Note that the first subscript m refers to the degrees of freedom of the sample variance in the numerator, and the second subscript n refers to the degrees of freedom of the sample variance in the denominator. The significance point $F_{10,20}(.05)$, illustrated in Figure 11.3, is found in Appendix X in the column labeled $m = 10$, in the horizontal section labeled $n = 20$, and in the row within the section labeled $\alpha = .05$ as 2.3479.

TABLE 11.3
Probability Curve of $F = s_1^2/s_2^2$ when $m = n_1 - 1 = 10$, $n = n_2 - 1 = 20$, and $\sigma_1^2/\sigma_2^2 = 1$

VALUE	HEIGHT OF PROBABILITY CURVE
0.0	.000
0.1	.015
0.2	.120
0.3	.311
0.4	.520
0.5	.688
0.6	.792
0.7	.833
0.8	.824
0.9	.779
1.0	.714
1.25	.525
1.5	.358
2.0	.153
2.5	.064
3.0	.027
3.5	.012

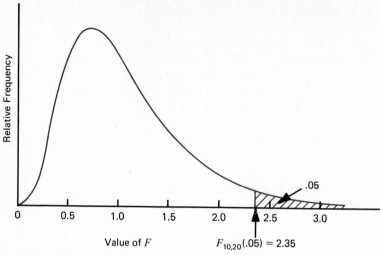

Figure 11.3 *Probability curve of $F = s_1^2/s_2^2$ when $m = n_1 - 1 = 10$, $n = n_2 - 1 = 20$, and $\sigma_1^2/\sigma_2^2 = 1$*

Note that significance points are given only for small values of α, that is, points in the upper tail. We have for any α

$$\alpha = \Pr\left\{\frac{s_1^2}{s_2^2} \geq F_{n_1,n_2}(\alpha)\right\}$$

$$= \Pr\left\{\frac{s_2^2}{s_1^2} \leq \frac{1}{F_{n_1,n_2}(\alpha)}\right\}$$

$$= 1 - \Pr\left\{\frac{s_2^2}{s_1^2} \geq \frac{1}{F_{n_1,n_2}(\alpha)}\right\}.$$

When we compare this equation with the definition of $F_{n_2,n_1}(1 - \alpha)$

$$\Pr\left\{\frac{s_2^2}{s_1^2} \geq F_{n_2,n_1}(1 - \alpha)\right\} = 1 - \alpha,$$

we see that

$$F_{m,n}(\alpha) = \frac{1}{F_{n,m}(1 - \alpha)}.$$

That is, if we interchange m and n, we replace α by $1 - \alpha$ and take the reciprocal. For example,

$$F_{10,20}(.95) = \frac{1}{F_{20,10}(.05)} = \frac{1}{2.7740} = .36049.$$

11.2.2 Testing the Hypothesis of Equality of Two Variances

A class of 32 students was randomly divided into an experimental set of size $n_1 = 11$ that received instruction in a new statistics unit and a control set of size $n_2 = 21$ that received the standard statistics instruction. All students were given a test of computational skill at the end of the course. Is the variability of this skill in the hypothetical population of students who might be given the new instruction the same as that of students who might be given the standard instruction? The standard deviation in the experimental population, σ_1, might be greater than the standard deviation in the control population, σ_2. The reason is that larger variability in the experimental population could result from the fact that not all students react to the experimental unit in the same way. Some may like it, while others dislike it; some may benefit from it, while others do not; diverting attention toward statistics and away from computation may affect the computational skills of some students and not others. (The interpretation would also depend on the relationship between the mean scores in the two groups.)

The sample standard deviations in this experiment were $s_1 = 3.05$ and $s_2 = 1.73$. The problem is to test $H: \sigma_1 \leq \sigma_2$ against the alternative: $\sigma_1 > \sigma_2$. Note that H and the alternative hypothesis can be stated in terms of the ratio of population variances as $H: \sigma_1^2/\sigma_2^2 \leq 1$ against the alternative: $\sigma_1^2/\sigma_2^2 > 1$. The ratio of sample variances is an estimate of the ratio of population variances, and one rejects H if the ratio of sample variances is sufficiently greater than 1. Here $s_1^2/s_2^2 = (3.05)^2/(1.73)^2 = 9.30/2.99 = 3.11$. This is the estimate of σ_1^2/σ_2^2. Is it sufficiently greater than 1 to warrant rejection of H? Since $3.11 > F_{10,20}(.05) = 2.3479$, H is rejected at the .05 level; in fact, the achieved level of significance is between .01 and .005. We conclude that the new statistics instruction results in a greater variability in computational skill.

Two-Tailed Test. The scores of $n_1 = 21$ undergraduates and $n_2 = 31$ graduate students in a statistics course gave respective sample variances of $s_1^2 = 478$ and $s_2^2 = 372$. The ratio of variances is $s_1^2/s_2^2 = 1.28$. Is this within the range of variability expected when the population variances are equal? That is, is the hypothesis of equality of population variances consistent with the data?

One tests $H: \sigma_1^2/\sigma_2^2 = 1$ against the alternative $\sigma_1^2/\sigma_2^2 \neq 1$. If the observed ratio $F = s_1^2/s_2^2$ is close to 1, H is accepted. That is, H is accepted if $a < F < b$, where the numbers a and b are chosen so that $\Pr(a < F < b$ when H is true$) = 1 - \alpha$, where α is the level of significance. We take $b = F_{m,n}(\alpha/2)$ and $a = F_{m,n}(1 - \alpha/2)$. In the example, $m = n_1 - 1 = 20$ and $n = n_2 - 1 = 30$. Suppose that $\alpha = .10$. Then $b = F_{20,30}(.05) = 1.9317$ and $a = F_{20,30}(.95) = 1/F_{30,20}(.05) = 1/2.0391 = 0.490$. The hypothesis H is accepted if F is between 0.490 and 1.9317. We have observed that $F = 1.28$, which is between these values. Hence, H is accepted.

Comparison of variances can be of interest in itself. Also, variances are sometimes compared before making a comparison of means, where assumptions are made about equality of variances. However, the F-test for equality of variances is sensitive to departures from normality; one should not make an F-test of equality of variances unless the sample distributions seem fairly close to normal.

11.2.3 Confidence Intervals for the Ratio of Two Variances

Regardless of whether or not H is true, the random variable $F^* = (s_1^2/\sigma_1^2) \div (s_2^2/\sigma_2^2)$ has an F-distribution when the samples are from normal distributions [because $(n_1 - 1)s_1^2/\sigma_1^2$ and $(n_2 - 1)s_2^2/\sigma_2^2$ have chi-square distributions with $n_1 - 1$ and $n_2 - 1$ degrees of freedom, respectively]. This fact can be used to construct confidence intervals for the ratio of population variances σ_1^2/σ_2^2. For example,

$$.90 = \Pr\ [F_{m,n}(.95) \le F^* \le F_{m,n}(.05)]$$

$$= \Pr\ \left[F_{m,n}(.95) \le \frac{s_1^2/\sigma_1^2}{s_2^2/\sigma_2^2} \le F_{m,n}(.05) \right]$$

$$= \Pr\ \left[\frac{1}{F_{m,n}(.95)} \ge \frac{s_2^2}{s_1^2} \times \frac{\sigma_1^2}{\sigma_2^2} \ge \frac{1}{F_{m,n}(.05)} \right]$$

$$= \Pr\ \left[\frac{1}{F_{m,n}(.95)} \times \frac{s_1^2}{s_2^2} \ge \frac{\sigma_1^2}{\sigma_2^2} \ge \frac{1}{F_{m,n}(.05)} \times \frac{s_1^2}{s_2^2} \right].$$

On the basis of observed sample variances s_1^2 and s_2^2, a 90% confidence interval for σ_1^2/σ_2^2 is

$$\left[\frac{1}{F_{m,n}(.05)} \times \frac{s_1^2}{s_2^2}, \frac{1}{F_{m,n}(.95)} \times \frac{s_1^2}{s_2^2} \right].$$

When the observed s_1^2/s_2^2 is 1.28, and $n_1 = 21$ and $n_2 = 31$ as in the previous example, the interval is $(1.28/1.9317, 1.28/0.490)$, which is $(0.663, 2.61)$. The confidence interval consists of those values of the ratio of population variances that are "acceptable" in a two-tailed test. For instance, the hypothesis $\sigma_1^2/\sigma_2^2 = 1$ would be accepted, but the hypothesis $\sigma_1^2/\sigma_2^2 = 3.1$ would be rejected.

The F-distribution was named for Sir Ronald A. Fisher, who in 1924 introduced the concepts and methods underlying the comparison of variances. (Actually Fisher used not F itself but another variable directly related to F. Tables for the sampling distribution of F were first constructed by George W. Snedecor.) Fisher was a greater contributor not only to statistical methodology but to the development of scientific methods of inquiry in general. Two books that are among Fisher's most important and influential works are listed in the references.

SUMMARY

The sum of squared deviations from the mean of a random sample of size n from a normal distribution with variance σ^2, divided by σ^2, has a chi-square distribution with $n - 1$ degrees of freedom. Hypothesis tests and confidence intervals for the population variance are based on this fact.

If s_1^2 is the variance of a random sample of n_1 observations from a normal distribution with variance σ_1^2 and s_2^2 is the variance of an independent random sample of n_2 observations from a normal distribution with variance σ_2^2, then the variable $(s_1^2/\sigma_1^2) \div (s_2^2/\sigma_2^2)$ is distributed according to the F-distribution, with degrees-of-freedom parameters $m = n_1 - 1$ and $n = n_2 - 1$. Inferences about the ratio of population variances are based on this fact.

APPENDIX 11A   The Family of Chi-Square Distributions

It was asserted in Section 11.1.1 that when the observations x_1, $x_2, \ldots, x_n$ are a random sample from a normal distribution with variance σ^2, the variable $w = (n - 1)s^2/\sigma^2 = \Sigma_{i=1}^n (x_i - \bar{x})^2/\sigma^2$ has the chi-square distribution with $n - 1$ degrees of freedom. We give now some explanation.

First, we note that if we define new variables $z_i = (x_i - \mu)/\sigma, i = 1, \ldots, n$, where μ is the mean of the normal distribution, then $z_1, \ldots, z_n$ are standard normal, $\bar{z} = (\bar{x} - \mu)/\sigma$ and $z_i - \bar{z} = (x_i - \bar{x})/\sigma$. (See Appendices 3C and 4C.) Then

$$\sum_{i=1}^n \frac{(x_i - \bar{x})^2}{\sigma^2} = \sum_{i=1}^n (z_i - \bar{z})^2.$$

This result verifies the assertion that the distribution of $\Sigma_{i=1}^n (x_i - \bar{x})^2/\sigma^2$ does not depend on σ^2.

The next assertion is that $\Sigma_{i=1}^n (z_i - z)^2$ has a sampling distribution identical to the sampling distribution of $\Sigma_{i=1}^{n-1} y_i^2$, where $y_1, \ldots, y_{n-1}$

are independent standard normal variables. As an example, take the case of $n = 2$. Then

$$\sum_{i=1}^{2} (z_i - \bar{z})^2 = \left[z_1 - \frac{1}{2} (z_1 + z_2) \right]^2 + \left[z_2 - \frac{1}{2} (z_1 + z_2) \right]^2$$

$$= \left[\frac{1}{2} (z_1 - z_2) \right]^2 + \left[\frac{1}{2} (z_2 - z_1) \right]^2$$

$$= \frac{1}{2} (z_1 - z_2)^2 = \left(\frac{z_1 - z_2}{\sqrt{2}} \right)^2,$$

but $y_1 = (z_1 - z_2)/\sqrt{2}$ is a normal variable with mean 0 and variance 1. The argument for general n uses the fact that

(11.3)
$$\sum_{i=1}^{n} (z_i - \bar{z})^2 = \sum_{i=1}^{n} z_i^2 - (\sqrt{n} \, \bar{z})^2.$$

More precisely, it is argued that the statistical behavior of $\Sigma_{i=1}^{n} z_i^2$ and $\sqrt{n} \, z_n$ is the same as that of $\Sigma_{i=1}^{n} y_i^2$ and y_n. Hence, the sampling distribution of (11.3) is that of $\Sigma_{i=1}^{n} y_i^2 - y_n^2 = \Sigma_{i=1}^{n-1} y_i^2$.

Finally, calculus is used to find the distribution of $\Sigma_{i=1}^{n-1} y_i^2$, which by definition is the χ^2-distribution with $n - 1$ degrees of freedom. The relative-frequency function of the chi-square distribution with f degrees of freedom is

$$\frac{1}{2^{f/2} \Gamma(f/2)} \, x^{f/2-1} e^{-x/2}$$

for positive x. This formula involves the gamma function $\Gamma(n)$. If n is a positive integer, then

$$\Gamma(n) = (n - 1)! = (n - 1) \times (n - 2) \times \cdots \times 2 \times 1$$

and

$$\Gamma(n + \tfrac{1}{2}) = (n - \tfrac{1}{2}) \times (n - \tfrac{3}{2}) \times \cdots \times \tfrac{3}{2} \times \tfrac{1}{2} \times \sqrt{\pi}.$$

The mean of the sampling distribution is f, the variance is $2f$, and the mode (highest value of the curve) is $f - 2$. More relative-frequency curves are given in Figure 11.4.

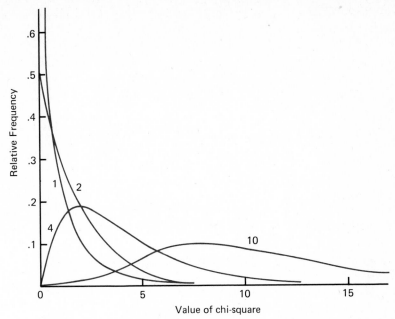

Figure 11.4 *Chi-square probability curves for 1, 2, 4, and 10 degrees of freedom*

APPENDIX 11B   Relations among Various Distributions

A variable having Student's t-distribution t_f is distributed as $z/\sqrt{w/f}$, where z has the standard normal distribution, w has a chi-square distribution with f degrees of freedom, and z and w are independent.

A variable having the F-distribution $F_{m,n}$ is distributed as $(u/m) \div (v/n)$, where u and v are independent chi-square variables with m and n degrees of freedom, respectively. The values in Table 11.3 are computed from the formula for the probability curve of the distribution $F_{m,n}$, given by

$$\frac{\Gamma[\tfrac{1}{2}(m+n)]}{\Gamma(\tfrac{1}{2}m)\Gamma(\tfrac{1}{2}n)} \left(\frac{m}{n}\right)^{m/2} \frac{x^{m/2-1}}{\left[1+\dfrac{m}{n}\,x\right]^{(m+n)/2}}$$

for positive x. The square of a variable having the t-distribution t_f has the F-distribution $F_{1,f}$.

EXERCISES

11.1 (Sec. 11.1)   From the table in Appendix IX, find the following percentage points.
  (a) $\chi_2^2(.05)$ (c) $\chi_4^2(.05)$
  (b) $\chi_2^2(.95)$ (d) $\chi_4^2(.95)$

11.2 (Sec. 11.1)   From the table in Appendix IX, find the following percentage points.
  (a) $\chi_2^2(.10)$ (c) $\chi_{20}^2(.10)$
  (b) $\chi_2^2(.90)$ (d) $\chi_{20}^2(.90)$

11.3 (Sec. 11.1)   A geneticist has some data from 1870 giving heights of human males who were then 30 years old. The standard deviation of these heights is 10 inches; that is, the variance is 100. One of the implications of the geneticist's theory is that the variance σ^2 of the distribution of heights of human males 30 years old in 1970 is 10% smaller than this. From a sample of size $n = 41$, the geneticist finds $s^2 = 92.1$. (Assume the distribution of heights is normal.)
  (a) Give a 99% confidence interval for the variance of the distribution of heights in 1970.
  (b) Do you reject at the 1% level the hypothesis that $\sigma^2 = 90$? Use a two-sided alternative.

11.4 (Sec. 11.1)   In the context of Exercise 11.3, do the following:
  (a) Give an 80% confidence interval for the variance of the distribution of heights in 1970.
  (b) Do you reject at the 20% level the hypothesis that $\sigma^2 = 90$? Use a two-sided alternative.

11.5 (Sec. 11.1.3)   Suppose that $s^2 = 7.96$ and $n = 10$, as in the example in Section 11.1. Construct an 80% confidence interval for the population variance. Compare the interval with the 95% interval given in the text.

11.6 (Sec. 11.1.3)   Suppose that $s^2 = 7.96$, as in Section 11.1, but $n = 20$ instead of 10. Construct a 95% confidence interval for the population variance. Compare the interval with the interval given in the text, which was constructed for $n = 10$.

11.7 (Sec. 11.1.3)   The standard deviation of a random sample of 20 observations from a normal distribution is 2.1. Construct an 80% confidence interval for the population standard deviation.

11.8 (Sec. 11.1.3)   The standard deviation of a random sample of 20 observations from a normal distribution is 2.1. Construct a 95% confidence interval for the population standard deviation.

11.9 (Sec. 11.1.3) The standard deviation of a random sample of 20 observations from a normal distribution is 2.1. Construct a 95% lower confidence interval for the population standard deviation, that is, an interval of the form (l, ∞).

11.10 (Sec. 11.1.3) The standard deviation of a random sample of 20 observations from a normal distribution is 2.1. Construct a 95% upper confidence interval for the population standard deviation, that is, an interval of the form $(0, u)$.

11.11 (Sec. 11.2) From the table in Appendix X, find the following percentage points.

  (a) $F_{10,20}(.05)$ (c) $F_{2,10}(.05)$
  (b) $F_{10,20}(.10)$ (d) $F_{10,2}(.05)$

11.12 (Sec. 11.2) From the table in Appendix X, find the following percentage points.

  (a) $F_{2,3}(.10)$ (c) $F_{2,4}(.10)$
  (b) $F_{2,3}(.05)$ (d) $F_{4,2}(.10)$

11.13 (Sec. 11.2) In Sample 1, $n_1 = 11$ and $s_1^2 = 18$. In Sample 2, $n_2 = 21$ and $s_2^2 = 36$. Assuming normality, test the hypothesis of equality of population variances against a two-sided alternative at the 5% level.

11.14 (Sec. 11.2) In Sample 1, $n_1 = 21$ and $s_1^2 = 30$. In Sample 2, $n_2 = 31$ and $s_2^2 = 10$. Assuming normality, test the hypothesis of equality of population variances against a two-sided alternative at the 10% level.

11.15 (Sec. 11.2) In Section 10.2 we considered the scores of $n_1 = 39$ undergraduates and $n_2 = 29$ graduate students in a statistics course. The sample variances were 409.2 and 478.1, respectively. We stated that the ratio of variances, 1.17, was well within the range of variability expected when the population variances are equal. Explain this by making the relevant test at the 10% level.

11.16 (Sec. 11.2) In Exercise 10.5 we recommended that separate variance estimates be used in making t-tests. To support this recommendation, test the hypothesis of equality of variances in the adult antique population and the adult contemporary population against the alternative that the variances are different. Show that the achieved level of significance in a two-tailed test is less than .005.

11.17 The *sample coefficient of variation* (defined in Exercise 4.22) is $s/\bar{x}$; the *population coefficient of variation* is σ/μ. How do you estimate this parameter? Estimate it for male cats and for female cats, using the statistics on male and female cats' weights in Table 4.11.

11.18   Compute the coefficients of variation for each variable in each population in Table 11.4. Comment on the results.

TABLE 11.4
Values of (μ, σ)

		POPULATION	
VARIABLE	A	B	C
x	(2, 4)	(3, 9)	(4, 16)
y	(3, 6)	(4, 12)	(5, 20)
z	(4, 8)	(5, 15)	(6, 24)
w	(5, 10)	(6, 18)	(7, 28)

11.19   Calculate $s/\bar{x}$ for each sample summarized in Table 10.16, and compare the four values. Suggest a reason for your findings.

MATHEMATICAL EXERCISES

11.20 (Sec. 11.1)   In Exercise 11.3, suppose that the sample size were not 41 but 5001. Give a 99% confidence interval for σ^2. [*Hint:* It follows from the Central Limit Theorem that χ^2_{5000} is approximately normal, with mean 5000 and variance 10,000.] Is the value 90 within the interval? Is the value 100?

11.21 (Sec. 11.1)   In Exercise 11.3, suppose the sample size were not 41 but 451. Give a 95% confidence interval for σ^2. [*Hint:* It follows from the Central Limit Theorem that χ^2_{450} is approximately normal, with mean 450 and variance 900.] Is the value 90 within the interval? Is the value 100?

11.22 (Sec. 11.2)   Construct a test of H: $\sigma_1^2 = 3\sigma_2^2$ against the alternative: $\sigma_1^2 \neq 3\sigma_2^2$.

11.23 (Sec. 11.2)   Construct a test of H: $\sigma_1 = 2\sigma_2$ against the alternative: $\sigma_1 \neq 2\sigma_2$.

11.24 (Sec. 11.1)   If z is distributed according to the standard normal distribution, what is the probability distribution of z^2?

11.25 (Sec. 11.1)   If x and y are independent and identically distributed according to the standard normal distribution, what is the probability distribution of $r^2 = x^2 + y^2$, the square of the length r in Figure 11.5?

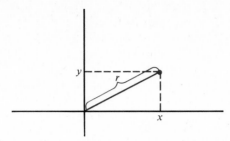

Figure 11.5 *Random horizontal and vertical coordinates*

11.26 The variance of the distribution of the sample variance is in general relatively complicated. However, when the sample is from a normal distribution, the variance of the sample variance is $2(\sigma^2)^2/(n-1)$. Compute this quantity for the following four cases: $n = 10$ and 28, $\sigma^2 = 9$ and 18.

11.27 Show that the median of $F_{n,n}$ is 1.

11.28 Show that if n is large, $\Pr(F_{m,n} \leq x)$ is approximately $\Pr(\chi^2_m \leq mx)$.

11.29 Show that, if t has a Student's distribution with f degrees of freedom, then t^2 is distributed according to $F_{1,f}$.

REFERENCES

Fisher, R. A. (1925). *Statistical Methods for Research Workers.* 1st ed. (12th ed., 1954). Oliver and Boyd, Edinburgh.

Fisher, R. A. (1935). *The Design of Experiments.* Oliver and Boyd, Edinburgh.

PART FIVE
STATISTICAL METHODS
FOR OTHER PROBLEMS

12 INFERENCE ON
CATEGORICAL DATA

INTRODUCTION

In this chapter we present some methods for treatment of categorical data. The methods involve the comparison of a set of observed frequencies with frequencies specified by some hypothesis to be tested. In Section 12.1 the hypothesis is that the categorical variable has a specific distribution. A test of such a hypothesis is called a test of *goodness of fit*.

The notion of independence of two categorical variables was developed in Chapter 5. In Section 12.2 we show how to test the hypothesis that two categorical variables are independent. The test statistics discussed have sampling distributions that are approximated by chi-square distributions. For this reason the tests are often called *chi-square tests*.

A measure of association for 2×2 tables was given in Chapter 5. In Section 12.3 we shall give several measures of association for two-way tables that are valid for larger tables.

12.1 TESTS OF GOODNESS OF FIT

One problem of statistics is to test whether a given distribution fits a set of data. A test of *goodness of fit* is based on comparison of an observed frequency distribution with the hypothesized distribution.

12.1.1 Two Categories—Dichotomous Data

We begin our discussion with the case of a variable defined by just two categories, which one may think of as Yes and No. Let p be the probability of Yes. Then $q = 1 - p$ is the probability of No, and the distribution of the variable is specified when the value of p is specified.

In Section 9.4.3 we considered a test of whether Black children were just as likely to choose a white as a nonwhite doll, that is, a test of the hypothesis H that the proportion who would choose a nonwhite doll is $p_0 = 1/2$. In a sample of 252 children, 83 chose a nonwhite doll. We developed a test based on consideration of the observed sample proportion $\hat{p} = 83/252 = .329$. Under the null hypothesis the standard deviation of a proportion in a random sample of 252 is $\sqrt{p_0 q_0}/\sqrt{n} = \sqrt{.5 \times .5}/\sqrt{252} = .0315$. Then $(\hat{p} - p_0)/(\sqrt{p_0 q_0}/\sqrt{n}) = (.329 - .5)/.0315 = -5.43$ is referred to the standard normal distribution for a two-tailed test (against a two-sided alternative). The null hypothesis is rejected at virtually any significance level; for example, it is rejected at 5%, since $|-5.43| > 1.96$. (The procedure is discussed in Section 9.4.)

An alternative method of carrying out this procedure is to use the fact that the square of a standard normal variable has a chi-square distribution with 1 degree of freedom. In the above example the observed value, $(-5.43)^2 = 29.5$, is to be compared with $1.96^2 = 3.84$ for a test at the 5% level. [In Appendix IX, $\chi_1^2(.05) = 3.843$.] In algebraic terms, when x is the number of Yes responses, p_0 the hypothesized value of the probability of Yes, and n the sample size, the criterion is

(12.1)
$$\left(\frac{\hat{p} - p_0}{\sqrt{p_0 q_0}/\sqrt{n}}\right)^2 = \frac{(\hat{p} - p_0)^2}{p_0 q_0/n} = n\frac{(\hat{p} - p_0)^2}{p_0 q_0}$$
$$= \frac{[n(\hat{p} - p_0)]^2}{n p_0 q_0} = \frac{(x - n p_0)^2}{n p_0 q_0}.$$

Of course, in terms of the No response, the null hypothesis is $1 - p = q = q_0 = 1 - p_0$, and the criterion is

(12.2)
$$\left(\frac{\hat{q} - q_0}{\sqrt{q_0 p_0}/\sqrt{n}}\right)^2 = \frac{(n - x - n q_0)^2}{n p_0 q_0}.$$

Since $n - x - n q_0 = n - x - n(1 - p_0) = -(x - n p_0)$, criterion (12.2) is identical to criterion (12.1). In fact, any weighted average of (12.1) and (12.2) is also equal to (12.1) and (12.2). We can multiply (12.1) by q_0 and (12.2) by p_0 to obtain the symmetric criterion

(12.3)
$$\frac{(x - n p_0)^2}{n p_0} + \frac{(n - x - n q_0)^2}{n q_0}.$$

The goodness-of-fit test is to compare the observed value of (12.3) with a value from the tables for chi-square with 1 degree of freedom.

In the example, $p_0 = q_0 = 1/2$, $n = 252$, $n p_0 = n q_0 = 126$, $x = 83$, and $n - x = 169$. The criterion is

$$\frac{(83-126)^2}{126} + \frac{(169-126)^2}{126} = \frac{(-43)^2}{126} + \frac{(43)^2}{126}$$

$$= 14.7 + 14.7$$

$$= 29.4,$$

which is within rounding error of $29.5 = (-5.43)^2$.

The purpose of developing (12.3) is to treat Yes and No (or nonwhite and white) symmetrically. This expression can be generalized for more than two categories. In (12.3) np_0 is the *expected* number of Yes responses under the null hypothesis, and nq_0 is the expected number of No responses. Each fraction in (12.3) is of the form

$$\frac{(\text{Observed Number} - \text{Expected Number})^2}{\text{Expected Number}}.$$

A symmetrical version of the null hypothesis is

$$H: p = p_0 \quad \text{and} \quad q = q_0.$$

12.1.2   An Arbitrary Number of Categories

Suppose that each trial can result in an outcome that falls into one of k categories. Let $p_1, p_2, \ldots, p_k$ denote the respective probabilities of these outcomes. Consider the null hypothesis

$$H: p_1 = p_{10}, p_2 = p_{20}, \ldots, p_k = p_{k0},$$

where $p_{10}, p_{20}, \ldots, p_{k0}$ are specified positive numbers adding to 1. Suppose that n trials are made. The expected number of outcomes in the ith category under the null hypothesis is $E_i = np_{i0}$; the actually observed number is denoted by O_i, $i = 1, \ldots, k$. The test statistic is

$$X^2 = \sum_{i=1}^{k} \frac{(O_i - E_i)^2}{E_i}.$$

This statistic has a sampling distribution that is approximately chi-square with $k - 1$ degrees of freedom. We reject H at level α if the observed test statistic exceeds $\chi^2_{k-1}(\alpha)$.

Note that $\sum_{i=1}^{k} E_i = n = \sum_{i=1}^{k} O_i$. The test statistic can also be written

$$X^2 = \sum_{i=1}^{k} \frac{O_i^2}{E_i} - n.$$

EXAMPLE   One might think that how long a person survives beyond a
birthday is a random event such that each day or week or month of
death (within a year) beyond the birthday is equally likely. In particular, in some specified population, it is expected that the probability of
the death of an individual occuring in the first month after a birthday
should be 1/12, etc. However, a person in a situation where death is
not unlikely who is looking forward to celebrating the next birthday
might stave off death, that is, reduce the probability of death shortly
before the birthday.

Phillips (1972) investigated the tendency of famous people to survive
until their next birthdays. Using a number of anthologies of famous
people, he noted that such people are least likely to die in the month before their birth month; there is a dip in the death rates corresponding to
the month before the birth month. Correspondingly there is an apparent rise in the death rate during the months immediately following the
birth month. Some data are given in Table 12.1. Are the observed frequencies 24, 31, etc., consistent with a probability of 1/12 for each
month? If the probability is in fact 1/12 for each month, then out of 348
people one would expect $348/12 = 29$ to die during each month. The expected value E of 29 is compared with the observed value O for each
month. The differences $O - E$ are shown in the table. To construct a
test statistic, each deviation $O - E$ is squared and divided by E. The resulting values $(O - E)^2/E$ are summed to obtain

TABLE 12.1
Number of Deaths Observed Before, During, and After the Birth Month

	NUMBER OF MONTHS BEFORE THE BIRTH MONTH						BIRTH MONTH	NUMBER OF MONTHS AFTER THE BIRTH MONTH				
	6	5	4	3	2	1		1	2	3	4	5
OBSERVED NUMBER OF DEATHS	24	31	20	23	34	16	26	36	37	41	26	34
EXPECTED NUMBER OF DEATHS	29	29	29	29	29	29	29	29	29	29	29	29
"EXCESS" NUMBER OF DEATHS	−5	+2	−9	−6	+5	−13	−3	+7	+8	+12	−3	+5
	$n = 348$ people						$n/12 = 29.0$					

SOURCE: Phillips (1972), p. 58.

$$\frac{(-5)^2}{29.0} + \frac{(+2)^2}{29.0} + \frac{(-9)^2}{29.0} + \frac{(-6)^2}{29.0} + \frac{(+5)^2}{29.0} + \frac{(-13)^2}{29.0} + \frac{(-3)^2}{29.0}$$

$$+ \frac{(+7)^2}{29.0} + \frac{(+8)^2}{29.0} + \frac{(+12)^2}{29.0} + \frac{(-3)^2}{29.0} + \frac{(+5)^2}{29.0} = 22.07.$$

There are $k = 12$ categories, and hence $k - 1 = 11$ degrees of freedom. We find $\chi^2_{11}(.025) = 21.920 < 22.07 < 24.724 = \chi^2_{11}(.01)$, and so at the .025 level we would reject the hypothesis that the hypothesized distribution is correct. The conclusion is that there is evidence that at least for famous people the death date has some relation to birth date; a possible explanation is that such a person looks forward to a birthday enough to affect health.

12.1.3 Choosing a Chi-Square Test

Lucky Louis has brought the dice for your weekly game. You wish to examine each of the dice, suspecting that either the dice have been loaded in favor of 6, or else that they are fair. You select one die and roll it 60 times, obtaining the results in Table 12.2. In formal terms the null hypothesis is $p_j = 1/6$, $j = 1, \ldots, 6$. The value of the chi-square statistic is

$$\frac{(5 - 10)^2}{10} + \frac{(10 - 10)^2}{10} + \frac{(10 - 10)^2}{10}$$

$$+ \frac{(9 - 10)^2}{10} + \frac{(9 - 10)^2}{10} + \frac{(17 - 10)^2}{10}$$

$$= \frac{25 + 0 + 0 + 1 + 1 + 49}{10}$$

$$= 7.6.$$

There are $6 - 1 = 5$ degrees of freedom. We have $7.6 < 9.236 = \chi^2_5(.10)$ from Appendix IX. This result is not particularly significant. However, this χ^2-test does not take into account the fact that you have one particu-

TABLE 12.2
Results of Rolling a Die

Face	1	2	3	4	5	6	Total
Frequency	5	10	10	9	9	17	60

(hypothetical data)

lar alternative in mind, namely, that the die may be loaded in favor of 6. Since a die is constructed so that the sum of the numbers of spots on opposite faces is seven, if 6 is favored, it must be the case that 1 will turn up less frequently. Accordingly, we categorize the outcome of the die into the three events: [1], [2, 3, 4, or 5], and [6]. The frequencies are given in Table 12.3. The value of the chi-square statistic for Table 12.3 is

$$\frac{(5 - 10)^2}{10} + \frac{(38 - 40)^2}{40} + \frac{(17 - 10)^2}{10} = \frac{25}{10} + \frac{4}{40} + \frac{49}{10}$$

$$= 2.5 + 0.1 + 4.9 = 7.5.$$

With $3 - 1 = 2$ degrees of freedom, this value is significant at the .05 level $[\chi^2_2(.025) = 7.378 < 7.5 < 9.210 = \chi^2_2(.01), .01 < P < .025]$.

These are various ways of categorizing the outcomes. For each there is a power function for the corresponding test, that is, the probability of rejecting the null hypothesis when it is not true; the power depends on the true probabilities of the outcomes. In the above example the test with 2 degrees of freedom is more powerful than that with 5 degrees of freedom when the discrepancy from the null hypothesis only involves the 1 and 6 faces.

TABLE 12.3
Results of Rolling a Die

FACE	1	2, 3, 4, 5	6	TOTAL
FREQUENCY	5	38	17	60

SOURCE: Table 12.2.

12.2   CHI-SQUARE TESTS OF INDEPENDENCE

12.2.1   Two-by-Two Tables

In Chapter 5 we studied frequency tables in the case in which individuals were classified simultaneously on two categorical variables. Independence in 2 × 2 tables was defined in Section 5.1.3 and in larger two-way tables in Section 5.2.2. In Section 12.2 we shall consider testing the null hypothesis that in a *population* two variables are independent on the basis of a *sample* drawn from that population. Even though the variables are independent in the population, they may not be (and in fact probably will not be) independent in a sample.

In Section 10.4 we also studied 2 × 2 tables. In particular, Table 10.5 presented data from one of Asch's experiments (discussed in Section 1.1.5); the numbers of subjects making one or more errors in the set of

TABLE 12.4
Favored Candidates in Two Polls

		DATE		Total
		First	Second	
CANDIDATE	Nixon	523	502	1025
	Kennedy	477	498	975
	Total	1000	1000	2000

trials with group pressure and in the control set and the numbers making no errors in each set constituted the entries. These data were used to test the null hypothesis that the probability of an individual's making one or more errors is the same for individuals under group pressure as for individuals not under group pressure. The null hypothesis is that the probability of error does not depend on whether group pressure is applied; this is a hypothesis of independence.

As another example, consider some data in Exercise 8.14. At two different dates a sample of 1000 registered voters was drawn; at each time each respondent was asked which of two potential candidates he or she would favor. The results are summarized in Table 12.4. In the underlying population, let the proportion favoring Nixon be p_1 at the time of the first poll and p_2 at the time of the second poll. The null hypothesis that the proportion favoring one candidate does not depend on the date of polling is $H: p_1 = p_2$. In this example the estimates of p_1 and p_2 are $\hat{p}_1 = .523$ and $\hat{p}_2 = .502$, respectively, based on sample sizes $n_1 = n_2 = 1000$. The estimate of the standard deviation of the difference between two sample proportions when the null hypothesis is true is

$$\sqrt{\hat{p}\hat{q}\left(\frac{1}{n_1} + \frac{1}{n_2}\right)} = \sqrt{.5125 \times .4875 \left(\frac{1}{1000} + \frac{1}{1000}\right)}$$

$$= .02235.$$

The test criterion (Section 10.4) is

(12.4) $$\frac{\hat{p}_1 - \hat{p}_2}{\sqrt{\hat{p}\hat{q}(1/n_1 + 1/n_2)}} = \frac{.021}{.02235} = .940,$$

which is to be referred to the standard normal tables. In this example a two-tailed test is called for, because a change could go either way; at the .05 level, the hypothesis would be rejected if (12.4) were greater in numerical value than 1.96.

An equivalent test procedure is to compare the square of (12.4) with the corresponding significance point of the chi-square distribution

TABLE 12.5

Expected Numbers for Table 12.4

		DATE		
		First	Second	Total
CANDIDATE	Nixon	$E_{11} = 512.5$	$E_{12} = 512.5$	1025
	Kennedy	$E_{21} = 487.5$	$E_{22} = 487.5$	975
	Total	$n_1 = 1000$	$n_2 = 1000$	2000

with 1 degree of freedom. Thus, $.940^2 = .884$ is to be compared with $1.96^2 = 3.84$.

The computation of the criterion can be put into the form $\Sigma(O - E)^2/E$, similar to the X^2-statistic in Section 12.1. The "expected" numbers are calculated on the basis of the hypothesis of independence being true. Then the estimate of the common value of $p_1 = p_2$ is $1025/2000 = .5125$. On the basis of this estimate of the proportion favoring Nixon, the expected numbers favoring Nixon on the two dates are $n_1\hat{p} = 1000 \times .5125 = 512.5$ and $n_2\hat{p} = 512.5$. Similarly, the expected numbers favoring Kennedy on the two dates are $n_1\hat{q} = 1000 \times .4875 = 487.5$ and $n_2\hat{q} = 487.5$. The marginal totals in Table 12.5 are, by construction, the same as the marginals in Table 12.4. The differences between observed numbers O and expected numbers E are given in Table 12.6. Note that the marginals are 0 and entries are ± 10.5. The X^2-criterion is

$$X^2 = \sum_{i,j=1}^{2} \frac{(O_{ij} - E_{ij})^2}{E_{ij}} = \frac{10.5^2}{512.5} + \frac{(-10.5)^2}{512.5} + \frac{(-10.5)^2}{487.5} + \frac{10.5^2}{487.5}$$

$$= 0.883.$$

It will be seen that this is the same value as that obtained before (except for rounding error). The purpose of writing X^2 as $\Sigma_{i,j}(O_{ij} - E_{ij})^2/E_{ij}$ is that this expression can be generalized to larger two-way tables.

TABLE 12.6

Values of $O - E$ for Table 12.4

		DATE		
		First	Second	Total
CANDIDATE	Nixon	$O_{11} - E_{11} = \quad 10.5$	$O_{12} - E_{12} = -10.5$	0
	Kennedy	$O_{21} - E_{21} = -10.5$	$O_{22} - E_{22} = \quad 10.5$	0
	Total	0	0	0

TABLE 12.7

Expected Numbers for a 2 × 2 Table

		B		Total
		B_1	B_2	
A	A_1	$E_{11} = \dfrac{(a + b)(a + c)}{n}$	$E_{12} = \dfrac{(a + b)(b + d)}{n}$	$a + b$
	A_2	$E_{21} = \dfrac{(c + d)(a + c)}{n}$	$E_{22} = \dfrac{(c + d)(b + d)}{n}$	$c + d$
Total		$a + c$	$b + d$	$n = a + b + c + d$

Let us now develop the general procedure for a 2 × 2 table in terms of the observed frequencies $O_{11} = a$, $O_{12} = b$, $O_{21} = c$, and $O_{22} = d$ as in Table 5.16. The expected number in the upper left-hand corner, for example, is $E_{11} = (a + b)(a + c)/n$; in Table 12.5, $512.5 = 1025 \times 1000/2000$. The calculations of E_{ij} are given in Table 12.7. The marginal totals are identical to those of the table of observed values. The table of differences $O_{ij} - E_{ij}$ will, therefore, have each marginal total equal to 0 as in Table 12.6, and the numerical values of the four entries are equal. The test criterion is

$$(12.5) \quad X^2 = \frac{(O_{11} - E_{11})^2}{E_{11}} + \frac{(O_{12} - E_{12})^2}{E_{12}}$$

$$+ \frac{(O_{21} - E_{21})^2}{E_{21}} + \frac{(O_{22} - E_{22})^2}{E_{22}}.$$

For random samples when the null hypothesis is true, X^2 has approximately a chi-square distribution with 1 degree of freedom. (The 1 degree of freedom is consistent with the fact that any one entry $O_{ij} - E_{ij}$ determines the other three entries because the marginals must be 0; see Exercises 5.3 and 5.4, for example.)

At the beginning of Section 5.1 double dichotomies were presented. Each individual in a sample or in a population is classified simultaneously on each of two dichotomies. Table 5.6, for example, gives the frequencies of female and male graduate and undergraduate students. Suppose that these frequencies refer to a sample (instead of a given class) from a large university. The dichotomy female and male and the dichotomy graduate and undergraduate are examples of two categorizations A_1 and A_2 and B_1 and B_2. If the numbers of individuals with property A_i and B_j is N_{ij} in a population (in the example, a university) of N individuals, the proportion of such persons is

(12.6)
$$p_{ij} = \frac{N_{ij}}{N}.$$

In Table 12.8 these proportions are displayed together with the marginal proportions

$$p_{i\cdot} = p_{i1} + p_{i2} = \frac{N_{i1} + N_{i2}}{N},$$

$$p_{\cdot j} = p_{1j} + p_{2j} = \frac{N_{1j} + N_{2j}}{N}.$$

Of course, the row marginals add to 1, as do the column marginals.

TABLE 12.8
Proportions in a Population

		B		
		B_1	B_2	*Total*
A	A_1	p_{11}	p_{12}	$p_{1\cdot}$
	A_2	p_{21}	p_{22}	$p_{2\cdot}$
	Total	$p_{\cdot 1}$	$p_{\cdot 2}$	1

If we draw an individual randomly from the population with the proportions in Table 12.8, the *probability* of drawing an individual from the pair of categories A_i and B_j is

(12.7)
$$\Pr(A_i \text{ and } B_j) = p_{ij}.$$

The probability of drawing an individual with characteristic A_i is $p_{i\cdot}$, and the probability of an individual with B_j is $p_{\cdot j}$. By (6.4), the A and B characteristics are independent (in the population or in the sense of probability) if $\Pr(A_i \text{ and } B_j) = \Pr(A_i) \times \Pr(B_j)$, that is, if

(12.8)
$$p_{ij} = p_{i\cdot}p_{\cdot j}.$$

The null hypothesis of independence is that (12.8) holds for $i = 1, 2$ and $j = 1, 2$. [In a 2×2 table, when (12.8) holds for one pair i, j, it holds for every pair.]

To test this null hypothesis of independence on the basis of a random sample of n with frequencies $O_{11} = a$, $O_{12} = b$, $O_{21} = c$, and $O_{22} = d$, we estimate the marginal probabilities as $\hat{p}_{1\cdot} = (a + b)/n$, $\hat{p}_{2\cdot} = (c + d)/n$,

$\hat{p}_{.1} = (a + c)/n$, and $\hat{p}_{.2} = (b + d)/n$. Under the null hypothesis the estimates of p_{ij} are

(12.9) $$\hat{p}_{ij} = \hat{p}_{i.} \times \hat{p}_{.j},$$

and the estimate of the expected number is $E_{ij} = n\hat{p}_{ij} = n\hat{p}_{i.} \times \hat{p}_{.j}$. The table of expected numbers is again Table 12.7, and the test criterion is again (12.5), which is compared with values of chi-square with 1 degree of freedom. Thus the test procedure for independence of two characteristics in a population based on a sample from the population is the same as the test procedure for independence of probabilities of a characteristic in two populations based on a sample from each population.

There is another sampling situation in which independence can be tested. In a bridge deck of 52 cards there are 12 face cards (King, Queen, and Jack in each of four suits). Consider a cross-classification of a card based on the partnership (North–South or East–West) to which it is dealt and on whether it is a face card or not. In the 2×2 table $O_{11} = a$ represents the number of face cards dealt to North and South, $O_{12} = b$ is the number of cards to North and South that are not face cards, etc. In this case the marginal totals are fixed: 26 cards dealt to each partnership $(a + b = c + d = 26)$, 12 face cards $(a + c = 12)$, and 40 nonface cards $(b + d = 40)$. Independence of partnership and type of card occurs when the deck has been well shuffled and dealt honestly. An alternative is that the dealer tends to give his or her partnership more than its fair share of face cards. The same test procedure is used. To summarize, in whichever of these three situations independence is to be tested (two samples, one sample, and fixed marginals), the X^2-criterion is calculated and compared with $\chi_1^2(\alpha)$.

In this special case of the 2×2 table, the X^2-criterion is

$$X^2 = \frac{n(ad - bc)^2}{(a + b)(c + d)(a + c)(b + d)},$$

which is $n\phi^2$, where ϕ is the measure of association presented in Section 5.1.3. (See Exercise 12.40.)

12.2.2 Two-Way Tables in General

Table 6.1 is a 12×3 table relating to the 1969 draft lottery. The 12 rows correspond to the months of the year; the 3 columns, to high-, medium-, or low-priority numbers. Under an assumption of adequate mixing of the 366 date-bearing capsules, about one-third of the days in each month should fall into each of the high, medium, and low categories. Table 12.9 shows the corresponding expected values; Table 12.10,

TABLE 12.9
Expected Values for Draft Lottery

| Month | PRIORITY NUMBERS | | | Total |
	1 to 122	123 to 244	245 to 366	
January	10.33	10.33	10.33	31
February	9.67	9.67	9.67	29
March	10.33	10.33	10.33	31
April	10.00	10.00	10.00	30
May	10.33	10.33	10.33	31
June	10.00	10.00	10.00	30
July	10.33	10.33	10.33	31
August	10.33	10.33	10.33	31
September	10.00	10.00	10.00	30
October	10.33	10.33	10.33	31
November	10.00	10.00	10.00	30
December	10.33	10.33	10.33	31
Total	122	122	122	366

SOURCE: Table 6.1.

the values of $O - E$. The value of the X^2-square statistic for testing independence of month and priority is the sum of the values of $(O - E)^2/E$, that is,

$$X^2 = \frac{(-1.33)^2}{10.33} + \frac{(-2.67)^2}{9.67} + \cdots + \frac{(-6.33)^2}{10.33} = 37.2.$$

An $r \times c$ table has $(r - 1)(c - 1)$ degrees of freedom. In this example, then, the number of degrees of freedom is $(12 - 1)(3 - 1) = 11 \times 2 = 22$. One finds from Appendix IX that $\chi^2_{22}(.025) = 36.781 < 37.2 < 40.289 = \chi^2_{22}(.01)$; so the achieved level of significance is between .01 and .025. The hypothesis of independence of month and priority is rejected; there is evidence that the three classes of priority numbers were not spread fairly among the different months.

In this example, the 366 dates of the year were classified simultaneously according to two classifications, month and priority number. The determination of the priority numbers was analogous to shuffling a deck of 366 cards, each card containing a date of the year; the first 122 cards were dealt to the priority category 1 to 122, the next 122 cards to category 123 to 244, and the last 122 cards to the third category. The mechanism is similar to dealing bridge hands; all marginal totals are specified in advance.

A situation that is more often met is drawing one sample from one population. Each individual may be cross-classified simultaneously

TABLE 12.10

Values of O − E for Draft Lottery

| | PRIORITY NUMBERS | | | |
Month	1 to 122	123 to 244	245 to 366	Total
January	−1.33	1.67	−0.33	0
February	−2.67	2.33	0.33	0
March	−5.33	−0.33	5.67	0
April	−2.00	−2.00	4.00	0
May	−1.33	−3.33	4.67	0
June	1.00	−3.00	2.00	0
July	1.67	−3.33	1.67	0
August	2.67	−3.33	0.67	0
September	0.00	5.00	−5.00	0
October	−1.33	4.67	−3.33	0
November	2.00	2.00	−4.00	0
December	6.67	−0.33	−6.33	0
Total	0	0	0	0

SOURCE: Tables 6.1 and 12.9.

into one of the categories $A_1, A_2, \ldots, A_r$ and into one of the categories $B_1, B_2, \ldots, B_c$. On the basis of a sample of n, one may test the hypothesis that these classifications are independent in the population, that is, that the corresponding variables A and B are independent. For example, we may classify an individual according to educational level (non-high school graduate, high school graduate, college graduate, post-college study) and political party affiliation (Democrat, Republican, Other). The hypothesis to be tested is that educational level and political party affiliation are independent. (It is assumed that the population is large relative to the sample.)

TABLE 12.11

Cross-Classification of Sample

		B			
		B_1	$B_2 \cdot \cdot \cdot B_c$		Total
	A_1	n_{11}	$n_{12} \cdot \cdot \cdot n_{1c}$		$n_{1.}$
	A_2	n_{21}	$n_{22} \cdot \cdot \cdot n_{2c}$		$n_{2.}$
A	$\vdots$	$\vdots$	$\vdots$ $\qquad$ $\vdots$		$\vdots$
	A_r	n_{r1}	$n_{r2} \cdot \cdot \cdot n_{rc}$		$n_{r.}$
	Total	$n_{.1}$	$n_{.2} \cdot \cdot \cdot n_{.c}$		n

An $r \times c$ table summarizing a sample of observations has the form of Table 12.11. Such a table is often referred to as a *contingency table*. Here we write n_{ij} instead of O_{ij} because the n individuals are assumed to be a random sample from the population of N individuals represented by Table 12.12, where we use N_{ij}.

In Table 12.11 the row totals are $n_{i\cdot} = n_{i1} + n_{i2} + \cdots + n_{ic}$, $i = 1, \ldots, r$, and the column totals are $n_{\cdot j} = n_{1j} + n_{2j} + \cdots + n_{rj}$, $j = 1, \ldots, c$; the notation in Table 12.12 is analogous. Table 12.13 is a table of population probabilities $\Pr(A_i \text{ and } B_j) = N_{ij}/N = p_{ij}$.

The classifications (or variables) A and B are said to be independent if $\Pr(A_i \text{ and } B_j) = \Pr(A_i) \times \Pr(B_j)$, that is, if (12.8) holds for $i = 1, \ldots, r$ and $j = 1, \ldots, c$. This is equivalent to

$$\frac{N_{ij}}{N} = \frac{N_{i\cdot}}{N} \times \frac{N_{\cdot j}}{N},$$

that is, $N_{ij} = N_{i\cdot}N_{\cdot j}/N$. This is equivalent to the definition of independence in Section 5.2.2. If independence holds in the population, then $p_{i\cdot}$ is estimated by $\hat{p}_{i\cdot} = n_{i\cdot}/n$, $p_{\cdot j}$ by $\hat{p}_{\cdot j} = n_{\cdot j}/n$, and $\hat{p}_{ij}$ by $\hat{p}_{i\cdot} \times \hat{p}_{\cdot j}$. The expected number in the ith row and jth column is $n\hat{p}_{i\cdot} \times \hat{p}_{\cdot j}$, which is

$$E_{ij} = \frac{n_{i\cdot}n_{\cdot j}}{n}.$$

Each E_{ij} is equal to its row total times its column total, divided by n. The χ^2-test is based on comparing each O_{ij} with its corresponding E_{ij}. We use the double summation notation

$$\sum_{i=1}^{r} \sum_{j=1}^{c} a_{ij} = a_{11} + a_{12} + \cdots + a_{1c} + a_{21} + a_{22} + \cdots + a_{2c}$$

$$+ \cdots + a_{r1} + a_{r2} + \cdots + a_{rc}.$$

TABLE 12.12

Cross-Classification of Population

		B_1	B_2	$\cdots$	B_c	Total
				B		
	A_1	N_{11}	N_{12}	$\cdots$	N_{1c}	$N_{1\cdot}$
	A_2	N_{21}	N_{22}	$\cdots$	N_{2c}	$N_{2\cdot}$
A	$\vdots$	$\vdots$	$\vdots$		$\vdots$	$\vdots$
	A_r	N_{r1}	N_{r2}	$\cdots$	N_{rc}	$N_{r\cdot}$
	Total	$N_{\cdot 1}$	$N_{\cdot 2}$	$\cdots$	$N_{\cdot c}$	N

TABLE 12.13

Table of Probabilities

		B				Total
		B_1	$B_2 \cdots B_c$			
A	A_1	p_{11}	$p_{12} \cdots p_{1c}$			$p_{1\cdot}$
	A_2	p_{21}	$p_{22} \cdots p_{2c}$			$p_{2\cdot}$
	$\vdots$	$\vdots$	$\vdots$		$\vdots$	$\vdots$
	A_r	p_{r1}	$p_{r2} \cdots p_{rc}$			$p_{r\cdot}$
	Total	$p_{\cdot 1}$	$p_{\cdot 2} \cdots p_{\cdot c}$			1

The test statistic is written

$$X^2 = \sum_{i=1}^{r} \sum_{j=1}^{c} \frac{(O_{ij} - E_{ij})^2}{E_{ij}}.$$

The sampling distribution of this statistic when the hypothesis of independence is true is approximated by the chi-square distribution χ_f^2 with $f = (r - 1)(c - 1)$ degrees of freedom. The approximation is good if each E_{ij} is at least 5. (The error of approximation is not large if only a few, say 20% or fewer, of the E_{ij}'s are less than 5.) At level α one rejects the hypothesis of independence if the value of the test statistic exceeds $\chi_f^2(\alpha)$, where $f = (r - 1)(c - 1)$.

The number of degrees of freedom f is $(r - 1)(c - 1)$ because that is the number of independent quantities among the rc quantities $O_{ij} - E_{ij}$, $i = 1, \ldots, r$, $j = 1, \ldots, c$. In a table of quantities $O_{ij} - E_{ij}$, all the marginals are zero. When $(r - 1)(c - 1)$ entries in the table, say all but those in the last row or column, are filled in, the omitted entries are determined by the fact that all the marginals are zero.

Another model for which this χ^2-test is appropriate is that in which a sample of size $n_{\cdot j}$ is drawn from the jth population of categories A_1, $\ldots$, A_r, $j = 1, \ldots, c$. The null hypothesis is that the sets of probabilities of these categories are the same in the c populations.

A handy computational formula for X^2 is

$$X^2 = \sum_{i=1}^{r} \sum_{j=1}^{c} \frac{O_{ij}^2}{E_{ij}} - n.$$

However, a table of $O_{ij} - E_{ij}$ is useful for analyzing a possible discrepancy from independence.

12.2.3 Choosing a Chi-Square Test

Just as in goodness-of-fit tests (Section 12.1), the investigator may exercise some judgment in assigning categories. From a table in which the categories are quite fine, the investigator may combine some categories or may delete some categories (which reduces the sample size). The powers of tests of independence will depend on the definitions of categories.

For example, Table 12.14 gives the voting records of professors from various colleges. Are the probabilities of voting Democratic the same in colleges of different kinds? The overall chi-square test statistic, with 2 degrees of freedom, is 12.98 ($P < .005$). Combining the first and second columns for comparison with the third produces Table 12.15; the value of the chi-square test statistic for this table, based on 1 degree of freedom, is 12.85 ($P < .005$). It should be noted that the value of X^2 for the 2×2 table is almost as great as for the 2×3 table; the lack of independence (that is, difference in Democrat probabilities) is due mainly to the difference between "Other" (that is, mixed sponsorship) and the combined category of purely private and purely public schools. This result is verified by observing that the percentages in the first two columns of Table 12.14 are about the same.

TABLE 12.14

2 × 3 Table: 1948 Voting Record of Professors from Different Types of Colleges

		TYPE OF SCHOOL			
		Public	*Private*	*Other*	
VOTE	*Democrat*	402 (71%)	493 (72%)	331 (63%)	1226 (69%)
	Not Democrat	164 (29%)	192 (28%)	195 (37%)	551 (31%)
		566 (100%)	685 (100%)	526 (100%)	1777 (100%)

SOURCE: Lazarsfeld and Thielens (1958), p. 28.

TABLE 12.15

2 × 2 Table: Public and Private Schools vs. Others

	PUBLIC OR PRIVATE	OTHER	
DEMOCRAT	895 (72%)	331 (63%)	1226 (69%)
NOT DEMOCRAT	356 (28%)	195 (37%)	551 (31%)
	1251 (100%)	526 (100%)	1777 (100%)

SOURCE: Table 12.14.

12.3 MEASURES OF ASSOCIATION

When two categorical variables are not independent, they are dependent or associated. How do we measure the degree of association? As we shall see, the appropriate definition of a measure of association depends to some extent on the purpose for which it is to be used. The measures to be developed apply to both samples and populations. The value of a measure of association in a sample usually varies statistically from the value of the measure in the population sampled. Even when a sample measure leads to rejection of the hypothesis of independence in the population (that is, is "statistically significant"), it may not indicate a practically significant degree of association. For instance, to predict one variable from another for a useful purpose may require a high degree of dependence.

12.3.1 The Phi Coefficient

One such measure for 2×2 tables was introduced in Section 5.1.3, "Interpretation of Frequencies." It is the *phi coefficient*,

$$\phi = \frac{ad - bc}{\sqrt{(a + b)(c + d)(a + c)(b + d)}},$$

where, as indicated in Table 5.16, a, b, c, and d are the entries in the cells. (Note that, when we use the Greek letter ϕ for the measure computed for a sample, we are breaking the rule of using Greek letters for population parameters and corresponding Roman letters for their sample analogs. However, in the literature on measures of association, Greek letters have been used for measures of association for samples as well as for populations.)

The difference between proportions based on the rows is

$$d_r = \frac{a}{a + b} - \frac{c}{c + d};$$

the difference between proportions based on the columns is

$$d_c = \frac{a}{a + c} - \frac{b}{b + d}.$$

Each of these could be a measure of association. To treat rows and columns symmetrically, we use the geometric mean $\sqrt{d_r d_c}$. This is ϕ.

Association is complete or perfect if a value of one variable implies the value of the other. In Table 12.16 every individual in category A_1 is in category B_1 and every individual in category A_2 is in category B_2. The

TABLE 12.16

Perfect Positive Association when A Is Ordered like B: $\phi = +1$

		B		
		B_1	B_2	Total
A	A_1	a	0	a
	A_2	0	d	d
	Total	a	d	$a + d$

combinations of A_1 and B_2 and of A_2 and B_1 never occur; that is, b and c are zero. We have

$$\phi = \frac{ad - 0}{\sqrt{adad}} = \frac{ad}{ad} = +1.$$

This is the case of *perfect positive association* when A and B are ordered similarly. In Table 12.17,

$$\phi = \frac{-bc}{\sqrt{bcbc}} = \frac{-bc}{bc} = -1.$$

This is the case of *perfect negative association*. The value of the coefficient ϕ is always between -1 and $+1$, and is zero when the variables are independent. As a rule of thumb, we may state the following in terms of the size (absolute value) of phi: a value of 0 to 1/3 is a "weak" association; a value of 1/3 to 2/3 is a "medium-sized" association; a value of 2/3 to 1 is a "strong" association.

TABLE 12.17

Perfect Negative Association when A Is Ordered like B: $\phi = -1$

		B		
		B_1	B_2	Total
A	A_1	0	b	b
	A_2	c	0	c
	Total	c	b	$b + c$

Given a measure of association and a corresponding test of the hypothesis of nullity of that association, one can make a simultaneous assessment of both "statistical significance" and "practical significance." The situation is similar to that discussed in Section 9.2.2, where we considered the relation of hypothesis tests to confidence intervals for a mean. The four cases that may arise are classified in Table 12.18.

TABLE 12.18
Strength of Association and Acceptance or Rejection of Hypothesis

| | | HYPOTHESIS OF INDEPENDENCE | |
		Accepted	*Rejected*
MEASURE OF	*Small*	(i)	(ii)
ASSOCIATION	*Large*	(iii)	(iv)

In case (i), the sample measure of association is small, that is, near zero. The hypothesis of independence is accepted with some conclusiveness, since the investigator has confidence that the value of the measure of association in the population is near zero. A confidence interval for it would include only values that are so close to zero as to be equivalent to zero in practical terms.

In case (ii), the sample measure of association is small, yet the hypothesis of independence is rejected. The investigator is confident that the population value is different from zero, yet this value may be so small that the relationship between variables cannot be exploited. If the sample size is large, there is a good chance that the value of the chi-square test statistic will lead to rejection of the hypothesis of independence, even though the association between variables is not large; this situation leads to case (ii).

In case (iii), the hypothesis of independence is accepted in spite of the fact that the sample measure of association is large. This large result may be due to a misleading sample and thus would not be reproducible. It may occur when the sample size is small, and hence the variability inherent in the sample measure of association is great.

In case (iv), the sample measure of association is large and the hypothesis of independence is rejected. The result is both statistically significant and practically significant. The result is reproducible and useful.

When the categories are ordered, it may make sense to assign numerical values to them, say $a_1 = 1$ for A_1 (High for A), $a_2 = -1$ for A_2 (Low for A), $b_1 = 1$ for B_1 (High for B), and $b_2 = -1$ for B_2 (Low for B). Then each of the n individuals is associated with one of the four pairs of numbers $(1, 1)$, $(1, -1)$, $(-1, 1)$, $(-1, -1)$. The coefficient of correlation r (Section 5.5.4) between the variables A and B then turns out to be $r = \phi$. In fact, the values of r are the same for all choices of a_1, a_2, b_1, and b_2, provided only that $a_1 > a_2$ and $b_1 > b_2$.

12.3.2   The Coefficient Based on Prediction

Another measure of association is based on the idea of using one variable to predict the other. Consider the cross-classification of income and occupation in Table 12.19. How well does occupational category predict income level? If one of the 202 persons represented in this table is selected at random, our best guess of the income level—if we don't know anything about the person—is to say that the person is in the high-income group because more of the people are in that group (117, compared with 85 for the low-income group); the high-income category is the *mode*. If we make this prediction for each of the 202 persons, we shall be right in 117 cases and wrong in 85 cases. However, if we take the person's occupational category into account, we can improve our prediction. If we know the person is a professional, our best guess is still that the person is in the high-income group, for 92 of the professionals are in the high-income group, compared with only 14 in the low-income group. On the other hand, if we know the person is not a professional, we should guess that the person is in the low-income group; in this case we would be correct 71 times and incorrect 25 times. Our total number of errors in predicting all 202 income levels for both nonprofessionals and professionals is 14 + 25 = 39, compared with 85 errors if we do not use the occupational category in making the prediction.

TABLE 12.19
Income and Occupation

		INCOME LEVEL		
		High	*Low*	*Total*
OCCUPATIONAL	*Professional*	92	14	106
CATEGORY	*Not professional*	25	71	96
	Total	117	85	202

A coefficient of association that measures the improvement in prediction of the column category due to using the row classification is

$$
(12.10) \quad \lambda_{c\cdot r} = \frac{\left(\begin{array}{c}\text{number of errors}\\\text{not using the rows}\end{array}\right) - \left(\begin{array}{c}\text{number of errors}\\\text{using the rows}\end{array}\right)}{\left(\begin{array}{c}\text{number of errors}\\\text{not using the rows}\end{array}\right)}
$$

$$
= \frac{\begin{array}{c}\text{reduction in errors when using}\\\text{the rows to predict columns}\end{array}}{\text{number of errors not using the rows}},
$$

the decrease in errors divided by the number of errors made when not using the rows. The letter λ is lowercase Greek *lambda*. The symbol $\lambda_{c \cdot r}$ may be read "lambda sub c dot r." The subscript $c \cdot r$ refers to predicting the column category using the row category. For the example, we have

$$\lambda_{c \cdot r} = \frac{85 - (14 + 25)}{85} = \frac{85 - 39}{85} = \frac{46}{85} = .54.$$

Another equivalent expression for this coefficient is

$$(12.11) \quad \lambda_{c \cdot r} = \frac{\left(\begin{array}{c} \text{number of} \\ \text{correct predictions} \\ \text{using rows} \end{array} \right) - \left(\begin{array}{c} \text{number of correct} \\ \text{predictions not} \\ \text{using rows} \end{array} \right)}{(\text{number}) - \left(\begin{array}{c} \text{number of correct predictions} \\ \text{not using rows} \end{array} \right)}.$$

This shows that $\lambda_{c \cdot r}$ can also be interpreted in terms of the increase in the number of correct predictions. For the example, this is

$$\lambda_{c \cdot r} = \frac{(92 + 71) - 117}{202 - 117} = \frac{163 - 117}{85} = \frac{46}{85} = .54.$$

Formula (12.10) or (12.11) can be used to calculate $\lambda_{c \cdot r}$ for larger two-way tables. In the special case of the 2 × 2 table with frequencies a, b, c, and d, (12.10) is

$$(12.12) \quad \lambda_{c \cdot r} = \frac{\min(a + c, b + d) - [\min(a, b) + \min(c, d)]}{\min(a + c, b + d)}.$$

Here $\min(a, b)$ denotes the minimum of the two numbers a and b; for example, $\min(117, 85) = 85$.

Formulas for the coefficient λ based on prediction of the row category from the column classification are obtained from (12.10) and (12.11) by interchanging "columns" and "rows." The formula for λ for the special case of a 2 × 2 table can be obtained from (12.12) by interchanging the symbols b and c.

The values of $\lambda_{r \cdot c}$ and $\lambda_{c \cdot r}$ range between 0 and 1. The value 1 is attained when there are no errors in using the cross-classification for predictions; for $\lambda_{c \cdot r}$ this happens when in each row all the frequencies but one are zero; similarly, $\lambda_{r \cdot c}$ equals 1 when each column contains only one frequency that is not zero. When two classifications are independent, the coefficient is zero, for in the case of independence the modal category is the same for each row, and the column classification gives no help in predicting. However, the coefficient can be zero even if the classifications are not independent. In Table 12.20,

$$\lambda_{c \cdot r} = \frac{(63 + 54) - 117}{85} = \frac{117 - 117}{85} = 0.$$

The modal category is high income for both professionals and nonprofessionals, so that $\lambda_{c \cdot r} = 0$; yet the proportion of professionals in the high-income group is $63/117 = .54$, which is not equal to $43/85 = .51$, the proportion of professionals in the low-income group.

TABLE 12.20
Income and Occupation

		INCOME LEVEL		
		High	*Low*	*Total*
OCCUPATIONAL	*Professional*	63	43	106
CATEGORY	*Not professional*	54	42	96
	Total	117	85	202

(hypothetical data)

A disadvantage of the coefficients of association based on predicting one variable from the other is that they do not treat the variables symmetrically. An alternative is to take a weighted average of the two coefficients:

$$\lambda = \frac{\left(\begin{array}{l}\text{reduction in errors}\\ \text{when using rows}\\ \text{to predict columns}\end{array}\right) + \left(\begin{array}{l}\text{reduction in}\\ \text{errors when}\\ \text{using columns}\\ \text{to predict rows}\end{array}\right)}{\left(\begin{array}{l}\text{number of errors}\\ \text{in predicting columns}\\ \text{without using rows}\end{array}\right) + \left(\begin{array}{l}\text{number of errors}\\ \text{in predicting rows}\\ \text{without using columns}\end{array}\right)}.$$

This is called the *coefficient of mutual association*. Let p be the reduction in errors when using rows to predict columns, q the number of errors in predicting columns without using rows, r the reduction in errors when using columns to predict rows, and s the number of errors in predicting rows without using columns. Then

$$\lambda_{r \cdot c} = \frac{p}{q}, \qquad \lambda_{c \cdot r} = \frac{r}{s}, \qquad \lambda = \frac{p + r}{q + s}.$$

An interpretation for λ can be made by supposing that on half the occasions we shall predict columns and on half the occasions rows. Then the average number of errors is $(q + s)/2$, and the average reduction in errors is $(p + r)/2$. Then we take as our coefficient of mutual association the ratio

$$\frac{\text{average reduction in errors}}{\text{average number of errors}}$$

and find that this is

$$\frac{(p + r)/2}{(q + s)/2} = \frac{p + r}{q + s} = \lambda.$$

The coefficient λ will be zero when and only when both p and r are zero, that is, when and only when both $\lambda_{r \cdot c}$ and $\lambda_{c \cdot r}$ are zero. In particular, $\lambda = 0$ in the case of independence. As with $\lambda_{r \cdot c}$ and $\lambda_{c \cdot r}$, the coefficient λ can be zero even in cases in which the classifications are not independent. For Table 12.20, $\lambda_{r \cdot c} = 0$ and $\lambda_{c \cdot r} = 0$, and so $\lambda = 0$, but the proportions based on the column totals are $63/117 = .54$ and $43/85 = .51$, which are not equal.

Next we illustrate the computation of the coefficients for a larger two-way table. How well can one predict hair color when one knows eye color? Refer to Table 5.21. The coefficient based on predicting columns from rows is

$\lambda_{c \cdot r}$

$$= \frac{(6800 - 2829) - [(2811 - 1768) + (3132 - 1387) + (857 - 438)]}{6800 - 2829}$$

$$= \frac{3971 - (1043 + 1745 + 419)}{3971}$$

$$= \frac{3971 - 3207}{3971} = \frac{764}{3971} = .192.$$

The coefficient based on predicting rows from columns is

$\lambda_{r \cdot c}$

$$= \frac{(6800 - 3132) - [(2829 - 1768) + (2632 - 1387) + (1223 - 746) + (116 - 53)]}{6800 - 3132}$$

$$= \frac{3668 - (1061 + 1245 + 477 + 66)}{3688}$$

$$= \frac{3668 - 2849}{3668} = \frac{819}{3668} = .223.$$

The coefficient of mutual association is

$$\lambda = \frac{819 + 764}{3668 + 3971}$$

$$= \frac{1583}{7639} = .207$$

The value of this index indicates there is a weak association between hair and eye color in this set of 6800 persons.

12.3.3   A Coefficient Based on Ordering

If both categorical variables are based on ordinal scales (that is, rankings), a measure of association may take this fact into account. We now develop a general measure for two-way tables.

Two-by-Two Tables.   In Table 5.45, 75 communities were cross-classified on rate of juvenile delinquency and population density. A measure of association for such data can be constructed as follows: Consider two of the 75 communities, A and B. A's population density may be classified as either high or low; B's population density may be either high or low. Thus A and B can be ranked according to population density. The population density of A is either higher than that of B (A High, B Low), equal to that of B (both High or both Low), or lower than that of B (A Low, B High). Similarly, A and B can be ranked according to rate of juvenile delinquency. There are nine possibilities if both orderings are considered (Table 12.21). The number of communities clas-

TABLE 12.21
Possible Orderings of Communities A and B on Two Variables

	Rate of Juvenile Delinquency	Population Density	Agreement
1	A's < B's	A's < B's	Yes
2	A's < B's	A's = B's	. . .
3	A's < B's	A's > B's	No
4	A's = B's	A's < B's	. . .
5	A's = B's	A's = B's	. . .
6	A's = B's	A's > B's	. . .
7	A's > B's	A's < B's	No
8	A's > B's	A's = B's	. . .
9	A's > B's	A's > B's	Yes

sified as Low for both variables is 30, and the number of communities classified as High for both variables is 36. This gives rise to 30×36 pairs of communities in which the orderings given by the two variables are the same. For each of the 30 we can select any one of the 36. The orderings disagree for 5×4 pairs. The value of the coefficient based on ordering, γ (lower-case Greek *gamma*), is

$$\gamma = \frac{30 \times 36 - 5 \times 4}{30 \times 36 + 5 \times 4}$$

$$= \frac{1080 - 20}{1080 + 20} = \frac{1060}{1100} = .96.$$

In general, the formula for γ in the case of a 2×2 table with ordered categories and with frequencies a, b, c, d is

$$\gamma = \frac{ad - bc}{ad + bc}.$$

Notice the familiar quantity $ad - bc$ in the numerator. If a pair of individuals is selected at random from among the $ad + bc$ pairs of individuals differing on both variables, then $ad/(ad + bc)$ is the probability that the chosen pair will be put in the same order by both variables, while $bc/(ad + bc)$ is the probability that the chosen pair will be put in opposite orders by the two variables. The coefficient γ is the difference between these two probabilities.

Larger Two-Way Tables. We note that a drawback of the chi-square test statistic and of the measure of association based on prediction is that they fail to take into account any ordering of the categories. Tables 12.22 and 12.23 have the same six frequencies, but the second and third columns of frequencies have been interchanged. The X^2-statistic and the measure based on prediction have the same values in

TABLE 12.22
Educational Level and Income

| | | INCOME | | | |
		Low	Medium	High	Total
EDUCATIONAL LEVEL	Low	30 (50%)	25 (42%)	5 (8%)	60 (100%)
	High	3 (8%)	17 (42%)	20 (50%)	40 (100%)
	Total	33 (33%)	42 (42%)	25 (25%)	100 (100%)

(hypothetical data)

TABLE 12.23
A 2 × 3 Table

		B			
		B_1	B_2	B_3	Total
A	A_1	30	5	25	60
	A_2	3	20	17	40
	Total	33	25	42	100

the two tables, yet if A_1 and A_2 and B_1, B_2, and B_3 are ordered, the positive association seems higher in Table 12.22. One method that distinguishes between the two tables is the coefficient based on ordering. The definitional formula for γ is, in words,

$$\gamma = \frac{\left(\begin{array}{l}\text{number of pairs of}\\ \text{individuals for}\\ \text{which both orderings}\\ \text{are the same}\end{array}\right) - \left(\begin{array}{l}\text{number of pairs of}\\ \text{individuals for}\\ \text{which the orderings}\\ \text{are different}\end{array}\right)}{\left(\begin{array}{l}\text{total number of pairs of individuals}\\ \text{for which neither ordering is a tie}\end{array}\right)}.$$

To illustrate the idea, we consider Table 12.24, with only six individuals. We list the 15 possible pairs of individuals in Table 12.25. The number of pairs with the same ordering is 5, and the number with different orderings is 1; hence the value of γ is

$$\gamma = \frac{5 - 1}{5 + 1} = \frac{4}{6} = .67.$$

TABLE 12.24
Data for Six Persons on Two Ordinal Variables

INDIVIDUAL	EDUCATION	INCOME
A	High	Low
B	Low	Low
C	High	High
D	High	Middle
E	Low	Middle
F	Low	Low

(hypothetical data)

TABLE 12.25
Ranking of Persons on Two Variables

Pairs of Persons	Relationship of Education of First Person to Education of Second Person	Relationship of Income of First Person to Income of Second Person	Same Order?	Opposite Order?
AB	>	=		
AC	=	<		
AD	=	<		
AE	>	<		×
AF	>	=		
BC	<	<	×	
BD	<	<	×	
BE	=	<		
BF	=	=		
CD	=	>		
CE	>	>	×	
CF	>	>	×	
DE	>	=		
DF	>	>	×	
EF	=	>		
			5	1

SOURCE: Table 12.24

However, the coefficient can be computed directly from cross-classification of the six persons in Table 12.26. The same reasoning as that given for the special case of a 2 × 2 table applies. Relative to the $n_{11} = 2$ persons in the (Low, Low) category, there are $n_{22} + n_{23} = 1 + 1 = 2$ persons in categories that are higher on both variables; this fact

TABLE 12.26
Cross-Classification of Six Persons

		INCOME			
		Low	Middle	High	Total
EDUCATION	Low	$n_{11} = 2$	$n_{12} = 1$	$n_{13} = 0$	3
	High	$n_{21} = 1$	$n_{22} = 1$	$n_{23} = 1$	3
	Total	3	2	1	6

SOURCE: Table 12.24

yields $2 \times 2 = 4$ pairs of persons that are ordered the same way on both variables. The $n_{12} = 1$ person in the (Low, Middle) category can be paired with the $n_{23} = 1$ person in the (High, High) category to yield $1 \times 1 = 1$ pair of persons ordered the same way on both variables. No other pairs of persons are ordered similarly. The $n_{12} = 1$ person can be paired with the $n_{21} = 1$ person in the (High, Low) category to form $1 \times 1 = 1$ pair of persons ordered differently on the two variables. This gives $4 + 1 = 5$ pairs of persons ordered the same way and 1 pair ordered differently on the two variables. The value of γ is $(5 - 1)/(5 + 1) = 4/6 = .67$. Table 12.27 is a worksheet for this calculation. At the bottom of column (4) is the sum of the numbers in that column; this sum is the number of pairs of individuals having the orderings the same. Similarly, the sum at the bottom of the column (5) is the number of pairs having different orderings. (The last three rows are unnecessary.)

TABLE 12.27
Worksheet for Calculations of γ for Table 12.26

(1)	(2)	(3)	(4)	(5)
	SUM OF FREQUENCIES BELOW AND TO THE RIGHT OF THIS	SUM OF FREQUENCIES BELOW AND TO THE LEFT OF THIS	PAIRS WITH SAME ORDERS	PAIRS WITH DIFFERENT ORDERS
FREQUENCY	FREQUENCY	FREQUENCY	(1) × (2)	(1) × (3)
$n_{11} = 2$	$1 + 1 = 2$	0	4	0
$n_{12} = 1$	1	1	1	1
$n_{13} = 0$	0	0	0	0
$n_{21} = 1$	0	0	0	0
$n_{22} = 1$	0	0	0	0
$n_{23} = 1$	0	0	0	0
			5	1

Returning to Table 12.22, we see that for each of the 30 persons in the category (Low, Low) there are $17 + 20 = 37$ persons in a category that is higher on both variables. This yields $30 \times 37 = 1110$ pairs of persons for which the variables give the same ordering (entered on the first line of Table 12.28). For each of the 25 persons in the category (Low, Medium), there are 20 persons who are ranked higher on both variables. This yields $25 \times 20 = 500$ pairs of individuals for which the ordering is the same for the variables. For each of these 25 persons in the category (Low, Medium), there are 3 persons who are ranked higher in

TABLE 12.28
Worksheet for Calculation of γ for Table 12.22

(1)	(2)	(3)	(4) = (1) × (2)	(5) = (1) × (3)
$n_{11} = 30$	$17 + 20 = 37$	0	1110	0
$n_{12} = 25$	20	3	500	75
$n_{13} = 5$	0	$3 + 17 = 20$	0	100
			1610	175
		(4) − (5):	1435	
		(4) + (5):	1785	
		γ:	.80	

education but lower in income. This yields $25 \times 3 = 75$ pairs for which the one variable gives a different ordering of the individuals than does the other variable. For each of the 5 persons in the (Low, High) category, there are $3 + 17 = 20$ persons who are ranked higher in education but lower in income. This yields $5 \times 20 = 100$ pairs with different orderings. The number of pairs with the same ordering is $1110 + 500 = 1610$. The number of pairs with different orderings is $75 + 100 = 175$. The difference is $1610 - 175 = 1435$. The total number of relevant pairs is $1610 + 175 = 1785$. The value of γ is $1435/1785 = .80$.

The value of γ for Table 12.23 is

$$\gamma = \frac{[30 \times (20 + 17) + 5 \times 17] - [5 \times 3 + 25 \times (3 + 20)]}{[30 \times (20 + 17) + 5 \times 17] + [5 \times 3 + 25 \times (3 + 20)]}$$

$$= \frac{[30 \times 37 + 5 \times 17] - [5 \times 3 + 25 \times 23]}{[30 \times 37 + 5 \times 17] + [5 \times 3 + 25 \times 23]}$$

$$= \frac{(1110 + 85) - (15 + 575)}{(1110 + 85) + (15 + 575)}$$

$$= \frac{605}{1785} = .34,$$

indicating a weaker but still positive association.

A general formula for γ for an $r \times c$ table is

$$\gamma = \frac{S - D}{S + D},$$

where

$$S = \sum_{i=1}^{r-1} \sum_{j=1}^{c-1} \left(n_{ij} \sum_{k=i+1}^{r} \sum_{l=j+1}^{c} n_{kl} \right)$$

$$D = \sum_{i=1}^{r-1} \sum_{j=2}^{c} \left(n_{ij} \sum_{k=i+1}^{r} \sum_{l=1}^{j-1} n_{kl} \right).$$

Tests of hypothesis could be based on the sampling distribution of γ. However, this distribution is unwieldy.

SUMMARY

A test of goodness of fit is a test of whether a specified distribution or family of distributions fits the distribution of a sample.

A test of independence is a test of whether the joint distribution of two variables is the product of their marginal distributions.

Tests of goodness of fit and independence are made by computing the frequencies expected under the respective hypotheses and comparing these with observed frequencies. The statistics used have distributions that are approximated by chi-square distributions.

The extent of the departure from independence is described by various measures of association.

APPENDIX 12A   The Continuity Adjustment

The continuity adjustment for the chi-square test statistic for independence in a 2×2 table consists of replacing $(ad - bc)^2$ by

$$\left(|ad - bc| - \frac{n}{2} \right)^2.$$

Note that this "makes it harder to reject." This continuity correction agrees with replacing $|\hat{p}_1 - \hat{p}_2|$ by

$$|\hat{p}_1 - \hat{p}_2| - \frac{1}{2} \left(\frac{1}{n_1} + \frac{1}{n_2} \right)$$

in the test statistic for comparing two proportions (Appendix 10A).

EXERCISES

12.1 (Sec. 12.1.1)   In 25 tosses of a coin, heads turned up 20 times. Test the null hypothesis that the coin is fair at level of significance .05.

12.2 (Sec. 12.1.1)   Do Exercise 9.8 by means of a chi-square test.

12.3 (Sec. 12.1.1)   Using the data in Exercise 8.14, test the null hypothesis that Nixon and Kennedy were equally favored in October 1959. Take $n = 1000$.

12.4 (Sec. 12.1.1)   Using the data in Exercise 8.14, test the null hypothesis that Nixon and Kennedy were equally favored in March 1960. Use $n = 1000$.

12.5 (Sec. 12.1.2)   Suppose that the data for a sample of size 200 were as in Table 12.29. Make a test of goodness of fit at the 1% level.

TABLE 12.29
Vote Intentions of 200 Faculty Members

Observed number	$O_1 = 94$	$O_2 = 90$	$O_3 = 16$
Hypothesized proportion	$p_{10} = .40$	$p_{20} = .45$	$p_{30} = .15$

(hypothetical data)

12.6 (Sec. 12.1.2)   A certain town has a large population that is 25% Blacks, 40% of Italian origin, and 35% from other ethnic backgrounds. The 40-member police force includes 3 Blacks and 21 Italians. Use a chi-square test at the 5% level to see whether a random recruitment model is satisfactory.

12.7 (Sec. 12.1.3)   Table 12.30 gives the number of wins for each of the first six post positions at Waterford Park, Chester, West Virginia, for the 118 races from the beginning of the season to March 11, 1969. Test at the 5% level the hypothesis that the probability of winning is the same for all post positions.

TABLE 12.30
Number of Wins, by Post Position

Post position	1	2	3	4	5	6
Number of wins	30	17	21	27	12	11

SOURCE: Pittsburgh Press, March 11, 1969.

12.8 (Sec. 12.1.3)   Test at the 10% level the hypothesis that the faces of the die have the same probability of turning up, based on the data in Table 12.31 for 100 rolls of the die.

12.9 (Sec. 12.1)   A die was tossed 300 times, giving the results in Table 12.32. Test at the 5% level the hypothesis that the die is fair.

TABLE 12.31
Results of 100 Rolls of a Die

Face	1	2	3	4	5	6	Total
Frequency	21	19	20	15	14	11	100

(hypothetical data)

TABLE 12.32
Results of Tossing a Die

Face	1	2	3	4	5	6	Total
Frequency	46	51	44	57	38	64	300

(hypothetical data)

12.10 (Sec. 12.1)   A random sample of 200 digits is tabulated in Table 12.33. Test the hypothesis that this random sample is from a uniform distribution.

TABLE 12.33
Distribution of a Random Sample of Digits

Digit	0	1	2	3	4	5	6	7	8	9	Total
Frequency	10	20	19	21	21	15	21	22	25	26	200

(hypothetical data)

12.11 (continuation of Exercise 12.9)   For the data in Table 12.32, do the following.
(a) Partition the table into Odd Faces vs. Even Faces.
(b) Compute the corresponding chi-square statistic.
(c) Test the hypothesis that the probability of an Odd Face is 1/2 at the 5% level.

12.12 (continuation of Exercise 12.10)   For the data in Table 12.33, do the following.
(a) Partition the table into 0 vs. Others.
(b) Compute the corresponding chi-square statistic.
(c) Test the hypothesis that the probability of 0 is .1.

12.13 (Sec. 12.2.1)   Consider the 25 students cross-tabulated in Table 5.6 as a sample from a large population, and test at the 5% level the hypothesis of independence between sex and university division in that population.

12.14 (Sec. 12.2.1)   Consider the 600 men cross-tabulated in Table 5.12 as a sample from a large population, and test at the 1% level the hypothesis of independence between use of the Brand X razor and use of the Brand X blades.

12.15 (Sec. 12.2.1)   Do Exercise 10.6 using a chi-square test statistic.

12.16 (Sec. 12.2.1)   Do Exercise 10.7 using a chi-square test statistic.

12.17 (Sec. 12.2)   Table 12.34 gives the distribution of a sample of household heads by income and age in Suburbs A and B of a metropolitan area.
   (a) Compute the age distribution in each suburb. Which suburb is "younger"?
   (b) Compute the income distribution in each suburb. Which suburb is "richer"?
   (c) Consider Table 12.34 as three separate 3 × 2 tables, one for each age group. Compute the chi-square statistic for each of these three tables. Are there any significant differences in income between the two suburbs, once one has controlled for age?

TABLE 12.34
Income and Age in Suburbs A and B

| | | \multicolumn{6}{c}{AGE} |
| | | *Below 25* | | *25–45* | | *Above 45* | |
		A	B	A	B	A	B
	Low	52	12	13	6	6	8
INCOME LEVEL	*Medium*	29	7	25	13	6	5
	High	33	7	12	6	25	26

(hypothetical data)

12.18 (Sec. 12.2)   As shown in Table 12.35, Blacks were 2% of all officers in the Armed Forces and 3% of those in Vietnam as of June 30, 1967. About 1000 in a total of 8000 Black officers were serving in Vietnam. Blacks were 10 percent of all enlisted men and 12 percent of those in Vietnam—51,000 in Vietnam in a total of 297,000 Black enlisted men in the Armed Forces.
   The percentages of Blacks in Vietnam were higher than would be expected on the basis of the overall percentages of Blacks among both officers and enlisted men. Could this be explained on the basis of chance alone? Study this question, as follows.
   (a) For Officers, make a 2 × 2 table, Outside Vietnam/Inside Vietnam vs. Blacks/Others. Compute the value of the chi-square statistic. Give the achieved level of significance.
   (b) Do the same for Enlisted Men.

TABLE 12.35

Negro Officers and Enlisted Men in the Armed Forces, June 20, 1967

	TOTAL	NEGROES	PERCENT NEGROES
	(Numbers in Thousands)		
Total	3365	305	9
Officers	384	8	2
Outside Vietnam	342	7	2
Inside Vietnam	43	1	3
Enlisted men	2981	297	10
Outside Vietnam	2536	246	10
Inside Vietnam	444	51	12

SOURCE: U.S. Department of Defense (*Social and Economic Conditions of Negroes in the United States*, October 1967, BLS Report No. 332, U.S. Department of Labor, Bureau of Labor Statistics, p. 84; Current Population Reports, Series P-23, No. 24, U.S. Bureau of the Census).

12.19 (Sec. 12.2)   Consider again the data on reading errors in Exercise 5.10.

(a) Is the hypothesis that the same error distribution obtains in the populations represented by the three groups tenable? (Do the chi-square test of independence.)

(b) Do a chi-square test to decide whether there are any significant differences for error categories other than "Don't Know," by computing the chi-square test statistic for Table 12.36.

TABLE 12.36

Summary of Data on Reading Errors

	SUBSTITUTIONS	DON'T KNOW	TOTAL
SQUIRRELS	30	53	83
BEARS	134	172	306
CATS	33	36	69
TOTAL	197	261	458

SOURCE: Table 5.64.

12.20 (Sec. 12.2.3)   Consider the data of Exercise 5.8 on opinions of freshmen, sophomores, and juniors in regard to increasing the representation of the freshman and sophomore classes in the student senate.

(a) Test the overall hypothesis that opinion is independent of college class.

(b) Compare the juniors with the lower classes. Do this by making a chi-square test on the data table of part (b) of Exercise 5.8.

12.21 (Sec. 12.2.3) An experiment was performed to test whether male moths' ability to steer toward a source of odorous sex pheromone depends on air movement. A male was considered to have steered successfully if, having started at one end of a flight tunnel, he flew through a hoop near the source at the other end. The results are summarized in Table 12.37. Test the hypothesis that flying through the hoop is independent of the pheromone plume.

TABLE 12.37
Numbers of Male Moths Flying Through Hoop and
Outside of Hoop for Three Test Conditions

	NUMBER FLYING THROUGH HOOP	NUMBER FLYING OUTSIDE HOOP
1. Pheromone plume in moving air	17	3
2. Pheromone plume in still air	16	4
3. No pheromone plume in still air (control)	3	17

SOURCE: Farkas and Shorey (1972). Copyright © 1972 by the American Association for the Advancement of Science.

12.22 (Sec. 12.2.3) In Table 12.38 a random sample of 250 persons is cross-classified according to level of education and opinion on lowering the voting age to 18. Suppose a political scientist had claimed that the only effect of education on such issues is that persons who have had ex-

TABLE 12.38
Opinion on Lowering the Voting Age

	IN FAVOR	AGAINST	No OPINION	TOTAL
ATTENDING COLLEGE	10 (40%)	13 (52%)	2 (8%)	25 (100%)
HIGH SCHOOL GRADUATE	21 (20%)	42 (40%)	42 (40%)	105 (100%)
GRADE SCHOOL ONLY	15 (12%)	60 (50%)	45 (38%)	120 (100%)
TOTAL	46 (18%)	115 (46%)	89 (36%)	250 (100%)

(hypothetical data)

posure to college are much less likely to have no opinion than others. (One might paraphrase such a hypothesis by saying that college-educated people are more opinionated.) Check this claim by testing the null hypothesis that education and opinion are independent.

12.23 (Sec. 12.3.1) Consider again the results of one of the early experiments in the Asch study on conformity under pressure in reporting comparative length of line segments, as summarized in Table 10.5.
(a) Compute the value of the chi-square test statistic for these data.
(b) Compute ϕ.
(c) Compute $n\phi^2$ in order to verify that it equals the chi-square test statistic. (It can be shown algebraically that this is true in general. Here you are asked only to verify it numerically.)

12.24 (Sec. 12.3.1) Consider the data in Table 12.39, a cross-tabulation of 140 persons by sex and by opinion regarding a political proposal.
(a) Compute the value of the chi-square test statistic for these data.
(b) Compute ϕ.
(c) Compute $n\phi^2$ in order to verify that it equals the chi-square test statistic.

TABLE 12.39
Cross-Classification According to Sex and Approval of Political Proposal

	APPROVE	DISAPPROVE	TOTAL
MEN	23	37	60
WOMEN	27	53	80
TOTAL	50	90	140

(hypothetical data)

12.25 (Sec. 12.3) The data of Table 12.40 are part of a study of the association between complications in pregnancy of mothers and behavior problems in children. The comparison is between mothers of children who had been referred by their teachers as behavior problems and mothers of children not so referred. For each mother it was recorded whether she had lost any children prior to the birth of the child. The birth order of the child was recorded and used as a control.
(a) Compute the phi coefficient for each of the three 2 × 2 tables.
(b) Compute the chi-square statistic for each of the three tables.

12.26 (Sec. 12.3) From Table 12.41, study the association of estrogen treatment with the condition of tumor, as follows. Study separately each of the four 2 × 2 tables for Estrogen Treatment and Condition of

Tumor; this controls for Age and Parity. (A nulliparous woman has never given birth; a parous woman has given birth at least once.)

(a) Compute the phi coefficient for each of the four tables.

(b) Compute the chi-square statistic for each of the four tables.

TABLE 12.40

Cross-Classification of Mothers by Referral of Child, Prior Infant Loss, and Birth Order of Child

BIRTH ORDER	BEHAVIORAL REFERRAL?	PRIOR INFANTS	
		Loss	*No Loss*
2	Yes	20	82
	No	10	54
3–4	Yes	26	41
	No	16	30
≥5	Yes	27	22
	No	14	23

SOURCE: Reproduced from W. G. Cochran, "Some Methods for Strengthening the Common Chi-Square Tests," *Biometrics* 10: 417–451, 1954, with the permission of the Biometric Society.

TABLE 12.41

Condition of Tumor, by Age, Parity, and Estrogen Treatment

		AGE < 50		AGE ≥ 50	
		Benign	*Malignant*	*Benign*	*Malignant*
NULLIPAROUS	*Estrogen*	9	14	9	14
	No Estrogen	64	27	24	38
PAROUS	*Estrogen*	30	9	14	14
	No Estrogen	189	71	39	93

SOURCE: Black and Leis (1972), Table I, p. 1602.

12.27 (Sec. 12.3.2)   Consider again the data of Table 5.35, "Classification of Social Scientists by Age, Productivity Score, and Party Vote in 1952."

(a) Calculate the tables for Productivity and Age for the Democrats and for the Others.

(b) Test the independence of Age and Productivity for the Democrats and for the Others.

12.28 (Sec. 12.3.2)   Consider again the data of Table 5.50, "Cross-Classification of 40 Years, by Yield of Hay, Rainfall, and Temperature."
(a) Calculate $\lambda_{\text{yield-temperature}}$ for the years of light rainfall and for the years of heavy rainfall.
(b) Test the independence of temperature and yield for the years of light rainfall and for the years of heavy rainfall.

12.29 (Sec. 12.3.3)   For the data in Table 5.35, study the association between age and productivity by calculating the coefficient based on ordering: (a) for the Democrats, (b) for the Others.

12.30 (Sec. 12.3.3)   For the data in Table 5.50, calculate the coefficient based on ordering: (a) for the years of light rainfall, (b) for the years of heavy rainfall.

12.31 (Sec. 12.3)   Analyze the data in Table 5.25 as follows.
(a) Compute the coefficient of association based on predicting the "leader's rating" from the "degree of change in understanding."
(b) Compute the coefficient of association based on predicting the "degree of change in understanding" from the "leader's rating."
(c) Compute the coefficient of mutual association, based on both types of prediction.
(d) Compute the coefficient of association based on ordering.
(e) Interpret the association in these data as measured by the coefficients in parts (a), (b), (c), and (d).

12.32 (Sec. 12.3)   Analyze the data in Table 5.23, as follows.
(a) Compute the coefficient of association based on predicting the "level of political interest" from the "level of education."
(b) Compute the coefficient of association based on predicting the "level of education" from the "level of political interest."
(c) Compute the coefficient of mutual association, based on both types of prediction.
(d) Compute the coefficient of association based on ordering.
(e) Interpret the association in these data as measured by the coefficients in parts (a), (b), (c), and (d).

12.33 (Sec. 12.3.1)   Calculate the phi coefficient for each of the three tables of Table 5.33. Call the coefficients ϕ_{RB} for Razors and Blades, ϕ_{RC} for Razors and Commercial, and ϕ_{CB} for Commercial and Blades.

12.34 (Sec. 12.3.1)   Calculate the phi coefficient for each of the three tables of Table 5.51. Call them ϕ_{TY} for Temperature and Yield, ϕ_{TR} for Temperature and Rainfall, and ϕ_{RY} for Rainfall and Yield.

12.35 (Sec. 12.1)   Test the hypothesis that the heights in Table 2.11 are a sample from a normal distribution with a mean of 52.5 inches and standard deviation of 2 inches, as follows.

(a) Standardize the endpoints of the class intervals by subtracting the mean of 52.5 and dividing the results by the standard deviation of 2. Make the first interval $-\infty$ to -3.0 (in standardized units) and the last 3.0 to $+\infty$.

(b) Use Appendix I to find the proportion of the standard normal distribution that lies between the standardized endpoints.

(c) Multiply each result of (b) by the sample size to obtain expected frequencies.

(d) Compare the expected frequencies with the observed frequencies by means of the chi-square statistic.

12.36 (Sec. 12.1)   Test the hypothesis that the heights in Table 7.19 are a sample from a normal distribution with a mean of 54.5 inches and a standard deviation of 2 inches. Use steps corresponding to (a) to (d) of Exercise 12.35.

12.37 (Sec. 12.1)   Table 12.42 resulted from a study of auto weight and safety. Test the hypothesis of independence between auto weight and accident frequency. How small is the achieved level of significance?

TABLE 12.42
Accident Frequency, by Weight Class

Auto Weight Class	Observed Accident Frequency	Registration Distribution
Under 3000 pounds	162	21.04%
3000–4000 pounds	318	46.13%
4000–5000 pounds	689	31.13%
Over 5000 pounds	35	1.70%
	1204	100.00%

SOURCE: Yu, Wrather, and Kozmetsky (1975), p. 8.

MATHEMATICAL EXERCISES

12.38 (Sec. 12.2.1)   Show that (12.5) is

$$\frac{n(ad - bc)^2}{(a + b)(c + d)(a + c)(b + d)}.$$

12.39 (Sec. 12.3.1)   Show that $\phi = \sqrt{d_r d_c}$.

12.40 (Sec. 12.3.1)   Show that the statistic

$$\frac{\hat{p}_1 - \hat{p}_2}{\sqrt{\hat{p}\hat{q}\left(\frac{1}{n_1} + \frac{1}{n_2}\right)}}$$

used for testing $H: p_1 = p_2$ is equal to $\sqrt{n}\phi$, and hence that its square is equal to $n\phi^2$, which is the chi-square test statistic.

12.41 (Sec. 12.3.3)   Show that, for the special case of a 2 × 2 table, $\gamma = 0$ if and only if the two variables are independent.

12.42 (Sec. 12.3)   Show that, for any given 2 × 2 table, ϕ is less than or equal to γ. [*Hint:* The numerators are both equal to $ad - bc$. Show that the denominator of ϕ is at least as large as that of γ by comparing their squares.]

12.43 (Sec. 12.3.1)   We define the *coefficient of partial association* between two variables, taking a third variable into account. In terms of the quantities of Exercise 12.34, the partial association between Temperature and Yield, taking Rainfall into account, is

$$\phi_{TY \cdot R} = \frac{\phi_{TY} - \phi_{TR}\phi_{RY}}{\sqrt{1 - \phi_{TR}^2}\sqrt{1 - \phi_{RY}^2}}.$$

Calculate this coefficient and interpret the result. (A partial ϕ coefficient ranges between -1 and $+1$, as does the ordinary ϕ coefficient, and the values -1, 0, and $+1$ have analogous interpretations in the two cases.)

12.44 (Sec. 12.3.1)   Calculate the partial coefficient of association

$$\phi_{RC \cdot B} = \frac{\phi_{RC} - \phi_{RB}\phi_{BC}}{\sqrt{1 - \phi_{RB}^2}\sqrt{1 - \phi_{BC}^2}},$$

using the coefficients computed in Exercise 12.33, and interpret the result.

12.45 (Sec. 12.3.1)   The *Cauchy-Schwarz inequality* states that, for any numbers $u_1, u_2, \ldots, u_n, v_1, v_2, \ldots, v_n$,

$$\left(\sum_{i=1}^{n} u_i v_i\right)^2 \le \left(\sum_{i=1}^{n} u_i^2\right)\left(\sum_{i=1}^{n} v_i^2\right).$$

This is equivalent to saying that the quantity

$$\frac{\sum_{i=1}^{n} u_i v_i}{\sqrt{\sum_{i=1}^{n} u_i^2 \sum_{i=1}^{n} v_i^2}}$$

is between -1 and $+1$, inclusive. In particular, given data $(x_i, y_i), i = 1, \ldots, n$, we can set $u_i = x_i - \bar{x}$ and $v_i = y_i - \bar{y}$ to obtain the fact that the correlation coefficient (Sec. 5.5)

$$r = \sum_{i=1}^{n} (x_i - \bar{x})(y_i - \bar{y}) \Big/ \sqrt{\sum_{i=1}^{n} (x_i - \bar{x})^2 \sum_{i=1}^{n} (y_i - \bar{y})^2}$$

is between -1 and $+1$, inclusive. (See also Section 14.5.) Use this fact to prove that ϕ is between -1 and $+1$, inclusive, as follows.

(a) In terms of the general notation for a 2×2 table, take $x_i = 1$ or 0 according as the ith individual is in category A_1 or A_2, and take $y_i = 1$ or 0 according as the ith individual is in category B_1 or B_2. Show that $\bar{x} = (a + b)/n$ and $\bar{y} = (a + c)/n$.

(b) Show that $x_i - \bar{x} = (c + d)/n$ or $-(a + b)/n$ according as the ith individual is in category A_1 or A_2. Show that $y_i - \bar{y} = (b + d)/n$ or $-(a + c)/n$ according as the ith individual is in category B_1 or B_2.

(c) Show that $\sum_{i=1}^{n}(x_i - \bar{x})^2 = (a + b)(c + d)/n$.

(d) Show that $\sum_{i=1}^{n}(y_i - \bar{y})^2 = (a + c)(b + d)/n$.

(e) Show that $\sum_{i=1}^{n}(x_i - \bar{x})(y_i - \bar{y}) = (ad - bc)/n$.

(f) Combine the results of (c), (d), and (e) to show that $\phi = r$, and hence that ϕ is between -1 and $+1$, inclusive.

12.46 (Sec. 12.1.3) Show that $\sum_{i=1}^{k}(O_i - E_i)^2/E_i = \sum_{i=1}^{k}O_i^2/E_i - n$.

12.47 (Sec. 12.2.2) Show that
$\sum_{i=1}^{r}\sum_{j=1}^{c}(O_{ij} - E_{ij})^2/E_{ij} = \sum_{i=1}^{r}\sum_{j=1}^{c}O_{ij}^2/E_{ij} - n$.

12.48 (Sec. 12.2) Show that $\sum_{i=1}^{r}\sum_{j=1}^{c}O_{ij}^2/E_{ij} - n$ can be computed as $n\sum_{i=1}^{r}(\sum_{j=1}^{c}O_{ij}^2/O_{.j})/O_{i.} - n$, where $O_{i.} = \sum_{j=1}^{c}O_{ij}$, $i = 1, 2, \ldots, c$, and $O_{.j} = \sum_{i=1}^{r}O_{ij}$, $j = 1, 2, \ldots, r$. This expression is useful for calculating the value of the chi-square statistic row by row. [*Hint*: $E_{ij} = n_i.n_{.j}/n = O_{i.}O_{.j}/n$.]

REFERENCES

Black, Maurice M., and Henry P. Leis (1972). "Mammary carcinogenesis: influence of parity and estrogens." *New York State Journal of Medicine* 72: 1601–1605.

Cochran, W. G. (1954). "Some methods for strengthening the common χ^2 tests." *Biometrics* 10: 417–451.

Farkas, S. R., and H. H. Shorey (1972). "Chemical trail-following by flying insects: a mechanism for orientation to a distant odor source." *Science* 178: 67–68.

Lazarsfeld, Paul F., and Wagner Thielens (1958). *The Academic Mind.* Free Press, Glencoe, Ill. Copyright © 1958 by The Free Press (Macmillan).

Phillips, David (1972). "Deathday and Birthday: An Unexpected Connection," from *Statistics: A Guide to the Unknown,* by Judith M. Tanur *et al.* (eds.), Holden-Day, San Francisco.

Yu, P. L., C. Wrather, and G. Kozmetsky (1975). "Auto weight and public safety, a statistical study of transportation hazards." Research Report 233, Center for Cybernetic Studies, University of Texas, Austin.

13 COMPARISON OF SEVERAL POPULATIONS

INTRODUCTION

Throughout this book we have stressed the basic statistical concept of *variability*. When some measurement, such as height or IQ, is made on several individuals, the values vary from person to person. This variability of a quantitative variable is often measured by the variance. If the set of individuals is stratified into more homogeneous groups, the variance of the measurements within the more homogeneous groups will be less than that of the entire group; in fact, that is what "more homogeneous" means. For example, the variance of the heights of pupils in an elementary school is usually greater than the variance of the heights of pupils in the first grade, the variance in the second grade, and the variance in each of the other grades. At the same time, the average height of pupils within a grade varies from grade to grade. As we shall see in this chapter, the facts that the within-grades variances are less than the overall variance and that the averages vary between grades are reciprocal; they correspond to two ways of looking at the same phenomenon. The total variability is made up of two components: the variability of individuals within groups (grades) and the variability of means between groups (grades).

The *analysis of variance* consists of statistical techniques for allocating variability to different sources and interpreting the allocation. The idea applies to both populations and samples. The analysis of variance of samples is used to make appropriate inferences about the populations; these involve tests of significance. In many studies the investigator is presented with data already classified into groups, such as states, countries, school grades, or religious affiliations. In other cases the statistician determines the classification into groups; for example, he or she may think that classification according to socioeconomic status is relevant to a measure of achievement. Sometimes the investigator constructs the groups by the assignment of individuals. In experimental studies the groups are defined by the various experimental conditions or "treatments," such as dosages of a drug. The scientist assigns individuals to the groups. (When possible, this assignment of subjects should be done randomly.) The groupings sometimes correspond to natural categories such as male and female. In other cases the classification may be made on the basis of intervals for a continuous variable, such as age.

EXAMPLE 1   An elementary school teacher wants to try out three different reading workbooks. At the end of the year the 18 children in the class will take a test in reading achievement. These test scores will be used to compare the workbooks.

EXAMPLE 2   In a study of television, viewing habits were related to the educational level of the viewer. Respondents (males over 25 years of age) were divided into 4 classifications according to the level of formal education attained. Each respondent's viewing was monitored for 3 weeks, and his daily average number of hours spent watching TV was computed. It is desired to decide whether these figures differ appreciably across educational levels, and if so, how the viewing depends on educational level.

EXAMPLE 3   In the study *The Academic Mind* by Lazarsfeld and Thielens, a total of 2451 social science faculty members from 165 of the larger American colleges and universities were interviewed in order to assess the impact of the McCarthy era on social science faculties. At each college, the number of "academic freedom incidents" was counted. These were incidents mentioned by more than one respondent as an attack on the academic freedom of the faculty. They ranged from small-scale matters, such as a verbal denunciation of a professor by a student group, to large-scale matters, including a Congressional investigation. It was of interest to examine whether and how the institutional basis of a school's support and control affected the number of "incidents" occurring there. Hence, each college was classified as publicly controlled, privately controlled, or controlled by some other institution. (Teachers' colleges and schools controlled by a religious institution were included in the "other" category.) The distributions of numbers of "incidents" in the different types of institutions were studied.

EXAMPLE 4   A theory states that the average IQ's of brunettes, of blonds, and of redheads are different, and that therefore three separate school systems should be set up, one for dark-haired children, one for blond children, and one for red-haired children, in order to be able to base instruction on IQ levels. A study of all the children in the United States aged 6 to 11 showed that on a certain set of intelligence tests, brunettes had an average score of 103, blonds of 97, and redheads of 100. The standard deviation of the scores of the brunettes was 15, and the

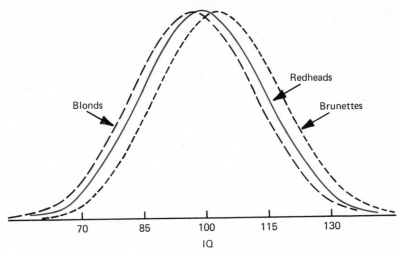

Figure 13.1 *Hypothetical distributions of IQ scores in three populations*

same standard deviation was found in the other groups. The graphs of the relative frequencies of the scores are given in Figure 13.1.

In this hypothetical example, the variability within each group is so large that the slight differences in average group scores are of no consequence. No differences in instructional methods would be based on such small differences in averages. If pupils were assigned to classes according to their hair color, there would still be large variation in IQ's within each class. Other characteristics, such as geographical location or income levels of parents, could potentially be of more importance.

An investigator may take samples from several populations in order to study differences between the populations. One question is whether there are differences between the averages of the populations. When we ask this question, what we really have in mind is whether there are differences large enough to be of importance. When we state a hypothesis that the means of several distributions are equal, this should be interpreted to mean that the differences are small enough to be considered negligible.

We formalize this question of whether the means of populations are different as a test of the hypothesis of equality of means based on a sample from each population. If there are two populations, and hence two samples, as in Chapter 10 ("Differences between Two Populations"), the test can be based on Student's t, the ratio of the difference between the sample means to an estimate of the standard deviation of that difference. The test used in the analysis of variance when there are more than two populations is a generalization of this procedure.

Problems connected with the use of *t*-tests to compare pairs of samples taken from a set of more samples are discussed in Section 13.3. When one does not want to rely on the population being approximately normal or the measurements being a ratio scale, one may use a rank test for comparing several samples, as given in Section 13.4, or a permutation test, as given in Appendix 13D.

The remaining material, Section 13.5 and Appendix 13C, deals with more complicated experimental designs. The analysis of variance refers to the allocation of the variability of the total sample (all groups combined) to differences among groups on the one hand and to differences among individuals in the same group on the other. The groups may correspond to a classification of individuals on two (or more) criteria simultaneously. This results in a so-called two-way (or higher-way) classification. The analysis of variance is then a decomposition of the total variability into separate parts, including one part for each criterion of classification, and a residual due to random errors and to factors not taken into account by the experimental design.

In order that the analysis of variance be appropriate, the mean should be a suitable measure of location. As we discussed earlier, this means that the distribution should be unimodal and (roughly) symmetric. The variance (or standard deviation) should be a suitable measure of variability, and, as in the case of two samples treated in Chapter 10, the variances in the different populations should be about the same. The formal mathematical assumptions for the tests of significance are that the distributions sampled be normal with equal variances. In practice this means that the distributions are unimodal, symmetric, and bell-shaped. The larger the samples, the less important the normality (due to the Central Limit Theorem). Information about these characteristics of the populations sampled can be obtained from histograms of the samples.

13.1 COMPARISON OF SEVERAL SAMPLES OF EQUAL SIZE

13.1.1 Three Samples of Equal Size

The ideas as well as the computations of the analysis of variance are easily introduced by the case of several samples of equal numbers of observations from the relevant populations. We are interested in comparing the variability of the group means with the variability within the groups. In particular, we may ask whether the variability of sample means indicates that the population means are different.

TABLE 13.1

Reading Achievement Scores of 18 Children Using Three Different Workbooks

	WORKBOOK 1	WORKBOOK 2	WORKBOOK 3
	2	9	4
	4	10	5
	3	10	6
	4	7	3
	5	8	7
	6	10	5
Sums	24	54	30
Sample means	4	9	5

Total of 3 samples: 108; mean of 3 samples: 6

(hypothetical data)

We consider data for Example 1. Table 13.1 gives the Reading Achievement Scores of 18 children using three different workbooks. The set of scores of the 6 children using Workbook 1 is considered as a sample from the hypothetical population of scores of all kindergarten children who might use that workbook. The scores are graphed in Figure 13.2.

The means of the three samples are 4, 9, and 5, respectively. Figure 13.2 shows these as the centers of the three samples; there is clearly variability from group to group. The variability in the entire pooled sample of 18 is shown by the last line.

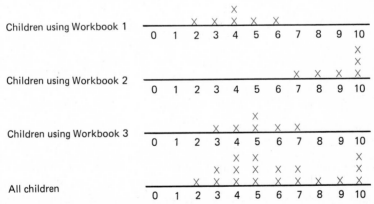

Figure 13.2 *Reading achievement scores, by workbook used, and for the combined sample*

TABLE 13.2

No Variation Within Groups

	GROUP		
	A	B	C
	3	5	8
	3	5	8
	3	5	8
	3	5	8
Means:	3	5	8

(hypothetical data)

In contrast to this rather typical allocation, we consider Tables 13.2 and 13.3 as illustrations of extreme cases. In Table 13.2 every observation in Group A is 3, every observation in Group B is 5, and every observation in Group C is 8. There is no variation within groups, but there is variation between groups. In Table 13.3 the mean of each group is 3. There is no variation among the group means, although there is variability within each group. Neither extreme can be expected to occur in an actual data set. In actual data, one needs to make an assessment of the relative sizes of the between-groups and within-groups variability. It is to this assessment that the term "analysis of variance" refers. We illustrate the method of assessment by calculating variances for the whole sample and the three separate samples in Table 13.1.

TABLE 13.3

No Variation Between Groups (Group Means Equal)

	GROUP		
	A	B	C
	3	3	1
	5	6	4
	1	2	3
	3	1	4
Means:	3	3	3

(hypothetical data)

The overall mean, say $\bar{x}$, is the sum of all the observations divided by the total number of observations: $\bar{x} = (2 + 4 + \cdots + 5)/18 = 108/18 = 6$. If we sum the observations in each group and then sum these three sums, we obtain $\bar{x} = (24 + 54 + 30)/18 = 108/18 = 6$. The sum of squared deviations of all 18 observations from the mean of the combined sample is a measure of variability of the combined sample. This sum of squared deviations is computed in Table 13.4. At the foot of each column is the sum of squared deviations of that group. The sum across groups is $34 + 62 + 16 = 112$. This is called the *Total Sum of Squares* and is denoted by SS(Total).

TABLE 13.4

Computation of the Sum of Squared Deviations from the Mean of the Combined Sample

	GROUP 1	GROUP 2	GROUP 3
	$(2 - 6)^2 = (-4)^2 = 16$	$(9 - 6)^2 = 3^2 = 9$	$(4 - 6)^2 = (-2)^2 = 4$
	$(4 - 6)^2 = (-2)^2 = 4$	$(10 - 6)^2 = 4^2 = 16$	$(5 - 6)^2 = (-1)^2 = 1$
	$(3 - 6)^2 = (-3)^2 = 9$	$(10 - 6)^2 = 4^2 = 16$	$(6 - 6)^2 = 0^2 = 0$
	$(4 - 6)^2 = (-2)^2 = 4$	$(7 - 6)^2 = 1^2 = 1$	$(3 - 6)^2 = (-3)^2 = 9$
	$(5 - 6)^2 = (-1)^2 = 1$	$(8 - 6)^2 = 2^2 = 4$	$(7 - 6)^2 = 1^2 = 1$
	$(6 - 6)^2 = 0^2 = 0$	$(10 - 6)^2 = 4^2 = 16$	$(5 - 6)^2 = (-1)^2 = 1$
Sum	34	62	16

SOURCE: Table 13.1

Next we measure the variability within samples. Table 13.5 shows the computation of the sum of squared deviations of each of the 18 scores from their respective group means. For each individual in each group, the table gives the deviation of that individual's score from the respective group mean and the square of the deviation. At the foot is

TABLE 13.5

Computation of Within-Groups Sum of Squares by Use of Deviations from Group Mean

	GROUP 1	GROUP 2	GROUP 3
	$(2 - 4)^2 = (-2)^2 = 4$	$(9 - 9)^2 = 0^2 = 0$	$(4 - 5)^2 = (-1)^2 = 1$
	$(4 - 4)^2 = 0^2 = 0$	$(10 - 9)^2 = 1^2 = 1$	$(5 - 5)^2 = 0^2 = 0$
	$(3 - 4)^2 = (-1)^2 = 1$	$(10 - 9)^2 = 1^2 = 1$	$(6 - 5)^2 = 1^2 = 1$
	$(4 - 4)^2 = 0^2 = 0$	$(7 - 9)^2 = (-2)^2 = 4$	$(3 - 5)^2 = (-2)^2 = 4$
	$(5 - 4)^2 = 1^2 = 1$	$(8 - 9)^2 = (-1)^2 = 1$	$(7 - 5)^2 = 2^2 = 4$
	$(6 - 4)^2 = 2^2 = 4$	$(10 - 9)^2 = 1^2 = 1$	$(5 - 5)^2 = 0^2 = 0$
Sum	10	8	10

SOURCE: Table 13.1

given the sum of squared deviations for each group. The sum over groups is $10 + 8 + 10 = 28$. This is called the *Sum of Squares Within Groups* and is denoted by SS(Within Groups).

Now let us consider the group means of 4, 9, and 5. The sum of squared deviations of the group means from the pooled mean of 6 is

$$(4 - 6)^2 + (9 - 6)^2 + (5 - 6)^2 = 4 + 9 + 1 = 14.$$

However, this sum is not comparable to the sum of squares within groups because the sampling variability of means is less than that of individual measurements. In fact, the mean of a sample of 6 observations has a sampling variance of 1/6, the sampling variance of a single observation. Hence, to put the sum of squared deviations of group means on a basis that can be compared with SS(Within Groups), we must multiply it by 6, the number of observations in each sample, to obtain $6 \times 14 = 84$. This is called the *Sum of Squares Between Groups* and is denoted by SS(Between Groups).

Now we have three sums that can be compared: SS(Between Groups), SS(Within Groups), and SS(Total). They are given in Table 13.6. Observe that addition of the first two sums of squares gives the last sum. This demonstrates what we mean by the allocation of the total variability to the variability due to differences between means of groups and the variability of individuals within (homogeneous) groups.

TABLE 13.6
Sums of Squares for Example of Reading Achievement Scores

SS(Between Groups)	84
SS(Within Groups)	28
SS(Total)	112

SOURCE: Table 13.1

In this example we notice that the variability between groups is a large proportion of the total variability. However, we have to adjust the numbers in Table 13.6 in order to take account of the number of pieces of information going into each sum of squares. That is, we want to use the sums of squares to calculate sample variances. The Sum of Squares Between Groups has 3 deviations about a mean. Therefore its number of degrees of freedom is $3 - 1 = 2$. The sample variance based on this sum of squares is

$$s_B^2 = \frac{\text{SS(Between Groups)}}{3 - 1} = \frac{84}{2} = 42.$$

The Sum of Squares Within Groups is made up of 3 sample sums of squares. Each involves 6 squared deviations, and hence, each has $6 - 1 = 5$ degrees of freedom. The 3 separate sample variances are

$$\frac{10}{5} = 2, \qquad \frac{8}{5} = 1.6, \qquad \frac{10}{5} = 2.$$

The estimate of variance is the average of these,

$$s_W^2 = \frac{2 + 1.6 + 2}{3} = \frac{5.6}{3} = 1.867.$$

This number can be calculated alternatively as

$$\frac{10 + 8 + 10}{5 + 5 + 5} = \frac{28}{15} = 1.867.$$

In this last expression the numerator is SS(Within Groups) and the denominator is the number of degrees of freedom, which is the sum of the numbers of degrees of freedom for the three sample variances.

The two estimates of variance s_B^2, measuring variability among groups, and s_W^2, measuring variability within groups, are now comparable. Their ratio is

$$\frac{s_B^2}{s_W^2} = \frac{42}{1.867} = 22.50.$$

The fact that s_B^2 is 22.5 times s_W^2 seems to indicate that the variability among groups is much greater than that within groups. However, we know that such a ratio computed for different triplets of random samples would vary from triplet to triplet, even if the population means were the same. We must take account of this sampling variability. This is done by referring to the *F*-tables. The calculated ratio of 22.50 is compared with a number from the *F*-tables depending on the desired significance level as well as on the number of degrees of freedom of s_B^2, which is 2 here, and the number of degrees of freedom of s_W^2, which is 15 here. The value in the *F*-table for a significance level of 1% is 6.36. Thus we would consider the calculated ratio of 22.50 as very significant. We conclude that there are real differences in average reading readiness due to the use of different workbooks.

The results of computation are set out in the Analysis of Variance Table (Table 13.7). The variances s_B^2 and s_W^2 are called "mean squares" because they are means of sets of squared deviations. In computing these "means" we divide not by the total number of squared devia-

TABLE 13.7
Analysis of Variance Table

Source of Variation	Sums of Squares	Degrees of Freedom	Mean Squares	F
Between groups	84	2	42	22.50
Within groups	28	15	1.867	
Total	112	17		

SOURCE: Table 13.1.

tions, but by the number of degrees of freedom associated with the respective sets of squared deviations.

Now let us treat the case of 3 groups of equal numbers of observations in general terms. We refer to the number of observations in each sample as r. Then the total number of observations is $n = 3r$. The samples are considered to have come from 3 populations, the first with some mean μ_1 and variance σ_1^2, the second with mean μ_2 and variance σ_2^2, and the third with mean μ_3 and variance σ_3^2. A hypothesis we may wish to test is $H: \mu_1 = \mu_2 = \mu_3$. It is assumed that it is reasonable to treat the population variances σ_1^2, σ_2^2, σ_3^2 as having a common value σ^2. Otherwise it would not be reasonable to reduce comparison of the groups to comparison of their means.

We denote the r observations in the first sample as $x_{11}, x_{12}, \ldots, x_{1r}$, those in the second sample as $x_{21}, x_{22}, \ldots, x_{2r}$, and those in the third sample as $x_{31}, x_{32}, \ldots, x_{3r}$. We can say this more compactly: x_{gi} is the ith observation in the gth group, $i = 1, \ldots, r$, $g = 1, 2, 3$. For instance, in Table 13.1, $x_{11} = 2$, $x_{12} = 4$, $\ldots$, $x_{21} = 9$, $\ldots$. (We read x_{21} as "x two one," not "x twenty-one.")

The mean of the gth group is

$$\bar{x}_g = \frac{x_{g1} + \cdots + x_{gr}}{r}, \qquad g = 1, 2, 3,$$

and the overall mean is

$$\bar{x} = \frac{\bar{x}_1 + \bar{x}_2 + \bar{x}_3}{3}.$$

We are interested in the deviation of each observation from its group mean, $x_{gi} - \bar{x}_g$; the deviation of the group mean from the overall mean, $\bar{x}_g - \bar{x}$; and the deviation of each observation from the overall mean.

We can write any observation x_{gi} as

$$x_{gi} = \bar{x} + (\bar{x}_g - \bar{x}) + (x_{gi} - \bar{x}_g), \qquad g = 1, 2, 3,$$
$$i = 1, 2, \ldots, r.$$

In words, this says

$$\begin{pmatrix} \text{Observation} \\ \text{on the } i\text{th unit} \\ \text{in the } g\text{th group} \end{pmatrix} = \begin{pmatrix} \text{overall} \\ \text{mean} \end{pmatrix} + \begin{pmatrix} \text{deviation of} \\ \text{group mean from} \\ \text{overall mean} \end{pmatrix} + \begin{pmatrix} \text{deviation of} \\ \text{observation} \\ \text{from group mean} \end{pmatrix}.$$

This can be written equivalently (by subtracting $\bar{x}$ from both sides) as

$$(x_{gi} - \bar{x}) = (\bar{x}_g - \bar{x}) + (x_{gi} - \bar{x}_g),$$

which shows how the deviation from the overall mean can be decomposed into the sum of a deviation of the group mean from the overall mean and the deviation of the observation from its group mean. It is an algebraic fact (see Appendix 13A at the end of the chapter) that the same decomposition holds for the sum of squared deviations, the numerator of the overall variance. This can be written in tabular form, as in Table 13.8. In words, the "Total" line of Table 13.8 says that

$$\begin{pmatrix} \text{Sum of} \\ \text{squared} \\ \text{deviations} \end{pmatrix} = \begin{pmatrix} \text{sum of squares} \\ \text{between groups} \end{pmatrix} + \begin{pmatrix} \text{sum of squares} \\ \text{within groups} \end{pmatrix},$$

$$\text{SS(Total)} = \text{SS(Between Groups)} + \text{SS(Within Groups)}.$$

TABLE 13.8
Sums of Squares

Group 1: $\displaystyle\sum_{i=1}^{r} (x_{1i} - \bar{x})^2 = r(\bar{x}_1 - \bar{x})^2 + \sum_{i=1}^{r} (x_{1i} - \bar{x}_1)^2$

Group 2: $\displaystyle\sum_{i=1}^{r} (x_{2i} - \bar{x})^2 = r(\bar{x}_2 - \bar{x})^2 + \sum_{i=1}^{r} (x_{2i} - \bar{x}_2)^2$

Group 3: $\displaystyle\sum_{i=1}^{r} (x_{3i} - \bar{x})^2 = r(\bar{x}_3 - \bar{x})^2 + \sum_{i=1}^{r} (x_{3i} - \bar{x}_3)^2$

Total: $\displaystyle\sum_{g=1}^{3}\sum_{i=1}^{r} (x_{gi} - \bar{x})^2 = r \sum_{g=1}^{3} (\bar{x}_g - \bar{x})^2 + \sum_{g=1}^{3}\sum_{i=1}^{r} (x_{gi} - \bar{x}_g)^2$

These properties of sums show why the technique is called the *analysis of variance:* the numerator of the overall variance, SS(Total), is decomposed into two parts, one a measure of differences *between* groups, the other a measure of differences *within* groups. In "analyzing" the variance, we compare the size of the sum of squares between groups with the sum of squares within groups. In this way we attempt to answer the question of whether the variability among individuals in different groups is more than would be accounted for by variability among individuals from the same group. Both terms in the above sum lead to an estimate of variance. If the variability of the group means is due merely to the variability of individuals, then these two estimates of variance should be about equal. If the estimate based on the sum of squares between groups is much larger than the estimate based on the within-groups sum of squares, then it appears that the variability of the group means cannot be explained merely in terms of the variability of individuals, and we decide that the three samples are from populations with different means.

A more precise explanation of this follows. Each of the three quantities

$$\frac{\sum_{i=1}^{r} (x_{gi} - \bar{x}_g)^2}{r - 1}, \qquad g = 1, 2, 3,$$

is an estimate of σ^2. Thus the average of these three quantities,

$$\frac{\sum_{g=1}^{3} \sum_{i=1}^{r} (x_{gi} - \bar{x}_g)^2}{3(r - 1)} = \frac{\text{SS(Within Groups)}}{3(r - 1)} = s_W^2,$$

is an estimate of σ^2, based on the observations in all three groups. The number of degrees of freedom associated with s_W^2 is $3(r - 1) = 3r - 3$.

Now consider the three group means, $\bar{x}_1$, $\bar{x}_2$, and $\bar{x}_3$. Under the null hypothesis, each group mean is the mean of a sample of r observations from a population with population mean μ, say, and population variance σ^2. We know from Section 8.1 that the means of random samples of r observations have a sampling distribution with a population mean equal to the population mean of the parent population, namely μ, and a population variance equal to the variance of the parent population divided by r. The sample variance of the group means,

$$\frac{\sum_{g=1}^{3} (\bar{x}_g - \bar{x})^2}{3 - 1},$$

is therefore an estimate of σ^2/r. Thus, to obtain an unbiased estimate of σ^2, we must multiply this by r, to obtain

$$\frac{r \sum_{g=1}^{3} (\bar{x}_g - \bar{x})^2}{3 - 1} = \frac{\text{SS(Between Groups)}}{2} = s_B^2.$$

If the null hypothesis is not true—that is, if the three population means are different—then we can expect the three sample means to be different (no matter how large the sample sizes are). This tends to inflate s_B^2 because the sample means are estimating three different population means: the variability of the sample means is greater than just the usual sampling variability of means from the same population. The statistic s_B^2 estimates σ^2 if the null hypothesis is true, but something bigger than σ^2 if the null hypothesis is false. On the other hand, whether or not the null hypothesis is true, s_W^2 estimates σ^2. The estimates s_B^2 and s_W^2 are compared by means of the F-ratio,

$$F = \frac{s_B^2}{s_W^2}.$$

Each of the mean squares s_B^2 and s_W^2 is a sum of squared deviations divided by the corresponding number of degrees of freedom. Thus an F-ratio is a ratio of two mean squares. We reject H if the calculated F is too large, for if the true means are unequal, then s_B^2 estimates a quantity larger than σ^2, whereas s_W^2 still estimates σ^2, so that F tends to be large. If the true means are equal, then the expected value of both numerator and denominator is σ^2.

The number of degrees of freedom associated with the numerator of F and the number of degrees of freedom associated with the denominator are 2 and $3r - 3$, respectively. To test the hypothesis of equality of group means at the α significance level, one compares the value of the calculated F-ratio with $F_{2,3r-3}(\alpha)$. This assumes that the populations from which the samples are drawn are normal. If they are only "roughly" normal, then this is an approximate test. The larger the sample size, the more assurance one has that the approximation is good. How seriously one should take the qualification of normality depends on the significance level and the sample size.

The results of the calculations are conveniently set out in a table with a standard format like Table 13.9. In the next section we shall discuss how the calculations can be done more efficiently.

13.1.2 Several Samples of Equal Size

Next we give the details of analysis of variance for the case in which there are k groups with r observations in each group. The obser-

TABLE 13.9

Analysis of Variance Table: Three Samples with r Observations in Each

Source of Variation	Sums of Squares	Degrees of Freedom	Mean Squares	F
Between groups	$r \sum_{g=1}^{3} (\bar{x}_g - \bar{x})^2$	2	s_B^2	s_B^2/s_W^2
Within groups	$\sum_{g=1}^{3} \sum_{i=1}^{r} (x_{gi} - \bar{x}_g)^2$	$3r - 3$	s_W^2	
Total	$\sum_{g=1}^{3} \sum_{i=1}^{r} (x_{gi} - \bar{x})^2$	$3r - 1$		

vations are x_{gi}, $g = 1, \ldots, k$, $i = 1, \ldots, r$. The total number of observations is $n = kr$. The samples are assumed to have come from k populations, the first with mean μ_1 and variance σ^2, the second with mean μ_2 and variance σ^2, ..., the kth with mean μ_k and variance σ^2. The hypothesis to be tested is

$$H: \mu_1 = \mu_2 = \cdots = \mu_k.$$

The sample means are $\bar{x}_1, \bar{x}_2, \ldots, \bar{x}_k$. The overall mean is $\bar{x} = (\bar{x}_1 + \bar{x}_2 + \cdots + \bar{x}_k)/k$. The sums of squares are

$$SS(\text{Total}) = \sum_{g=1}^{k} \sum_{i=1}^{r} (x_{gi} - \bar{x})^2,$$

$$SS(\text{Between Groups}) = r \sum_{g=1}^{k} (\bar{x}_g - \bar{x})^2,$$

$$SS(\text{Within Groups}) = \sum_{g=1}^{k} \sum_{i=1}^{r} (x_{gi} - \bar{x}_g)^2.$$

The mean squares are

$$s_B^2 = \frac{SS(\text{Between Groups})}{k - 1},$$

$$s_W^2 = \frac{SS(\text{Within Groups})}{n - k},$$

where the number of degrees of freedom associated with SS(Within Groups) is $r - 1$ for each of k groups, or $k(r - 1) = kr - k = n - k$ altogether. Note that the degrees of freedom add in the same way as the sums of squares:

$$SS(\text{Total}) = SS(\text{Between Groups}) + SS(\text{Within Groups})$$

and

$$n - 1 = k - 1 + k(r - 1),$$

$n = kr$.

Computational Formulas. The formulas given above for the sums of squares are definitional formulas, not usually used for computation. The usual computational formulas are given below.

FORMULA 1

$$SS(\text{Total}) = \sum_{g=1}^{k}\sum_{i=1}^{r} x_{gi}^2 - \frac{\left(\sum_{g=1}^{k}\sum_{i=1}^{r} x_{gi}\right)^2}{n},$$

where $n = kr$. (This formula is Rule 9 of Appendix 4C in terms of summations on two indices; see Appendix 13A for details.)

We illustrate the use of Formula 1 for Example 1; some of the computations appear in Table 13.10. We have

$$SS(\text{Total}) = 760 - \frac{(108)^2}{18} = 760 - \frac{11,664}{18}$$

$$= 760 - 648 = 112.$$

FORMULA 2

$$SS(\text{Between Groups}) = \frac{\sum_{g=1}^{k}\left(\sum_{i=1}^{r} x_{gi}\right)^2}{r} - \frac{\left(\sum_{g=1}^{k}\sum_{i=1}^{r} x_{gi}\right)^2}{n}.$$

TABLE 13.10

Use of Computational Formula for SS(Total), Illustrated for Example 1

	Group 1			Group 2			Group 3	
	x	x^2		x	x^2		x	x^2
	2	4		9	81		4	16
	4	16		10	100		5	25
	3	9		10	100		6	36
	4	16		7	49		3	9
	5	25		8	64		7	49
	6	36		10	100		5	25
	24	106		54	494		30	160
	$\sum\limits_{i=1}^{6} x_{1i}$	$\sum\limits_{i=1}^{6} x_{1i}^2$		$\sum\limits_{i=1}^{6} x_{2i}$	$\sum\limits_{i=1}^{6} x_{2i}^2$		$\sum\limits_{i=1}^{6} x_{3i}$	$\sum\limits_{i=1}^{6} x_{3i}^2$

$$\sum_{g=1}^{3}\sum_{i=1}^{6} x_{gi} = 108 \qquad\qquad \sum_{g=1}^{3}\sum_{i=1}^{6} x_{gi}^2 = 760$$

(This is Rule 9 of Appendix 4C, as

$$r \sum_{g=1}^{k} (\bar{x}_g - \bar{x})^2 = r \left[\sum_{g=1}^{k} \bar{x}_g^2 - k\bar{x}^2 \right];$$

see Appendix 13A for details.)

For Example 1, we have

$$\text{SS(Between Groups)} = \frac{(24)^2 + (54)^2 + (30)^2}{6} - 648$$

$$= \frac{576 + 2916 + 900}{6} - 648$$

$$= \frac{4392}{6} - 648 = 732 - 648 = 84.$$

Proofs of the computational formulas are given in Appendix 13A. The computational formulas are convenient in that they avoid computing the deviations from the means before squaring. They are good for use on a desk calculator as well as in a computer program. Care must be taken to carry enough significant digits, however, as digits are lost in the subtraction. (See Section 4.5.1.)

13.2 COMPARISON OF SEVERAL SAMPLES OF UNEQUAL SIZE

In Example 2 (Table 13.11) there are four educational levels, so $k = 4$. Let n_g be the number of observations in group g, $g = 1, \ldots, k$. Here we have $n_1 = 3, n_2 = 4, n_3 = 3, n_4 = 2$. The size of the total sample is $n = n_1 + n_2 + \cdots + n_k$; here this is $n = 3 + 4 + 3 + 2 = 12$.

TABLE 13.11
Average Number of Hours of TV Viewing per Day

	$g = 1$ Not a High School Graduate	$g = 2$ High School Graduate Only	$g = 3$ College Graduate Only	$g = 4$ Post- College Study
	$x_{11} = 5.0$ $x_{12} = 6.0$ $x_{13} = 4.0$	$x_{21} = 5.0$ $x_{22} = 4.0$ $x_{23} = 3.0$ $x_{24} = 4.0$	$x_{31} = 3.0$ $x_{32} = 4.0$ $x_{33} = 2.0$	$x_{41} = 1.5$ $x_{42} = .5$
$\sum_{i=1}^{n_g} x_{gi}$:	$\sum_{i=1}^{3} x_{1i} = 15.0$	$\sum_{i=1}^{4} x_{2i} = 16.0$	$\sum_{i=1}^{3} x_{3i} = 9.0$	$\sum_{i=1}^{2} x_{4i} = 2.0$
n_g:	3	4	3	2
$\bar{x}_g$:	5.0	4.0	3.0	1.0

(hypothetical data)

The observations in the sample are x_{gi}, $g = 1, \ldots, k$, $i = 1, \ldots, n_g$. The group means are

$$\bar{x}_g = \frac{1}{n_g} \sum_{i=1}^{n_g} x_{gi}, \quad g = 1, \ldots, k.$$

The overall mean $\bar{x}$ is simply the total of all observations divided by n,

$$\bar{x} = \frac{\sum_{g=1}^{k} \sum_{i=1}^{n_g} x_{gi}}{n}.$$

Note that $\sum_{i=1}^{n_g} x_{gi} = n_g \bar{x}_g$, so that

$$\bar{x} = \frac{\sum_{g=1}^{k} n_g \bar{x}_g}{n};$$

the overall mean is a weighted average of the group means, the weights being proportional to the sample sizes. When each n_g is equal to r, this reduces to

$$\bar{x} = \frac{\sum_{g=1}^{k} r\bar{x}_g}{n} = \frac{r \sum_{g=1}^{k} \bar{x}_g}{kr} = \frac{\sum_{g=1}^{k} \bar{x}_g}{k}.$$

The terms in the decomposition SS(Total) = SS(Between Groups) + SS(Within Groups) are

$$\text{SS(Total)} = \sum_{g=1}^{k} \sum_{i=1}^{n_g} (x_{gi} - \bar{x})^2,$$

$$\text{SS(Between Groups)} = \sum_{g=1}^{k} n_g(\bar{x}_g - \bar{x})^2,$$

$$\text{SS(Within Groups)} = \sum_{g=1}^{k} \sum_{i=1}^{n_g} (x_{gi} - \bar{x}_g)^2.$$

(See Appendix 13A for the proof of the decomposition.)
   The estimate of variance based on SS(Within Groups) is

$$s_W^2 = \frac{\text{SS(Within Groups)}}{n - k}.$$

The denominator is $n - k$ because the deviations $x_{g1} - \bar{x}_g$, $x_{g2} - \bar{x}_g$, $\ldots$, $x_{g,n_g} - \bar{x}_g$ of the observations in group g from their mean have $n_g - 1$ degrees of freedom. Summing over groups, we obtain $\sum_{g=1}^{k}(n_g - 1) = \sum_{g=1}^{k} n_g - k = n - k$ degrees of freedom altogether. The estimate of variance based on SS(Between Groups) is

$$s_B^2 = \frac{\text{SS(Between Groups)}}{k - 1},$$

the denominator being $k - 1$ because we are computing a variance from the k numbers $\bar{x}_1, \bar{x}_2, \ldots, \bar{x}_k$.
   Table 13.12 is the analysis of variance table. The first basic computations are the overall sum of the measurements $\sum_{g=1}^{k}\sum_{i=1}^{n_g} x_{gi}$ and the overall sum of squares of the measurements

$$\sum_{g=1}^{k} \sum_{i=1}^{n_g} x_{gi}^2.$$

Then the total sum of squares of deviations is calculated as

TABLE 13.12
Analysis of Variance Table

Source of Variation	Sum of Squares	Degrees of Freedom	Mean Squares	F
Between groups	$\sum_{g=1}^{k} n_g(\bar{x}_g - \bar{x})^2$	$k - 1$	s_B^2	s_B^2/s_W^2
Within groups	$\sum_{g=1}^{k} \sum_{i=1}^{n_g} (x_{gi} - x_g)^2$	$n - k$	s_W^2	
Total	$\sum_{g=1}^{k} \sum_{i=1}^{n_g} (x_{gi} - \bar{x})^2$	$n - 1$		

$$SS(\text{Total}) = \sum_{g=1}^{k} \sum_{i=1}^{n_g} x_{gi}^2 - \frac{1}{n} \left(\sum_{g=1}^{k} \sum_{i=1}^{n_g} x_{gi} \right)^2.$$

The sum of squares of deviations of group means multiplied by the group size is usually computed as

$$SS(\text{Between Groups}) = \sum_{g=1}^{k} \frac{\left(\sum_{i=1}^{n_g} x_{gi} \right)^2}{n_g} - \frac{1}{n} \left(\sum_{g=1}^{k} \sum_{i=1}^{n_g} x_{gi} \right)^2.$$

Finally the sum of squares of deviations about the group means is calculated as

$$SS(\text{Within Groups}) = SS(\text{Total}) - SS(\text{Between Groups}).$$

The proofs of these formulas are given in Appendix 13A.

We illustrate the computations with Example 2 (Table 13.13). We have

$$SS(\text{Total}) = 174.5 - \frac{(42)^2}{12} = 174.5 - 147 = 27.5,$$

$$SS(\text{Between Groups}) = \frac{15^2}{3} + \frac{16^2}{4} + \frac{9^2}{3} + \frac{2^2}{2} - \frac{(42)^2}{12}$$

$$= 75 + 64 + 27 + 2 - 147$$

$$= 168 - 147 = 21,$$

$$SS(\text{Within Groups}) = SS(\text{Total}) - SS(\text{Between Groups})$$

$$= 27.5 - 21 = 6.5.$$

TABLE 13.13

Computations for Analysis of Variance for Hours of TV Viewing

	$g = 1$ NOT A HIGH SCHOOL GRADUATE		$g = 2$ HIGH SCHOOL GRADUATE ONLY		$g = 3$ COLLEGE GRADUATE ONLY		$g = 4$ POST-COLLEGE STUDY	
	x	x^2	x	x^2	x	x^2	x	x^2
	5	25	5	25	3	9	1.5	2.25
	6	36	4	16	4	16	.5	0.25
	4	16	3	9	2	4		
			4	16				
Sums:	15	77	16	66	9	29	2	2.5
n_g:	3		4		3		2	

$$15 + 16 + 9 + 2 = 42$$
$$77 + 66 + 29 + 2.5 = 174.5$$

SOURCE: Table 13.11

The analysis of variance table appears as Table 13.14. The mean squares are $6.5/8 = .812$ and $21/3 = 7$; the ratio is $7/.812 = 8.62$. Since $F_{3,8}(.01) = 7.59$, even at the 1% level we would reject the hypothesis that the four group means are equal. The conclusion is that the average amount of TV viewing differs according to the educational level. This conclusion suggests further study of why and how education affects the number of hours persons spend in front of TV sets.

In Section 10.2 we developed Student's t-test for the null hypothesis of the equality of the means of two populations when the variances are

TABLE 13.14

Analysis of Variance Table for Hours of TV Viewing

SOURCE OF VARIATION	SUM OF SQUARES	DEGREES OF FREEDOM	MEAN SQUARES	F
Between educational levels	21.0	3	7.00	8.62
Within educational levels	6.5	8	0.812	
Total	27.5	11		

SOURCE: Table 13.13

considered equal, but unknown; this is the problem considered here for $k = 2$. It is shown in Appendix 13B that the statistic t^2 is algebraically identical to the F-ratio when $k = 2$. The reader can verify from Appendices III and X that $t_f^2(\alpha/2) = F_{1,f}(\alpha)$. Thus a two-sided t-test at significance level α is exactly equivalent to an F-test at significance level α, when $k = 2$.

13.3 MULTIPLE COMPARISONS AND CONFIDENCE REGIONS

The F-test gives information about all the means $\mu_1, \ldots, \mu_k$ simultaneously. In this section we consider inferences about differences of pairs of means. Instead of simply concluding that some of $\mu_1, \ldots, \mu_k$ are different, we may conclude that specific pairs μ_g, μ_h are different.

The variance of the difference between two means, say $\bar{x}_1$ and $\bar{x}_2$, is $\sigma^2(1/n_1 + 1/n_2)$, which is estimated as $s_W^2(1/n_1 + 1/n_2)$. The corresponding estimated standard deviation is $s_W\sqrt{1/n_1 + 1/n_2}$. If one were interested simply in determining whether the first two population means differed, one would test the null hypothesis that $\mu_1 = \mu_2$ at significance level α by the use of a t-test, rejecting the null hypothesis if $|\bar{x}_1 - \bar{x}_2|/(s_W\sqrt{1/n_1 + 1/n_2})$ exceeds $t(\alpha/2)$, where the number of degrees of freedom for the t-value is the number of degrees of freedom of s_W. However, now we want to consider each possible difference $\mu_g - \mu_h$; that is, we want to test all the null hypothesis H_{gh}: $\mu_g = \mu_h$, $g \neq h$, g, $h = 1, \ldots, k$. There are $k(k - 1)/2$ such hypotheses.

If, indeed, all the μ's were equal, so that there were no real differences, the probability that any particular *one* of the pairwise differences in absolute value would exceed the relevant t-value is α. Hence the probability that *at least one* of them would exceed the t-value, would, of course, be greater than α. When many differences are tested, the probability that some will appear to be "significant" is greater than the nominal significance level α when all the null hypotheses are true. How can one eliminate this false significance? It can be shown (see Appendix 13C) that, if m comparisons are to be made and the *overall* Type I error probability is to be at most α, it is sufficient to use α/m for the significance level of the individual tests. By overall Type I error we mean concluding $\mu_g \neq \mu_h$ for at least one pair g, h when actually $\mu_1 = \mu_2 = \cdots = \mu_k$.

We illustrate with Example 1 (Tables 13.1 and 13.7). Here $s_W^2 = 1.867$, based on 15 degrees of freedom ($s_W = 1.366$). Since all the sample sizes are 6, the value with which to compare each difference $\bar{x}_g - \bar{x}_h$ is $t(\alpha^*/2)$

$s_W\sqrt{1/6 + 1/6} = t(\alpha^*/2) \times 1.366 \times \sqrt{1/3} = t(\alpha^*/2) \times .789$, where a^* is to be the level of the individual tests. The number of comparisons to be made for $k = 3$ is $k(k - 1)/2 = 3 = m$. If we want the overall Type I error probability to be at most .03, then it suffices to choose the level α^* to be $.03/3 = .01$. The corresponding percentage point of Student's t with 15 degrees of freedom is $t_{15}(.01/2) = t_{15}(.005) = 2.947$. The value with which to compare $\bar{x}_g - \bar{x}_h$ is $.789 \times 2.947 = 2.33$. In Table 13.1 the means are $\bar{x}_1 = 4$, $\bar{x}_2 = 9$, and $\bar{x}_3 = 5$. The difference $\bar{x}_2 - \bar{x}_1 = 9 - 4 = 5$ is significant; so is the difference $\bar{x}_2 - \bar{x}_3 = 9 - 5 = 4$. The difference $\bar{x}_3 - \bar{x}_1 = 5 - 4 = 1$ is not significant. The conclusion is that μ_2 is different from both μ_1 and μ_3, but μ_1 and μ_3 may be equal; Workbook 2 appears to be superior.

Other techniques are available for multiple comparisons; some are discussed in Appendix 13C. The reader is also referred to the section on comparisons in Snedecor and Cochran (1967).

Confidence Regions. The confidence-interval procedure described in Section 10.2 can similarly be modified to obtain confidence intervals holding simultaneously on $\mu_g - \mu_h$ for all pairs g, h. With confidence at least $1 - \alpha$, the following inequalities hold:

$$\bar{x}_g - \bar{x}_h - t_{n-k}\left(\frac{\alpha^*}{2}\right) s_W \sqrt{\frac{1}{n_g} + \frac{1}{n_h}} < \mu_g - \mu_h$$

$$< \bar{x}_g - \bar{x}_h + t_{n-k}\left(\frac{\alpha^*}{2}\right) s_W \sqrt{\frac{1}{n_g} + \frac{1}{n_h}}$$

for $g \neq h$, $g, h = 1, \ldots, k$, if $\alpha^* = \alpha/m$.

Comparing Several Groups with a Control Group. In some experiments one group is untreated and each of the other groups is treated. Then it is of interest to compare each of the treated groups in an experimental condition with the control group. Has a particular treatment made a difference? If the control group is indexed by 0, the null hypotheses are H_g: $\mu_g = \mu_0$, $g = 1, \ldots, k$. With $k + 1$ treatments, one counted as a control, the tests should be done at level $\alpha^* = \alpha/k$.

13.4 A RANK TEST FOR COMPARISON OF SEVERAL DISTRIBUTIONS

In the previous section we compared the distributions of a variable in several populations by focusing on means and variances of the samples, and we developed tests that rely on the assumption that the variables have normal distributions and the same variances in the populations. One can also compare locations of distributions by means of ranks. This approach, which can be applied even when the data are ordinal, has the same relationship to the analysis of variance (Sections 13.1 and 13.2) as the rank sum test of Section 10.6.2 has to the two-sample *t*-test. Specifically, the hypothesis to be tested is that the distributions in the populations are the same against the alternative that they differ with respect to location.

The test procedure was developed by Kruskal and Wallis (1952); we illustrate with some data from their article, given in Table 3.26 (Daily Outputs of Three Bottle-Cap Machines). In Table 13.15 the values for each machine in Table 3.26 are replaced by the ranks of those values in the data for all three machines. The lowest value received rank 1; the highest, rank 12.

TABLE 13.15
Ranks in the Combined Sample

	RANKS	SUM OF RANKS	SAMPLE SIZE
Machine A	5, 9, 1, 6, 3	24	5
Machine B	4, 2, 8	14	3
Machine C	10, 7, 11, 12	40	4
Sum		78	12

SOURCE: Table 3.26

The test applies the analysis of variance formulas to these ranks, replacing x_{gi} by r_{gi}, the rank in the whole sample of the *i*th object in the *g*th group. Since the sum of the integers $1, \ldots, n$ is $n(n + 1)/2$ (Exercise 10.28), the mean of the ranks over the combined sample is $(n + 1)/2$. Then

$$\text{SS(Between Groups)} = \sum_{g=1}^{k} \frac{\left(\sum_{i=1}^{n_g} r_{gi} \right)^2}{n_g} - n \frac{(n + 1)^2}{4}.$$

The sum of squared deviations of the ranks around the mean is $n(n + 1)(2n + 1)/6 - n(n + 1)^2/4$ (Exercise 10.28), and this, divided by $n - 1$ (the degrees of freedom), is $n(n + 1)/12$. This quantity is MS(Total); it serves as an estimate of variance when the null hypothesis is true. (MS stands for "mean square.") The analog of $(k - 1)F_{k-1,n-k}$ is

$$H = \cfrac{\displaystyle\sum_{g=1}^{k} \left(\sum_{i=1}^{n_g} r_{gi}\right)^2 \Big/ n_g - n(n+1)^2/4}{n(n+1)/12}$$

$$= \frac{12}{n(n+1)} \sum_{g=1}^{k} \frac{\left(\displaystyle\sum_{i=1}^{n_g} r_{gi}\right)^2}{n_g} - 3(n+1).$$

The null hypothesis of no difference between groups is equivalent to random sampling (that is, random allocation of the ranks to the k groups). Under this hypothesis, the sampling distribution of H is approximately χ^2 with $k-1$ degrees of freedom when $n_1, \ldots, n_k$ are large.

In our example,

$$H = \frac{12}{12 \times 13} \left(\frac{24^2}{5} + \frac{14^2}{3} + \frac{40^2}{4}\right) - 3 \times (12 + 1) = 5.66.$$

We have $\chi_2^2(.10) = 4.605 < 5.66 < 5.992 = \chi_2^2(.05)$. So the P-value associated with 5.66 is between .05 and .10. The sample sizes are rather small for this approximate distribution, but we give this example to illustrate the procedure in terms of easy arithmetic. There are special tables for small sample-size cases.

13.5 COMPARISONS FOR ONE CLASSIFICATION WHEN THE DATA ARE CLASSIFIED TWO WAYS

13.5.1 The F-Test

Three methods of treating beer cans are being compared by a panel of 5 people. Each person samples beer from each type of can and scores the beer with a number (integer) between 0 and 6, 6 indicating a strong metallic taste and 0 meaning no metallic taste. It is obvious that different people will use the scale somewhat differently, and we shall take this into account when we compare the different types of can. The data are reported in Table 13.16. This is an example of a situation in which the investigator has data pertaining to k *treatments* ($k = 3$ types of can) in b *blocks* ($b = 5$ persons). We let x_{gj} denote the observation corresponding to the gth treatment and the jth block, $\bar{x}_{g\cdot}$ denote the mean of the b observations for the gth treatment, $\bar{x}_{\cdot j}$ the mean of the k observations in the jth block, and $\bar{x}_{\cdot\cdot}$ the overall mean of all $n = kb$ observations. A plan such as this for allocating individuals to experimental con-

TABLE 13.16
Scores of Three Types of Can on "Metallic" Scale

TYPE OF CAN	Adams	Clark	PERSON Davis	Jones	Smith	SUMS
A	6	5	6	4	3	24
B	2	3	2	2	1	10
C	6	4	4	4	3	21
Sums	14	12	12	10	7	55

(hypothetical data)

ditions is called an *experimental design*. When this particular design is used, the three types of can are presented to the individuals in random order.

This design differs basically from the design treated in Section 13.1. In that case the group could also be defined by the treatment applied, but the units of observation are different in the different groups. Here the units of observations in the different groups are the same. The purpose of using the same individuals here is to balance out the effects of variation from person to person.

An experimental design of this type is called a *randomized blocks design*. In agricultural experiments* the k treatments might correspond, for example, to k different fertilizers; the field would be divided into blocks of presupposed similar fertility; and every fertilizer was used in each block so that differences in fertility of the soil in different parts of the field (blocks) would not bias the comparison of the fertilizers. Each block would be subdivided into k subblocks, called "plots." The k fertilizers would be randomly assigned to the plots in each block; hence the name, "randomized blocks."

Table 13.17 shows the pattern of a data set resulting from a randomized blocks design; it is a two-way table with single measurements as entries (instead of frequencies as in Section 12.2). In the example persons correspond to blocks and cans to treatments. The observation x_{gj} is called the response to treatment g in block j.

The treatment mean $\bar{x}_{g\cdot}$ estimates the population mean μ_g for treatment g (averaged out over persons). An objective may be to test the hypothesis that treatments make no difference,

$$H: \mu_1 = \mu_2 = \cdots = \mu_k.$$

* Many of the basic ideas of "design of experiments" and "analysis of variance" were developed by R. A. Fisher at the Rothamstead Agricultural Experiment Station, Harpenden, England; some of the terminology reflects this origin.

TABLE 13.17
Randomized Blocks Design

TREATMENTS		BLOCKS		
	1	2	$\cdots$	b
1	x_{11}	x_{12}	$\cdots$	x_{1b}
2	x_{21}	x_{22}	$\cdots$	x_{2b}
.	.	.		.
.	.	.		.
.	.	.		.
k	x_{k1}	x_{k2}	$\cdots$	x_{kb}

Each observation x_{gj} can be written as a sum of meaningful terms by means of the identity

$$x_{gj} = \bar{x}.. + (\bar{x}_{g.} - \bar{x}..) + (\bar{x}._{j} - \bar{x}..) + (x_{gj} - \bar{x}_{g.} - \bar{x}._{j} + \bar{x}..).$$

In words, the

$$\begin{pmatrix} \text{Observed value for} \\ g\text{th treatment in} \\ j\text{th block} \end{pmatrix} = \begin{pmatrix} \text{overall} \\ \text{mean} \end{pmatrix} + \begin{pmatrix} \text{deviation} \\ \text{due to } g\text{th} \\ \text{treatment} \end{pmatrix} + \begin{pmatrix} \text{deviation} \\ \text{due to} \\ j\text{th block} \end{pmatrix} + (\text{residual}).$$

The "residual" is

$$x_{gj} - [\bar{x}.. + (\bar{x}_{g.} - \bar{x}..) + (\bar{x}._{j} - \bar{x}..)],$$

which is the difference between the observation and $\bar{x}.. + (\bar{x}_{g.} - \bar{x}..) + (\bar{x}._{j} - \bar{x}..)$, obtained by taking into account the overall mean, the effect of the gth treatment, and the effect of the jth block. Algebra shows that the corresponding decomposition is true for sums of squares:

$$\sum_{g=1}^{k} \sum_{j=1}^{b} (x_{gj} - \bar{x}..)^2 = b \sum_{g=1}^{k} (\bar{x}_{g.} - \bar{x}..)^2 + k \sum_{j=1}^{b} (\bar{x}._{j} - \bar{x}..)^2$$

$$+ \sum_{g=1}^{k} \sum_{j=1}^{b} (x_{gj} - \bar{x}_{g.} - \bar{x}._{j} + \bar{x}..)^2;$$

that is,

$$\text{SS(Total)} = \text{SS(Treatments)} + \text{SS(Blocks)} + \text{SS(Residuals)}.$$

The number of degrees of freedom of SS(Total) is $kb - 1 = n - 1$, the number of observations less 1 for the mean $\bar{x}...$ The number of degrees of freedom of SS(Treatments) is $k - 1$, the number of means $\bar{x}_g$. less 1 for the mean $\bar{x}.. = \Sigma_{g=1}^k \bar{x}_g./k$. Similarly, the number of degrees of freedom of SS(Blocks) is $b - 1$. There remain as the number of degrees of freedom for SS(Residuals)

$$kb - 1 - (k - 1) - (b - 1) = (k - 1)(b - 1).$$

(Note that the calculation corresponds to the two-way tables in Section 12.2.)

There is a hypothetical model behind the analysis. It is assumed that in repeated experiments the measurement for the gth treatment in the jth block would be the sum of a constant pertaining to the treatment, namely μ_g, a constant pertaining to the jth block, and a random "error" term with a variance of σ^2. The mean square for residuals,

$$\text{MS(Residuals)} = \frac{\text{SS(Residuals)}}{(k - 1)(b - 1)},$$

is an unbiased estimate of σ^2 regardless of whether the μ_g's differ (that is, whether there are true effects due to treatments). If there are no differences in the μ_g's,

$$\text{MS(Treatments)} = \frac{\text{SS(Treatments)}}{k - 1}$$

is an unbiased estimate of σ^2 (whether or not there are true effects due to blocks). If there are differences among the μ_g's, then MS(Treatments) will tend to be larger than σ^2. One tests H by means of

$$F = \frac{\text{MS(Treatments)}}{\text{MS(Residuals)}}.$$

When H is true, F is distributed as $F_{k-1,(k-1)(b-1)}$. One rejects H if F is sufficiently large, that is, if F exceeds $F_{k-1,(k-1)(b-1)}(\alpha)$. Table 13.18 is the analysis of variance table.

The computational formulas are

$$\text{SS(Total)} = \sum_{g=1}^{k} \sum_{j=1}^{b} x_{gj}^2 - \frac{1}{kb} \left(\sum_{g=1}^{k} \sum_{j=1}^{b} x_{gj} \right)^2,$$

$$\text{SS(Treatments)} = \frac{1}{b} \sum_{g=1}^{k} \left(\sum_{j=1}^{b} x_{gj} \right)^2 - \frac{1}{kb} \left(\sum_{g=1}^{k} \sum_{j=1}^{b} x_{gj} \right)^2,$$

TABLE 13.18
Analysis of Variance Table for Randomized Blocks Design

Source of Variation	Sum of Squares	Degrees of Freedom	Mean Square	F
Treatments	$b \sum_{g=1}^{k} (\bar{x}_{g \cdot} - \bar{x}_{\cdot \cdot})^2$	$k - 1$	MS(Treatments)	$\dfrac{\text{MS(Treatments)}}{\text{MS(Residuals)}}$
Blocks	$k \sum_{j=1}^{b} (\bar{x}_{\cdot j} - \bar{x}_{\cdot \cdot})^2$	$b - 1$	MS(Blocks)	$\dfrac{\text{MS(Blocks)}}{\text{MS(Residuals)}}$
Residuals	$\sum_{g=1}^{k} \sum_{j=1}^{b} (x_{gj} - \bar{x}_{g \cdot} - \bar{x}_{\cdot j} + \bar{x}_{\cdot \cdot})^2$	$(k - 1)(b - 1)$	MS(Residuals)	
Total	$\sum_{g=1}^{k} \sum_{j=1}^{b} (x_{gj} - \bar{x}_{\cdot \cdot})^2 = \text{SS(Total)}$	$n - 1$		

$$SS(\text{Blocks}) = \frac{1}{k} \sum_{j=1}^{b} \left(\sum_{g=1}^{k} x_{gj} \right)^2 - \frac{1}{kb} \left(\sum_{g=1}^{k} \sum_{j=1}^{b} x_{gj} \right)^2,$$

$$SS(\text{Residuals}) = SS(\text{Total}) - SS(\text{Treatments}) - SS(\text{Blocks}).$$

For the data in Table 13.16 we have

$$\sum_{i=1}^{k} \sum_{j=1}^{b} x_{gj}^2 = 6^2 + 5^2 + \cdots + 3^2 = 237,$$

$$\frac{1}{kb} \left(\sum_{i=1}^{k} \sum_{j=1}^{b} x_{gi} \right)^2 = \frac{55^2}{15} = 201.67,$$

$$SS(\text{Total}) = 237 - 201.67 = 35.33,$$

$$SS(\text{Treatments}) = \frac{24^2 + 10^2 + 21^2}{5} - 201.67$$

$$= 223.40 - 201.67 = 21.73,$$

$$SS(\text{Blocks}) = \frac{14^2 + 12^2 + 12^2 + 10^2 + 7^2}{3} - 201.67$$

$$= 211 - 201.67 = 9.33,$$

$$SS(\text{Residuals}) = 35.33 - 21.73 - 9.33 = 4.27.$$

The analysis of variance table is Table 13.19. The null hypothesis of no difference in metallic taste of types of can is rejected.

The roles of cans and persons can be interchanged. To test the hypothesis that there are no differences in scoring among persons (in the hypothetical population of repeated experiments), one uses the ratio of MS(Blocks) to MS(Residuals) and rejects the null hypothesis if that ratio is greater than an F-value for $b - 1$ and $(b - 1)(k - 1)$ degrees of freedom. The value here of 4.38 is referred to the table with 4 and 8 de-

TABLE 13.19
Analysis of Variance Table for "Metallic" Scale

SOURCE OF VARIATION	SUM OF SQUARES	DEGREES OF FREEDOM	MEAN SQUARE	F
Cans	21.73	2	10.87	20.4 ($P < .001$)
Persons	9.33	4	2.33	4.38
Residual	4.27	8	0.533	
Total	35.33	14		

SOURCE: Table 13.16

grees of freedom, for which the 5% point is 3.8379; it is barely significant.

The two-way design discussed here has the advantage that in studying the effects of treatments, the effects of blocks are balanced out. In the agricultural experiments there is no interest in the blocks. In the experiment we have discussed, however, a psychologist might be interested in variation among individuals (blocks). Variation due to either treatments or blocks, or both, can be studied in this design.

13.5.2 The Rank Test

To do a rank test for the randomized blocks design, one ranks the observations within each block. This procedure eliminates differences between blocks (or individuals), since the rank is within a block (as contrasted to the procedure of Section 13.4). The observation x_{gj} is replaced by r_{gj}, the rank of x_{gj} among $x_{1j}, x_{2j}, \ldots, x_{kj}$, the observations in the jth block. When the individuals are asked to rank the items, the r_{gj} are the raw data. The test statistic is

$$ W = \frac{12}{k(k+1)} \frac{1}{b} \sum_{g=1}^{k} \left(\sum_{j=1}^{b} r_{gj} \right)^2 - 3b(k+1). $$

[This is SS(Treatments) for ranks divided by MS(Residuals); the latter is $k(k+1)/12$, the variance of $1, \ldots, k$.] The distribution of W is approximated by χ^2_{k-1} when the null hypothesis is true. The significance test based on W was developed in 1937 by the now well-known American economist Milton Friedman. It is referred to as the *Friedman rank test*.

Consider again the data in Table 13.16. The data for each person are ranked, ties being assigned the mean of the ranks tied for, in Table 13.20. The value of W is

TABLE 13.20
Rank Scores of Three Types of Can

TYPE OF CAN			PERSON			SUMS
	Adams	Clark	Davis	Jones	Smith	
A	2.5	3	3	2.5	2.5	13.5
B	1	1	1	1	1	5.0
C	2.5	2	2	2.5	2.5	11.5

SOURCE: Table 13.17

$$W = \frac{12}{3(3+1)} \frac{1}{5} (13.5^2 + 5^2 + 11.5^2) - 3 \times 5 \times (3+1) = 7.9.$$

Referring to the table of percentage points of chi-square in Appendix IX, we find from the line for $3 - 1 = 2$ degrees of freedom that 7.9 falls between the tabled values of 7.378 and 9.210, which correspond to P-values of .025 and .01. The hypothesis of no differences between cans is rejected.

This value of W is not "corrected for ties." The correction consists of dividing the value of W by the quantity t, defined as

$$t = 1 - \frac{\Sigma_i(i-1)i(i+1)t_i}{b(k-1)k(k+1)},$$

where t_i is the number of ties involving i observations. In the example we have only two-way ties, so $t_2 = 3$ and $t = 1 - (1 \times 2 \times 3 \times 3)/(5 \times 2 \times 3 \times 4) = 17/20$. The corrected value of W is $W/t = 7.9/(17/20) = 9.3$, giving a P-value just under .01.

13.5.3 Two-Way Classification with Repeated Observations

We shall sketch the ideas of a two-way classification with several observations in each cell by means of an example. A full treatment is beyond the scope of this book.

Street and Carroll (1972) discuss the preliminary evaluation of a new food product, called H, on the basis of a consumer taste test. The aim was to compare the tastes of the liquid and solid forms of H and of a competitive product already on the market, called C. Each product was tasted by 25 men and 25 women, with a total of 150 people participating in the experiment. After tasting the product, each person marked a ballot using a pictorial scale with seven drawings of a woman's face showing a range of feelings about the product. The pictures corresponded to terms ranging from "excellent" and "very good," to "very poor" and "terrible." Each picture was assigned a score in the sequence $+3, +2, +1, 0, -1, -2, -3$, with $+3$ being most acceptable. Table 13.21 gives the mean scores of 25 persons for the 6 different pairings of sex and product. The differences in means between male tasters and female tasters is not great. The females gave smaller means than the males for Liquid H and Solid H but gave a larger mean for C. The means for Liquid H and Solid H do not differ much, but the means for C are considerably lower.

If we denote the mean score of male tasters for Product C as $\bar{x}_{MC}$ and the score of the ith male taster of Product C as x_{MCi}, then $\bar{x}_{MC} = \Sigma_{i=1}^{25} x_{MCi}/25$ and

TABLE 13.21

Data from Palatability Evaluation of Product H:
Means of Scores From 25 Tasters in Each Group

Sex	Product Tasted			Means
	C	*Liquid H*	*Solid H*	
Male	0.20	1.24	1.08	0.84
Female	0.24	1.12	1.04	0.80
Means	0.22	1.18	1.06	0.82

SOURCE: Street and Carroll (1972), p. 226.

$$s_{MC}^2 = \frac{\sum_{i=1}^{25} (x_{MCi} - \bar{x}_{MC})^2}{24}$$

is an estimate, having 24 degrees of freedom, of the variance of male tasters for Product C. Under the assumption that the population variances for all combinations of sex of tasters and product are the same, there are 5 more such estimates of the "error" variance. The average of the 6 estimates is called the "error mean square" and is recorded as 1.35 in Table 13.22. (The error sum of squares is the sum of the 6 numerators, and the degrees of freedom is the sum of the 6 denominators, namely 144.)

The sum of squares for row means is computed from Table 13.21 in the usual way and multiplied by 75 (the number of tasters) to obtain the Sex Sum of Squares. Similarly, the sum of squares for column means is

TABLE 13.22

Analysis of Variance Table for Palatability Evaluation

Source of Variation	Sum of Squares	Degrees of Freedom	Mean Square	F
Sex	0.06	1	0.06	0.04
Product	27.36	2	13.7	10.1
Interaction	0.16	2	0.08	0.06
Error	194.56	144	1.35	
Total	222.14	149		

SOURCE: Street and Carroll (1972), p. 226.

multiplied by 50 (the number of times each product was tasted) to obtain the Product Sum of Squares.

As was noted, female tasters rate C slightly higher than male tasters, but rate Liquid H and Solid H lower. A measure of these differential effects, called "interactions," is calculated in Table 13.21 as the sum of squares for residuals, defined in Section 13.5; it has to be multiplied by 25 (the number of tasters of each sex of each product).

For each source of variability there is an *F*-test based on the ratio of the mean square for that source to the error mean square. The degrees of freedom for each *F* agree with the degrees of freedom in the numerator and denominator mean squares. In Table 13.22 the one *F* of three that is significant is that for product effects; the conclusion is that there is a real difference in tasters among the products. (A finer analysis shows that this conclusion is due to the difference between C and H.)

SUMMARY

When more than two population means are to be compared, analysis of variance is used. The dispersion among the sample means, measured by their variance (called a "mean square" in the context of analysis of variance), is compared with the dispersion within groups. If the former variance is significantly greater than the latter, one rejects the hypothesis of equality of population means. The *F*-table provides critical values for testing whether the ratio of two variances can be attributed to chance.

The analysis of variance is extended to data classified on the basis of two factors. The analysis of variance of ranks (ordinal data) is presented.

APPENDIX 13A   Formulas Relating to Sums of Squares

FORMULA 1   [Computational formula for SS(Total) in one-way analysis of variance]

$$SS(Total) = \sum_{g=1}^{k} \sum_{i=1}^{n_g} x_{gi}^2 - \frac{\left(\sum_{g=1}^{k} \sum_{i=1}^{n_g} x_{gi} \right)^2}{n}.$$

PROOF: The proof is along the lines of the proof of Rule 8 in Appendix 4C. By definition,

$$\text{SS(Total)} = \sum_{g=1}^{k} \sum_{i=1}^{n_g} (x_{gi} - \bar{x})^2$$

$$= \sum_{g=1}^{k} \sum_{i=1}^{n_g} x_{gi}^2 - 2\bar{x} \sum_{g=1}^{k} \sum_{i=1}^{n_g} x_{gi} + n\bar{x}^2$$

$$= \sum_{g=1}^{k} \sum_{i=1}^{n_g} x_{gi}^2 - 2 \frac{\displaystyle\sum_{g=1}^{k} \sum_{i=1}^{n_g} x_{gi}}{n} \sum_{g=1}^{k} \sum_{i=1}^{n_g} x_{gi}$$

$$+ n \left(\frac{\displaystyle\sum_{g=1}^{k} \sum_{i=1}^{n_g} x_{gi}}{n} \right)^2$$

$$= \sum_{g=1}^{k} \sum_{i=1}^{n_g} x_{gi}^2 - \frac{\left(\displaystyle\sum_{g=1}^{k} \sum_{i=1}^{n_g} x_{gi} \right)^2}{n}.$$

FORMULA 2   [Computational formula for SS(Between Groups) in one-way analysis of variance]

$$\text{SS(Between Groups)} = \sum_{g=1}^{k} \frac{\left(\displaystyle\sum_{i=1}^{n_g} x_{gi} \right)^2}{n_g} - \frac{1}{n} \left(\sum_{g=1}^{k} \sum_{i=1}^{n_g} x_{gi} \right)^2.$$

PROOF: By definition,

$$\text{SS(Between Groups)} = \sum_{g=1}^{k} n_g (\bar{x}_g - \bar{x})^2$$

$$= \sum_{g=1}^{k} n_g \bar{x}_g^2 - 2\bar{x} \sum_{g=1}^{k} n_g \bar{x}_g + n\bar{x}^2$$

$$= \sum_{g=1}^{k} n_g \bar{x}_g^2 - n\bar{x}^2$$

$$= \sum_{g=1}^{k} n_g \left(\frac{\displaystyle\sum_{i=1}^{n_g} x_{gi}}{n_g} \right)^2 - n \left(\frac{\displaystyle\sum_{g=1}^{k} \sum_{i=1}^{n_g} x_{gi}}{n} \right)^2$$

since $n = \sum_{g=1}^{k} n_g$ and $\bar{x} = \sum_{g=1}^{k} n_g \bar{x}_g / n$.

FORMULA 3   [Decomposition of SS(Total)]

SS(Total) = SS(Between Groups) + SS(Within Groups),

that is,

$$\sum_{g=1}^{k} \sum_{i=1}^{n_g} (x_{gi} - \bar{x})^2 = \sum_{g=1}^{k} n_g(\bar{x}_g - \bar{x})^2 + \sum_{g=1}^{k} \sum_{i=1}^{n_g} (x_{gi} - \bar{x}_g)^2.$$

PROOF: We first write

$$(x_{gi} - \bar{x}) = (\bar{x}_g - \bar{x}) + (x_{gi} - \bar{x}_g);$$

all we have done here is add and subtract $\bar{x}_g$ and rearrange items. Squaring, we obtain

$$(x_{gi} - \bar{x})^2 = (\bar{x}_g - \bar{x})^2 + (x_{gi} - \bar{x}_g)^2 + 2(\bar{x}_g - \bar{x})(x_{gi} - \bar{x}_g).$$

Summing over i, $i = 1, \ldots, n_g$, we obtain

$$\sum_{i=1}^{n_g} (x_{gi} - \bar{x})^2 = \sum_{i=1}^{n_g} (\bar{x}_g - \bar{x})^2 + \sum_{i=1}^{n_g} (x_{gi} - \bar{x}_g)^2$$

$$+ 2(\bar{x}_g - \bar{x}) \sum_{i=1}^{n_g} (x_{gi} - \bar{x}_g)$$

$$= n_g(\bar{x}_g - \bar{x})^2 + \sum_{i=1}^{n_g} (x_{gi} - \bar{x}_g)^2 + 0,$$

the cross-product term being 0 because the sum of deviations $\sum_{i=1}^{n_g}(x_{gi} - \bar{x}_g)$ is 0. When the above terms are added over g, $g = 1, \ldots, k$, Formula 3 results.

FORMULA 4   [Expected mean square between groups] The population mean of the sampling distribution of s_B^2 is

$$\sigma^2 + \frac{1}{k - 1} \sum_{g=1}^{k} n_g(\mu_g - \bar{\mu})^2,$$

where $\bar{\mu} = \sum_{g=1}^{k} n_g \mu_g / n$. The population mean of s_B^2 is called the "expected mean square" between groups.

PROOF: Any observation x_{gi} can be written as

$$x_{gi} = \mu_g + e_{gi},$$

where each e_{gi} is a random error with mean 0 and variance σ^2 and all the e's are statistically independent. Defining $\bar{e}_g$ and $\bar{e}$ in a manner analogous to $\bar{x}_g$ and $\bar{x}$, we have

$$\bar{x}_g - \bar{x} = (\mu_g - \bar{\mu}) + (\bar{e}_g - \bar{e}).$$

Squaring and summing, we obtain

$$\text{SS(Between Groups)} = \sum_{g=1}^{k} n_g(\bar{x}_g - \bar{x})^2$$

$$= \sum_{g=1}^{k} n_g(\mu_g - \bar{\mu})^2 + \sum_{g=1}^{k} n_g(\bar{e}_g - \bar{e})^2 + 2\sum_{g=1}^{k} n_g(\mu_g - \bar{\mu})(\bar{e}_g - \bar{e}).$$

Thus,

$$E[\text{SS(Between Groups)}] = \sum_{g=1}^{k} n_g(\mu_g - \bar{\mu})^2 + \sum_{g=1}^{k} n_g E[(\bar{e}_g - \bar{e})^2],$$

since $E[\bar{e}_g - \bar{e}] = 0 - 0 = 0$. Straightforward algebra shows that $E[(\bar{e}_g - \bar{e})^2] = \sigma^2(1/n_g - 1/n)$, and summing this expression over g, one sees that the second term on the right-hand side is $(k - 1)\sigma^2$. Thus

$$E[s_B^2] = \frac{E[\text{SS(Between Groups)}]}{k - 1} = \frac{1}{k - 1}\sum_{g=1}^{k} n_g(\mu_g - \bar{\mu})^2 + \sigma^2.$$

APPENDIX 13B   Relation between Student's t and the F-Ratio

If we had only 2 groups ($k = 2$), then we would be testing the hypothesis $H: \mu_1 = \mu_2$. The test statistic is Student's t,

$$t = \frac{\bar{x}_1 - \bar{x}_2}{s\sqrt{\dfrac{1}{n_1} + \dfrac{1}{n_2}}},$$

where

$$s^2 = \frac{1}{n_1 + n_2 - 2}\left[\sum_{i=1}^{n_1} (x_{1i} - \bar{x}_1)^2 + \sum_{i=1}^{n_2} (x_{2i} - \bar{x}_2)^2\right],$$

the number of degrees of freedom being $n_1 + n_2 - 2$. We now show that $t^2 = F_{1,n_1+n_2-2}$. First note that $s^2 = s_W^2$. Also, we have

$$\text{SS(Between Groups)} = \sum_{g=1}^{2} n_g(\bar{x}_g - \bar{x})^2 = n_1\left(\bar{x}_1 - \frac{n_1\bar{x}_1 + n_2\bar{x}_2}{n}\right)^2$$

$$+ n_2\left(\bar{x}_2 - \frac{n_1\bar{x}_1 + n_2\bar{x}_2}{n}\right)^2$$

$$= n_1 \left(\frac{n\bar{x}_1 - n_1\bar{x}_1 - n_2\bar{x}_2}{n} \right)^2$$

$$+ n_2 \left(\frac{n\bar{x}_2 - n_1\bar{x}_1 - n_2\bar{x}_2}{n} \right)^2$$

$$= n_1 \left[\frac{n_2(\bar{x}_1 - \bar{x}_2)}{n} \right]^2 + n_2 \left[\frac{n_1(\bar{x}_2 - \bar{x}_1)}{n} \right]^2$$

$$= \frac{n_1 n_2^2 + n_1^2 n_2}{n^2} (\bar{x}_1 - \bar{x}_2)^2$$

$$= \frac{n_1 n_2}{n} (\bar{x}_1 - \bar{x}_2)^2 = \frac{n_1 n_2}{n_1 + n_2} (\bar{x}_1 - \bar{x}_2)^2$$

$$= \frac{(\bar{x}_1 - \bar{x}_2)^2}{\dfrac{1}{n_1} + \dfrac{1}{n_2}}.$$

Thus,

$$F = \frac{s_B^2}{s_W^2} = \frac{(\bar{x}_1 - \bar{x}_2)^2}{\left(\dfrac{1}{n_1} + \dfrac{1}{n_2} \right) s^2} = t^2.$$

APPENDIX 13C More on Multiple Comparisons and Confidence Regions

Confidence Regions Corresponding to an F-Test.   When $\bar{x}_g$ is the mean of a random sample of n_g observations from the normal distribution with mean μ_g and variance σ^2, then the sampling distribution of $\bar{x}_g - \mu_g$ is normal with mean 0 and variance σ^2, $g = 1, \ldots, k$. Thus, if $\bar{\mu} = \Sigma_{g=1}^k n_g \mu_g / n$,

$$\sum_{g=1}^k n_g [(\bar{x}_g - \mu_g) - (\bar{x} - \bar{\mu})]^2 / \sigma^2 = \sum_{g=1}^k n_g [(\bar{x}_g - \bar{x}) - (\mu_g - \bar{\mu})]^2 / \sigma^2$$

has a χ^2-distribution with $k - 1$ degrees of freedom, and

$$\frac{\sum_{g=1}^k n_g [(\bar{x}_g - \bar{x}) - (\mu_g - \bar{\mu})]^2}{(k - 1)s_W^2}$$

has an F-distribution with $k - 1$ and $n - k$ degrees of freedom. A confidence region for $\mu_1, \ldots, \mu_k$ based on observed $\bar{x}_1, \ldots, \bar{x}_k$ and s_W^2 at confidence level $1 - \alpha$ is

(13.1) $\sum_{g=1}^{k} n_g[(\mu_g - \bar{\mu}) - (\bar{x}_g - \bar{x})]^2 \leq (k-1)s_W^2 F_{k-1,n-k}(\alpha).$

A way of using this method is to try out sets of values of $\mu_1, \ldots, \mu_k$ (computing the corresponding $\bar{\mu}$) and see whether they satisfy the inequality.

Multiple Comparisons.   The above procedure is difficult to apply, but a more useful method can be derived. It follows from the fact that the correlation r lies between -1 and 1 (Appendix 14C) that

$$\left| \sum_{j=1}^{k} a_j b_j \right| \leq \sqrt{\sum_{j=1}^{k} a_j^2} \sqrt{\sum_{j=1}^{k} b_j^2}.$$

Now let $a_j = c_j/\sqrt{n_j}$ and $b_j = \sqrt{n_j}[(\mu_j - \bar{\mu}) - (\bar{x}_j - \bar{x})]$. Then (13.1) implies

$$\left| \sum_{j=1}^{k} c_j[(\mu_j - \bar{\mu}) - (\bar{x}_j - \bar{x})] \right| \leq \sqrt{\sum_{j=1}^{k} \frac{c_j^2}{n_j}} \sqrt{k-1} \, s_W \sqrt{F_{k-1,n-k}(\alpha)}.$$

If $\sum_{j=1}^{k} c_j = 0$, then the left-hand side of the above expression is

$$\left| \sum_{j=1}^{k} c_j(\mu_j - \bar{x}_j) \right|.$$

In this case

(13.2) $\sum_{j=1}^{k} c_j \bar{x}_j - \sqrt{\sum_{j=1}^{k} \frac{c_j^2}{n_j}} \sqrt{k-1} s_W \sqrt{F_{k-1,n-k}(\alpha)} \leq \sum_{j=1}^{k} c_j \mu_j$

$$\leq \sum_{j=1}^{k} c_j \bar{x}_j + \sqrt{\sum_{j=1}^{k} \frac{c_j^2}{n_j}} \sqrt{k-1} s_W \sqrt{F_{k-1,n-k}(\alpha)}.$$

Given observed $\bar{x}_1, \ldots, \bar{x}_k$ and s_W^2, one can assert with confidence $1 - \alpha$ that (13.2) holds simultaneously for all sets $c_1, c_2, \ldots, c_k$ satisfying $\sum_{j=1}^{k} c_j = 0$. In particular, it holds for $c_g = 1$, $c_h = -1$, and $c_j = 0$, $j \neq g$, $j \neq h$. Then (13.2) is

$$\bar{x}_g - \bar{x}_h - \sqrt{\frac{1}{n_g} + \frac{1}{n_h}} \sqrt{k-1} s_W \sqrt{F_{k-1,n-k}(\alpha)} \leq \mu_g - \mu_h$$

$$\leq \bar{x}_g - \bar{x}_h + \sqrt{\frac{1}{n_g} + \frac{1}{n_h}} \sqrt{k-1} s_W \sqrt{F_{k-1,n-k}(\alpha)}.$$

These inequalities are similar to those in Section 13.4, but $t_{n-k}(\alpha^*/2)$ for $\alpha^* = \alpha/m$ is replaced by $\sqrt{(k-1)}F_{k-1,n-k}(\alpha)$.

As an illustration, consider Example 1 (Tables 13.1 and 13.7), which was treated in Section 13.3. Here $k = 3$; $n_g = 6$; $g = 1, 2, 3$; $\bar{x}_1 = 4$; $\bar{x}_2 = 9$; $\bar{x}_3 = 5$; and $s_W = 1.366$. For $\alpha = .03$, $F_{2,15}(.03) = 4.546$, and the term to be added to and subtracted from $\bar{x}_g - \bar{x}_h$ is $\sqrt{1/6 + 1/6} \times \sqrt{2 \times 4.546} \times 1.366 = 2.38$. The intervals are

$$2.62 \le \mu_2 - \mu_1 \le 7.38,$$

$$-1.38 \le \mu_3 - \mu_1 \le 3.38,$$

$$1.62 \le \mu_2 - \mu_3 \le 6.38.$$

Note that $\mu_3 - \mu_1 = 0$ is included, but $\mu_2 - \mu_1 = 0$ and $\mu_2 - \mu_3 = 0$ are excluded (in agreement with Section 13.4).

Probability Inequality. In Section 13.3 the procedure depended on

$$\Pr\left(|t_i| \ge t(\alpha^*/2) \text{ for some } i = 1, \ldots, m\right)$$

$$\le m\Pr\left(|t_i| \ge t(\alpha^*/2) \text{ for any given } i\right).$$

Note that the right-hand side does not depend on i. Let A_i be the event $|t_i| \ge t(\alpha^*/2)$, $i = 1, \ldots, m$. Then the inequality is

$$\Pr\left(A_1 \text{ or } A_2 \text{ or } \ldots \text{ or } A_m\right) \le \Pr\left(A_1\right) + \Pr\left(A_2\right) + \cdots + \Pr\left(A_m\right).$$

In case $m = 2$ the inequality follows directly from Section 6.4.3. We shall treat the case $m = 3$ by considering the event

$$A_1 \text{ or } A_2 \text{ or } A_3 = A_1 \text{ or } [\bar{A}_1 \text{ and } (A_2 \text{ or } A_3)]$$

$$= A_1 \text{ or } (\bar{A}_1 \text{ and } A_2) \text{ or } (\bar{A}_1 \text{ and } \bar{A}_2 \text{ and } A_3).$$

The three sets on the right-hand side are mutually exclusive and exhaustive of A_1 or A_2 or A_3. Hence,

$$\Pr\left(A_1 \text{ or } A_2 \text{ or } A_3\right) = \Pr\left(A_1\right) + \Pr\left(\bar{A}_1 \text{ and } \bar{A}_2\right) + \Pr\left(\bar{A}_1 \text{ and } \bar{A}_2 \text{ and } A_3\right).$$

From $\Pr\left(\bar{A}_1 \text{ and } A_2\right) \le \Pr\left(A_2\right)$ and $\Pr\left(\bar{A}_1 \text{ and } \bar{A}_2 \text{ and } A_3\right) \le \Pr\left(A_3\right)$, the desired inequality follows.

APPENDIX 13D   Randomization (Permutation) Tests

In Appendix 10B we discussed randomization tests for the locations of two populations. Here we consider randomization tests for the locations of several populations. For definiteness suppose there are three populations, with sample sizes n_1, n_2, and n_3.

Suppose we take SS(Between Groups) as a test statistic. We want to know whether the observed value of this statistic is large enough to reject the null hypothesis that the location parameters are equal. The answer involves consideration of an appropriate sampling distribution. To make a permutation test, one generates this sampling distribution from the data themselves.

Suppose the numerical values of all $n_1 + n_2 + n_3$ observations were written on slips of paper and put into a bowl. Then n_1 were drawn out, and then n_2 from among the remaining observations. This would give us fictitious samples of sizes n_1, n_2, and n_3. There are many possible such fictitious samples. For each we compute the value of the statistic SS(Between Groups), generating a sampling distribution of this statistic. If the observed value is larger than 95% of the fictitious values, then at the 5% level the hypothesis of no differences in location would be rejected.

There are usually many possible permutations. To conduct a test it is not always necessary to consider all possible permutations.

(1) It is sometimes easy to determine that the observed value is in the upper $100\alpha\%$.

(2) It is sometimes easy to determine that the observed value is *not* in the upper $100\alpha\%$.

(3) One can sample randomly from the population of all possible permutations and use the distribution of the test statistic based on this sampling of permutations.

EXERCISES

13.1 (Sec. 13.1.1)   In order to compare three brands of beer, 15 people were randomly divided into three groups of 5 each. Each person scored the beer tasted with a number (integer) between 0 and 6, 6 indicating excellent taste and 0 meaning extremely bad taste. The data are given in Table 13.23. Test at the 5% level the hypothesis of equality of the three corresponding population means.

13.2 (Sec. 13.1.1)   Professor X can drive to the university along four different routes. Table 13.24 shows the numbers of minutes needed to

TABLE 13.23
Scores of Three Brands of Beer on Taste Scale

Beer A		Beer B		Beer C	
Adams	6	Abrams	2	Carter	6
Clark	5	Brown	3	Edwards	4
Davis	6	Garcia	2	Flynn	4
Jones	4	Murphy	2	Gross	4
Smith	3	Thomas	1	Peters	3

(hypothetical data)

make the trip on seven different occasions for each route. Is there reason to think that in the long run it will matter which route Professor X chooses? Test at the 5% level the hypothesis that there is no difference in the true mean time it takes the professor to drive to the university along the four different routes.

TABLE 13.24
Times (Minutes) by Four Different Routes

	Sum	Mean
Route 1: 39, 38, 43, 37, 33, 40, 36	266	38
Route 2: 34, 41, 44, 45, 35, 29, 31	259	37
Route 3: 32, 46, 47, 42, 48, 51, 49	315	45
Route 4: 53, 55, 54, 52, 58, 50, 56	378	54

(hypothetical data)

13.3 (Sec. 13.1.2) Test the hypothesis that there is no difference in the true mean number of trials in the populations from which the samples in Table 13.25 were drawn.

TABLE 13.25
Number of Trials before Perfect Recall of a Series of Words
(Serial Presentation)

English majors:	2, 3, 2, 1
Math majors:	1, 5, 3, 3
Psychology majors:	3, 5

(hypothetical data)

13.4 (See Example 3 of the Introduction.) Figure 13.3 gives the distribution of the number of academic freedom incidents by type of control

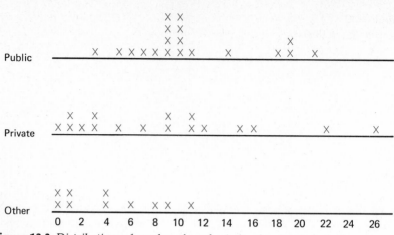

Figure 13.3 *Distributions of number of academic freedom incidents in three types of colleges (Source: Lazarsfeld and Thielens (1958))*

of college. Only 46 "moderately large" schools (student bodies of 2500 to 9000) are included to improve comparability. Table 13.26 gives some statistics derived from the above distributions. These are to be used in making a test of the hypothesis that the mean number of incidents is the same for the three populations from which the schools were drawn.

(a) Compute s_B^2 and s_W^2.

(b) Test at the 1% level the hypothesis that the population means are identical.

(c) What conclusion do you draw?

TABLE 13.26

Summary Statistics for Distributions of Number of Academic Freedom Incidents in Three Types of College

TYPE OF CONTROL	SUM	SUM OF SQUARES	NUMBER OF COLLEGES
Public	207	2350	19
Private	153	2287	17
Other	44	336	10

SOURCE: Figure 13.3

13.5 (Sec. 13.4)   The operating costs of three makes of cars, nine of each make, are given in Table 13.27. Use the analysis of variance of ranks to test the null hypothesis of no differences at the 5% level.

13.6 (Sec. 13.4)   Use the analysis of variance of ranks to make a 5%-level test of equality of locations for the data of Exercise 13.2.

TABLE 13.27

Operating Costs per Mile of Three Makes of Automobiles

	Cost (Cents per Mile)	
C	F	P
7.85	9.40	4.86
6.90	8.30	5.96
4.00	9.10	6.08
4.56	6.62	9.88
6.98	4.26	6.30
8.50	9.37	4.92
4.76	5.36	6.68
6.04	4.72	4.77
6.52	7.86	4.54

(hypothetical data)

13.7 (Sec. 13.5.1)   Consider the "data" of Table 13.28.

(a) At the 5% level, make the F-test of equality of population (treatment) means.

(b) What is the value of SS(Residual) if the 7 is changed to a 6?

TABLE 13.28

Data in a Randomized Blocks Design

Treatment	Block			
	1	2	3	4
1	1	2	3	4
2	2	3	4	5
3	3	4	5	7

(hypothetical data)

13.8 (Sec. 13.5.1)   At the 5% level make the F-test of equality of population (treatment) means for the "data" in Table 13.29.

TABLE 13.29

Data in a Randomized Blocks Design

Treatment	Block		
	1	2	3
1	1	4	9
2	4	9	16
3	9	16	25

(hypothetical data)

13.9 (Sec. 13.5.2)   Four subjects were asked to rank each of three beers. Compare the beers, using the results in Table 13.30.

TABLE 13.30
Beer-Ranking Experiment

BEER	SUBJECT			
	1	2	3	4
A	1	1	1	1
B	2	3	2	3
C	3	2	3	2

(hypothetical data)

13.10 (Sec. 13.5.2)   Five subjects were asked to rank each of four paintings. Compare their preferences, using the rankings given in Table 13.31.

TABLE 13.31
Picture-Ranking Experiment

PAINTING	SUBJECT				
	1	2	3	4	5
A	1	1	2	3	2
B	2	2	1	1	1
C	4	3	3	2	4
D	3	4	4	4	3

(hypothetical data)

*****13.11** (Sec. 13.5.3)   Table 13.32 gives the results of two repetitions of the beer-can assessment discussed in Section 13.5. Compute F-statistics for persons, cans, and interaction. Locate the achieved levels of significance.

TABLE 13.32
Scores of Three Types of Can on Two "Metallic Scale" Replicates

CAN	PERSONS				
	Adams	*Clark*	*Davis*	*Jones*	*Smith*
A	6, 6	4, 6	6, 6	5, 3	2, 4
B	1, 3	2, 4	3, 1	1, 3	0, 2
C	6, 6	3, 5	4, 4	3, 5	2, 4

(hypothetical data)

*13.12 (Sec. 13.5.3)   Table 13.33 gives the weights of sets of three to-matoes grown under four different conditions (in all, a dozen tomatoes were weighed). Compute *F*-statistics for nitrogen, phosphate, and interaction. Locate the achieved levels of significance.

TABLE 13.33

Weights of 12 Tomatoes (ounces)

Nitrogen	Phosphate	
	Absent	*Present*
Absent	5, 4, 6	9, 7, 11
Present	6, 7, 8	15, 12, 18

(hypothetical data)

MATHEMATICAL EXERCISES

13.13 (Sec. 13.3)   It is desired to make *m* comparisons. Suppose that all population means are equal, so that there are no real differences. Let A_i, $i = 1, \ldots, m$, be the event that the *i*th comparison is significant. The overall Type I error probability is the probability that at least one A_i occurs. Show that if each individual test is made at level α/m, this probability is $\leq \alpha$.

13.14 (Sec. 13.4)   Show that the Kruskal-Wallis test is equivalent to an *F*-test replacing measurements by ranks. [*Hint:* The Kruskal-Wallis test statistic is analogous to SS(Between Groups)/SS(Total). Use SS(Total) = SS(Between Groups) + SS(Within Groups) and simplify the inequality defining the rejection procedure.]

13.15 (Sec. 13.1)   When using a pretreatment/posttreatment design, one usually needs to check for uniformity of experimental groups with respect to pretest scores. Then the emphasis is not so much on whether one accepts or rejects the hypothesis of equality of population means, but rather on determining whether these means are really close together. One approach to this problem is to construct a confidence set for the population means. Such a set is the set consisting of all values $\mu_1, \mu_2, \ldots, \mu_k$ sufficiently close to the sample means $\bar{x}_1, \bar{x}_2, \ldots, \bar{x}_k$ in the sense that they satisfy the inequality

$$\frac{r[(\bar{x}_1 - \mu_1)^2 + (\bar{x}_2 - \mu_2)^2 + \cdots + (\bar{x}_k - \mu_k)^2]/k}{s_W^2} < F_{k,k(r-1)}(\alpha),$$

where $1 - \alpha$ is the confidence coefficient. Suppose that $k = 3$, $r = 6$, $\bar{x}_1 = 5$, $\bar{x}_2 = 6$, $\bar{x}_3 = 7$, and $s_W^2 = 10$. Take $\alpha = .10$.

(a) Is the point $(\mu_1, \mu_2, \mu_3) = (5, 5, 9)$ in the confidence set?
(b) Is the point $(\mu_1, \mu_2, \mu_3) = (5, 7, 6)$ in the confidence set?

REFERENCES

Friedman, M. (1937). "The use of ranks to avoid the assumption of normality in the analysis of variance." *Journal of the American Statistical Association* 32: 675–701.

Kruskal, W. H., and W. Allen Wallis (1952). "Use of ranks in one-criterion analysis of variance." *Journal of the American Statistical Association* 47: 583–621.

Lazarsfeld, Paul F., and Wagner Thielens (1958). *The Academic Mind.* Free Press, Glencoe, Ill. Copyright © 1958 by The Free Press (Macmillan).

Snedecor, G. W., and W. G. Cochran (1967). *Statistical Methods.* 6th ed., Iowa State University Press, Ames.

Street, Elisabeth, and Mavis B. Carroll (1972). "Preliminary evaluation of a new food product." In *Statistics: A Guide to the Unknown,* Judith M. Tanur *et al.* (eds.), Holden-Day, San Francisco.

14 SIMPLE REGRESSION AND CORRELATION

INTRODUCTION

In this chapter we study the statistical relationship between two quantitative variables, just as in Chapters 5 and 12 we studied relationships between two qualitative variables. In many instances one variable may have a direct effect on the other or may be used to predict the other. Age may affect blood pressure; rainfall influences crop yield; SAT scores may predict college grade averages; and parents' heights may predict offspring's heights. In a statistical relationship, when the first variable is used to predict or explain the determination of the second, there is some variation of the second variable around the determination from the first. First we shall develop ways of expressing statistical relationships, then methods for estimating them from observations on the pairs of variables. We shall also measure and interpret variability around the predicted or "explained" values.

In Sections 5.1 and 12.3 we considered measures of association between two categorical variables. In this chapter we develop further the correlation coefficient, the measure of association between two quantitative variables introduced in Section 5.5. It is a measure of relationship that is symmetric in the two variables.

14.1 FUNCTIONAL RELATIONSHIP

A variable y is said to be a *function* of a variable x if to any value of x there corresponds one and only one value of y; that is, if we know that $x = x_0$, a specified numerical value of x, then we know that $y = y_0$, a specific value of y. We symbolize a functional relationship by writing $y = f(x)$, where f represents the function. For example, if f stands for the function "squaring," then $y = f(x)$ is $y = x^2$. Table 14.1 gives five values of x and y when y is x^2. The variable x is called the *independent variable;*

TABLE 14.1
Some Values of the Function $y = x^2$

x	-1	0	$\frac{1}{2}$	1	2
y	1	0	$\frac{1}{4}$	1	4

the variable y is called the *dependent variable* because it is considered to depend on x. If x is the height from which a ball is dropped and y is the time the ball takes to fall to the ground, then y is functionally related to x because the law of gravity determines y in terms of x.

Figure 14.1 is a graph of a functional relationship. To the value x_0 there corresponds a value of y, which is labeled y_0.

A *linear relationship* is a functional relationship of the form $y = a + bx$, where a and b are numbers. For example, $y = 3 + 2x$ and $y = 23.7 + 37.4x$ are linear relationships. Figure 14.2 is the graph of a linear functional relationship; it is a straight line. The number a is called the *intercept* because it is the height at which the line intercepts the y axis, that is, the value of y when $x = 0$. Table 14.2 is a table of four values of x and y when y is linearly related to x. The variable y increases by b as x increases by 1. An example of a linear relationship is that between Centigrade and Fahrenheit temperatures, $F = 32 + 1.8C$.

A relationship that is not linear, but is nevertheless smooth, can be *approximated* as a straight line, as in Figure 14.3. Although the real rela-

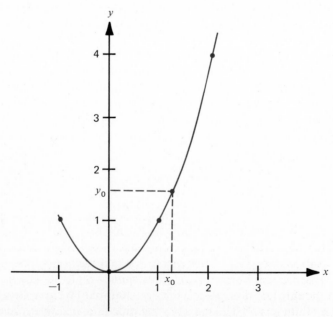

Figure 14.1 *Graph of the functional relationship* $y = x^2$

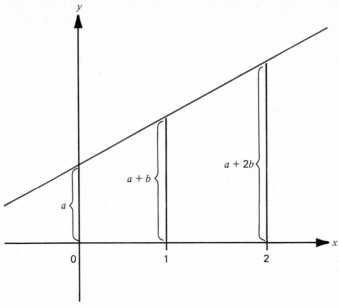

Figure 14.2 *Graph of the linear relationship* $y = a + bx$

tionship is not linear, the linear description may be adequate over a particular interval of interest.

TABLE 14.2
Some Values of a Linear Function

x	y
0	a
0.5	$a + 0.5b$
1	$a + b$
2	$a + 2b$

14.2 STATISTICAL RELATIONSHIP

Often the relationship between x and y is not an exact, mathematical relationship, but rather several y values corresponding to a given x value scatter about a value that depends on the x value. For example, although not all persons of the same height have exactly the same weight, their weights bear some relation to that height. On the average, people who are 6 feet tall are heavier than those who are 5 feet

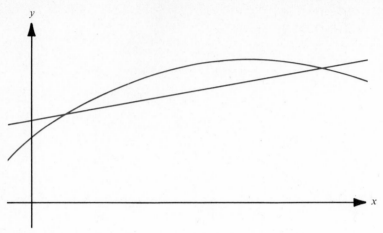

Figure 14.3 *A curve approximated by a straight line*

tall; the mean weight in the population of 6-footers exceeds the mean weight in the population of 5-footers.

The model is as follows: For every value of x, there is a corresponding population of y values. The population mean of y corresponding to x is denoted by $\mu(x)$ and is called the *regression function*. Often it is a reasonable working assumption to take this function to be a linear function of x, a function of the form $\mu(x) = \alpha + \beta x$. The quantities α and β are unknown parameters, to be estimated from observations on pairs (x, y). The variable x may be uncontrolled, as in the case of observing a sample of n individuals with their heights x and weights y, or it may be controlled, as in a designed experiment in which persons are trained as typists for different lengths of time x, and one measures the corresponding values of y, words per minute. Often it is reasonable to assume that the population of y values at an x value has variance σ^2 for all values of x. (This property is called *homoscedasticity*.)

The statistical problems connected with the model have to do with inferences about α, β, and σ^2. One would like to be able to estimate these parameters and test hypotheses about them. Of special interest is the hypothesis $H: \beta = 0$, for, if it is true, then $\mu(x) = \alpha + 0 \times x = \alpha$, the same number for all values of x, and this means that the mean value of y does not depend on x.

14.3 LEAST-SQUARES ESTIMATES

The data consist of n pairs of numbers $(x_1, y_1), (x_2, y_2), \ldots, (x_n, y_n)$; they can be plotted as a *scatter plot*, such as Figure 14.4 (or Figure

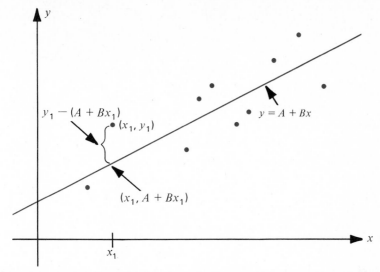

Figure 14.4 *Scatter plot. Deviation from the line $y = A + Bx$ is shown for (x_1, y_1).*

5.11). We shall estimate the parameters α and β by the *method of least squares*.

We want to find a line—that is, a formula of the type $\hat{y}_i = A + Bx_i$—that will give values $\hat{y}_i$ that are as close as possible to the data values y_i. Suppose that trial values A and B have been chosen and the corresponding estimates $\hat{y}_1, \hat{y}_2, \ldots, \hat{y}_n$ of the values $y_1, y_2, \ldots, y_n$ have been computed. To decide whether the values A and B are good choices, we determine the distance between the $\hat{y}$'s and the y's. The differences $y_i - \hat{y}_i$ are squared and summed, giving the criterion

$$\sum_{i=1}^{n} (y_i - \hat{y}_i)^2.$$

Since $\hat{y}_i = A + Bx_i$, this is

$$\sum_{i=1}^{n} [y_i - (A + Bx_i)]^2.$$

The *principle of least squares* says to take as the estimates of α and β the values of A and B that minimize this criterion. The quantity $y_i - (A + Bx_i)$ is the vertical distance from the line $y = A + Bx$ to the point (x_i, y_i). (See Figure 14.4.) The criterion is the sum of the squares of these vertical distances. The least-squares estimates, which will be denoted by a and b, are the values of A and B that minimize the criterion. It is shown in Appendix 14A that these values are

(14.1)
$$b = \frac{s_{xy}}{s_x^2}$$

and

(14.2)
$$a = \bar{y} - b\bar{x},$$

where

$$s_{xy} = \frac{\sum\limits_{i=1}^{n} (x_i - \bar{x})(y_i - \bar{y})}{n - 1}$$

is the covariance between x and y (see Section 5.5.4) and s_x^2 is the variance of x. The statistics a and b are the *least-squares estimates*.

The *estimated regression line*, the estimate of the function $\mu(x)$, is $\hat{y} = a + bx$. It is interesting to write this using $\bar{y}$ instead of a: $\hat{y} = a + bx = (\bar{y} - b\bar{x}) + bx = \bar{y} + b(x - \bar{x})$. The predicted value of y_i is thus

$$\hat{y}_i = \bar{y} + b(x_i - \bar{x}).$$

Note that the point $(\bar{x}, \bar{y})$ is on the regression line. That is, when $x = \bar{x}$, then $\hat{y} = \bar{y}$.

The vertical distance from the data point (x_i, y_i) to the line $\hat{y} = \bar{y} + b(x - \bar{x})$ is $y_i - [\bar{y} + b(x_i - \bar{x})]$. (See Figure 14.4.) The estimate of the variance σ^2 depends on these distances and is

$$s_{y\cdot x}^2 = \frac{\sum\limits_{i=1}^{n} (y_i - \hat{y}_i)^2}{n - 2}.$$

[The conventional symbol $s_{y\cdot x}^2$, which is the estimate of variance, should not be confused with s_{xy}, which is the covariance between x and y.] The numerator of $s_{y\cdot x}^2$ is

$$\sum_{i=1}^{n} [y_i - \bar{y} - b(x_i - \bar{x})]^2 = \sum_{i=1}^{n} [(y_i - \bar{y})^2 - 2b(y_i - \bar{y})(x_i - \bar{x})$$

$$+ b^2(x_i - \bar{x})^2]$$

$$= (n - 1)(s_y^2 - 2bs_{xy} + b^2 s_x^2)$$

$$= (n - 1)\left[s_y^2 - 2\frac{s_{xy}^2}{s_x^2} + \left(\frac{s_{xy}}{s_x^2}\right)^2 s_x^2\right]$$

$$= (n - 1) \left(s_y^2 - \frac{s_{xy}^2}{s_x^2} \right),$$

where $(n - 1)s_y^2 = \Sigma_{i=1}^n (y_i - \bar{y})^2$; we have used the definition of b in (14.1). The estimate of σ^2 can be written

(14.3)
$$s_{y \cdot x}^2 = \left(s_y^2 - \frac{s_{xy}^2}{s_x^2} \right) \frac{n - 1}{n - 2}.$$

The divisor $n - 2$ is the number of degrees of freedom, deduced as the number of observations minus the number of coefficients estimated. Appendix 14B gives some other computational formulas for the *residual sum of squares*, $\Sigma_{i=1}^n (y_i - \hat{y}_i)^2$, the numerator of $s_{y \cdot x}^2$. Computations will be illustrated in the following example.

EXAMPLE   The energy crisis has raised many serious questions about energy policies, such as whether nuclear reactors should play a more prominent role in producing power. It is necessary to investigate the possibility that radioactive contamination poses health hazards. Since World War II, plutonium has been produced at the Hanford, Washington, facility of the Atomic Energy Commission. Over the years, appreciable quantities of radioactive wastes have leaked from their open-pit storage areas into the nearby Columbia River, which flows through parts of Oregon to the Pacific. As part of the assessment of the conse-

TABLE 14.3
Radioactive Contamination and Cancer Mortality

COUNTY	INDEX OF EXPOSURE	CANCER MORTALITY PER 100,000 PERSON-YEARS
Clatsop	8.34	210.3
Columbia	6.41	177.9
Gilliam	3.41	129.9
Hood River	3.83	162.3
Morrow	2.57	130.1
Portland	11.64	207.5
Sherman	1.25	113.5
Umatilla	2.49	147.1
Wasco	1.62	137.5

SOURCE: Fadeley (1965).

quences of this contamination on human health, investigators calculated, for each of the nine Oregon counties having frontage on either
the Columbia River or the Pacific Ocean, an "index of exposure." This
index of exposure was based on several factors, including distance from
Hanford and average distance of the population from water frontage.
The cancer mortality rate, cancer mortality per 100,000 person-years
(1959–1964), was also determined for each of these nine counties. Table
14.3 shows the index of exposure and the cancer mortality rate for the
nine counties.

Figure 14.5 is a scatter plot of the data in Table 14.3. Each point corresponds to one county. The straight line in the figure is the estimated regression line of y (cancer mortality) on x (index of exposure). Summary
statistics needed for the computation are given in Table 14.4.

The numerator of the sample covariance is

$$(n-1)s_{xy} = \sum_{i=1}^{n} (x_i - \bar{x})(y_i - \bar{y}) = \sum_{i=1}^{n} x_i y_i - n\bar{x}\bar{y}$$

$$= \sum_{i=1}^{n} x_i y_i - \frac{\left(\sum_{i=1}^{n} x_i\right)\left(\sum_{i=1}^{n} y_i\right)}{n}$$

$$= 7439.37 - \frac{41.56 \times 1416.1}{9}$$

$$= 900.13.$$

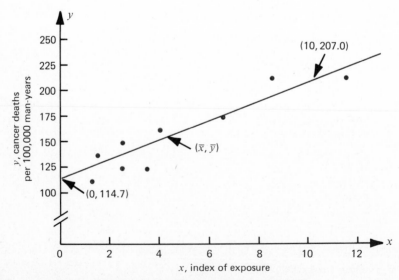

Figure 14.5 *Scatter plot of cancer mortality data of Table 14.3*

TABLE 14.4
Computations for Regression Example

$$n = 9$$

$\sum_{i=1}^{n} x_i = 41.56$	$\sum_{i=1}^{n} y_i = 1{,}416.1$
$\bar{x} = 4.618$	$\bar{y} = 157.34$
$\sum_{i=1}^{n} x_i^2 = 289.4222$	$\sum_{i=1}^{n} y_i^2 = 232{,}498.97$

$$\sum_{i=1}^{n} x_i y_i = 7439.37$$

$$\sum_{i=1}^{n} (x_i - \bar{x})^2 = 97.5074 \qquad \sum_{i=1}^{n} (x_i - \bar{x})(y_i - \bar{y}) = 900.13 \qquad \sum_{i=1}^{n} (y_i - \bar{y})^2 = 9683.50$$

SOURCE: Table 14.3.

The numerator of the variance of x is

$$(n - 1)s_x^2 = \sum_{i=1}^{n} (x_i - \bar{x})^2$$

$$= \sum_{i=1}^{n} x_i^2 - \frac{\left(\sum_{i=1}^{n} x_i\right)^2}{n}$$

$$= 289.4222 - \frac{(41.56)^2}{9} = 97.5074.$$

The sample regression coefficient is

$$b = \frac{s_{xy}}{s_x^2} = \frac{(n-1)s_{xy}}{(n-1)s_x^2} = \frac{900.13}{97.507} = 9.23.$$

The estimate of the intercept is

$$a = \bar{y} - b\bar{x} = 157.34 - 9.23 \times 4.62 = 114.7.$$

The estimated regression line is

$$\hat{y} = 114.7 + 9.23x.$$

One concludes that a zero exposure rate ($x = 0$) would suggest a death rate of about 115 and that, as the exposure index increases by one unit, the death rate increases by about 9.23 per 100,000.

Two points are required to draw the line in Figure 14.5. One of these can be taken to be (0, *a*), which is (0, 114.7). Another point may be obtained by substituting into the equation of the estimated regression line a large value of *x* and computing the corresponding value of $\hat{y}$. For example, $x = 10$ gives $\hat{y} = 114.7 + 9.23 \times 10 = 207.0$. The corresponding point, (10, 207.0), is shown in Figure 14.5.

The residual sum of squares, which is a sum of squares of deviations, is computed from

$$\sum_{i=1}^{n} (y_i - \hat{y}_i)^2 = (n - 1)s_y^2 - \frac{[(n-1)s_{xy}]^2}{(n-1)s_x^2}$$

$$= 9683.50 - \frac{(900.13)^2}{97.507} = 1374.06.$$

The estimate of the variance is

$$s_{y \cdot x}^2 = \frac{1374.06}{9 - 2} = \frac{1374.06}{7} = 196.294;$$

the corresponding standard deviation is $s_{y \cdot x} = \sqrt{196.294} = 14.01$. This standard deviation is used in statistical inference.

14.4 SAMPLING VARIABILITY; STATISTICAL INFERENCE

Suppose we drew repeated samples with the same set of *x*'s; these samples would be of the form

$$(x_1, y_1), \ldots, (x_n, y_n),$$

$$(x_1, y_1'), \ldots, (x_n, y_n'),$$

$$(x_1, y_1''), \ldots, (x_n, y_n''),$$

and so forth. For example, we could think of $x_1, \ldots, x_n$ as being fixed numbers of hours of training in typing; each sample arises from *n* different people who give rise to *n* different words-per-minute scores.

For each sample we could compute the value of *b*, the estimate of β. This procedure would generate the sampling distribution of *b*. The mean of this sampling distribution is β; that is, b is an unbiased estimate of β. The standard deviation of this distribution, the standard deviation or *standard error* of b, denoted by σ_b, is

$$\sigma_b = \frac{\sigma}{\sqrt{\sum_{i=1}^{n} (x_i - \bar{x})^2}}.$$

The more spread out the x's are, the smaller is σ_b. Since $s_{y \cdot x}^2$ estimates σ^2, the estimate of the standard error of b is

$$s_b = \frac{s_{y \cdot x}}{\sqrt{\sum_{i=1}^{n} (x_i - \bar{x})^2}}.$$

If the y's are normally distributed, b has a normal distribution, and $(b - \beta)/\sigma_b$ is normally distributed with mean 0 and variance 1. This provides a basis for testing hypotheses and forming confidence intervals for β. In particular, if the null hypothesis is $H\colon \beta = \beta_0$, where β_0 is specified, and σ is known, the observed $(b - \beta_0)/\sigma_b$ is referred to the normal tables. For example, one would reject this hypothesis at the 5% level if the absolute value of the computed ratio were greater than 1.96. (The procedures here are valid if n is fairly large, even if normality is not a good approximation.)

Usually σ is unknown, and use must be made of the estimates s_b. Since $s_{y \cdot x}^2$ has $n - 2$ degrees of freedom under random sampling, the ratio $(b - \beta)/s_b$ has as its sampling distribution the t-distribution with $n - 2$ degrees of freedom.

As pointed out earlier, the y's do not depend on the x's if $\beta = 0$; hence, a hypothesis of particular interest is $H\colon \beta = 0$. At significance level α the null hypothesis is rejected against the two-sided alternative if

$$\left| \frac{b}{s_b} \right| \geq t_{n-2} \left(\frac{\alpha}{2} \right).$$

In the example, $s_b = 14.01/\sqrt{97.5074} = 1.419$. The ratio $b/s_b = 6.50$ is significant at any reasonable level $[t_7(.005) = 3.499]$.

The power of the test of $H\colon \beta = 0$ depends on the ratio of the true values of β and σ_b, which we call $\delta\colon \delta = \beta/\sigma_b$. The power curve for a two-sided test is shown in Figure 14.6. [The level of significance should not be confused with the intercept α in $\mu(x) = \alpha + \beta x$.]

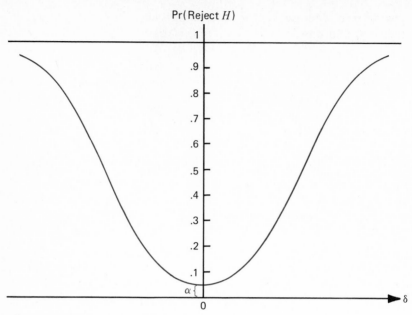

Figure 14.6 *Power curve* $(\delta = \beta/\sigma_b)$

When σ^2 is unknown, a confidence region for β with confidence coefficient $1 - \alpha$ is

$$b - s_b t_{n-2}\left(\frac{\alpha}{2}\right) < \beta < b + s_b t_{n-2}\left(\frac{\alpha}{2}\right).$$

In the example, the interval has endpoints $9.23 \pm .467 \times 3.499 = 9.23 \pm 1.63 = 7.60, 10.86$. Thus, one has confidence that β is close to $b = 9.23$.

14.5   THE CORRELATION COEFFICIENT: A MEASURE OF LINEAR RELATIONSHIP

In Section 5.5.4 the *correlation coefficient*

$$r = \frac{s_{xy}}{s_x s_y}$$

was introduced as a measure of association between two quantitative variables. It is a statistic describing one aspect of n pairs of measurements on two variables x and y. Although the measure is symmetric in the two variables, it can be interpreted in terms of how well one variable y can be predicted from the other variable x.

In Chapter 12 we defined a measure of association for categorical variables that took the form

$$\frac{\left(\begin{array}{c}\text{Number of errors}\\\text{in prediction of } y\\\text{without using } x\end{array}\right) - \left(\begin{array}{c}\text{Number of errors}\\\text{in prediction of } y\\\text{using } x\end{array}\right)}{\left(\begin{array}{c}\text{Number of errors in prediction}\\\text{of } y \text{ without using } x\end{array}\right)}.$$

With quantitative variables, the notion that replaces "number of errors" is "sum of squared deviations" of the predicted values of y from the actual values. This gives rise to a measure of association for quantitative variables, defined by

$$\frac{\left(\begin{array}{c}\text{Sum of squares of}\\\text{deviations in predicting}\\y \text{ without using } x\end{array}\right) - \left(\begin{array}{c}\text{Sum of squares of}\\\text{deviations in predicting}\\y \text{ using } x\end{array}\right)}{\left(\begin{array}{c}\text{Sum of squares of deviations in}\\\text{predicting } y \text{ without using } x\end{array}\right)}.$$

Without using x_i, the best prediction for y_i is the mean, $\bar{y}$, in the sense that the sum of squares of deviations about the number v, $\sum_{i=1}^{n}(y_i - v)^2$, is made a minimum by choosing $\bar{y}$ as the value of v. Using x_i, the predicted value of y_i is $\hat{y}_i = a + bx_i = \bar{y} + b(x_i - \bar{x})$, and the sum of squared deviations is $\sum_{i=1}^{n}(y_i - \hat{y}_i)^2$. Thus the measure of association is

$$\frac{\sum_{i=1}^{n}(y_i - \bar{y})^2 - \sum_{i=1}^{n}(y_i - \hat{y}_i)^2}{\sum_{i=1}^{n}(y_i - \bar{y})^2} = \frac{(n-1)s_y^2 - (n-1)(s_y^2 - s_{xy}^2/s_x^2)}{(n-1)s_y^2}$$

$$= 1 - 1 + \frac{s_{xy}^2}{s_y^2 s_x^2} = r^2,$$

the square of the correlation between x and y. Thus the square of the correlation coefficient can be interpreted as the proportional reduction in the sum of squared deviations achieved by using the best linear function of x to predict y.

Note that $0 \le r^2 \le 1$. If $r = 0$, x is of no help in predicting y. If $r^2 = 1$, x predicts y exactly; each point (x_i, y_i) falls on the regression line. The correlation is positive or negative according to whether the regression line has a positive or negative slope (runs in a southwest to northeast direction or a northwest to southeast direction). Table 14.5 gives some guidelines for interpretation of r.

TABLE 14.5
Guidelines for Interpretation of Size of Correlation Coefficient

r^2	$|r|$	INTERPRETATION
0	0	"no correlation"
0–.25	0–.5	"weak correlation"
.25–.64	.5–.8	"medium correlation"
.64–1	.8–1	"strong correlation"
1	1	"perfect correlation"

For the cancer mortality data in our example, we compute $r =$ 900.13/$\sqrt{97.507 \times 9683.52}$ = .926, which is a very high value of the correlation coefficient.

An example of a two-way frequency table of pairs of quantitative variables is Table 14.10, which gives the frequencies of average heights of pairs of parents and heights of grown children. For example, there were 48 children who had a height of 69.2 inches (within .5 inch) and whose parents had an average height of 68.5 inches. A three-dimensional histogram can be constructed by raising a column over each square with height equal to the frequency. For instance, over the square from 68 to 69 in one direction and 68.7 to 69.7 in the other, erect a height of 48. The frequency table can be described by such a collection of columns. A particular smooth surface fitted over this is a bivariate normal distribution. This is a theoretical relative frequency surface. The normal distribution of one variable x depends on its mean μ_x and its standard deviation σ_x, and the normal distribution of the other variable y depends on its mean μ_y and its standard deviation σ_y. The *bivariate* normal distribution depends on these four parameters and the population correlation coefficient ρ, which is the analog of r for the theoretical distribution.

The mode of the normal distribution (that is, the maximum value of the relative frequency function) is the point (μ_x, μ_y). The distribution of y values for a fixed value of x is a normal distribution with a mean that bears a linear relation to x and a variance not depending on x. Similarly, the distribution of x values for a fixed value of y is normal with a mean linearly related to y and a variance independent of y. Another remarkable property of the bivariate normal distribution is that x and y are independent if and only if $\rho = 0$.

Now consider testing the null hypothesis $H: \rho = 0$ when x and y have a bivariate normal distribution. This is equivalent to the hypothesis that x and y are statistically independent. If the alternative is $\rho \neq 0$, the hypothesis H will be rejected if $|r|$ is sufficiently large, that is, if $|r| > c$, where c is chosen so that the test has level α. It is shown in Appendix 14C that the statistic

$$\sqrt{n-2}\,\frac{r}{\sqrt{1-r^2}}$$

is distributed according to t_{n-2} under the null hypothesis. The condition $|r| > c$ is equivalent to the condition

$$\sqrt{n-2}\,\frac{|r|}{\sqrt{1-r^2}} > t_{n-2}\left(\frac{\alpha}{2}\right).$$

It follows from Appendix 14C that this test is equivalent to the test that the regression of y on x is a constant and to the test that the regression of x on y is a constant.

14.6   RANK CORRELATION

The *rank correlation coefficient* is the correlation coefficient defined in terms of the ranks of the two variables. It is thus appropriate for data defined only on an ordinal scale. If the observations are the pairs $(x_1, y_1), \ldots, (x_n, y_n)$, the observed x_i is replaced by s_i, its rank in $x_1, \ldots, x_n$, and y_i is replaced by t_i, its rank in $y_1, \ldots, y_n$. Then the rank correlation coefficient is

$$r_s = \frac{\sum_{i=1}^{n} s_i t_i - \left(\sum_{i=1}^{n} s_i\right)\left(\sum_{i=1}^{n} t_i\right)\big/n}{\sqrt{\sum_{i=1}^{n} s_i^2 - \left(\sum_{i=1}^{n} s_i\right)^2\big/n}\,\sqrt{\sum_{i=1}^{n} t_i^2 - \left(\sum_{i=1}^{n} t_i\right)^2\big/n}}.$$

When we make use of the facts that

$$\sum_{i=1}^{n} s_i = \sum_{i=1}^{n} t_i = \sum_{i=1}^{n} i = \frac{n(n+1)}{2},$$

$$\sum_{i=1}^{n} s_i^2 = \sum_{i=1}^{n} t_i^2 = \sum_{i=1}^{n} i^2 = \frac{n(n+1)(2n+1)}{6},$$

we obtain

$$r_s = \frac{12\sum_{i=1}^{n} s_i t_i}{n(n^2-1)} - 3\,\frac{n+1}{n-1}.$$

This statistic is known as *Spearman's rank correlation coefficient.*

EXAMPLE   The ranks of the values of the index of exposure and the ranks of the values of cancer mortality from Table 14.3 are given in Table 14.6. Here $n = 9$ and $\Sigma_{i=1}^{9} s_i t_i = 275$. The value of the rank correlation is $12 \times 275/720 - 30/8 = 5/6 = .833$.

TABLE 14.6
Ranks of Index of Exposure and Cancer Mortality

COUNTY	RANK OF INDEX OF EXPOSURE	RANK OF CANCER MORTALITY
Clatsop	8	9
Columbia	7	7
Gilliam	5	2
Hood River	6	6
Morrow	4	3
Portland	9	8
Sherman	1	1
Umatilla	3	5
Wasco	2	4

SOURCE: Table 14.3.

Another way of computing r_s is

$$r_s = 1 - \frac{6 \sum_{i=1}^{n} d_i^2}{n(n^2 - 1)},$$

where $d_i = s_i - t_i$, $i = 1, \ldots, n$. The formula is verified in Appendix 14D. The reader can use this procedure on the ranks in Table 14.6 to confirm that when the association is positive, this method is numerically easier.

The rank correlation can be used to test the hypothesis that x and y are independent in the population sampled. The distribution of r_s for any given n can be calculated. Tables are available in Beyer (1968), for example. If n is not small, one can use the fact that the coefficient is approximately normally distributed with variance $1/(n - 1)$. Thus, a test at level α is made by rejecting the hypothesis of no association if the observed value of r_s lies outside the interval $(-z(\alpha/2)/\sqrt{n - 1}, z(\alpha/2)/\sqrt{n - 1})$. For the example with $\alpha = .05$, $z(\alpha/2)/\sqrt{n - 1} = 1.96/\sqrt{8} = 1.96/2.828 = 0.771$. Since the observed result of 0.833 is greater than 0.771, one rejects the hypothesis of no association.

SUMMARY

The standard method for estimating the parameters in the regression of y on x is the *method of least squares*.

The *correlation coefficient* is a measure of linear relationship between x and y. The square of the correlation coefficient is the proportionate reduction in variability achieved by using a linear function of x to predict y.

APPENDIX 14A   Derivation of the Least-Squares Estimates

The principle of least squares says to choose as the estimate of the parameters α and β the values of A and B that minimize the least-squares criterion $q = \sum_{i=1}^{n}[y_i - (A + Bx_i)]^2$. We can write x_i and y_i in terms of their deviations from the sample means, as $y_i = (y_i - \bar{y}) + \bar{y}$ and $x_i = (x_i - \bar{x}) + \bar{x}$. Then

$$q = \sum_{i=1}^{n} \{[(y_i - \bar{y}) + \bar{y}] - A - B[(x_i - \bar{x}) + \bar{x}]\}^2$$

$$= \sum_{i=1}^{n} \{[(y_i - \bar{y}) - B(x_i - \bar{x})] - [A - (\bar{y} - B\bar{x})]\}^2.$$

Upon expanding the square, one obtains

(14.4) $$q = \sum_{i=1}^{n} [(y_i - \bar{y}) - B(x_i - \bar{x})]^2 + n[A - (\bar{y} - B\bar{x})]^2$$

because the cross-product term is zero:

$$\sum_{i=1}^{n} [(y_i - \bar{y}) - B(x_i - \bar{x})][A - (\bar{y} - B\bar{x})]$$

$$= [A - (\bar{y} - B\bar{x})] \sum_{i=1}^{n} [(y_i - \bar{y}) - B(x_i - \bar{x})] = 0,$$

since

$$\sum_{i=1}^{n} [(y_i - \bar{y}) - B(x_i - \bar{x})] = \sum_{i=1}^{n} (y_i - \bar{y}) - B \sum_{i=1}^{n} (x_i - \bar{x}) = 0;$$

the sums of deviations are zero. From (14.4) we see that, whatever the value of B, to make q a minimum we must choose A to be $\bar{y} - B\bar{x}$, for this makes the second term in (14.4) equal to zero, whereas otherwise it would be positive. With this choice of A, (14.4) is $\sum_{i=1}^{n}[(y_i - \bar{y}) - B(x_i - \bar{x})]^2$, which can be written as

$$\sum_{i=1}^{n} (y_i - \bar{y})^2 - 2B \sum_{i=1}^{n} (x_i - \bar{x})(y_i - \bar{y}) + B^2 \sum_{i=1}^{n} (x_i - \bar{x})^2.$$

A quadratic form

$$cB^2 + 2dB + e = c \left(B + \frac{d}{c} \right)^2 + e - \frac{d^2}{c}$$

is minimized by taking $B = -d/c$ when c is positive. Here $c = \sum_{i=1}^{n} (x_i - \bar{x})^2$ is positive, $d = -\sum_{i=1}^{n}(x_i - \bar{x})(y_i - \bar{y})$, and $e = \sum_{i=1}^{n}(y_i - \bar{y})^2$. Thus, the quadratic is minimized by taking B as

$$-[-\sum_{i=1}^{n}(x_i - \bar{x})(y_i - \bar{y})]/\sum_{i=1}^{n}(x_i - \bar{x})^2 = b,$$

as was to be shown.

APPENDIX 14B Computational Formulas for the Residual Sum of Squares

The estimate of the variance σ^2 is $s_{y \cdot x}^2 = \sum_{i=1}^{n}(y_i - \hat{y}_i)^2/(n - 2)$. The numerator, the *residual sum of squares*, is

$$\sum_{i=1}^{n} (y_i - \hat{y}_i)^2 = \sum_{i=1}^{n} [y_i - (a + bx_i)]^2$$

$$= \sum_{i=1}^{n} [y_i - \bar{y} - b(x_i - \bar{x})]^2$$

$$= \sum_{i=1}^{n} (y_i - \bar{y})^2 - 2b \sum_{i=1}^{n} (y_i - \bar{y})(x_i - \bar{x})$$

$$+ b^2 \sum_{i=1}^{n} (x_i - \bar{x})^2$$

$$= \sum_{i=1}^{n} (y_i - \bar{y})^2 - b^2 \sum_{i=1}^{n} (x_i - \bar{x})^2$$

$$= \sum_{i=1}^{n} (y_i - \bar{y})^2 - b \sum_{i=1}^{n} (x_i - \bar{x})(y_i - \bar{y})$$

$$= \sum_{i=1}^{n} (y_i - \bar{y})^2 - \frac{\left[\sum_{i=1}^{n} (x_i - \bar{x})(y_i - \bar{y}) \right]^2}{\sum_{i=1}^{n} (x_i - \bar{x})^2}$$

$$= \sum_{i=1}^{n} y_i^2 - \frac{\left(\sum_{i=1}^{n} y_i \right)^2}{n}$$

$$- \frac{\left[\sum_{i=1}^{n} x_i y_i - \left(\sum_{i=1}^{n} x_i \right) \left(\sum_{i=1}^{n} y_i \right) \Big/ n \right]^2}{\left[\sum_{i=1}^{n} x_i^2 - \left(\sum_{i=1}^{n} x_i \right)^2 \Big/ n \right]}.$$

The last is the best computing procedure.

APPENDIX 14C Formulas for the Statistic for Testing Independence

The least-squares estimate b can be written as

$$b = \frac{s_{xy}}{s_x^2} = \frac{s_{xy}}{s_x s_y} \frac{s_y}{s_x} = r \frac{s_y}{s_x}.$$

Also,

$$(n - 2)s_{y \cdot x}^2 = (n - 1)s_y^2(1 - r^2).$$

The statistic for testing independence can thus be written as

$$\frac{b}{s_{y \cdot x} \Big/ \sqrt{\sum_{i=1}^{n} (x_i - \bar{x})^2}} = \frac{\sqrt{n - 2} \; r s_y / s_x}{\sqrt{n - 1} \; s_y \sqrt{1 - r^2} / (\sqrt{n - 1} \; s_x)}$$

$$= \sqrt{n - 2} \frac{r}{\sqrt{1 - r^2}}.$$

APPENDIX 14D Derivation of the Rank Correlation Coefficient Formula

Since $d_i = s_i - t_i$, and $(s_1, \ldots, s_n)$ and $(t_1, \ldots, t_n)$ are rearrangements of $(1, \ldots, n)$, we have

$$\sum_{i=1}^{n} d_i^2 = \sum_{i=1}^{n} (s_i - t_i)^2$$

$$= \sum_{i=1}^{n} s_i^2 - 2 \sum_{i=1}^{n} s_i t_i + \sum_{i=1}^{n} t_i^2$$

$$= 2 \sum_{i=1}^{n} i^2 - 2 \sum_{i=1}^{n} s_i t_i$$

$$= \frac{n(n + 1)(2n + 1)}{3} - 2 \sum_{i=1}^{n} s_i t_i.$$

Then from Section 14.6 we obtain

$$r_s = \frac{12}{n(n^2 - 1)} \sum_{i=1}^{n} s_i t_i - 3 \frac{n + 1}{n - 1}$$

$$= \frac{12}{n(n^2 - 1)} \left[\frac{n(n + 1)(2n + 1)}{6} - \frac{1}{2} \sum_{i=1}^{n} d_i^2 \right] - 3 \frac{n + 1}{n - 1}$$

$$= \frac{2(2n + 1)}{n - 1} - \frac{6}{n(n^2 - 1)} \sum_{i=1}^{n} d_i^2 - 3 \frac{n + 1}{n - 1},$$

which simplifies to the second expression for r_s in Section 14.6.

APPENDIX 14E   Other Intervals in Regression

Confidence Region for the True Regression Line.   The estimate of $\mu(x) = \alpha + \beta x$ is $a + bx$. A confidence interval for $\mu(x)$ with confidence coefficient $1 - \alpha$ is obtained by adding to and subtracting from the estimate $a + bx$ the quantity

$$t_{n-2}\left(\frac{\alpha}{2}\right) s_{y \cdot x} \sqrt{\frac{1}{n} + \frac{(x - \bar{x})^2}{\sum_{i=1}^{n} (x_i - \bar{x})^2}}.$$

This follows from the fact that the variance of the point estimate $a + bx$ is

$$\text{Var}(a + bx) = \sigma^2 \left[\frac{1}{n} + \frac{(x - \bar{x})^2}{\sum_{i=1}^{n} (x_i - \bar{x})^2} \right]$$

and the variable $[(a + bx) - (\alpha + \beta x)]/s_{a+bx}$ is distributed as Student's t with $n - 2$ degrees of freedom.

Prediction Interval for a New y Value.   A new observation y corresponding to a specified value x is equal to its mean $\mu(x)$ plus random error due to observing a new individual value. The estimate of y is the estimate of its mean $\mu(x)$. This estimate is $a + bx$. The associated prediction interval is obtained by adding to and subtracting from the estimate $a + bx$ the quantity

$$t_{n-2}\left(\frac{\alpha}{2}\right) s_{y \cdot x} \sqrt{1 + \frac{1}{n} + \frac{(x - \bar{x})^2}{\sum\limits_{i=1}^{n} (x_i - \bar{x})^2}}.$$

In the radioactivity example, the estimated mean value corresponding to $x = 9$ is $a + 9b = 114.7 + 9 \times 9.23 = 197.8$. A 95% confidence interval is obtained by computing $1/n + (x - \bar{x})^2/\sum_{i=1}^{n}(x_i - \bar{x})^2 = 1/9 + (9 - 4.62)^2/97.5074 = 0.308$ and $\sqrt{0.308} = 0.555$. The interval has the limits $157.3 - 2.365 \times 14.01 \times 0.555 = 157.3 - 18.4 = 138.9$ and $157.3 + 18.4 = 175.7$; the interval is $(138.9, 175.7)$. It could as well be reported as $(139, 176)$ or even $(140, 180)$. A 95% prediction interval for a new observation corresponding to $x = 9$ is obtained by computing $1 + 1/n + (x - \bar{x})^2/\sum_{i=1}^{n}(x_i - \bar{x})^2 = 1.308$ and $\sqrt{1.308} = 1.144$. The interval has limits $157.3 - 2.365 \times 14.01 \times 1.144 = 157.3 - 37.9 = 119.4$ and $157.3 + 37.9 = 195.2$. The interval is $(119, 195)$. The interpretation is that, if the exposure rate x is 9 for a county, then one expects the death rate y for that county to be within these limits.

EXERCISES

14.1 (Sec. 14.2)   Make a two-way frequency table for the data of Table 14.7.

TABLE 14.7
(x, y) Pairs

x	1	1	2	4	4	5	5	7	7	8
y	1	3	3	1	5	3	5	3	5	5

14.2 (Sec. 14.2)   Make a two-way frequency table for the data of Table 14.8.

TABLE 14.8
(x, y) Pairs

x	2	2	3	3	3	4	4	4	4	5	5	5	5	5
y	1	2	3	4	5	6	7	8	9	9	9	9	9	9

TABLE 14.9

Heights and Weights of 41 Five-Year-Old Boys

Boy	Height	Weight	Boy	Height	Weight
1	42	39	21	40	40
2	46	53	22	40	40
3	39	29	23	38	37
4	39	37	24	41	36
5	36	31	25	36	39
6	38	36	26	36	36
7	40	43	27	37	32
8	36	37	28	33	24
9	41	42	29	40	38
10	34	30	30	41	37
11	37	35	31	39	36
12	39	34	32	41	34
13	40	35	33	41	41
14	37	38	34	39	42
15	33	29	35	41	39
16	40	35	36	42	35
17	39	39	37	44	44
18	41	47	38	42	45
19	36	34	39	42	42
20	39	41	40	44	41
			41	43	40

SOURCE: Subset of a data set from Isserlis (1915), p. 62.

14.3 (Sec. 14.2)   The data of Table 14.9 consist of the heights (in inches) and weights (in pounds) of 41 five-year-old boys.

(a) Tabulate the bivariate frequency distribution of these data; use the class intervals 23.5–28.5, 28.5–33.5, . . . , 48.5–53.5 for weight and 32.5–35.5, 35.5–38.5, . . . , 44.5–47.5 for height.

(b) Find the marginal distributions of height and of weight.

(c) What is the mode of the height distribution? Of the weight distribution?

(d) Divide the entries in the body of the table, the row totals, and the column totals by 41 to obtain, in terms of relative frequency, the bivariate height-weight distribution, the univariate height distribution, and the univariate weight distribution.

14.4 (Sec. 14.2)   In the study from which "regression" takes its name, Galton obtained the heights of 205 couples together with the heights of their grown children. The frequencies of heights of children and their parents are given in Table 14.10. (These data are for both males and

TABLE 14.10
Number of Grown Children of Various Heights Born to 205 Couples of Various Heights (Units: Inches)

Heights of Grown Children	Average Height of the Two Parents											Totals	Medians of Rows
	Below 64.5	64.5	65.5	66.5	67.5	68.5	69.5	70.5	71.5	72.5	Above 72.5		
Below 62.2	1	1	1			1		1				5	
62.2		1		3	3							7	
63.2	2	4	9	3	5	7	1	1				32	66.3
64.2	4	4	5	5	14	11	16		1			59	67.8
65.2	1	1	7	2	15	16	4	1	3			48	67.9
66.2	2	5	11	17	36	25	17	1	4			117	67.7
67.2	2	5	11	17	38	31	27	3	3			138	67.9
68.2	1		7	14	28	34	20	12	5	1		120	68.3
69.2	1	2	7	13	38	48	33	18	10	2		167	68.5
70.2			5	4	19	21	25	14	4	1		99	69.0
71.2			2		11	18	20	7	9	2	1	64	69.0
72.2			1		4	4	11	4	2	7		41	70.0
73.2						3	4	3	2	2		17	
Above 73.2							5	3		4	3	14	
Total Number of:													
Children	14	23	66	78	211	219	183	68	43	19	4	928	
Couples	1	5	12	20	33	49	41	22	11	6	5	205	
Medians of Columns		65.8	66.7	67.2	67.6	68.2	68.9	69.5	69.9	72.2			

source: Galton (1885).

females. It was desirable to adjust for the sex difference in height. Since the mean height for males was 8% greater than that for females, all female heights were multiplied by 1.08 to make them equivalent, on the average, to male heights.)

(a) Give the relative frequencies of average heights of the 205 couples (that is, the marginal distribution).

(b) Give the relative frequencies of heights of the 928 children.

(c) Give the conditional distribution of the heights of the children, given that the average height of the two parents was 68.5 inches.

(d) Plot the medians of the average height of the two parents against the height of the children.

(e) Plot the medians of the height of the children against the average height of the two parents.

In (d) and (e), draw on the graphs the 45° line $y = x$, the line representing equality of height of children and the average height of their two parents.

14.5 (Sec. 14.3)   Find the regression line for the data of Table 14.11.

TABLE 14.11
Data Set

x	1	2	5	8	9
y	5	2	3	4	1

14.6 (Sec. 14.3)   Find the regression line for the data of Table 14.12.

TABLE 14.12
Data Set

x	1	2	5	8	9
y	1	4	3	2	5

14.7 (Sec. 14.3)   Figure 14.7 gives two population regression lines showing the relationship between hours of study for the final examination and average grade on the examination. Line CD is for an art course; line EF is for a psychology course. What do these lines imply about the relationships between the variables in the two cases?

14.8 (Sec. 14.3)   In an investigation of hours of study and grade in two large lecture courses in history, the data and sample regression lines shown in Figure 14.8 were obtained. The observed sample regression lines are very nearly the same in the two courses. Relatively how reliable are conclusions about the population regression line in the two cases? Compare the usefulness of x in predicting y in the two cases.

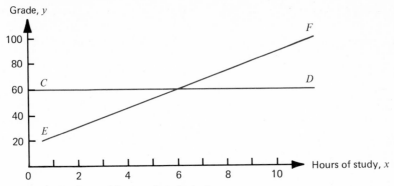

Figure 14.7 *Grades and hours of study in two courses*

14.9 (Sec. 14.5)   For each of the following examples, state from your general knowledge whether the correlation between the two variables is positive, negative, or zero.

(a) Severity of earthquakes and amplitude of change in a recording pen on a seismograph

(b) Golf scores and number of hours spent practicing

(c) The length of dogs' tails and their owners' waist measurements

(d) Heights of fathers and their sons

(e) Bowling scores and number of hours spent practicing

14.10 (Sec. 14.5)   For each of the following examples, state from your general knowledge whether the correlation between the two variables is positive, negative, or zero.

(a) Height and weight

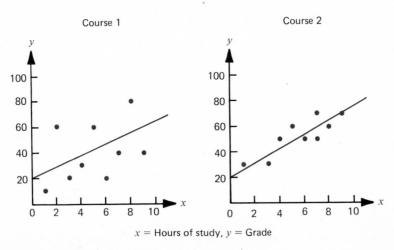

Figure 14.8 *Grades and hours of study in two courses*

(b) Amount of traffic and time needed to reach destination
(c) Education and income
(d) Price of hogs in the United States and price of tea in China
(e) Grade in physics and grade in mathematics

14.11 (Sec. 14.5)   Table 14.13 gives five (x, y) pairs. Treating these data as a sample, calculate:
(a) The variance of x
(b) The standard deviation of x
(c) The variance of y
(d) The standard deviation of y
(e) The covariance of x and y
(f) The coefficient of correlation between x and y

TABLE 14.13
Data Set

x	1	2	3	4	5
y	4	5	3	1	2

14.12 (Sec. 14.5)   Table 14.14 gives five (x, y) pairs. Treating these data as a sample, calculate:
(a) The variance of x
(b) The standard deviation of x
(c) The variance of y
(d) The standard deviation of y
(e) The covariance of x and y
(f) The coefficient of correlation between x and y

TABLE 14.14
Data Set

x	1	2	3	4	5
y	5	4	3	1	2

14.13 (Sec. 14.5)   Consider the data of Table 14.15.
(a) Make a scatter plot of the data.
(b) Note that the three points fall on a straight line. Verify that the value of the correlation coefficient is 1.

14.14 (Sec. 14.5)   Consider the data of Table 14.16.
(a) Make a scatter plot of the data.
(b) Note that the four points fall on a straight line. Verify that the value of the correlation coefficient is 1.

TABLE 14.15
Three Collinear Points

x	1	2	3
y	4	7	10

TABLE 14.16
Four Collinear Points

x	1	2	3	4
y	4	7	10	13

14.15 (Sec. 14.5)   For the data in Table 3.20, compute the correlation between the number of hogs and the number of chicks.

14.16 (Sec. 14.5)   For the data in Table 3.20, one can see without computation that the correlation between the number of children and the number of toys is $+1$. How?

14.17 (Sec. 14.5)   Scores of 20 husbands and wives on a test of conformity are given in Table 14.17.
(a) Make a scatter plot of the data. What does the plot indicate about the magnitude (high, medium, low) and direction of the correlation?
(b) Compute the value of r.

14.18 (Sec. 14.5)   A student of Southern race relations wished to test quantitatively the suspicion that lynching, in this century, was essentially a "survival" phenomenon, that is, a practice that came into being because of the absence of effective political organization, in frontier days, and outlived its manifest function. The student felt that the incidence of lynching in an area should be negatively related to that area's experience with due process of law. For each of 13 states (the 11 Confederate states plus Kentucky and Oklahoma), the student obtained the value of y, the number of lynchings per 10,000 Black population (for the period 1930–1939), and x, the number of decades (as of 1920) since the passing of the frontier (statehood or the colonial equivalent). The data are plotted in Figure 14.9. Summary statistics are given in Table 14.18.
(a) Compute the least-squares estimates a and b.
(b) Test at the 5% level the hypothesis that there is no association between the variables in the (hypothetical) population from which these observations are a sample against the alternative hypothesis that there is a negative relation.
(c) Compute r.

TABLE 14.17

Scores of 20 Husbands and Wives on a Test of Conformity

COUPLE	HUSBAND'S SCORE, x	WIFE'S SCORE, y
1	9	8
2	19	31
3	13	18
4	8	7
5	6	10
6	25	28
7	6	8
8	15	16
9	4	8
10	3	4
11	12	21
12	22	26
13	14	12
14	13	17
15	12	9
16	20	16
17	12	12
18	18	18
19	7	6
20	15	24

(hypothetical data)

14.19 (Sec. 14.5)   The following example was suggested by results reported in Key and Munger (1959). Table 14.19 shows the Republican percentage of the total vote for President in 1920 and 1952 for a random sample of 20 of the 92 counties in Indiana.

(a) Present these data in the form of a scatter plot, to show the relation between the 1920 and 1952 votes. (This is what Key and Munger call "traditionalism" in voting patterns.)

(b) In order to predict the 1952 vote (y) from the 1920 vote (x) on a

TABLE 14.18

Summary Statistics (Figure 14.9)

$$n = 13$$

$\Sigma x_i = \ \ 169$	$\Sigma y_i = 26$
$\Sigma x_i^2 = 2797$	$\Sigma y_i^2 = 71$

$$\Sigma x_i y_i = 248$$

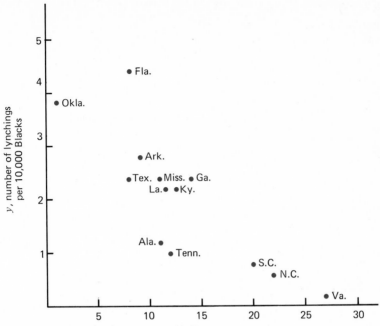

Figure 14.9 *Scatter plot for Exercise* 14.18

county basis, one can consider a regression equation. Find the least-squares estimates a and b. Draw in the estimated regression line on the scatter plot. To simplify the computations, summary statistics are given in Table 14.19.

(c) Compute the standard error of estimate $s_{y \cdot x}$.

(d) Compute a 99% confidence interval for the slope of the population regression line β.

(e) Test the hypothesis $H: \beta = 0$ against the alternative $\beta \neq 0$ at the 1% level.

(f) Compare the total sum of squares, $\Sigma(y_i - \bar{y})^2$, with the residual sum of squares, $\Sigma(y_i - \hat{y}_i)^2$.

14.20 (Sec. 14.4)   For the data of Table 5.74, find the regression of percent registered to vote on mean income. Test the significance of the relationship.

14.21 (Sec. 14.5)   Table 3.28 gives values of x = number of hours' increase in sleep due to use of Drug A and y = number of hours' increase in sleep due to use of Drug B, for each of ten patients. Compute the coefficient of correlation between x and y.

14.22 (Sec. 14.5)   Table 14.20 gives the IQ's of monozygotic (identical) twins raised apart.

TABLE 14.19

Republican Percentage of Total Presidential Vote in 1920 and 1952 for a Random Sample of 20 of the 92 Counties of Indiana

COUNTY	1920	1952	COUNTY	1920	1952
Brown	36.8	51.8	LaPorte	65.0	60.1
Clark	48.4	48.9	Marion	54.9	60.7
Clay	48.2	53.9	Noble	60.8	66.4
Dearborn	50.7	55.0	Porter	72.7	69.1
Decatur	57.7	65.7	St. Joseph	56.5	50.1
Fountain	54.1	61.6	Switzerland	50.1	48.9
Greene	48.6	53.8	Washington	46.9	55.8
Harrison	51.5	54.6	Wells	47.6	57.6
Jasper	66.9	72.6	White	58.6	67.9
Johnson	45.4	60.7	Whitley	52.5	61.1

SOURCE: Key and Munger (1959).

$$n = 20$$
$$\Sigma x_i = 1{,}073.9 \qquad \Sigma y_i = 1{,}176.3$$
$$\Sigma x_i^2 = 58{,}976.39 \qquad \Sigma y_i^2 = 70{,}103.07$$
$$\Sigma x_i y_i = 63{,}947.86$$

(a) Compute the correlation r between $x =$ IQ of first born and $y =$ IQ of second born.

(b) Compute r^2. The percentage $100r^2\%$ is sometimes called *geneticity*. The complementary percentage $100(1 - r^2)\%$ is then taken as a measure of the contribution of environmental (as distinct from genetic) effects.

14.23 (Sec. 14.5) Suppose that a manufacturing process were performed at each of five different temperatures and the strength of the resulting product measured. The data might have appeared as in Table 14.21.

(a) Compute the residual sum of squares $\Sigma(y_i - \hat{y}_i)^2$ that results when $\hat{y}_i = 0.3x_i - 87$.

(b) Compute the residual sum of squares that results when $\hat{y}_i = 0.3x_i - 86$.

(c) When $\hat{y}_i = a + bx_i$, where a and b are the least-squares estimates, then $\Sigma(y_i - \hat{y}_i)^2 = \Sigma(y_i - \bar{y})^2 - [\Sigma(x_i - \bar{x})(y_i - \bar{y})]^2/\Sigma(x_i - \bar{x})^2$. Use the results in Table 14.22 to compute this value. Note that it is considerably less than the results of (a) and (b).

14.24 (Sec. 14.5) Table 14.23 gives aptitude and achievement scores for 12 students.

TABLE 14.20

IQ's of Monozygotic Twins Raised Apart
(n = 34 pairs of twins)

CASE	FIRST BORN	SECOND BORN
Sm 1	22	12
Sm 2	36	34
Sm 4	13	10
Sm 5	30	25
Sm 6	32	28
Sm 7	26	17
Sm 8	20	24
Sm 10	30	26
Sm 11	29	35
Sm 12	26	20
Sm 13	28	22
Sm 14	21	27
Sm 15	13	4
Sf 5	32	33
Sf 7	30	34
Sf 9	27	24
Sf 10	32	18
Sf 11	27	28
Sf 12	22	23
Sf 13	15	9
Sf 14	24	33
Sf 15	6	10
Sf 16	23	21
Sf 17	38	27
Sf 18	33	26
Sf 19	16	28
Sf 20	27	25
Sf 22	4	2
Sf 23	19	9
Sf 24	41	40
Sf 25	40	38
Sf 26	12	9
Sf 27	13	22
Sf 29	29	30

SOURCE: Shields (1962).

TABLE 14.21

Temperature and Strength of Product

x, TEMPERATURE	y, STRENGTH
480	57
500	63
520	71
540	75
560	82

(hypothetical data)

TABLE 14.22

Summary Statistics for Data of Table 14.21

		$n = 5$			
$\Sigma x_i =$	2,600			$\Sigma y_i =$	348
$\Sigma x_i^2 =$	1,356,000			$\Sigma y_i^2 =$	24,608
		$\Sigma x_i y_i = 182,200$			
$\bar{x} =$	520			$\bar{y} =$	69.6
$s_x^2 =$	1,000			$s_y^2 =$	96.8
		$s_{xy} =$	310		

SOURCE: Table 14.21.

(a) Compute the line of regression of y on x.

(b) Construct a 95% confidence interval for β.

(c) Using the fact that $\mathrm{Var}(a) = \sigma^2[1/n + \bar{x}^2/\Sigma_{i=1}^{n}(x_i - \bar{x})^2]$, construct a 95% confidence interval for α.

(d) Construct a 95% confidence interval for the mean value of y.

(e) Construct a 95% confidence interval for the mean value of the population of y's corresponding to $x = 700$.

(f) Compute r.

(g) Test $H: \rho = 0$ against the alternative $\rho \neq 0$.

14.25 (Sec. 14.3)   For the data on auto noise pollution of Exercise 5.22, compute the coefficients a and b in the linear regression of y on x^2.

14.26 (Sec. 14.4)   A benchmark study of the relationship between environment and IQ is the study by Klineberg (1935). He studied the influence of the urban environment on intelligence-test scores. The subjects were 12-year-old boys and girls in schools in Harlem. Table 14.24 shows some of the data for boys.

(a) Plot mean IQ score against number of years' residence, using 1.5, 3.5, 5.5, 7.5, 10, and 12 years as the points along the horizontal axis at which to plot.

TABLE 14.23

Aptitude and Achievement Scores of 12 Students

x, APTITUDE	*y,* ACHIEVEMENT
440	250
400	440
370	260
600	660
400	410
560	650
540	400
410	420
550	570
550	680
530	470
610	550

(hypothetical data)

TABLE 14.24

Intelligence Test Score and Length of Residence

	NUMBER OF YEARS OF RESIDENCE IN NEW YORK					
	1 or 2	*3 or 4*	*5 or 6*	*7 or 8*	*9, 10, or 11*	*12**
Number of children	56	35	41	33	44	308
Mean score	64.21	66.86	72.32	83.58	84.64	86.93
Standard deviation	29.5	29.5	33.2	29.1	30.4	28.9

SOURCE: Study by George Lapidus of 517 twelve-year-old boys, reported in Klineberg (1935).
* Born in New York

(b) Estimate the regression coefficient, using the formula

$$b = \frac{\Sigma n_j x_j \bar{y}_j - (\Sigma n_j x_j)(\Sigma n_j \bar{y}_j)/(\Sigma n_j)}{\Sigma n_j x_j^2 - (\Sigma n_j x_j)^2/(\Sigma n_j)},$$

where $n_1 = 56$, $n_2 = 35$, etc.,
$x_1 = 1.5$, $x_2 = 3.5$, etc.,
$\bar{y}_1 = 64.21$, $\bar{y}_2 = 66.86$, etc.,

and Σ denotes $\Sigma_{j=1}^6$.

(c) Estimate the intercept, using the formula $a = \bar{y} - b\bar{x}$, where $\bar{y} = \Sigma n_j \bar{y}_j / \Sigma n_j$ and $\bar{x} = \Sigma n_j x_j / \Sigma n_j$.

(d) Compute $s_{y \cdot x}^2$ according to the formula $s_{y \cdot x}^2 = \Sigma(n_j - 1)s(j)^2/$

$\Sigma(n_j - 1)$, where $s(1) = 29.5$, $s(2) = 29.5$, $s(3) = 33.2$, etc. Make a 95% confidence interval for β.

MATHEMATICAL EXERCISES

14.27 (Sec. 14.5)   Show that, when $n = 2$, $r = +1$ or -1.

14.28 (Sec. 14.5)   Show that the correlation coefficient can be written as $r = \Sigma x_i' y_i' / (n - 1)$, where $x_i' = (x_i - \bar{x})/s_x$ and $y_i' = (y_i - \bar{y})/s_y$ are the so-called "standard scores" corresponding to x_i and y_i.

14.29 (Sec. 14.3)   Show that $\Sigma(x_i - \bar{x})(y_i - \bar{y}) = \Sigma(x_i - \bar{x})y_i$.

14.30 (Sec. 14.4)   Show that $\text{Var}(b) = \sigma^2 / \Sigma(x_i - \bar{x})^2$.

14.31 (Sec. 14.4)   Using the result of Exercise 14.30 (or otherwise), show that $\text{Var}(a) = \sigma^2[1/n + \bar{x}^2 / \Sigma(x_i - \bar{x})^2]$.

REFERENCES

Beyer, William H., ed. (1972). *Handbook of Tables for Probability and Statistics*, 2nd ed. Chemical Rubber Co., Cleveland, Ohio.

Fadeley, Robert C. (1965). "Oregon malignancy pattern physiographically related to Hanford, Washington, radioisotope storage." *Journal of Environmental Health* 27: 883–897.

Galton, Francis (1885). "Regression towards mediocrity in hereditary stature." *Journal of the Anthropological Institute* 15: 246–263.

Isserlis, L. (1915). "On the partial correlation ratio. Part II: Numerical." *Biometrika* 11: 50–66.

Key, V. O., and Frank Munger (1959). "Social determinism and electoral decision: the case of Indiana." In *American Voting Behavior*, Eugene L. Burdick and A. J. Brodbeck, eds. Free Press, Glencoe, Ill.

Klineberg, Otto (1935). *Negro Intelligence and Selective Migration*. Columbia University Press, New York.

Shields, James (1962). *Monozygotic Twins*. Oxford University Press, London.

15 MULTIPLE REGRESSION AND PARTIAL CORRELATION

INTRODUCTION

In this chapter we consider explaining and predicting one variable from several explanatory variables (*multiple regression*). We shall develop a measure of association between the one variable and the set of explanatory variables in terms of the effectiveness of the latter in predicting the former (*multiple correlation coefficient*). A partial correlation coefficient measures association between two variables after taking account of the effect of a third.

In Sections 5.3 and 5.4 we gave many examples of studies involving three categorical variables. Now we shall give several examples of studies with three or more quantitative variables. In each there is one "dependent" variable y and two or more explanatory variables x_1, x_2, ..., x_p. In each study the data consist of the measurements of the $p + 1$ variables in n units of observation.

EXAMPLE 1 [Conner, Gibbs, and Reynolds (1973)] The object of this study was to assess the effect of water frontage on land values. The data are as follows.

$n = 316$ sales of residential lots in the Kissimmee River Basin of Florida,

y = price per front foot,

x_1 = year in which the sale occurred (1966 = 1, ..., 1970 = 5),

x_2 = lot size (acres),

x_3 = the presence ($x_3 = 1$) or absence ($x_3 = 0$) of lake frontage,

x_4 = the presence ($x_4 = 1$) or absence ($x_4 = 0$) of canal frontage,

x_5 = miles to the nearest paved road,

x_6 = the number of utilities available in the subdivision,

x_7 = the percentage of wood homes in the subdivision in which the sale occurred.

Interest centers on x_3 and x_4; the other x's are variables whose effects must be simultaneously estimated to obtain estimates of the effects of water frontage alone. From their data, the authors obtained the following regression equation, using the seven independent variables in an additive way:

$$\hat{y} = 10.32 + 1.47x_1 - 1.08x_2 + 32.41x_3$$
$$+ 7.41x_4 - 1.34x_5 + 3.77x_6 - 0.05x_7.$$

The coefficient of x_3, 32.41, is interpreted as the dollar value added to the price per front foot by lake frontage. The coefficient of x_4, 7.41, is the value added by canal frontage. The square of the multiple correlation coefficient is 0.68. We shall see that this means that the seven x's explain 68% of the variability of y.

EXAMPLE 2 [Galle, Gove, and McPherson (1972)]   For each of $n = 75$ community districts in and around Chicago, these sociologists studied y = rate of juvenile delinquency in relation to x_1 = population density and x_2 = socioeconomic status. It was found that, even when x_2 is taken into account, x_1 plays a significant role in explaining the variation in y. The conclusions held up for other dependent variables measuring "social pathology." (See Section 5.4.3.)

EXAMPLE 3 [Kelley, Ayres, and Bowen (1967)]   Multiple regression was used to study voter registration rates y in United States cities with 1960 populations greater than 100,000. The explanatory variables were factors affecting the value of the vote, measured by x_1 = closeness of recent elections, factors affecting the costs of registration, measured by x_2 = closing date for registration, x_3 = provisions regarding literacy tests, and x_4 = times and places of registration; and socioeconomic characteristics, measured by x_5 = age (percent between 20 and 34), x_6 = race (percent white), and x_7 = education (median school years completed). A linear function of the seven explanatory variables accounted for 80% of the variation in y from city to city.

15.1   REGRESSION ON TWO EXPLANATORY VARIABLES

A statistical relationship between one dependent variable y and two explanatory variables x_1 and x_2 exists when, for fixed values of x_1 and x_2, there is a population of y values such that the mean of the y values is a function of x_1 and x_2, say $\mu(x_1, x_2)$. If the so-called *regression function* is linear, it can be written

$$\mu(x_1, x_2) = \alpha + \beta_1 x_1 + \beta_2 x_2.$$

In many situations it is reasonable to believe that a linear function is an adequate approximation over the range of x_1 and x_2 of relevance. It is further assumed that for given values of x_1 and x_2 the values of y in the population have a variance σ^2 whose value does not depend on x_1 and x_2. If the parameters α, β_1, and β_2 were known as well as the values of x_1 and x_2, the prediction of y would be $\alpha + \beta_1 x_1 + \beta_2 x_2$.

15.1.1 The Least-Squares Estimates

The data consist of n triplets of numbers (x_{11}, x_{21}, y_1), (x_{12}, x_{22}, y_2), $\ldots$, (x_{1n}, x_{2n}, y_n). The first problem of inference is to estimate the parameters α, β_1, β_2, and σ^2. The principle of least squares is applied again; the estimates of α, β_1, and β_2 are the values of A, B_1, and B_2 that minimize $\Sigma_{i=1}^{n}[y_i - (A + B_1 x_{1i} + B_2 x_{2i})]^2$. The statistics needed from the sample are $\bar{x}_1$, $\bar{x}_2$, $\bar{y}$,

$$S_{11} = \sum_{i=1}^{n} (x_{1i} - \bar{x}_1)^2 = \sum_{i=1}^{n} x_{1i}^2 - \frac{\left(\sum_{i=1}^{n} x_{1i}\right)^2}{n},$$

$$S_{12} = \sum_{i=1}^{n} (x_{1i} - \bar{x}_1)(x_{2i} - \bar{x}_2)$$

$$= \sum_{i=1}^{n} x_{1i} x_{2i} - \frac{\left(\sum_{i=1}^{n} x_{1i}\right)\left(\sum_{i=1}^{n} x_{2i}\right)}{n},$$

$$S_{22} = \sum_{i=1}^{n} (x_{2i} - \bar{x}_2)^2 = \sum_{i=1}^{n} x_{2i}^2 - \frac{\left(\sum_{i=1}^{n} x_{2i}\right)^2}{n},$$

$$S_{y1} = \sum_{i=1}^{n} (y_i - \bar{y})(x_{1i} - \bar{x}_1)$$

$$= \sum_{i=1}^{n} y_i x_{1i} - \frac{\left(\sum_{i=1}^{n} y_i\right)\left(\sum_{i=1}^{n} x_{1i}\right)}{n},$$

$$S_{y2} = \sum_{i=1}^{n} (y_i - \bar{y})(x_{2i} - \bar{x}_2)$$

$$= \sum_{i=1}^{n} y_i x_{2i} - \frac{\left(\sum_{i=1}^{n} y_i\right)\left(\sum_{i=1}^{n} x_{2i}\right)}{n},$$

$$S_{yy} = \sum_{i=1}^{n} (y_i - \bar{y})^2 = \sum_{i=1}^{n} y_i^2 - \frac{\left(\sum_{i=1}^{n} y_i\right)^2}{n}.$$

The least-squares estimates of the coefficients β_1 and β_2 are the solutions b_1 and b_2 to the equations

(15.1)

$$S_{11}b_1 + S_{12}b_2 = S_{y1},$$

$$S_{12}b_1 + S_{22}b_2 = S_{y2}.$$

The least-squares estimate of α is

(15.2)

$$a = \bar{y} - b_1\bar{x}_1 - b_2\bar{x}_2.$$

(See Appendix 15A.) An explicit solution to equations (15.1) is

$$b_1 = \frac{S_{y1}S_{22} - S_{y2}S_{12}}{S_{11}S_{22} - S_{12}^2},$$

$$b_2 = \frac{S_{y2}S_{11} - S_{y1}S_{12}}{S_{11}S_{22} - S_{12}^2}.$$

In order to advise students whether or not to take calculus, Dr. Jones is considering using her own pretest and a standard mathematics test. Suppose that a sample of $n = 53$ students yields the following data:
y = final numerical grade in calculus ($\bar{y} = 600$),
x_1 = score on Dr. Jones's pretest ($\bar{x}_1 = 600$),
x_2 = score on standard math test ($\bar{x}_2 = 500$).
Table 15.1 gives the sums of squares and cross-products.

TABLE 15.1
Sums of Squares and Cross-Products

$S_{11} = 400,000$	$S_{12} = 200,000$	$S_{y1} = 360,000$
	$S_{22} = 500,000$	$S_{y2} = 240,000$
		$S_{yy} = 480,000$

(hypothetical data)

The estimating equations for b_1 and b_2 are

$$400,000b_1 + 200,000b_2 = 360,000,$$

$$200,000b_1 + 500,000b_2 = 240,000.$$

The solution is $b_1 = 0.825$ and $b_2 = 0.150$. Then $a = 700 - 0.825 \times 600 - 0.150 \times 500 = 130$. The regression equation is $\hat{y} = 130 + 0.825x_1 + 0.150x_2$.

The estimate of the variance σ^2 is based on the sum of squares of deviations of the dependent variable from the regression or predicted values:

(15.3)
$$\sum_{i=1}^{n} (y_i - \hat{y}_i)^2 = \sum_{i=1}^{n} [y_i - (a + b_1 x_{1i} + b_2 x_{2i})]^2$$

$$= \sum_{i=1}^{n} [y_i - \bar{y} - b_1(x_{1i} - \bar{x}_1) - b_2(x_{2i} - \bar{x}_2)]^2$$

$$= S_{yy} - S_{11}b_1^2 - 2S_{12}b_1b_2 - S_{22}b_2^2$$

$$= S_{yy} - S_{y1}b_1 - S_{y2}b_2.$$

The last two expressions are derived in Appendix 15A. The number of degrees of freedom of the above sum of squares of residuals is $n - 3$ because 3 constants are determined in order to minimize this sum. The estimate of σ^2 is then

$$s_{y \cdot 12}^2 = \frac{\sum_{i=1}^{n} (y_i - \hat{y}_i)^2}{n - 3}.$$

The numerator can be calculated by any of the expressions in (15.3). In the illustration,

$$s_{y \cdot 12}^2 = \frac{1}{50}(480{,}000 - 360{,}000 \times 0.825 - 240{,}000 \times 0.150)$$

$$= \frac{147{,}000}{50} = 2{,}940.$$

The estimated standard deviation is 54.2.

15.1.2 Statistical Inference

The estimates b_1 and b_2 are unbiased estimates of β_1 and β_2, respectively. The estimated standard errors (standard deviations) of b_1 and b_2 are, respectively,

$$s_{b_1} = s_{y \cdot 12} \sqrt{\frac{S_{22}}{S_{11}S_{22} - S_{12}^2}}, \qquad s_{b_2} = s_{y \cdot 12} \sqrt{\frac{S_{11}}{S_{11}S_{22} - S_{12}^2}}.$$

If the distribution of values of y for given values of x_1 and x_2 is normal, then b_1 and b_2 have normal sampling distributions. In any case, if n is large, the sampling distribution is approximately normal. Statistical inference is then based on $(b_j - \beta_j)/s_b$ having a t-distribution with $n - 3$ degrees of freedom. A confidence interval for β_j at confidence level $1 - \alpha$ is

$$b_j - s_{b_j}t_{n-3}\left(\frac{\alpha}{2}\right) < \beta_j < b_{j_j} + s_{b_j}t_{n-3}\left(\frac{\alpha}{2}\right).$$

A test of the null hypothesis $\beta_j = \beta_0$, a specified value, is carried out by referring $(b_j - \beta_0)/s_{b_j}$ to the table of the t-distribution with $n - 3$ degrees of freedom. A particularly important null hypothesis is that one of the β_j's is zero, say β_2. The null hypothesis $H: \beta_2 = 0$ implies that x_2 is not needed in the regression function, that x_2 is of no use in predicting y, that y does not depend on x_2. These statements, it should be noted, are on the basis that x_1 is included in the regression function. The test at significance level α and against two-sided alternatives is to reject H if

$$|b_2| > s_{b_2} \times t_{n-3}\left(\frac{\alpha}{2}\right).$$

In our illustration, $n - 3 = 50$,

$$s_{b_2} = 54.2 \times \sqrt{\frac{400,000}{400,000 \times 500,000 - (200,000)^2}}$$

$$= 54.2 \times .001581 = .0857,$$

and $b_2 = 0.150$. If the significance level is .05, we use $t_{50}(.025) = 2.010$. Then the critical value is $.0857 \times 2.010 = .1723$, and the hypothesis is accepted. The sample does not provide Dr. Jones with evidence that the standard mathematics test adds any further accuracy to the prediction of grades in calculus.

15.1.3 More Explanatory Variables

If there are more than two explanatory variables, the investigator is likely to resort to a "canned" computer program that is able to handle a large number of explanatory variables. The ideas developed in this chapter for two explanatory variables hold for an arbitrary number, although the algebra is more complicated and the numerical computations more laborious.

15.1.4 Nonlinear Regression

Some nonlinear functions can be accommodated by the format of this section. In the case of one explanatory variable x, the linear form $\alpha + \beta x$ may not be an adequate approximation to the regression function $\mu(x)$. One might try a function of the form $\alpha + \beta_1 x + \beta_2 x^2$ or $\alpha + \beta_1 x + \beta_2 x^2 + \beta_3 x^3$. These are *polynomial* regression functions. Note that if, in the regression function of the form $\alpha + \beta_1 x_1 + \beta_2 x_2$, one takes the first explanatory variable x_1 to be x and the second explanatory variable x_2 to be x^2, then one obtains the polynomial regression function $\alpha + \beta_1 x + \beta_2 x^2$. Thus, polynomial regression can be viewed as a special case of multiple regression.

15.2 MULTIPLE CORRELATION

We can ask how useful x_1 and x_2 are in predicting y. One measure of this is the proportion by which the sum of squared errors of prediction is reduced by using x_1 and x_2. This proportion is

$$\frac{\sum_{i=1}^{n}(y_i - \bar{y})^2 - \sum_{i=1}^{n}(\hat{y} - \bar{y}_i)^2}{\sum_{i=1}^{n}(y_i - \bar{y})^2} = \frac{S_{yy} - (S_{yy} - S_{11}b_1^2 - 2S_{12}b_1b_2 - S_{22}b_2^2)}{S_{yy}}$$

$$= \frac{S_{11}b_1^2 + 2S_{12}b_1b_2 + S_{22}b_2^2}{S_{yy}}$$

by use of (15.3). The square root of this proportion is the multiple correlation coefficient

$$(15.4) \qquad R_{y\cdot12} = \sqrt{\frac{S_{11}b_1^2 + 2S_{12}b_1b_2 + S_{22}b_2^2}{S_{yy}}}.$$

This is the correlation between y_i and $\hat{y}_i$, $i = 1, \ldots, n$. Of all linear combinations $A + B_1 x_{1i} + B_2 x_{2i}$, $\hat{y}_i = a + b_1 x_{1i} + b_2 x_{2i}$ not only minimizes the sum of squared deviations but also maximizes the correlation between y_i and such a linear combination. Thus the multiple correlation is the maximum of all possible correlations between y_i and linear combinations of x_{1i} and x_{2i}.

The multiple correlation ranges between 0 and 1. A value of 0 occurs only if y_i is uncorrelated with x_{1i} and with x_{2i} separately. (This hardly ever occurs in practice.) A multiple correlation of 1 occurs when $\hat{y}_i = y_i$, $i = 1, \ldots, n$.

A hypothesis that may be of interest is whether x_1 and x_2 are of any value in predicting or explaining y. In our linear model, a formal statement of the null hypothesis is $H: \beta_1 = \beta_2 = 0$. The test criterion depends on comparing the reduction in the sum of squares of errors of prediction by using x_1 and x_2 with the "error variance" $s_{y \cdot 12}^2$. It is

$$\frac{S_{11}b_1^2 + 2S_{12}b_1b_2 + S_{22}b_2^2}{2s_{y \cdot 12}^2} = \frac{R_{y \cdot 12}^2}{1 - R_{y \cdot 12}^2} \times \frac{n - 3}{2}.$$

The numerator has 2 degrees of freedom, and the denominator has $n - 3$ degrees of freedom. At significance level α the hypothesis is rejected if the ratio exceeds $F_{2,n-3}(\alpha)$. In our illustration, this ratio is

$$\frac{333,000}{2 \times 2940} = 56.6,$$

which is significant at virtually any significance level.

15.3 PARTIAL CORRELATION

In Section 5.4 we studied the effect of one categorical variable on two others. We saw that in some cases a third variable accounts for most of the association between two others. For instance, it was found (Tables 5.40 to 5.44) that the association between level of education and intention to vote was accounted for by the intervening variable of political interest. The higher the level of education, the greater the political interest, and the greater the political interest, the greater the intention to vote. Controlling for political interest virtually eliminates the association between level of education and intention to vote.

In the case of quantitative variables, we may also be interested in the association between two variables after we have taken out the effect of another variable on these two. The partial correlation coefficient measures this residual association.

Consider a large sample of triplets of measurements (x_{1i}, x_{2i}, y_i), $i = 1, \ldots, n$. On the basis of a set of intervals for x_1, x_2, and y, one can construct a three-way frequency table. For observations in which x_{1i} is in a given interval, there is a two-way frequency table for pairs (x_{2i}, y_i). If the correlation between x_{2i} and y_i is about the same in each of these tables (irrespective of x_{1i}), then this common correlation is the conditional correlation between x_2 and y, given x_1. (In a *trivariate* normal distribution this is the case.)

Usually there are not enough observations to permit the construction of a three-way frequency table in which there are enough observations in each x_1 layer to estimate the correlation between x_2 and y. Instead it is *assumed* that in the population the mean of y is a linear function of x_1, the mean of x_2 is a linear function of x_1, and the (conditional) variation of y and x_2 and the (conditional) correlation between y and x_2 do not depend on x_1. Under these assumptions the regression of y on x_1 is calculated as $a + bx_1$ (according to Chapter 14), and the regression of x_2 on x_1 is calculated as $a' + b'x_1$. From these regressions we obtain the predicted values (based on x_1).

$$\hat{y}_i = a + bx_{1i}, \qquad \hat{x}_{2i} = a' + b'x_{1i}, \qquad i = 1, 2, \ldots, n.$$

The *residuals* are

$$y_i - \hat{y}_i, \quad x_{2i} - \hat{x}_{2i}, \qquad i = 1, 2, \ldots, n.$$

These residuals are the parts of y and x_2 not explained by x_1. The sample *partial correlation coefficient*, denoted by the symbol $r_{y2 \cdot 1}$, is the simple correlation between $y - \hat{y}$ and $x_2 - \hat{x}_2$:

$$r_{y2 \cdot 1} = r_{(y - \hat{y})(x_2 - \hat{x}_2)}$$

$$= \frac{\displaystyle\sum_{i=1}^{n} (y_i - \hat{y}_i)(x_{2i} - \hat{x}_{2i})}{\sqrt{\displaystyle\sum_{i=1}^{n} (y_i - \hat{y}_i)^2} \sqrt{\displaystyle\sum_{i=1}^{n} (x_{2i} - \hat{x}_{2i})}} .$$

Here we do not need to indicate the subtraction of means of residuals in the sums, because they are zero.

Straightforward (but tedious) algebra shows that

(15.5)
$$r_{y2 \cdot 1} = \frac{r_{yx_2} - r_{yx_1} r_{x_1 x_2}}{\sqrt{1 - r_{yx_1}^2} \sqrt{1 - r_{x_1 x_2}^2}} .$$

This formula is the one used for computation. It is also very important for analysis. In terms of simple correlation of quantitative variables, Figure 15.1 is analogous to Figures 5.3, 5.6, and 5.7 for categorical variables.

The simple correlation coefficients for the illustration of Section 15.1 are $r_{yx_1} = .822$, $r_{yx_2} = .490$, and $r_{x_1 x_2} = .447$. The partial correlation between y and x_2, after taking x_1 into account, is

$$r_{y2 \cdot 1} = \frac{.490 - .822 \times .447}{\sqrt{1 - .822^2}\sqrt{1 - .447^2}} = \frac{.1226}{.5094} = .241.$$

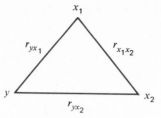

Figure 15.1 *Correlations among three variables*

This is a relatively small value ($r^2_{y2 \cdot 1} = .058$). The conclusion is that after we have taken account of the knowledge of the pretest, there is not much further association between grades in calculus and the standard math test.

The interpretation of partial correlations is similar to the interpretation of the effects of a third categorical variable as discussed in Section 5.4. The partial correlation may be numerically small; this feature may be due to x_1 having a causal effect on both y and x_2 (Section 5.4.2). It is possible that the partial correlation may be similar to the simple correlation. Rate of juvenile delinquency y may be positively related to population density x_2 even after the effect of social class x_1 has been removed (Section 5.4.3). The sign of the partial correlation coefficient may be opposite to that of the simple correlation coefficient (Section 5.4.4). It is also possible that the partial correlation is different from zero when the simple correlation is near zero (Section 5.4.5).

EXAMPLE   Hooker (1907) considered the correlations among rainfall x_1, temperature x_2, and yield of hay y over a 20-year period in England ($i = 1, 2, \ldots, n = 20$ years). The correlations are indicated on Figure 15.2. One might conclude that high temperature tends to cause low yield. However, $r_{y2 \cdot 1} = 0.097 > 0$. Having adjusted for rainfall, we find that yield and temperature are positively related. We conclude that high temperature increases yield, but usually the temperature is low when

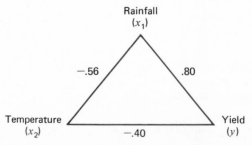

Figure 15.2 *Triangle diagram showing pairwise correlations*

there is much rain and high when there is little rain. Further, $r_{y·12} = 0.76$. We conclude that high rainfall increases yield.

The associated causal diagram is shown in Figure 15.3. The interpretation is that, among years when the rainfall is light, the yield is apt to be greater when the temperature is high than when it is low, and the same is true among years when the rainfall is heavy. It would be a mistake to assume from the fact that $r_{yx_2} = -.40$ that decreasing the temperature *causes* a larger yield. It is just that low temperature *occurs naturally* with high rainfall, which gives a larger yield. In a greenhouse, *both* high temperature *and* high humidity could be achieved.

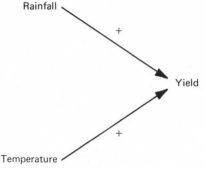

Figure 15.3 *Causal diagram*

Statistical Inference. In a trivariate statistical population, the partial correlation coefficient between y and x_2 adjusted for x_1 is denoted by $\rho_{y2·1}$ and is calculated in a manner analogous to the sample coefficient. The analog of (15.5) holds. In a trivariate *normal* population, $\rho_{y2·1} = 0$ corresponds to the independence of y and x_2 in a population in which the x_1 value is fixed.

To test the null hypothesis H: $\rho_{y2·1} = 0$, one uses the sample coefficient $r_{y2·1}$. When the null hypothesis is true and the sampling distribution of y, x_1, and x_2 is trivariate normal,

(15.6)
$$\sqrt{n-3}\ \frac{r_{y2·1}}{\sqrt{1-r_{y2·1}^2}}$$

has a *t*-distribution with $n-1$ degrees of freedom. The null hypothesis is rejected if the calculated value exceeds the suitable t. In our illustration, $r_{y2·1} = .239$ and (15.6) is $\sqrt{50} \times .241/\sqrt{1-.241^2} = 1.76$. This is not significant at the 5% level, because $t_{50}(.025) = 2.010$. The conclusion is that the data are consistent with the hypothesis that, given x_1, there is no further association between y and x_2. Note that this is the same conclusion that was arrived at by testing $\beta_2 = 0$. Although the computations are carried out differently, the two tests are identical.

SUMMARY

The relation of a dependent variable to a set of independent variables is stated as the mean of the dependent variable being a linear function of the independent variables. From a set of data, the coefficients of the linear function are estimated by the method of least squares. The multiple correlation coefficient is a measure of association between the dependent variable and the independent variables. It is the simple correlation between the dependent variable and the estimated linear function of the independent variables. Tests of hypotheses concerning the coefficients of the independent variables and confidence intervals are based on t- and F-distributions.

The partial correlation coefficient between two variables given a third is the correlation between the residuals of the two variables from their regressions on the third.

APPENDIX 15A   Derivation of the Least-Squares Estimates

The criterion to be minimized is

$$q(A, B_1, B_2) = \sum_{i=1}^{n} [y_i - (A + B_1 x_{1i} + B_2 x_{2i})]^2$$

$$= \sum_{i=1}^{n} [(y_i - B_1 x_{1i} - B_2 x_{2i}) - A]^2.$$

The minimum with respect to A is achieved for the value

$$A^* = \frac{1}{n} \sum_{i=1}^{n} (y_i - B_1 x_{1i} - B_2 x_{2i})$$

$$= \bar{y} - B_1 \bar{x}_1 - B \bar{x}_2.$$

Then we minimize with respect to B_1 and B_2 the quantity

$$q(A^*, B_1, B_2) = \sum_{i=1}^{n} [y_i - \bar{y} - B_1(x_{1i} - \bar{x}_1) - B_2(x_{2i} - \bar{x}_2)]^2.$$

We shall show that $q(A^*, B_1, B_2)$ is minimized when $B_1 = b_1$ and $B_2 = b_2$, where b_1 and b_2 satisfy the simultaneous linear equations

$$S_{11}b_1 + S_{12}b_2 = S_{y1},$$

$$S_{12}b_1 + S_{22}b_2 = S_{y2}.$$

LEMMA Let (u_1, v_1, w_1), (u_2, v_2, w_2), $\ldots$, (u_n, v_n, w_n) be n triplets of numbers. Suppose that $\hat{c}$ and $\hat{d}$ satisfy

(15.7)
$$\sum_{i=1}^{n} v_i^2 \hat{c} + \sum_{i=1}^{n} v_i w_i \hat{d} = \sum_{i=1}^{n} v_i u_i,$$

(15.8)
$$\sum_{i=1}^{n} v_i w_i \hat{c} + \sum_{i=1}^{n} w_i^2 \hat{d} = \sum_{i=1}^{n} w_i u_i.$$

Then

(15.9) $\displaystyle\sum_{i=1}^{n} [u_i - (cv_i + dw_i)]^2$

$$= \sum_{i=1}^{n} [u_i - (\hat{c}v_i + \hat{d}w_i)]^2 + \sum_{i=1}^{n} [(\hat{c} - c)v_i + (\hat{d} - d)w_i]^2.$$

PROOF: We have

$$\sum_{i=1}^{n} [u_i - (cv_i + dw_i)]^2$$

$$= \sum_{i=1}^{n} [u_i - (\hat{c}v_i + \hat{d}w_i) + (\hat{c} - c)v_i + (\hat{d} - d)w_i]^2$$

$$= \sum_{i=1}^{n} [u_i - (\hat{c}v_i + \hat{d}w_i)]^2 + \sum_{i=1}^{n} [(\hat{c} - c)v_i + (\hat{d} - d)w_i]^2$$

$$+ 2 \sum_{i=1}^{n} [u_i - (\hat{c}v_i + \hat{d}w_i)][(\hat{c} - c)v_i + (\hat{d} - d)w_i].$$

The last term is 2 times

$$\sum_{i=1}^{n} [u_i - (\hat{c}v_i + \hat{d}w_i)][(\hat{c} - c)v_i + (\hat{d} - d)w_i]$$

$$= (\hat{c} - c) \sum_{i=1}^{n} [u_i - (\hat{c}v_i + \hat{d}w_i)]v_i$$

$$+ (\hat{d} - d) \sum_{i=1}^{n} [u_i - (\hat{c}v_i + \hat{d}w_i)]w_i$$

$$= (\hat{c} - c) \left[\sum_{i=1}^{n} u_i v_i - \left(\hat{c} \sum_{i=1}^{n} v_i^2 + \hat{d} \sum_{i=1}^{n} w_i v_i \right) \right]$$

$$+ (\hat{d} - d) \left[\sum_{i=1}^{n} u_i w_i - \left(\hat{c} \sum_{i=1}^{n} v_i w_i + \hat{d} \sum_{i=1}^{n} w_i^2 \right) \right].$$

The two terms in brackets are 0 by virtue of (15.7) and (15.8).

A consequence of the lemma is that (15.9) is a minimum for $c = \hat{c}$ and $d = \hat{d}$ because then the second term on the right-hand side is 0.

THEOREM   *The least-squares estimates are given by the solution to the simultaneous equations* (15.1) *and* (15.2).

PROOF: In the lemma, take $c = B_1$, $d = B_2$, $\hat{c} = b_1$, and $\hat{d} = b_2$. Take $u_i = y_i - \bar{y}$, $v_i = x_{1i} - \bar{x}_1$, and $w_i = x_{2i} - \bar{x}_2$. Then application of the lemma shows that the solutions b_1 and b_2 of the first two equations minimize the criterion. The relation for a was proved above.

APPENDIX 15B   Formula for the Residual Sum of Squares with Two Independent Variables

We have for the residual sum of squares

$$\Sigma(y_i - \hat{y}_i)^2 = \Sigma[y_i - (a + b_1 x_{1i} + b_2 x_{2i})]^2$$
$$= \Sigma[(y_i - \bar{y}) - b_1(x_{1i} - \bar{x}_1) - b_2(x_{2i} - \bar{x}_2)]^2$$
$$= S_{yy} - 2(b_1 S_{y1} + b_2 S_{y2}) + b_1^2 S_{11}$$
$$+ 2b_1 b_2 S_{12} + b_2^2 S_{22}.$$

The equations for the estimates b_1 and b_2 are

[1]: $$S_{11}b_1 + S_{12}b_2 = S_{y1},$$
[2]: $$S_{12}b_1 + S_{22}b_2 = S_{y2}.$$

These equations give the following relations:

$[1'] = [1] \times b_1$: $\qquad\qquad S_{11}b_1^2 + S_{12}b_1b_2 = S_{y1}b_1$

$[2'] = [2] \times b_2$: $\qquad\qquad S_{12}b_1b_2 + S_{22}b_2^2 = S_{y2}b_2$

$[1'] + [2']$: $\qquad\qquad S_{11}b_1^2 + 2b_1b_2S_{12} + S_{22}b_2^2 = b_1S_{y1} + b_2S_{y2} = T,$

say. Hence, the residual sum of squares is $S_{yy} - 2T + T = S_{yy} - T = S_{yy} - b_1S_{y1} - b_2S_{y2}$. The quantity T is the reduction in the residual sum of squares due to regression on x_1 and x_2. It is often denoted by $SS(x_1 \text{ and } x_2)$ or $SS(\text{Regression})$.

EXERCISES

15.1 (Sec. 15.1)   An avid gardener wishes to predict the weight of berries to be obtained from each plant. The gardener has the following data on 40 plants from the previous year:

   x_1 = height of plant in full bloom (mean of 4.5 inches),
   x_2 = amount of fertilizer applied after blooming (mean of 1 pound),
   y = weight of berries produced by plant (mean of 2.0 pounds).
The summary statistics are

$$S_{11} = 10 \qquad S_{12} = 0 \qquad S_{y1} = 15$$
$$S_{22} = 40 \qquad S_{y2} = 30$$
$$S_{yy} = 90.$$

(a) Find the coefficients a, b_1, and b_2 in the regression equation $\hat{y} = a + b_1x_1 + b_2x_2$.
(b) Test at the 5% level the hypothesis that the yield y is independent of height and amount of fertilizer.

15.2 (Sec. 15.1)   A model for a quadratic time trend is that the mean of y_t is $\alpha + \beta_1 t + \beta_2 t^2$. Suppose the time series $y_1, y_2, \ldots, y_{10}$ is observed. Set up the equations for estimating β_1 and β_2 on the basis of these data. (The equations will be in terms of the algebraic symbols $y_1, y_2, \ldots, y_{10}$.) The required sums of powers of t are given in Table 15.2.

TABLE 15.2
Sums of Powers

p	1	2	3	4
$\sum_{t=1}^{10} t^p$	55	385	3025	25,333

15.3 (Sec. 15.1) The relevant data are given in Table 15.3.

TABLE 15.3
Multiple Regression Example

YEAR	MEAN CONGRES-SIONAL VOTE FOR PARTY OF CURRENT PRESIDENT IN LAST 8 ELECTIONS	NATIONWIDE CONGRESSIONAL VOTE FOR PARTY OF CURRENT PRESIDENT	STANDARD-IZED VOTE LOSS y_i	GALLUP POLL RATING OF PRESIDENT AT TIME OF ELECTION x_{1i}	CURRENT YEARLY CHANGE IN REAL DISPOSABLE INCOME PER CAPITA x_{2i}
1946	Dem. 52.57%	45.27%	7.30%	32%	−$36
1950	Dem. 52.04%	50.04%	2.00%	43%	$99
1954	Rep. 49.79%	47.46%	2.33%	65%	−$12
1958	Rep. 49.83%	43.91%	5.92%	56%	−$13
1962	Dem. 51.63%	52.42%	−0.79%	67%	$60
1966	Dem. 53.06%	51.33%	1.73%	48%	$96
1970	Rep. 46.66%	45.68%	0.98%	56%	$69

	x_1	x_2	y
Mean	52.42857	37.57143	2.78143
Standard deviation	12.39432	56.42948	2.83397
Sum of squared deviations	921.71429	19,105.72	48.18829

$$r_{x_1x_2} = .07156, \quad r_{yx_1} = -.63214, \quad r_{yx_2} = -.74510$$
$$S_{12} = 300.29654, \quad S_{y1} = -133.2240, \quad S_{y2} = -714.93591$$

SOURCE: Tufte (1974), Table 4-1, page 142.

(a) Graph the values of x_{1i}, x_{2i}, and y_i, with $i = 1, 2, \ldots, 7$ along the horizontal axis and x_{1i}, x_{2i}, and y_i along the vertical axis. Use different scales for the three variables and use different colors (or marks).

(b) Make scatter plots for each pair of variables, (x_1, x_2), (x_1, y), and (x_2, y).

(c) Calculate the least-squares estimates a, b_1, and b_2 of the parameters in the regression $\alpha + \beta_1 x_1 + \beta_2 x_2$.

(d) Find the estimate of the variance $s^2_{y \cdot 12}$.

(e) Find the multiple correlation coefficient $R_{y \cdot 12}$ and its square $R^2_{y \cdot 12}$.

15.4 (Sec. 15.1) Use the data from Exercise 15.3.

(a) Calculate the partial correlations $r_{y1 \cdot 2}$ and $r_{y2 \cdot 1}$.

(b) Calculate the *t*-statistic

$$\sqrt{n - 3} \frac{r_{y2 \cdot 1}}{\sqrt{1 - r^2_{y2 \cdot 1}}}.$$

(c) Calculate the estimated standard deviation of b_2.

(d) Give the 95% confidence interval for β_2.

(e) Give the analysis of variance table (effect of x_1, additional effect of

x_2, error, and total). Include sums of squares and numbers of degrees of freedom.

(f) Calculate $R_{y\cdot12}$ and

$$F = \frac{n-3}{2} \frac{R^2_{y\cdot12}}{1 - R^2_{y\cdot12}}.$$

(g) What conclusions do you draw from this analysis?

(h) "Predict" the vote for 1970 and compare with the actual vote. (The Gallup prediction was 47.0%.)

15.5 (Sec. 15.1)   Use the data from Exercise 15.3 and the results of Exercise 15.4 to give a 95% confidence interval for $\sigma^2_{y\cdot12}$. [*Note:* $(n-3)s^2_{y\cdot12}/\sigma^2_{y\cdot12}$ has a sampling distribution that is chi-square with $n-3$ degrees of freedom.]

15.6 (Sec. 15.1)   For the data for the second grade in Table 2.22, Exercise 2.19, compute the regression equation for predicting final reading-achievement score from initial reading-achievement score and verbal IQ. Is the coefficient of initial reading-achievement score significant at the 5% significance level? Is the coefficient of verbal IQ significant at the 5% level?

15.7 (Sec. 15.3)   Consider the data on noise pollution from autos in Table 5.67. Decide whether the regression function $\alpha' + \beta'x^2 + \gamma'x^3$ gives a significantly better fit than the regression function $\alpha + \beta x^2$. (The term in x is omitted due to physical considerations.) Do this by testing the null hypothesis that $\gamma' = 0$ at the 5% significance level.

15.8 (Sec. 15.3)   The observed correlations among three variables, $w =$ score on test of memory for *words*, $f =$ score on test of memory for meaning*ful* symbols, and $l =$ score on test of memory for meaning*less* symbols, in a sample of 140 are as given in (the upper portion of) the correlation matrix in Table 15.4.

(a) Compute $r_{fl\cdot w}$, the partial correlation between f and l, taking w into account.

(b) Would you say that the correlation between the memory for

TABLE 15.4

Correlations among Three Variables

	f	l
w	.67	.46
f		.69

SOURCE: Kelley (1928).

meaning*f*ul symbols and the memory for meaning*l*ess symbols is explained by their correlations with the memory for *w*ords? Why or why not?

15.9 (Sec. 15.3) The variables y, x, and t have the correlations given in Table 15.5. Would you say that the relationship between y and t can be explained in terms of their relationship with x?

TABLE 15.5
Correlations among Three Variables

	t	x
y	.60	.80
t		.75

MATHEMATICAL EXERCISES

15.10 (Sec. 15.3) Prove that introducing an additional x into the regression equation cannot increase the residual sum of squares.

15.11 (Sec. 15.3) When an additional x is introduced into a regression equation, how great must the reduction in the residual sum of squares be in order for there also to be a reduction in residual *mean* square, that is, in variance?

15.12 (Sec. 15.2) Prove that $R_{y\cdot12}$, given by (15.4), is the correlation between y_i and $\hat{y}_i$, $i = 1, \ldots, n$.

REFERENCES

Conner, J. R., K. C. Gibbs, and J. E. Reynolds (1973). "The effects of water frontage on recreational property values." *Journal of Leisure Research* 5: 26–38.

Galle, Omer R., Walter R. Gove, and J. Miller McPherson (1972). "Population density and pathology: What are the relations for man?" *Science* 176: 23–30.

Hooker, R. H. (1907). "The correlation of the weather and crops." *Journal of the Royal Statistical Society* 70: 1–42.

Kelley, Stanley, Jr., Richard E. Ayres, and William G. Bowen (1967). "Registration and voting: Putting first things first." *American Political*

Science Review 61: 359–379. (Reprinted in *The Quantitative Analysis of Social Problems*. Edward R. Tufte, ed. Addison-Wesley, Reading, Mass., 1970, pages 250–283. See also Tufte, Edward R. (1974), *Data Analysis for Politics and Policy*. Prentice-Hall, Englewood Cliffs, N.J.)

Kelley, T. L. (1928). *Crossroads in the Mind of Man*. Stanford University Press, Stanford, Calif.

Tufte, Edward R. (1974), *Data Analysis for Politics and Policy*. Prentice-Hall, Englewood Cliffs, N.J.

16 SAMPLING FROM POPULATIONS: SAMPLE SURVEYS

INTRODUCTION

Much empirical data arises from experiments, in which the investigator interacts in some way with the units of observation and actually influences the conditions of the units leading to the measurements. Many other sets of data result from simply *observing*, that is, making a *survey*. It is to such investigations that we now turn our attention. Usually one cannot observe every individual in the population, and often this would not even be desirable, for many individuals are similar. One does not need to eat the whole bowl to learn how the soup tastes; a spoonful will suffice, provided that the soup has been adequately stirred. The "spoonful" is a *sample* from the bowl (*population*), and "stirring" corresponds to drawing a random sample.

Sample surveys are especially useful in business and economics, market research, sociology, and industrial quality control. A survey often involves an interview, a questionnaire, or some sort of inspection. Care must be taken in working out the details of administration of the interview, questionnaire, or inspection. The planning of a survey should involve two kinds of professionals working together—a subject-matter specialist (economist, sociologist, production engineer) and a statistician.

Suppose that one wants to draw a random sample from among the households in a county to find out how many hours a day children watch television. Individual children are the "units of observation" (Section 2.1.4). Households are the "sampling units." The list of households on the county tax roll is a "frame." The "population sampled" consists of the children in the county who live in households on the tax roll. The "target population" is the set of all children who live in the county.

In general, the *target population* is the population about which it is desired to make inferences. Any difference between the *population sampled* and the target population is a potential source of bias. The *frame* is a physical list of the sampled population. Each *sampling unit* contains one, more than one, or no units of observation.

Many problems would arise in carrying out the survey of TV-viewing habits. How should the survey be designed? For how many days should each child be observed? If it is decided to monitor the TV viewing of the children for two weeks, which two weeks should be chosen? Should different two-week periods be chosen for different sub-samples of children? How can one ensure that the parents keep an accurate record of how long their children watch TV? If there is more than one child in a household, should all of them be included in the sample? Should some inducement to participate be offered?

When a survey involves an expenditure of time or effort on the part of those selected to be in the sample, there are almost always some individuals who refuse to respond. When the sampling unit is a household, an interviewer may find no one home when he or she calls a household designated to be in a sample. Such failures to obtain information from each unit meant to be sampled can introduce some bias into the results when those not responding differ systematically from those responding. Bias resulting from this source is termed *nonresponse bias*. It means, in effect, that the population sampled differs from the target population, inasmuch as the population sampled consists only of those individuals in the target population who are willing to respond or can be induced to do so. If, with respect to the characteristic of interest in the study, the population of persons who do not respond differs from those who do, there is a nonresponse bias. A statistical method for adjusting for nonresponse bias will be discussed briefly. In order for any such adjustment to be possible, however, there must be a second try, in which at least some of those who did not respond the first time do in fact respond.

Another type of bias is *interviewer bias*, which results when different persons in the sample are questioned by different interviewers. This may occur because interviewers interpret the instructions given by the director of the survey in slightly different ways, or simply because people react in different ways to different interviewers. Care should be taken to minimize these effects. Sometimes statistical methods, such as analysis of variance, can be employed after the fact to assess the extent of interviewer bias and adjust for it.

Our emphasis in this chapter will be on explanation of several sampling methods, and we shall focus attention on the problem of assessing the accuracy and precision of estimators resulting from the various methods. For such assessment to be possible, the sample drawn must be a *probability sample,* a sample drawn in such a way that the investigator knows the probability that any individual unit of observation will be included in the sample. We shall discuss several sampling methods: simple random sampling; stratified random sampling; cluster sampling; systematic sampling with a random start; and systematic subsampling with random starts. In each case, some random device plays a role in determining which members of the population shall be included in the sample.

When sampling involves interviewing people selected in a probability sample, the amount of work involved can be tremendous. W. E. Deming (1950, page 10), a leading sampling expert, puts it this way: "A probability-sample will send the interviewer through mud and cold, over long distances, up decrepit stairs, to people who do not welcome an interviewer; but such cases occur only in their correct proportions. Substitutions are not permitted: the rules are ruthless."

16.1   SIMPLE RANDOM SAMPLING

A *simple random sample* of size n is a sample of n drawn in such a way that all sets of n units in the population have the same chance of being the sample. (There are C_n^N such sets, where N is the population size and C_n^N was defined in Section 7.5.1. The fact that all these sets are equally likely to be chosen implies in particular that all units have the same chance of being included.) A *frame,* a list of all units with one of the numbers from 1 to N assigned to each, is constructed. A random sample can then be drawn.

The basic ideas of drawing a random sample were presented in Section 6.8. If the sample size is not very large, random numbers can be taken from a table of random numbers such as Appendix V. Since the frame of N units is finite, the sampling is done without replacement. A computer can be used to obtain a stream of numbers that behave like random numbers (often called pseudorandom numbers). The set of n random numbers determines the n sampling units to be included in the sample. Characteristics of the sampled units are then obtained by interview, written questionnaire, or direct measurement.

In the case of a numerical variable, let the numbers $y_1, y_2, \ldots, y_N$ represent the values of the variable for the N individuals in the population. The population mean μ is $\sum_{k=1}^{n} y_k / N$ (page 318), and the population variance is $\sigma^2 = \sum_{k=1}^{n} (y_k - \mu)^2 / N$. In a finite population, the total, such as total energy used, may be of primary interest rather than the mean, such as average amount of energy used. The total $\sum_{k=1}^{n} y_k$ is, of course, N times the mean, that is, $N\mu$.

Let the values of the variable for the n individuals in the sample be $x_1, x_2, \ldots, x_n$, where x_i is the value of the variable for the ith individual *in the sample,* $i = 1, 2, \ldots, n$. (The symbol x_1 bears no special relation to

y_1; if $n = 3$ and the sample of three individuals consists of the individuals numbered 13, 17, and 8 in the frame, then $x_1 = y_{13}$, $x_2 = y_{17}$, and $x_3 = y_8$.) The mean of the population μ is estimated by the mean of the sample $x = \Sigma_{i=1}^n x_i / n$. Then the total $N\mu$ is estimated by $N\bar{x}$. Note that

$$N\bar{x} = \frac{N}{n} \sum_{i=1}^n x_i;$$

this is the sample total $\Sigma_{i=1}^n x_i$ scaled up to population size by multiplying by N/n.

The sample mean $\bar{x}$ is an unbiased estimate of the population mean μ (page 318); that is, $\mu_{\bar{x}} = \mu$ (whether sampling is with or without replacement). The variance of the sample mean for sampling without replacement (Section 8.7.1) is

$$\sigma_{\bar{x}}^2 = \frac{N-n}{N-1} \frac{\sigma^2}{n}.$$

The finite population correction factor for the variance $(N-n)/(N-1)$ is approximately $(N-n)/N$, or $1 - n/N$. When the fraction sampled n/N is small, this correction is negligible.

The statistic $\bar{x}$ has a sampling distribution that is approximated by a normal distribution when n is large. Tests and confidence intervals are based on this fact.

Adjustment for Nonresponse Bias. Often a nonresponse can be converted into a response. If no one in a household is at home at the time the interviewer calls, the interviewer can return (a "call-back"). If a questionnaire mailed to a respondent is not returned, the respondent can be tried by telephone. Typically, the second attempt is more expensive than the first (for example, the households may be more scattered), and the response on the second round may not be complete.

Suppose that, of a sample of 100 taken in March to estimate the unemployment rate in the county, there were 10 nonrespondents. Of the 90 respondents, 5 were unemployed, and the unemployment rate was estimated as $\hat{p}_1 = 5/90 = .056$. This is an estimate of the unemployment rate *in the population of respondents*. The 10 nonrespondents were revisited in April and asked if they were employed in March; 8 of them responded, 3 saying that they were unemployed in March. The estimate of the unemployment in the population of nonrespondents is $\hat{p}_2 = 3/8 = .375$. The overall unemployment rate p is $.9p_1 + .1p_2$, in terms of the rates in the population of respondents and the population of nonrespondents. The corresponding estimate is $.9\hat{p}_1 + .1\hat{p}_2 =$ $.9 \times .056 + .1 \times .375 = .088$. This is considerably higher than the initial estimate of 5.6%, due to the high rate of unemployment among the nonrespondents.

16.2   STRATIFIED RANDOM SAMPLING

Randomness in drawing a sample, which is essential to obtaining unbiased estimates, results in sampling variability. In some situations the variability can be reduced without introducing bias by using other information about the population. Suppose that a sample of engineers employed in a large corporation is to be drawn to estimate the mean salary of all engineers. An individual's salary depends heavily on his or her corporate function—whether the position is supervisory or nonsupervisory. If the listing of engineers (a frame) is such that the function of each is identified, it is possible to draw a random sample of each type of engineer and estimate the average salary of each type of engineer in the entire corporation. A weighted average of these two estimates yields an estimate of the average salary of all engineers. The sampling variability of this procedure is less than that of simple random sampling because the variability within the set of supervisory engineers and within the set of nonsupervisory engineers is less than the variability among all the engineers.

A *stratum* is a subpopulation. A set of *strata* is a collection of subsets of individuals in the population such that each individual belongs to one and only one such subset. To use "stratified sampling," it is essential that a frame be available for each stratum; this implies that the sizes of the strata are known.

16.2.1   Case of Two Strata

To begin the development of the theory of stratified random sampling, let us assume two strata: supervisory and nonsupervisory. Let N denote the size of the total population of engineers and μ the mean salary in this population. Suppose that some N_1 of these N have supervisory positions; let μ_1 denote the mean income in this stratum. The other $N_2 (=N - N_1)$ engineers have nonsupervisory positions, and their mean income is μ_2. The parameter μ is equal to $(N_1\mu_1 + N_2\mu_2)/N$. The parameters μ, μ_1, and μ_2 are, of course, unknown to the investigator, but N, N_1, and N_2 are known from information on the sampling frame (the list of engineers).

For the purpose of explaining the principles, suppose that the engineers are numbered in such a way that the numbers $1, 2, \ldots, N_1$ denote the supervisory engineers and the numbers $N_1 + 1,$

$N_1 + 2, \ldots, N_1 + N_2$ denote the nonsupervisory engineers. Here y_i is the income of the ith engineer. The mean for supervisory engineers is

$$\mu_1 = \frac{y_1 + y_2 + \cdots + y_{N_1}}{N_1},$$

the mean for nonsupervisory engineers is

$$\mu_2 = \frac{y_{N_1+1} + y_{N_1+2} + \cdots + y_{N_1+N_2}}{N_2}.$$

The overall mean is

$$\mu = \frac{y_1 + y_2 + \cdots + y_{N_1} + y_{N_1+1} + y_{N_1+2} + \cdots + y_{N_1+N_2}}{N_1 + N_2}$$

$$= \frac{N_1\mu_1 + N_2\mu_2}{N}.$$

That is, the mean of all engineers is the weighted mean of the two types of engineers,

(16.1) $$\mu = \frac{N_1}{N} \mu_1 + \frac{N_2}{N} \mu_2.$$

The population has been *stratified* by occupational position (supervisory vs. nonsupervisory). To estimate μ by stratified random sampling, one takes a random sample of specified size n_1 from the N_1 supervisory engineers and a random sample of size n_2 from the N_2 nonsupervisory engineers. The estimates of the strata means μ_1 and μ_2 are the sample means $\hat{\mu}_1 = \bar{x}_1$ and $\hat{\mu}_2 = \bar{x}_2$ of the samples from the two strata. As an estimate of μ, we take the weighted average corresponding to (16.1), namely,

(16.2) $$\hat{\mu} = \frac{N_1}{N} \hat{\mu}_1 + \frac{N_2}{N} \hat{\mu}_2 = \frac{N_1}{N} \bar{x}_1 + \frac{N_2}{N} \bar{x}_2.$$

The mean of the sampling distribution of $\hat{\mu}$, with $E[x]$ as an alternative symbol for μ_x, is

$$E[\hat{\mu}] = E \left[\frac{N_1}{N} \bar{x}_1 + \frac{N_2}{N} \bar{x}_2 \right]$$

$$= E \left[\frac{N_1}{N} \bar{x}_1 \right] + E \left[\frac{N_2}{N} \bar{x}_2 \right]$$

$$= \frac{N_1}{N} \mu_1 + \frac{N_2}{N} \mu_2 = \mu,$$

by (16.1); thus $\hat{\mu}$ is an unbiased estimate of μ. Note that, in order to form the estimate, one has to know the weights N_1/N and N_2/N

From Section 8.7 we know that

$$\text{Var}(\bar{x}_g) = \frac{N_g - n_g}{N_g - 1} \frac{\sigma_g^2}{n_g}$$

for $g = 1, 2$, where $\text{Var}(\bar{x}_g)$ means $\sigma_{\bar{x}_g}^2$ and σ_g^2 is the variance within the gth stratum. That is, $\sigma_1^2 = \Sigma_{k=1}^{N_1}(y_k - \mu_1)^2/N_1$ and $\sigma_2^2 = \Sigma_{k=N_1+1}^{N_1+N_2}(y_k - \mu_2)^2/N_2$. The variance of the estimate $\hat{\mu}$ is

(16.3) $\text{Var}(\hat{\mu}) = \text{Var}\left(\dfrac{N_1}{N} \bar{x}_1 + \dfrac{N_2}{N} \bar{x}_2\right)$

$$= \text{Var}\left(\frac{N_1}{N} \bar{x}_1\right) + \text{Var}\left(\frac{N_2}{N} \bar{x}_2\right)$$

$$= \left(\frac{N_1}{N}\right)^2 \text{Var}(\bar{x}_1) + \left(\frac{N_2}{N}\right)^2 \text{Var}(\bar{x}_2)$$

$$= \left(\frac{N_1}{N}\right)^2 \frac{N_1 - n_1}{N_1 - 1} \frac{\sigma_1^2}{n_1} + \left(\frac{N_2}{N}\right)^2 \frac{N_2 - n_2}{N_2 - 1} \frac{\sigma_2^2}{n_2}.$$

If the fractions sampled, n_1/N_1 and n_2/N_2, are small, we can ignore the finite population correction and write

(16.4) $$\text{Var}(\hat{\mu}) \doteq \left(\frac{N_1}{N}\right)^2 \frac{\sigma_1^2}{n_1} + \left(\frac{N_2}{N}\right)^2 \frac{\sigma_2^2}{n_2}.$$

The estimate $\hat{\mu}$ is unbiased no matter what n_1 and n_2 are (as long as each is positive). Since the variance, however, depends on the sample sizes as well as population sizes, n_1 and n_2 will be selected to make (16.3) or (16.4) small. When the sample sizes are taken so that $n_1/n_2 = N_1/N_2$, the procedure is called *proportional stratified sampling*. If the total sample size is n, the stratum sample sizes are $n_1 = (N_1/N)n$ and $n_2 = (N_2/N)n$, and the approximate variance (16.4) is

$$\text{Var}(\hat{\mu}_{\text{prop}}) = \frac{1}{n} \left(\frac{N_1}{N} \sigma_1^2 + \frac{N_2}{N} \sigma_2^2\right),$$

which is $1/n$ times the same weighted average of the stratum variances that μ is of the stratum means. An analysis of variance, which is detailed in Section 16.2.2, shows that $\text{Var}(\hat{\mu}_{\text{prop}})$ is never greater than σ^2/n, where σ^2 is the variance of the whole population, and is less than σ^2/n when $\mu_1 \neq \mu_2$. Thus proportional stratified random sampling is preferred over simple random sampling.

Figure 16.1 suggests the distributions of salaries in the two strata and the entire population. The variability in each of the strata is less than the variability in the whole population, and the weighted average of the strata variabilities is less than the variability of the whole population. The example in Section 14.1 shows how the weighted average of the variances of sets of measurements is smaller than the variance of all the measurements.

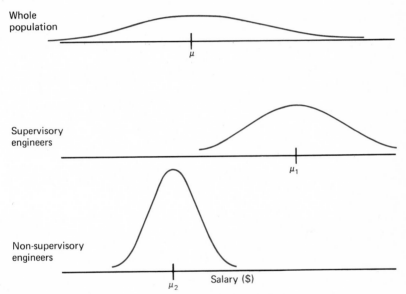

Figure 16.1 *Distributions in strata and in whole population*

If the variances of the strata are known (while the means of the strata are unknown), the allocation of the sample can be improved. The basic idea is that, if one stratum has a very small variance, even if the sample from that stratum is quite small, the sample mean from that stratum will have a small variance. It turns out that the (approximate) variance (16.4) is minimized subject to $n_1 + n_2 = n$, fixed, by taking

$$n_1 = \frac{\sigma_1 N_1}{\sigma_1 N_1 + \sigma_2 N_2} \, n, \qquad n_2 = \frac{\sigma_2 N_2}{\sigma_1 N_1 + \sigma_2 N_2} \, n.$$

That is, n_i is proportional to $\sigma_i N_i$. For instance, if $n = 70$, $N_1 = 2N_2$, and $\sigma_1 = 3\sigma_2$, then $\sigma_1 N_1 = 6\sigma_2 N_2$, and $n_1 = 60$ and $n_2 = 10$. Stratified random sampling with this allocation of total sample size is called *Neyman sampling,* because it was developed by the famous Polish-American statistician Jerzy Neyman. Neyman sampling is an improvement over proportional sampling if the variances in the strata are different (and known).

16.2.2 More than Two Strata

The efficiency of stratification may be increased by increasing the number of strata. The ideas presented for the case of two strata hold for more strata.

In the case of k strata, let y_{gi} be the value of the variable for the ith individual in the gth stratum, $i = 1, 2, \ldots, N_g$, $g = 1, 2, \ldots, k$. The population mean is written in this notation as

$$\mu = \frac{1}{N} \sum_{g=1}^{k} \sum_{i=1}^{N_g} y_{gi};$$

the population variance is

$$\sigma^2 = \frac{1}{N} \sum_{g=1}^{k} \sum_{i=1}^{N_g} (y_{gi} - \mu)^2.$$

The mean in the gth stratum is

$$\mu_g = \frac{1}{N_g} \sum_{i=1}^{N_g} y_{gi};$$

the variance in the gth stratum is

$$\sigma_g^2 = \frac{1}{N_g} \sum_{i=1}^{N_g} (y_{gi} - \mu_g)^2.$$

Making a decomposition like that used in analysis of variance (Section 13.2), we can write the population variance in terms of the stratum means and variances as

$$N\sigma^2 = \sum_{g=1}^{k} \sum_{i=1}^{N_g} (y_{gi} - \mu)^2$$

$$= \sum_{g=1}^{k} \sum_{i=1}^{N_g} (y_{gi} - \mu_g)^2 + \sum_{g=1}^{k} N_g(\mu_g - \mu)^2$$

$$= \sum_{g=1}^{k} N_g\sigma_g^2 + \sum_{g=1}^{k} N_g(\mu_g - \mu)^2.$$

TABLE 16.1

Decomposition of Population Sum of Squares in Terms of Stratum Means and Variances

SOURCE OF VARIATION	SUM OF SQUARES
Between Strata	$\sum_{g=1}^{k} N_g(\mu_g - \mu)^2$
Within Strata	$\sum_{g=1}^{k} \sum_{i=1}^{N_g} (y_{gi} - \mu_g)^2 = \sum_{g=1}^{k} N_g \sigma_g^2$
Population	$\sum_{g=1}^{k} \sum_{i=1}^{N_g} (y_{gi} - \mu)^2 = N\sigma^2$

The indicated decomposition is shown in an analysis of variance table for the population in Table 16.1.

A random sample of size n_g is taken from the gth stratum, yielding a sample mean of $\bar{x}_g, g = 1, \ldots, k$. The estimate for the mean based on sampling from k strata is the weighted average of the estimates of the strata means,

$$\hat{\mu} = \sum_{g=1}^{k} \frac{N_g}{N} \bar{x}_g.$$

Note that the weights are the population proportions. The variance of $\hat{\mu}$ is

$$\text{Var}(\hat{\mu}) = \sum_{g=1}^{k} \left(\frac{N_g}{N}\right)^2 \frac{N_g - n_g}{N_g - 1} \frac{\sigma_g^2}{n_g},$$

which, ignoring the finite population correction factors, is approximately

(16.5) $$\text{Var}(\hat{\mu}) \doteq \sum_{g=1}^{k} \left(\frac{N_g}{N}\right)^2 \frac{\sigma_g^2}{n_g}.$$

Proportional sampling occurs when an overall sample of size n is allocated to the strata in proportion to the strata sizes, that is, $n_g = (N_g/N)n$. Then the approximate variance (16.5) is

$$\text{Var}(\hat{\mu}_{\text{prop}}) = \frac{1}{n} \sum_{g=1}^{k} \frac{N_g}{N} \sigma_g^2.$$

This is the Within Strata sum of squares in Table 16.1 divided by $n \times N$. Since the variance of the estimate based on simple random sampling is the Population sum of squares divided by $n \times N$, the reduction in variance due to proportional sampling is the Between Strata sum of squares divided by $n \times N$, that is,

$$\frac{1}{n} \sum_{g=1}^{k} \frac{N_g}{N} (\mu_g - \mu)^2.$$

This decrease in the variability of the estimate is made large by choosing homogeneous strata, in which case the strata means will be very different. (We note that for proportional sampling, $\hat{\mu} = \Sigma_{g=1}^{n} n_g \bar{x}_g / n$, which is the mean of all the observations.)

If the strata variances are known and some are different, a further increase in efficiency can be made. The optimal allocation of a total sample number n (Neyman sampling) is proportional to $N_g \sigma_g$:

$$n_g = \frac{N_g \sigma_g}{N \bar{\sigma}} n, \qquad g = 1, \ldots, k,$$

where $\bar{\sigma} = \Sigma_{g=1}^{k}(N_g/N)\sigma_g$, the weighted average of strata standard deviations. The variance of the estimate, say $\hat{\mu}_{\text{Neyman}}$, is

$$\text{Var}(\hat{\mu}_{\text{Neyman}}) = \sum_{g=1}^{k} \left(\frac{N_g}{N}\right)^2 \frac{\sigma_g^2}{n\sigma_g N_g/(N\bar{\sigma})}$$

$$= \frac{\bar{\sigma}}{n} \sum_{g=1}^{k} \frac{N_g}{N} \sigma_g = \frac{\bar{\sigma}^2}{n}.$$

The saving over proportional sampling is

$$\frac{1}{n} \left(\sum_{g=1}^{k} \frac{N_g}{N} \sigma_g^2 - \bar{\sigma}^2\right) = \frac{1}{n} \sum_{g=1}^{k} \frac{N_g}{N} (\sigma_g - \bar{\sigma})^2.$$

This shows that such a saving is due to the variation among standard deviations of the strata.

Usually the strata variances are not known exactly; they may be guessed on the basis of studies made earlier in time or based on related characteristics. For any method of stratified random sampling, the variance σ_g^2 can be estimated by the sample variance s_g^2 of the n_g observations from that stratum (if $n_g \geq 2$). Substituting s_g^2 for σ_g^2 into (16.5) or the preceding exact expression yields an estimate of the variance of the estimate of μ.

16.3 CLUSTER SAMPLING

Cluster sampling refers to sampling "clusters" of potential respondents and then sampling respondents in the clusters in the sample. To determine the total number of unemployed in a city, for example, one might consider city blocks as the clusters and households as the "respondents." A sample of city blocks is taken, using a map of the city to number the blocks. In each block the households are enumerated, and a random sample of households is taken in each block in the sample. The *total* number of unemployed in a sampled block is estimated from the sample of households in that block. In turn, the total number of unemployed in the city is estimated from these estimated block totals. An unbiased estimate of the population total results from this procedure. An unbiased estimate of the population mean is obtained by dividing the estimated total by the number of units in the population.

Sampling variability arises from two sources: the sampling of clusters and the sampling of units within clusters. The formula for the variance of an estimate depends on the rule for sample size within the sampled cluster. For example, the sample sizes may be the same in all clusters or they may be a fixed proportion of the cluster sizes. It is beyond the scope of this book to develop these formulas. See Cochran (1963), Chapter 9, for example.

What are the advantages of cluster sampling? When the clusters represent geographically compact sets of units, as in the above illustration, with cluster sampling the interviewers may spend more time in interviewing than traveling. Also, a frame for clusters (for instance, blocks) may be available, making enumeration of clusters feasible, while preparation of a frame for the entire population of units is not practical.

In a sense the clusters are strata, since each individual or unit belongs to one and only one cluster. The greater flexibility here results because clusters (or strata) are themselves sampled.

16.4 SYSTEMATIC SAMPLING WITH A RANDOM START

The idea of systematic sampling is to take every tenth name on a list, or check every fifth car passing a toll booth, or review every twentieth file folder in a drawer. The method is appealing because it is easy to carry out and it spreads the "sample" out through the population. However, it is clearly not random. To add an element of randomness—which is necessary to obtain unbiased estimates—one may select the starting point at random.

The systematic sampling procedure begins with the construction of a frame, as before. Suppose we want a sample of size n out of a population of N, and suppose that $N/n = l$, approximately an integer. The number l is called the *sampling interval*. If we want a sample of 300 families out of 3000, the sampling interval is 10. We choose an integer at random from $1, 2, \ldots, l$. Suppose we get 7. Then we choose the individuals numbered $7, 7 + l, 7 + 2l, \ldots, 7 + (n - 1)l$ as our sample; our observations will be

$$x_1 = y_7, x_2 = y_{7+l}, \ldots, x_n = y_{7+(n-1)l}.$$

The estimate of the population mean is the sample mean $\bar{x} = \sum_{i=1}^{n} x_i / n$.

The sample mean is unbiased (because the start is random), but the precision of $\bar{x}$ depends on how the characteristic under observation varies as we go through the frame. If the population is the 365 days of the year, the frame is the calendar. When the sampling interval is 7 and $n = 52$, we get a systematic sample that is based on the same day of the week over the entire year. The method is good for estimating the average hours of daylight per day over the year, but poor for estimating the average hours of work per day over the year.

Suppose again that we are sampling households and that the frame lists houses in the following order: Avenue A from west to east; Avenue B from west to east on the north side of the street, Avenue B from west to east on the south side of the street, etc. The blocks are oblong and narrow between avenues, so that all houses face on avenues. Suppose that the ordering in the frame corresponds to the cost of the houses; house 1 is least expensive and house 36 most expensive (Figure 16.2). Then systematic sampling gives us a sample of households that is varied and representative as far as cost of house (and consequently family income) is concerned. In this case, systematic sampling performs roughly like stratified sampling, where the strata are defined in terms of income.

On the other hand, if the variation in the population is related to the sampling interval, then systematic sampling can be much less precise than simple random sampling; at worst it can be equivalent to having only a sample of size 1. If the corner houses are most expensive, and our sampling interval is 3, then we get a very nonrepresentative sample: either all corner houses or all middle-of-the-block houses.

Let us denote by m the starting number, the integer chosen at random from $1, 2, \ldots, l$. The quantity m is a random variable that takes on the values $1, 2, \ldots, l$, each with probability $1/l$. Only l different samples are possible. These are

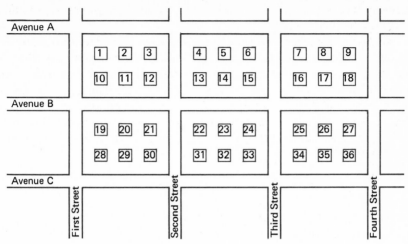

Figure 16.2 *Frame for systematic sampling*

$$x_1 = y_m, \, x_2 = y_{m+l}, \, \ldots, \, x_n = y_{m+(n-1)l},$$

where m ranges over $1, 2, \ldots, l$. Although all individuals have the same chance of being included in the sample, not all possible sets of n have the same chance: l sets have probability $1/l$ each. All the other sets have zero probability. The possible values of $\bar{x}$ are

$$\bar{y}_1 = \frac{y_1 + y_{1+l} + \cdots + y_{1+(n-1)l}}{n},$$

$$\bar{y}_2 = \frac{y_2 + y_{2+l} + \cdots + y_{2+(n-1)l}}{n},$$

(16.6)

$$\vdots$$

$$\bar{y}_l = \frac{y_l + y_{2l} + \cdots + y_{nl}}{n};$$

each of these values has probability $1/l$.

In effect, the observed $\bar{x}$ is a random sample of 1 from the population of l values, $\bar{y}_1, \ldots, \bar{y}_l$. The mean of the sampling distribution of $\bar{x}$ is

$$E(\bar{x}) = \frac{1}{l} \sum_{h=1}^{l} \bar{y}_h = \frac{1}{nl} \sum_{k=1}^{nl} y_k,$$

which is μ if $nl = N$ (and is approximately μ in any case). The variance is $\Sigma_{h=1}^{l}(\bar{y}_h - \mu)^2/l$, which is unknown. It cannot be estimated from the sample because the sample is effectively a sample of 1 from the population (16.6). The investigator cannot assess variability, carry out tests of significance, or construct confidence intervals.

16.5 SYSTEMATIC SUBSAMPLING WITH RANDOM STARTS

A simple modification of systematic sampling makes it possible for one to estimate an appropriate variance. The point is to carry out the procedure with several random starts.

Suppose that the population is of size $N = 3000$ and a sample of $n = 300$ is wanted. With systematic sampling we would compute the sampling interval as $3000/300 = 10$ and observe every 10th individual, randomly selecting a starting number from 1, 2, ... , 10. Suppose that, instead of taking *one* sample of size $n = 300$, we take, for example, 5 *subsamples* of size $n/5 = 300/5 = 60$. For samples of size 60, the sampling interval is $N/60 = 3000/60 = 50$. We randomly obtain five different starting numbers from 1, 2, ... , 50. Suppose that the first starting number is 17. Then the first subsample of 60 consists of the individuals numbered 17, 67, 117, ... , 2967, and we observe the values $y_{17}, y_{67}, y_{117}, \ldots, y_{2967}$. Suppose that the second starting number is 23. The second subsample gives the values $y_{23}, y_{73}, y_{123}, \ldots, y_{2973}$. Continuing in this way, we get five systematic subsamples. Let the means of the subsamples be denoted by $\bar{x}_1, \bar{x}_2, \bar{x}_3, \bar{x}_4, \bar{x}_5$. Then the mean of all $5 \times 60 = 300$ observations can be written as the mean of the subsample means,

$$\bar{x} = \frac{\bar{x}_1 + \bar{x}_2 + \cdots + \bar{x}_5}{5}.$$

This statistic is an unbiased estimate of the population mean. The sample variance of subsample means,

$$s_{\bar{x}_i}^2 = \frac{\displaystyle\sum_{i=1}^{5} (\bar{x}_i - \bar{x})^2}{4},$$

is an unbiased estimate of the variance $\sigma_{\bar{x}_i}^2$ of subsample means. The subsample means constitute a sample of size 5 from a population of subsample means. This population is of size 50, for the possible values of the subsample means are

$$\frac{y_1 + y_{51} + \cdots + y_{2951}}{60},$$

$$\frac{y_2 + y_{52} + \cdots + y_{2952}}{60},$$

.
.
.

$$\frac{y_{50} + y_{100} + \cdots + y_{3000}}{60}.$$

Hence the variance of $\bar{x}$ is

$$\sigma_{\bar{x}}^2 = \frac{\sigma_{\bar{x}_i}^2}{5} \frac{50 - 5}{50 - 1},$$

which is estimated by

$$s_{\bar{x}}^2 = \frac{s_{\bar{x}_i}^2}{5} \frac{50 - 5}{50 - 1}.$$

Here the "population" is a population of subsample means; this population is usually approximately normal, and inferences can be based on the normal distribution.

SUMMARY

Usually, for budgetary or other reasons, an investigator knows that only a sample of relatively limited size is affordable. Then the design of the survey becomes especially important.

Often an investigator partitions the population into subpopulations, or *strata*. Making the allocation of observations proportional to the strata sizes is called *proportional sampling*.

Proportional sampling is always better than simple random sampling from the population as a whole. If the strata variances are known, the best method of stratified sampling is *Neyman sampling*, in which the sample sizes are proportional to the products of the strata sizes and the strata standard deviations.

Systematic sampling has also been discussed.

APPENDIX 16A   Dichotomous Variables

Often one is interested in a Yes-No characteristic, as in the case of inspection of manufactured articles, which can be classified as either defective or not. Here each value y_k is either 1 or 0 according as the kth item is defective or not, $k = 1, 2, \ldots, N$. If A articles from among the population of N are defective, then the proportion of defective articles is $p = A/N$. The probability that exactly d articles in a random sample of n are defective is

(16.7)
$$\frac{C_d^A C_{n-d}^{N-A}}{C_n^N}.$$

Here d runs over the range $\max\{0, n - (N - A)\}, \ldots, \min\{A, n\}$; (16.7) is the *hypergeometric probability distribution*. If N is large, n is less than A, and n/N is small, (16.7) is approximated by the binomial distribution

$$C_d^n p^d (1 - p)^{n-d}, \qquad d = 0, 1, \ldots, n.$$

EXERCISES

16.1 (Sec. 16.1)   Define the events $A = (1, 2, 3)$, $B = (4, 5, 6)$, and $C = (7, 8, 9)$. If the procedure used in constructing the table in Appendix V is random, the (conditional) probability of each event is 1/3, given that the digit is not 0. Use as your "sample" the first 15 nonzero digits in the line of the table, numbered according to the day of the month of your birthday. For example, if your birthday is on the thirteenth of some month, use the thirteenth line of the table.
  (a) Calculate the chi-square statistic.
  (b) Test the null hypothesis that the (conditional) probabilities are 1/3 at the 5% level.

16.2 (Sec. 16.1)   Show that for simple random sampling of a sample of size $n = 2$ without replacement from a population of size $N = 3$, the variance of the sample mean is

$$\frac{\sigma^2}{4} = \frac{3 - 2}{3 - 1} \frac{\sigma^2}{2} = \frac{N - n}{N - 1} \frac{\sigma^2}{n}.$$

16.3 (Introduction)   In the TV-viewing example, if more than one child in a household is included in the sample, do you think it would be valid to consider the corresponding observations as statistically independent? Why or why not?

16.4 (Sec. 16.2)   From (16.4), deduce the conditions under which n_1 and n_2 would be chosen to be equal.

16.5 (Sec. 16.2)   [*Note:* The assumption is made throughout this exercise that sampling (from a finite population) is done with replacement, so that there need be no correction for finite population. It is assumed that the stratum sizes are known, so that proportional random sampling can be used.]

Suppose an investigator wishes to estimate the income of a population of engineers by drawing a sample. The investigator believes that a stratified sample would be better than a simple random sample, but is not sure whether it would be better to stratify by occupational position (supervisory vs. nonsupervisory) or by department (production vs. development). Table 16.2 shows a hypothetical *population* of 10 incomes (to the nearest thousand dollars) in two departments for supervisory and nonsupervisory engineers. Your computations in answer to the following questions illustrate the answer to this investigator's question.

TABLE 16.2

Incomes of Engineers ($1000's)

	PRODUCTION	DEVELOPMENT
Supervisory	20, 20, 17	23
Nonsupervisory	15, 12, 12	8, 7, 6

(hypothetical data)

(a) Compute the standard deviation of the sampling distribution of the mean income (to the nearest dollar) for samples of size 5 in proportional stratified sampling from this population: (i) stratified by occupational position, and (ii) stratified by department.
(b) Show (algebraically) what you would use as an estimate of mean income in the population on the basis of the above samples.
(c) Compute the standard deviation of the sampling distribution of the mean income for simple random sampling.
(d) Describe the information about stratified sampling that is illustrated by your computations.
(e) Graph on one line the incomes of supervisory engineers and on a parallel line the incomes of nonsupervisory engineers. Graph on a pair of parallel lines the incomes of production and development engineers, respectively.

16.6 (Sec. 16.2)   An investigator is interested in the average amount a large garage charges for automobile repairs. The repair bills from the garage are categorized according to the size of car on which the repair was made, giving the breakdown in Table 16.3.

TABLE 16.3
Repair Bills for Autos

Category	Number of Receipts	Average Charge	Standard Deviation within Category
Large cars (8 cylinders or more)	100	$37.50	$14.00
Small cars (less than 8 cylinders)	150	25.25	11.00

(hypothetical data)

(a) What is the average repair bill from the garage?

(b) Suppose that the investigator does not know the average charge and variance within each category, but only the number of receipts in each category. The investigator randomly selects 10 receipts from "large cars" and finds that their total is $480, then randomly selects 10 receipts from "small cars" and finds that their total is $250. What is the best estimate of the overall average repair bill?

(c) What is the variance of this estimate? (The investigator cannot compute this, but since we have more information, we can.)

16.7 (Sec. 16.4)   An investigator proposed a systematic sample of residential apartment buildings for a study of neighborhood incomes. Since all buildings had one family per floor and the same number of floors, the proposal was to choose a floor randomly, then to sample the *same* floor in each building. (Note that we can think of the families as listed in order with a serial number, so that our sample includes every third family, which makes the sampling interval equal to 3.)

(a) Assuming a random starting point for this sampling, list all the different possible samples of 3 that could be drawn by systematic sampling in the two situations shown in Table 16.4.

(b) Compute the standard error of the mean (to the nearest dollar) for samples of 3 in the above two situations.

(c) Define the sample mean whose standard error is computed in (b).

(d) In Situations A and B of Table 16.4, compute the standard error of the mean, assuming simple random sampling from the population, for samples of size 3.

(e) Discuss your conclusions in regard to the efficiency of systematic sampling.

TABLE 16.4

Incomes ($1000s) of Residents on Each Floor for Three Apartment Buildings

SITUATION A

	Building		
	1	2	3
First floor	4	10	8
Second floor	4	10	8
Third floor	4	10	8

SITUATION B

	Building		
	1	2	3
First floor	10	10	10
Second floor	8	8	8
Third floor	4	4	4

(hypothetical data)

16.8 (Sec. 16.4)   Table 16.5 is a table of hours of study time per day for a Stanford student on 28 successive days.

TABLE 16.5

Hours of Study Time per Day for 28 Successive Days

4 6 5 4 3 1 2 6 6 4 3 5 2 2 4 6 5 5 3 2 3 6 6 6 4 5 3 1

$$\bar{x} = 4, \ \Sigma x_i^2 = 520, \ \Sigma(x_i - \bar{x})^2 = 72$$

(a) What is the variance of the estimate of the mean number of hours studied per day, using systematic sampling with a random start and a sampling interval of 7 days (sample size of 4)?

(b) What is the variance of the estimate of the mean hours studied per day, using random sampling with sample size 4?

(c) Compare the results in parts (a) and (b), and explain.

MATHEMATICAL EXERCISES

16.9 (Sec. 16.2)   Let $\hat{\mu}_{\text{prop}}$ be the estimate based on proportional stratified random sampling. Show that

$$(16.8) \quad \text{Var}(\bar{x}) - \text{Var}(\hat{\mu}_{\text{prop}}) \doteq \frac{1}{N}(b - 1) \sum_{g=1}^{k} \frac{N_g}{N}(\mu_g - \mu)^2,$$

where there are k strata and $b = N_1/n_1 = N_2/n_2 = \cdots = N_k/n_k = N/n$.

Note that the difference is never negative, so that proportional sampling is always at least as good as simple random sampling. Also, note that the difference involves only the stratum means and not the stratum variances. Finally, note that if the means are not very different, the summation is small, and it may not be worth the trouble to stratify. [*Hint:* You can find this exercise worked out in the text except for the finite population correction factor.]

16.10 (Sec. 16.2) Let $\hat{\mu}_{\text{Neyman}}$ be the stratified-sampling estimate based on Neyman sampling. Show that

$$(16.9) \quad \text{Var}(\hat{\mu}_{\text{prop}}) - \text{Var}(\hat{\mu}_{\text{Neyman}}) = \frac{1}{n} \sum_{\substack{g=1 \\ g<h}}^{k} \sum_{h=1}^{k} \frac{N_g}{N} \frac{N_h}{N} (\sigma_g - \sigma_h)^2.$$

Note in particular that this expression does not involve the means.

16.11 (Sec. 16.2) Combine (16.8) and (16.9) to show that

$$(16.10) \quad \text{Var}(\bar{x}) - \text{Var}(\hat{\mu}_{\text{Neyman}}) = \frac{1}{n} \sum_{\substack{g=1 \\ g<h}}^{k} \sum_{h=1}^{k} \frac{N_g}{N} \frac{N_h}{N} (\sigma_g - \sigma_h)^2$$

$$+ \frac{1}{N} (b-1) \sum_{g=1}^{k} \frac{N_g}{N} (\mu_g - \mu)^2.$$

This is the maximum possible improvement achievable by using stratified sampling instead of simple random sampling. Note that the more disparate the means and standard deviations, the larger the improvement.

16.12 (Sec. 16.2) Suppose that $\sigma_1 N_1$ is larger than $\sigma_2 N_2$. Show that $\text{Var}(\hat{\mu})$ can be as large as

$$(16.11) \qquad\qquad \left(\frac{N_1}{N}\right)^2 \sigma_1^2 + \left(\frac{N_2}{N}\right)^2 \frac{\sigma_2^2}{n-1}$$

if the choice of n_1 and n_2 is indiscriminate. Show that (16.11) can be larger than $\text{Var}(\bar{x})$, even when σ_1 and σ_2 are both smaller than σ.

16.13 (Sec. 16.2) Suppose that $N_1 = N_2 = N/2$, $\sigma_1^2 = 400$, and $\sigma_2^2 = 40,000$.

(a) Show that if $n = 20$, the optimal allocation, that is, the allocation that minimizes (16.4), is $n_1 = 20/11 = 1.82$ and $n_2 = 18.18$. This suggests rounding to $n_1 = 2$, $n_2 = 18$.

(b) Show that it is better to round to $n_1 = 1, n_2 = 19$. (Note, however, that then one could not estimate σ_1 and, hence, verify the *a priori* information.)

16.14 (Sec. 16.2) Note that (16.4) can be written as

$$\text{Var}(\hat{\mu}) \doteq \frac{a_1^2}{n_1} + \frac{a_2^2}{n_2},$$

where $a_1 = N_1\sigma_1/N$ and $a_2 = N_2\sigma_2/N$. The total sample size $n_1 + n_2$ is required by budgetary considerations to be equal to a fixed number n; thus $n_2 = n - n_1$, and if we increase n_1 we must decrease n_2 by the same amount. Suppose that we set n_1 at a trial value v. Compute the change in $\text{Var}(\hat{\mu})$ due to increasing n_1 from v to $v + 1$. Hence deduce the optimal values of n_1 and n_2, as in (16.4).

16.15 (continuation of 16.3) Let x_1 and x_2 represent the total amount of time two children spend watching TV over the two-week period. We observe $x_1' = x_1 + e_1$ and $x_2' = x_2 + e_2$, where e_1 and e_2 are errors of observation introduced by the parents in recording the data. Do you think it is reasonable to assume that x_1 and x_2 are statistically independent? That e_1 and e_2 are statistically independent? That x_1 and x_2 are statistically independent, given that the two children are in the same family? That e_1 and e_2 are statistically independent, given that the two children are in the same family?

REFERENCES

Cochran, William G. (1963). *Sampling Techniques*. 2nd ed. Wiley, New York.

Deming, W. E. (1950). *Some Theory of Sampling*. Wiley, New York. (Reprinted by Dover, New York, 1966.)

APPENDICES

APPENDIX I / THE CUMULATIVE DISTRIBUTION FUNCTION OF THE STANDARD NORMAL DISTRIBUTION

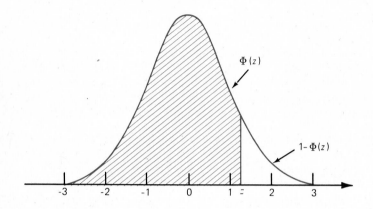

INTEGER AND FIRST DECIMAL OF z	SECOND DECIMAL OF z									
	.00	.01	.02	.03	.04	.05	.06	.07	.08	.09
0.0	.5000	.5040	.5080	.5120	.5160	.5199	.5239	.5279	.5319	.5359
0.1	.5398	.5438	.5478	.5517	.5557	.5596	.5636	.5675	.5714	.5753
0.2	.5793	.5832	.5871	.5910	.5948	.5987	.6026	.6064	.6103	.6141
0.3	.6179	.6217	.6255	.6293	.6331	.6368	.6406	.6443	.6480	.6517
0.4	.6554	.6591	.6628	.6664	.6700	.6736	.6772	.6808	.6844	.6879
0.5	.6915	.6950	.6985	.7019	.7054	.7088	.7123	.7157	.7190	.7224
0.6	.7257	.7291	.7324	.7357	.7389	.7422	.7454	.7486	.7517	.7549
0.7	.7580	.7611	.7642	.7673	.7704	.7734	.7764	.7794	.7823	.7852
0.8	.7881	.7910	.7939	.7967	.7995	.8023	.8051	.8078	.8106	.8133
0.9	.8159	.8186	.8212	.8238	.8264	.8289	.8315	.8340	.8365	.8389
1.0	.8413	.8438	.8461	.8485	.8508	.8531	.8554	.8577	.8599	.8621
1.1	.8643	.8665	.8686	.8708	.8729	.8749	.8770	.8790	.8810	.8830
1.2	.8849	.8869	.8888	.8907	.8925	.8944	.8962	.8980	.8997	.9015
1.3	.9032	.9049	.9066	.9082	.9099	.9115	.9131	.9147	.9162	.9177
1.4	.9192	.9207	.9222	.9236	.9251	.9265	.9279	.9292	.9306	.9319

$\Phi(z)$ for $z = 0.02$ is found in the row labelled 0.0 and the column labelled .02. Thus, for $z = 0.02$, $\Phi(z) = \Phi(0.02) = .5080$. $\Phi(z) = 1 - \Phi(-z)$; thus for $z = -0.02$, $\Phi(-0.02) = 1 - \Phi(0.02) = 1 - .5080 = .4920$.

Appendix I / The Cumulative Distribution Function of the
Standard Normal Distribution

INTEGER AND FIRST DECIMAL OF z	SECOND DECIMAL OF z									
	.00	.01	.02	.03	.04	.05	.06	.07	.08	.09
1.5	.9332	.9345	.9357	.9370	.9382	.9394	.9406	.9418	.9429	.9441
1.6	.9452	.9463	.9474	.9484	.9495	.9505	.9515	.9525	.9535	.9545
1.7	.9554	.9564	.9573	.9582	.9591	.9599	.9608	.9616	.9625	.9633
1.8	.9641	.9649	.9656	.9664	.9671	.9678	.9686	.9693	.9699	.9706
1.9	.9713	.9719	.9726	.9732	.9738	.9744	.9750	.9756	.9761	.9767
2.0	.9772	.9778	.9783	.9788	.9793	.9798	.9803	.9808	.9812	.9817
2.1	.9821	.9826	.9830	.9834	.9838	.9842	.9846	.9850	.9854	.9857
2.2	.9861	.9864	.9868	.9871	.9875	.9878	.9881	.9884	.9887	.9890
2.3	.9893	.9896	.9898	.9901	.9904	.9906	.9909	.9911	.9913	.9916
2.4	.9918	.9920	.9922	.9925	.9927	.9929	.9931	.9932	.9934	.9936
2.5	.9938	.9940	.9941	.9943	.9945	.9946	.9948	.9949	.9951	.9952
2.6	.9953	.9955	.9956	.9957	.9959	.9960	.9961	.9962	.9963	.9964
2.7	.9965	.9966	.9967	.9968	.9969	.9970	.9971	.9972	.9973	.9974
2.8	.9974	.9975	.9976	.9977	.9977	.9978	.9979	.9979	.9980	.9981
2.9	.9981	.9982	.9982	.9983	.9984	.9984	.9985	.9985	.9986	.9986
3.0	.9987	.9987	.9987	.9988	.9988	.9989	.9989	.9989	.9990	.9990

APPENDIX II / PERCENTILES OF THE STANDARD NORMAL DISTRIBUTION

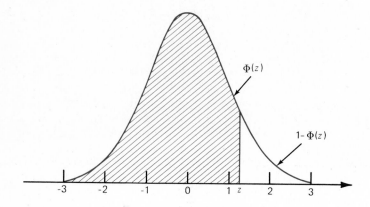

First Decimal of $\Phi(z)$	Second Decimal of $\Phi(z)$									
	.00	.01	.02	.03	.04	.05	.06	.07	.08	.09
.5	.000	.025	.050	.075	.100	.126	.151	.176	.202	.228
.6	.253	.279	.305	.332	.358	.385	.412	.440	.468	.496
.7	.524	.553	.583	.613	.643	.674	.706	.739	.772	.806
.8	.842	.878	.915	.954	.994	1.036	1.080	1.126	1.175	1.227
.9	1.282	1.341	1.405	1.476	1.555	1.645	1.751	1.881	2.054	2.326

$$.975 = \Phi(1.960) \qquad .995 = \Phi(2.576)$$

The 90th percentile, which is the value of z for which $\Phi(z) = .90$, is found in the row labelled .9 and the column labelled .00. Thus, for $\Phi(z) = .90$, $z = 1.282$.

For percentages under 50 use the relationship $\Phi(z) = 1 - \Phi(-z)$. For example, to find the 25th percentile, let z equal this unknown quantity. Then $\Phi(z) = .25$, so $1 - \Phi(-z) = .25$. Thus $\Phi(-z) = .75$, and from the table $-z = .674$. Therefore, $z = -.674$; the 25th percentile is $-.674$.

APPENDIX III / PERCENTAGE POINTS OF STUDENT'S t-DISTRIBUTIONS: VALUES OF $t_f(\alpha)$ FOR VARIOUS α AND f

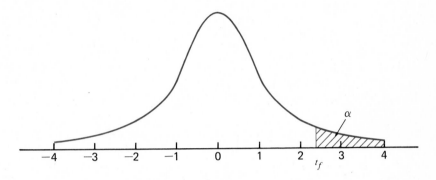

f	SIGNIFICANCE LEVEL FOR ONE-TAILED TEST α				
	.10	.05	.025	.01	.005
1	3.078	6.314	12.706	31.821	63.657
2	1.886	2.920	4.303	6.964	9.925
3	1.638	2.353	3.182	4.541	5.841
4	1.533	2.132	2.776	3.747	4.604
5	1.476	2.015	2.571	3.365	4.032
6	1.440	1.943	2.447	3.143	3.707
7	1.415	1.895	2.365	2.998	3.499
8	1.397	1.860	2.306	2.896	3.355
9	1.383	1.833	2.262	2.821	3.250
10	1.372	1.812	2.228	2.764	3.169
11	1.363	1.796	2.201	2.718	3.106
12	1.356	1.782	2.179	2.681	3.054
13	1.350	1.771	2.160	2.650	3.012
14	1.345	1.761	2.145	2.624	2.977
15	1.341	1.753	2.132	2.602	2.947
16	1.337	1.746	2.120	2.584	2.921
17	1.333	1.740	2.110	2.567	2.898
18	1.330	1.734	2.101	2.552	2.878
19	1.328	1.729	2.093	2.540	2.861
20	1.325	1.725	2.086	2.528	2.845
21	1.323	1.721	2.080	2.518	2.831
22	1.321	1.717	2.074	2.508	2.819
23	1.320	1.714	2.069	2.500	2.807

Appendix III / Percentage Points of Student's t-Distributions: Values of $t_f(\alpha)$ for Various α and f

f	SIGNIFICANCE LEVEL FOR ONE-TAILED TEST α				
	.10	.05	.025	.01	.005
24	1.318	1.711	2.064	2.492	2.797
25	1.316	1.708	2.060	2.485	2.788
26	1.315	1.706	2.056	2.479	2.779
27	1.314	1.703	2.052	2.473	2.771
28	1.312	1.701	2.048	2.467	2.763
29	1.311	1.699	2.045	2.462	2.756
30	1.310	1.697	2.042	2.457	2.750
40	1.303	1.684	2.021	2.423	2.704
60	1.296	1.671	2.000	2.390	2.660
120	1.289	1.658	1.980	2.358	2.617
∞	1.282	1.645	1.960	2.326	2.576

For a one-tailed test α is the significance level of the test against the alternative that the mean is positive. For a two-tailed test refer to the column headed by 1/2 of the desired significance level; for example, if the significance level is 5% use the percentage point in the column headed .025.

APPENDIX IV / BINOMIAL PROBABILITIES

		PROBABILITY OF HEAD p							
n	r	.01	.05	.10	1/6	.25	1/3	.49	.50
2	0	.9801	.9025	.8100	.6944	.5625	.4444	.2601	.2500
	1	.0198	.0950	.1800	.2778	.3750	.4444	.4998	.5000
	2	.0001	.0025	.0100	.0278	.0625	.1111	.2401	.2500
3	0	.9703	.8574	.7290	.5787	.4219	.2963	.1327	.1250
	1	.0294	.1354	.2430	.3472	.4219	.4444	.3823	.3750
	2	.0003	.0071	.0270	.0694	.1406	.2222	.3674	.3750
	3	.0000	.0001	.0010	.0046	.0156	.0370	.1176	.1250
4	0	.9606	.8145	.6561	.4823	.3164	.1975	.0677	.0625
	1	.0388	.1715	.2916	.3858	.4219	.3951	.2600	.2500
	2	.0006	.0135	.0486	.1157	.2109	.2963	.3747	.3750
	3	.0000	.0005	.0036	.0154	.0469	.0988	.2400	.2500
	4	.0000	.0000	.0001	.0008	.0039	.0123	.0576	.0625
5	0	.9510	.7738	.5905	.4019	.2373	.1317	.0345	.0312
	1	.0480	.2036	.3280	.4019	.3955	.3292	.1657	.1562
	2	.0010	.0214	.0729	.1608	.2637	.3292	.3185	.3125
	3	.0000	.0011	.0081	.0322	.0879	.1646	.3060	.3125
	4	.0000	.0000	.0004	.0032	.0146	.0412	.1470	.1562
	5	.0000	.0000	.0000	.0001	.0010	.0041	.0282	.0312
6	0	.9415	.7351	.5314	.3349	.1780	.0878	.0176	.0156
	1	.0571	.2321	.3543	.4019	.3560	.2634	.1014	.0938
	2	.0014	.0305	.0984	.2009	.2966	.3292	.2436	.2344
	3	.0000	.0021	.0146	.0536	.1318	.2195	.3121	.3125
	4	.0000	.0001	.0012	.0080	.0330	.0823	.2249	.2344
	5	.0000	.0000	.0001	.0006	.0044	.0165	.0864	.0938
	6	.0000	.0000	.0000	.0000	.0002	.0014	.0138	.0156
7	0	.9321	.6983	.4783	.2791	.1335	.0585	.0090	.0078
	1	.0659	.2573	.3720	.3907	.3115	.2048	.0604	.0547
	2	.0020	.0406	.1240	.2344	.3115	.3073	.1740	.1641
	3	.0000	.0036	.0230	.0781	.1730	.2561	.2786	.2734
	4	.0000	.0002	.0026	.0156	.0577	.1280	.2676	.2734
	5	.0000	.0000	.0002	.0019	.0115	.0384	.1543	.1641
	6	.0000	.0000	.0000	.0001	.0013	.0064	.0494	.0547
	7	.0000	.0000	.0000	.0000	.0001	.0005	.0068	.0078

Appendix IV / Binomial Probabilities

		PROBABILITY OF HEAD p							
n	r	.01	.05	.10	1/6	.25	1/3	.49	.50
8	0	.9227	.6634	.4305	.2326	.1001	.0390	.0046	.0039
	1	.0746	.2793	.3826	.3721	.2670	.1561	.0352	.0312
	2	.0026	.0515	.1488	.2605	.3115	.2731	.1183	.1094
	3	.0001	.0054	.0331	.1042	.2076	.2731	.2273	.2188
	4	.0000	.0004	.0046	.0260	.0865	.1707	.2730	.2734
	5	.0000	.0000	.0004	.0042	.0231	.0683	.2098	.2188
	6	.0000	.0000	.0000	.0004	.0038	.0171	.1008	.1094
	7	.0000	.0000	.0000	.0000	.0004	.0024	.0277	.0312
	8	.0000	.0000	.0000	.0000	.0000	.0002	.0033	.0039
9	0	.9135	.6302	.3874	.1938	.0751	.0260	.0023	.0020
	1	.0830	.2985	.3874	.3489	.2253	.1171	.0202	.0176
	2	.0034	.0629	.1722	.2791	.3003	.2341	.0776	.0703
	3	.0001	.0077	.0446	.1302	.2336	.2731	.1739	.1641
	4	.0000	.0006	.0074	.0391	.1168	.2048	.2506	.2461
	5	.0000	.0000	.0008	.0078	.0389	.1024	.2408	.2461
	6	.0000	.0000	.0001	.0010	.0087	.0341	.1542	.1641
	7	.0000	.0000	.0000	.0001	.0012	.0073	.0635	.0703
	8	.0000	.0000	.0000	.0000	.0001	.0009	.0153	.0176
	9	.0000	.0000	.0000	.0000	.0000	.0001	.0016	.0020
10	0	.9044	.5987	.3487	.1615	.0563	.0173	.0012	.0010
	1	.0914	.3151	.3874	.3230	.1877	.0867	.0114	.0098
	2	.0042	.0746	.1937	.2907	.2816	.1951	.0494	.0439
	3	.0001	.0105	.0574	.1550	.2503	.2601	.1267	.1172
	4	.0000	.0010	.0112	.0543	.1460	.2276	.2130	.2051
	5	.0000	.0001	.0015	.0130	.0584	.1366	.2456	.2461
	6	.0000	.0000	.0001	.0022	.0162	.0569	.1966	.2051
	7	.0000	.0000	.0000	.0002	.0031	.0163	.1080	.1172
	8	.0000	.0000	.0000	.0000	.0004	.0030	.0389	.0439
	9	.0000	.0000	.0000	.0000	.0000	.0003	.0083	.0098
	10	.0000	.0000	.0000	.0000	.0000	.0000	.0000	.0010

$$\text{Pr (exactly } r \text{ heads in } n \text{ tosses)} = C_r^n p^r q^{n-r}, \qquad r = 0, 1, 2, \ldots, n,$$

$$C_r^n = \frac{n!}{r!(n-r)!}.$$

For values of p greater than .50, use the rule Pr (exactly r heads in n tosses if heads probability is p) = Pr (exactly $n - r$ heads in n tosses if heads probability is $1 - p$).

APPENDIX V / RANDOM DIGITS

50960	40163	81961	00843	24550	36522	39452	79917	98586	17521
72630	45181	52058	76764	18220	83033	57036	86600	71728	72333
60589	62357	03386	99275	88123	26521	20050	02695	26609	34169
54589	97714	44600	45043	52743	03018	37700	82690	63322	01915
35443	54474	41248	38339	20225	64986	17386	92055	13455	02781
02760	79500	83272	98876	31455	82773	86285	08893	14852	23231
34909	34768	90680	30766	62491	31826	20327	81683	21028	17206
40078	15933	04669	70936	21735	38584	26359	38503	95990	23622
90600	49418	84750	57037	33133	91190	41517	42265	94794	85845
45868	96616	95195	86485	23985	45892	47622	11304	75337	54682
06969	56737	61956	43382	94856	49786	75302	71354	47964	50766
89343	79183	89631	51299	69278	35013	14400	76360	33694	14115
42214	46401	85574	35076	00934	70478	84657	62188	42880	47766
85200	83665	55576	37988	84408	53648	06898	04419	32694	88302
21909	68892	32437	55567	87705	22638	87875	24268	88587	25071
66555	36076	03095	26490	53387	66699	56518	11847	43201	80755
45764	52347	43037	59347	11985	42296	15462	24165	34020	36122
78490	31181	43566	56600	74896	61496	76828	09063	50937	13891
37710	73871	97251	11926	68385	78055	17524	84727	79283	33652
26690	80042	97961	66031	10880	01264	55734	40502	62151	66529
72845	34959	14587	56200	31529	84536	90864	78110	60105	43172
61946	33833	19507	06006	84126	38045	92093	49265	68888	43544
03578	32398	90583	40531	33545	74017	51434	22521	78657	16986
25833	81557	40853	85116	97095	55739	02178	91182	55061	40148
33262	49673	38264	52881	20295	38399	26633	38876	21231	17876
87840	72801	26879	27046	12486	14732	16496	46441	53447	26679
40517	16564	44137	76289	35586	05925	69187	54135	09813	19009
77240	23861	44528	03035	29244	16032	05958	75696	02818	70855
71823	83501	11356	57658	08014	31502	44422	79503	53656	74133
27271	69648	48332	97709	60814	15250	56055	09976	21694	34299
07183	68954	42201	22979	09485	59191	00142	71063	85767	65472
88217	39439	92319	58339	31969	84814	90869	34498	38823	44718
94483	32185	34424	53284	43915	49574	07406	55179	60883	05521
88773	25562	96265	70398	98538	26428	84011	85880	66962	37190
92175	22746	58292	37741	92953	32708	68923	33573	11934	27532

SOURCE: The RAND Corporation (1955), p. 186.

APPENDIX VI / SQUARE ROOTS

FIRST DIGIT	SECOND DIGIT									
	0	1	2	3	4	5	6	7	8	9
0	0.000	1.000	1.414	1.732	2.000	2.236	2.449	2.646	2.828	3.000
1	3.162	3.317	3.464	3.606	3.742	3.873	4.000	4.123	4.243	4.359
2	4.472	4.583	4.690	4.796	4.899	5.000	5.099	5.196	5.292	5.385
3	5.477	5.568	5.657	5.745	5.831	5.916	6.000	6.083	6.164	6.245
4	6.325	6.403	6.481	6.557	6.633	6.708	6.782	6.856	6.928	7.000
5	7.071	7.141	7.211	7.280	7.348	7.416	7.483	7.550	7.616	7.681
6	7.746	7.810	7.874	7.937	8.000	8.062	8.124	8.185	8.246	8.307
7	8.367	8.426	8.485	8.544	8.602	8.660	8.718	8.775	8.832	8.888
8	8.944	9.000	9.055	9.110	9.165	9.220	9.274	9.327	9.381	9.434
9	9.487	9.539	9.592	9.644	9.695	9.747	9.798	9.849	9.899	9.950

For example, $\sqrt{12}$ is in the row labelled 1 and the column labelled 2, that is, 3.464; $\sqrt{.12} = .1 \times \sqrt{12} = .3464$; $\sqrt{1200} = 10 \times \sqrt{12} = 34.64$.

FIRST DIGIT	SECOND AND THIRD DIGITS									
	00	10	20	30	40	50	60	70	80	90
0	0.000	3.162	4.472	5.477	6.325	7.071	7.746	8.367	8.944	9.487
1	10.000	10.488	10.954	11.402	11.832	12.247	12.649	13.038	13.416	13.784
2	14.142	14.491	14.832	15.166	15.492	15.811	16.125	16.432	16.733	17.029
3	17.321	17.607	17.889	18.166	18.439	18.708	18.974	19.235	19.494	19.748
4	20.000	20.248	20.494	20.736	20.976	21.213	21.448	21.679	21.909	22.136
5	22.361	22.583	22.804	23.022	23.238	23.452	23.664	23.875	24.083	24.290
6	24.495	24.698	24.900	25.100	25.298	25.495	25.690	25.884	26.077	26.268
7	26.458	26.646	26.833	27.019	27.203	27.386	27.568	27.749	27.928	28.107
8	28.284	28.460	28.636	28.810	28.983	29.155	29.326	29.496	29.665	29.833
9	30.000	30.166	30.332	30.496	30.659	30.822	30.984	31.145	31.305	31.464

Examples: $\sqrt{120} = 10.954$; $\sqrt{1.20} = .1 \times \sqrt{120} = 1.0954$; $\sqrt{12,000} = 10 \times \sqrt{120} = 109.54$

APPENDIX VII / COMPUTATION OF SQUARE ROOTS

We illustrate the method by computing $\sqrt{150}$ although its value is in Appendix VI.

STEP 1   Make a guess of the square root.

$$\sqrt{150} \doteq \sqrt{144} = 12.$$

STEP 2   Divide the number by the guess.

$$150 \div 12 = 12.50$$

STEP 3   Average this quotient and the guess. The result is a better guess than the original guess.

$$\frac{12 + 12.5}{2} = 12.25$$

$$\sqrt{150} \doteq 12.25$$

Repeat Steps 2 and 3 until the needed accuracy is obtained. (The number of accurate significant figures nearly doubles with each repetition.)

$$150 \div 12.25 = 12.244898$$
$$\frac{12.25 + 12.244898}{2} = 12.247449$$

$$\sqrt{150} \doteq 12.247449$$

The approximation 12.25 is accurate to two decimal places.

APPENDIX VIII / INTERPOLATION

The function $\Phi(z)$, for example, is defined for all values of z, but can only be tabulated for some values. However, by interpolation it is possible to derive approximate values of $\Phi(z)$ for values of z which are not tabulated.

As an example, consider $\Phi(0.024)$. In Appendix I we find $\Phi(0.02) = .5080$ and $\Phi(0.03) = .5120$; the second value is .0040 more than the first and it results from an increase of .01 in z. We can expect that an increase of .001 in z would result in an increase of about .0004 in $\Phi(z)$. Since 0.024 is 4/10 of the way from 0.02 to 0.03, we expect $\Phi(0.024)$ to be about 4/10 of the way from $\Phi(0.02)$ to $\Phi(0.03)$; that is,

$$\Phi(0.024) \doteq \Phi(0.02) + \frac{4}{10}[\Phi(0.03) - \Phi(0.02)]$$
$$= .5080 + .4 \times .0040$$
$$= .5080 + .0016$$
$$= .5096.$$

APPENDIX IX / PERCENTAGE POINTS OF CHI-SQUARE DISTRIBUTIONS: VALUES OF $\chi_f^2(\alpha)$ FOR VARIOUS α AND f

Suppose that w is distributed according to the chi-square distribution with f degrees of freedom, denoted by χ_f^2. Then the upper αth percentage point, $\chi_f^2(\alpha)$, is defined by $\Pr[w > \chi_f^2(\alpha)] = \alpha$. For example, if $f = 5$ and $\alpha = .05$, then $\chi_5^2(.05) = 11.070$.

The following table is for $\alpha = .005, .01, .025, .05, .10, .90, .95, .975, .99, .995$ and $f = 1, 2, \ldots, 40$. For large values of f, one may use the approximation [Kendall and Stuart (1969), page 371]

$$\chi_f^2(\alpha) \doteq f \left(1 - \frac{2}{9f} + z(\alpha) \sqrt{\frac{2}{9f}}\right)^3,$$

where $z(\alpha)$ is the upper αth percentage point of the standard normal distribution. For example, $\chi_{60}^2(.05) \doteq 60[1 - .00370 + 1.645 \times .06086]^3 = 79.080$.

Appendix IX / Percentage Points of Chi-Square Distributions: Values of $\chi_f^2(\alpha)$ for Various α and f

	SIGNIFICANCE LEVEL, α									
f	.995	.99	.975	.95	.90	.10	.05	.025	.01	.005
1	0.000	0.000	0.001	0.004	0.016	2.706	3.843	5.025	6.637	7.882
2	0.010	0.020	0.051	0.103	0.211	4.605	5.992	7.378	9.210	10.597
3	0.072	0.115	0.216	0.352	0.584	6.251	7.815	9.348	11.344	12.837
4	0.207	0.297	0.484	0.711	1.064	7.779	9.488	11.143	13.277	14.860
5	0.412	0.554	0.831	1.145	1.610	9.236	11.070	12.832	15.085	16.748
6	0.676	0.872	1.237	1.635	2.204	10.645	12.592	14.440	16.812	18.548
7	0.989	1.239	1.690	2.167	2.833	12.017	14.067	16.012	18.474	20.276
8	1.344	1.646	2.180	2.733	3.490	13.362	15.507	17.534	20.090	21.954
9	1.735	2.088	2.700	3.325	4.168	14.684	16.919	19.022	21.665	23.587
10	2.156	2.558	3.247	3.940	4.865	15.987	18.307	20.483	23.209	25.188
11	2.603	3.053	3.816	4.575	5.578	17.275	19.675	21.920	24.724	26.755
12	3.074	3.571	4.404	5.226	6.304	18.549	21.026	23.337	26.217	28.300
13	3.565	4.107	5.009	5.892	7.041	19.812	22.362	24.735	27.687	29.817
14	4.075	4.660	5.629	6.571	7.790	21.064	23.685	26.119	29.141	31.319
15	4.600	5.229	6.262	7.261	8.547	22.307	24.996	27.488	30.577	32.799
16	5.142	5.812	6.908	7.962	9.312	23.542	26.296	28.845	32.000	34.267
17	5.697	6.407	7.564	8.682	10.085	24.769	27.587	30.190	33.408	35.716
18	6.265	7.015	8.231	9.390	10.865	25.989	28.869	31.526	34.805	37.156
19	6.843	7.632	8.906	10.117	11.651	27.203	30.143	32.852	36.190	38.580
20	7.434	8.260	9.591	10.851	12.443	28.412	31.410	34.170	37.566	39.997
21	8.033	8.897	10.283	11.591	13.240	29.615	32.670	35.478	38.930	41.399
22	8.643	9.542	10.982	12.338	14.042	30.813	33.924	36.781	40.289	42.796
23	9.260	10.195	11.688	13.090	14.848	32.007	35.172	38.075	41.637	44.179
24	9.886	10.856	12.401	13.848	15.659	33.196	36.415	39.364	42.980	45.558
25	10.519	11.523	13.120	14.611	16.473	34.381	37.652	40.646	44.313	46.925
26	11.160	12.198	13.844	15.379	17.292	35.563	38.885	41.923	45.642	48.290
27	11.807	12.878	14.573	16.151	18.114	36.741	40.113	43.194	46.962	49.642
28	12.461	13.565	15.308	16.928	18.939	37.916	41.337	44.461	48.278	50.993
29	13.120	14.256	16.147	17.708	19.768	39.087	42.557	45.722	49.586	52.333
30	13.787	14.954	16.791	18.493	20.599	40.256	43.773	46.979	50.892	53.672
31	14.457	15.655	17.538	19.280	21.433	41.422	44.985	48.231	52.190	55.000
32	15.134	16.362	18.291	20.072	22.271	42.585	46.194	49.480	53.486	56.328
33	15.814	17.073	19.046	20.866	23.110	43.745	47.400	50.724	54.774	57.646
34	16.501	17.789	19.806	21.664	23.952	44.903	48.602	51.966	56.061	58.964
35	17.191	18.508	20.569	22.465	24.796	46.059	49.802	53.203	57.340	60.272
36	17.887	19.233	21.336	23.269	25.643	47.212	50.998	54.437	58.619	61.581
37	18.584	19.960	22.105	24.075	26.492	48.363	52.192	55.667	59.891	62.880
38	19.289	20.691	22.878	24.884	27.343	49.513	53.384	56.896	61.162	64.181
39	19.994	21.425	23.654	25.695	28.196	50.660	54.572	58.119	62.426	65.473
40	20.706	22.164	24.433	26.509	29.050	51.805	55.758	59.342	63.691	66.766

SOURCE: Zehna and Barr (1970), Table 2, pages 11–14.

APPENDIX X / PERCENTAGE POINTS OF F-DISTRIBUTIONS: VALUES OF $F_{m,n}(\alpha)$ FOR VARIOUS m, n, AND α

The parameters m and n denote the numbers of degrees of freedom in the numerator and denominator, respectively. Suppose that F is distributed according to $F_{m,n}$. Then Pr $[F > F_{m,n}(\alpha)] = \alpha$. For example, if $m = 2$, $n = 9$, and $\alpha = .05$, then $F_{2,9}(.05) = 4.2565$.

The following table can be used for values corresponding to $1 - \alpha$ by means of the identity $F_{n,m}(\alpha) = 1/F_{m,n}(1 - \alpha)$.

Interpolation on m is illustrated in the following example. (Interpolation on n follows a similar pattern.) Suppose that one wishes to find the upper 5% point of $F_{17,30}$. The table contains values for 15 and 20, but not for 17.

$$F_{20,30}(.05) = 1.9317 \qquad F_{17,30}(.05) = ? \qquad F_{15,30}(.05) = 2.0148$$

$$F_{17,30}(.05) \doteq 1.9317 + \frac{20 - 17}{20 - 15} \times (2.0148 - 1.9317)$$

$$= 1.9317 + .6 \times .0831$$

$$= 1.9816.$$

To find, for example, $F_{17,33}(.05)$, one has to interpolate simultaneously on m and n.

$$F_{15,30}(.05) = 2.0148 \qquad F_{15,50}(.05) = 1.8714$$

$$F_{17,30}(.05) \doteq 1.9816 \qquad F_{17,50}(.05) \doteq 1.8365$$

$$F_{20,30}(.05) = 1.9317 \qquad F_{20,50}(.05) = 1.7841$$

$$F_{17,33}(.05) \doteq 1.8365 + \frac{50 - 33}{50 - 30} \times (1.9816 - 1.8365) = 1.9598.$$

Appendix X / Percentage Points of F-Distributions: Values of $F_{m,n}(\alpha)$ for Various m, n, and α

α	n	1	2	3	4	m 5	6	7	8	9
.100		39.9	49.5	53.6	55.8	57.2	58.2	58.9	59.4	59.9
.050		161.4	199.5	215.7	224.6	230.2	234.0	236.8	238.9	240.5
.025	1	647.8	799.5	864.2	899.6	921.9	937.1	948.2	956.7	963.3
.010		4,052.2	4,999.5	5,403.3	5,624.5	5,763.6	5,849.0	5,928.3	5,981.2	6,022.5
.005		16,211.	20,000.	21,615.	22,499.	23,056.	23,438.	23,716.	23,925.	24,091.
.100		8.5263	9.0000	9.1618	9.2434	9.2926	9.3255	9.3491	9.3668	9.3805
.050		18.513	19.000	19.164	19.274	19.297	19.329	19.354	19.371	19.385
.025	2	38.506	39.000	39.167	39.249	39.299	39.331	39.355	39.373	39.388
.010		98.502	98.999	99.165	99.250	99.299	99.332	99.356	99.375	99.389
.005		198.50	199.00	199.17	199.25	199.30	199.33	199.36	199.38	199.39
.100		5.5383	5.4624	5.3908	5.3426	5.3092	5.2847	5.2662	5.2517	5.2400
.050		10.128	9.5521	9.2766	9.1172	9.0135	8.9406	8.8867	8.8452	8.8123
.025	3	17.443	16.044	15.439	15.101	14.885	14.735	14.624	14.540	14.473
.010		34.116	30.816	29.457	28.710	28.237	27.911	27.672	27.489	27.345
.005		55.552	49.799	47.467	46.195	45.392	44.839	44.434	44.126	43.883
.100		4.5448	4.3246	4.1909	4.1072	4.0506	4.0098	3.9790	3.9549	3.9357
.050		7.7086	6.9443	6.5914	6.3882	6.2561	6.1631	6.0942	6.0410	5.9988
.025	4	12.218	10.649	9.9792	9.6045	9.3645	9.1973	9.0741	8.9796	8.9047
.010		21.198	18.000	16.694	15.977	15.522	15.207	14.976	14.799	14.659
.005		31.333	26.284	24.259	23.154	22.456	21.974	21.622	21.352	21.139
.100		4.0604	3.7797	3.6195	3.5202	3.4530	3.4045	3.3679	3.3393	3.3163
.050		6.6079	5.7861	5.4094	5.1922	5.0503	4.9503	4.8759	4.8183	4.7725
.025	5	10.007	8.4336	7.7636	7.3879	7.1464	6.9777	6.8531	6.7572	6.6810
.010		16.258	13.274	12.060	11.392	10.967	10.672	10.456	10.289	10.158
.005		22.785	18.314	16.530	15.556	14.940	14.513	14.200	13.961	13.772
.100		3.7760	3.4633	3.2888	3.1808	3.1075	3.0546	3.0145	2.9830	2.9577
.050		5.9874	5.1433	4.7571	4.5337	4.3874	4.2839	4.2067	4.1468	4.0990
.025	6	8.8131	7.2598	6.5988	6.2272	5.9876	5.8198	5.6955	5.5996	5.5234
.010		13.745	10.925	9.7796	9.1483	8.7459	8.4661	8.2600	8.1017	7.9761
.005		18.635	14.544	12.917	12.027	11.464	11.073	10.786	10.566	10.392
.100		3.5894	3.2574	3.0741	2.9605	2.8833	2.8274	2.7849	2.7516	2.7247
.050		5.5915	4.7374	4.3468	4.1203	3.9715	3.8660	3.7870	3.7257	3.6767
.025	7	8.0727	6.5415	5.8898	5.5226	5.2852	5.1186	4.9949	4.8993	4.8232
.010		12.246	9.5466	8.4513	7.8466	7.4604	7.1914	6.9928	6.8401	6.7188
.005		16.236	12.404	10.882	10.050	9.5220	9.1553	8.8853	8.6781	8.5138
.100		3.4579	3.1131	2.9238	2.8064	2.7264	2.6683	2.6241	2.5893	2.5612
.050		5.3177	4.4590	4.0662	3.8379	3.6875	3.5806	3.5005	3.4381	3.3881
.025	8	7.5709	6.0595	5.4160	5.0526	4.8173	4.6517	4.5286	4.4333	4.3572
.010		11.259	8.6491	7.5910	7.0061	6.6318	6.3707	6.1776	6.0289	5.9106
.005		14.688	11.042	9.5965	8.8051	8.3018	7.9520	7.6942	7.4959	7.3386

SOURCE: Zehna and Barr (1970)
m = degrees of freedom for numerator
n = degrees of freedom for denominator

Appendix X / Percentage Points of F-Distributions: Values of $F_{m,n}(\alpha)$ for Various m, n, and α

α	n	m						
		10	15	20	30	50	100	200
.100		60.2	61.2	61.7	62.3	62.7	63.0	63.2
.050		241.9	245.9	248.0	250.1	251.8	253.0	253.7
.025	1	968.6	984.9	993.1	1,001.4	1,008.1	1,013.2	1,015.7
.010		6,055.9	6,157.3	6,208.6	6,260.7	6,302.5	6,334.0	6,349.9
.005		24,225.	24,629.	24,837.	25,043.	25,211.	25,338.	25,401.
.100		9.3916	9.4247	9.4413	9.4579	9.4712	9.4812	9.4862
.050		19.396	19.429	19.445	19.463	19.476	19.486	19.491
.025	2	39.396	39.432	39.447	39.465	39.476	39.488	39.491
.010		99.399	99.432	99.450	99.466	99.478	99.491	99.495
.005		199.40	199.43	199.45	199.46	199.48	199.49	199.49
.100		5.2304	5.2003	5.1845	5.1681	5.1546	5.1443	5.1390
.050		8.7855	8.7029	8.6602	8.6166	8.5810	8.5539	8.5402
.025	3	14.419	14.253	14.167	14.080	14.010	13.956	13.929
.010		27.229	26.872	26.690	26.505	25.354	26.240	26.183
.005		43.686	43.085	42.778	42.466	42.213	42.022	41.925
.100		3.9199	3.8704	3.8443	3.8174	3.7952	3.7782	3.7695
.050		5.9644	5.8578	5.8026	5.7459	5.6995	5.6641	5.6461
.025	4	8.8439	8.6565	8.5599	8.4613	8.3808	8.3195	8.2885
.010		14.546	14.198	14.020	13.838	13.690	13.577	13.520
.005		20.967	20.438	20.167	19.891	19.667	19.497	19.411
.100		3.2974	3.2380	3.2066	3.1741	3.1471	3.1263	3.1157
.050		4.7351	4.6188	4.5581	4.4957	4.4444	4.4051	4.3851
.025	5	6.6192	6.4277	6.3286	6.2269	6.1436	6.0800	6.0478
.010		10.051	9.7222	9.5527	9.3794	9.2378	9.1299	9.0754
.005		13.618	13.146	12.904	12.656	12.454	12.300	12.222
.100		2.9369	2.8712	2.8363	2.8000	2.7697	2.7463	2.7343
.050		4.0600	3.9381	3.8742	3.8082	3.7537	3.7117	3.6904
.025	6	5.4613	5.2687	5.1684	5.0652	4.9804	4.9154	4.8824
.010		7.8741	7.5590	7.3958	7.2285	7.0915	6.9867	6.9336
.005		10.250	9.8140	9.5888	9.3583	9.1697	9.0257	8.9529
.100		2.7025	2.6322	2.5947	2.5555	2.5226	2.4971	2.4841
.050		3.6365	3.5107	3.4445	3.3758	3.3189	3.2749	3.2525
.025	7	4.7611	4.5678	4.4667	4.3624	4.2763	4.2101	4.1764
.010		6.6200	6.3143	6.1554	5.9920	5.8577	5.7547	5.7024
.005		8.3803	7.9678	7.7540	7.5345	7.3544	7.2165	7.1466
.100		2.5380	2.4642	2.4246	2.3830	2.3481	2.3208	2.3068
.050		3.3472	3.2184	3.1503	3.0794	3.0204	2.9747	2.9513
.025	8	4.2951	4.1012	3.9995	3.8940	3.8067	3.7393	3.7050
.010		5.8143	5.5151	5.3591	5.1981	5.0654	4.9633	4.9114
.005		7.2107	6.8142	6.6082	6.3961	6.2215	6.0875	6.0194

Appendix X / Percentage Points of F-Distributions: Values of $F_{m,n}(\alpha)$ for Various m, n, and α

α	n	1	2	3	4	m 5	6	7	8	9
.100		3.3603	3.0064	2.8129	2.6927	2.6106	2.5509	2.5053	2.4694	2.4403
.050		5.1174	4.2565	3.8625	3.6331	3.4817	3.3738	3.2927	3.2296	3.1789
.025	9	7.2093	5.7147	5.0781	4.7181	4.4844	4.3197	4.1970	4.1020	4.0260
.010		10.561	8.0215	6.9919	6.4221	6.0569	5.8018	5.6129	5.4671	5.3511
.005		13.614	10.107	8.7171	7.9559	7.4712	7.1338	6.8849	6.6933	6.5411
.100		3.2850	2.9245	2.7277	2.6053	2.5216	2.4606	2.4140	2.3772	2.3473
.050		4.9646	4.1028	3.7083	3.4780	3.3258	3.2172	3.1355	3.0717	3.0204
.025	10	6.9367	5.4564	4.8256	4.4683	4.2361	4.0721	3.9498	3.8549	3.7790
.010		10.044	7.5594	6.5523	5.9943	5.6363	5.3858	5.2001	5.0567	4.9424
.005		12.826	9.4270	8.0808	7.3428	6.8724	6.5446	6.3025	6.1159	5.9676
.100		3.0732	2.6952	2.4898	2.3614	2.2730	2.2081	2.1582	2.1185	2.0862
.050		4.5431	3.6823	3.2874	3.0556	2.9013	2.7905	2.7066	2.6408	2.5876
.025	15	6.1995	4.7650	4.1528	3.8043	3.5764	3.4147	3.2934	3.1987	3.1227
.010		8.6831	6.3589	5.4170	4.8932	4.5556	4.3183	4.1415	4.0045	3.8948
.005		10.798	7.7007	6.4760	5.8029	5.3721	5.0708	4.8472	4.6743	4.5364
.100		2.9747	2.5893	2.3801	2.2489	2.1582	2.0913	2.0697	1.9985	1.9649
.050		4.3512	3.4928	3.0984	2.8661	2.7109	2.5990	2.5140	2.4471	2.3928
.025	20	5.8715	4.4613	3.8587	3.5147	3.2891	3.1283	3.0074	2.9128	2.8365
.010		8.0960	5.8489	4.9382	4.4307	4.1027	3.8714	3.6987	3.5644	3.4567
.005		9.9439	6.9865	5.8177	5.1743	4.7616	4.4721	4.2569	4.0900	3.9564
.100		2.8807	2.4887	2.2761	2.1422	2.0492	1.9803	1.9269	1.8841	1.8490
.050		4.1709	3.3158	2.9223	2.6896	2.5336	2.4205	2.3343	2.2662	2.2107
.025	30	5.5675	4.1821	3.5894	3.2499	3.0265	2.8667	2.7460	2.6513	2.5746
.010		7.5625	5.3903	4.5097	4.0179	3.6990	3.4735	3.3045	3.1726	3.0665
.005		9.1797	6.3547	5.2388	4.6234	4.2276	3.9492	3.7416	3.5801	3.4505
.100		2.8087	2.4120	2.1967	2.0608	1.9660	1.8954	1.8405	1.7963	1.7598
.050		4.0343	3.1826	2.7900	2.5572	2.4004	2.2864	2.1992	2.1299	2.0733
.025	50	5.3403	3.9749	3.3902	3.0544	2.8326	2.6736	2.5530	2.4579	2.3808
.010		7.1706	5.0566	4.1993	3.7195	3.4077	3.1864	3.0202	2.8900	2.7850
.005		8.6258	5.9016	4.8259	4.2316	3.8486	3.5785	3.3764	3.2189	3.0920
.100		2.7564	2.3564	2.1394	2.0019	1.9057	1.8339	1.7778	1.7324	1.6949
.050		3.9361	3.0873	2.6955	2.4626	2.3053	2.1906	2.1025	2.0323	1.9748
.025	100	5.1786	3.8284	3.2496	2.9166	2.6961	2.5374	2.4168	2.3215	2.2439
.010		6.8953	4.8239	3.9837	3.5127	3.2059	2.9877	2.8233	2.6943	2.5898
.005		8.2407	5.5892	4.5424	3.9634	3.5895	3.3252	3.1271	2.9722	2.8472
.100		2.7308	2.3293	2.1114	1.9732	1.8763	1.8038	1.7470	1.7011	1.6630
.050		3.8884	3.0411	2.6497	2.4168	2.2592	2.1441	2.0556	1.9849	1.9269
.025	200	5.1004	3.7578	3.1820	2.8503	2.6304	2.4720	2.3513	2.2558	2.1780
.010		6.7633	4.7128	3.8810	3.4143	3.1100	2.8933	2.7298	2.6012	2.4971
.005		8.0572	5.4412	4.4085	3.8368	3.4674	3.2059	3.0097	2.8560	2.7319

Appendix X / Percentage Points of F-Distributions: Values of $F_{m,n}(\alpha)$ for Various m, n, and α

α	n	m						
		10	15	20	30	50	100	200
.100		2.4163	2.3396	2.2983	2.2547	2.2180	2.1892	2.1744
.050		3.1373	3.0061	2.9365	2.8637	2.8028	2.7556	2.7313
.025	9	3.9639	3.7694	3.6669	3.5604	3.4719	3.4034	3.3684
.010		5.2565	4.9621	4.8080	4.6486	4.5167	4.4150	4.3631
.005		6.4172	6.0325	5.8318	5.6248	5.4539	5.3223	5.2553
.100		2.3226	2.2435	2.2007	2.1554	2.1171	2.0869	2.0713
.050		2.9782	2.8450	2.7740	2.6996	2.6371	2.5884	2.5634
.025	10	3.7168	3.5217	3.4185	3.3110	3.2214	3.1517	3.1161
.010		4.8492	4.5581	4.4054	4.2469	4.1155	4.0137	3.9617
.005		5.8467	5.4707	5.2740	5.0705	4.9022	4.7722	4.7058
.100		2.0593	1.9722	1.9243	1.8728	1.8284	1.7929	1.7743
.050		2.5437	2.4034	2.3275	2.2468	2.1780	2.1234	2.0950
.025	15	3.0602	2.8621	2.7559	2.6437	2.5488	2.4739	2.4352
.010		3.8049	3.5222	3.3719	3.2141	3.0814	2.9772	2.9235
.005		4.4235	4.0698	3.8826	3.6867	3.5225	3.3941	3.3279
.100		1.9367	1.8449	1.7938	1.7382	1.6896	1.6501	1.6292
.050		2.3479	2.2033	2.1242	2.0391	1.9656	1.9066	1.8755
.025	20	2.7737	2.5731	2.4645	2.3486	2.2493	2.1699	2.1284
.010		3.3682	3.0880	2.9377	2.7785	2.6430	2.5353	2.4792
.005		3.8470	3.5020	3.3178	3.1234	2.9586	2.8282	2.7603
.100		1.8195	1.7223	1.6673	1.6065	1.5522	1.5069	1.4824
.050		2.1646	2.0148	1.9317	1.8409	1.7609	1.6950	1.6597
.025	30	2.5112	2.3072	2.1952	2.0739	1.9681	1.8816	1.8354
.010		2.9791	2.7002	2.5487	2.3860	2.2450	2.1307	2.0700
.005		3.3440	3.0057	2.8230	2.6278	2.4594	2.3234	2.2514
.100		1.7291	1.6269	1.5681	1.5018	1.4409	1.3885	1.3590
.050		2.0261	1.8714	1.7841	1.6872	1.5995	1.5249	1.4835
.025	50	2.3168	2.1090	1.9933	1.8659	1.7520	1.6558	1.6029
.010		2.6981	2.4190	2.2652	2.0976	1.9490	1.8248	1.7567
.005		2.9875	2.6531	2.4702	2.2717	2.0967	1.9512	1.8719
.100		1.6632	1.5566	1.4943	1.4227	1.3548	1.2934	1.2571
.050		1.9267	1.7675	1.6764	1.5733	1.4772	1.3917	1.3416
.025	100	2.1793	1.9679	1.8486	1.7148	1.5917	1.4833	1.4203
.010		2.5033	2.2230	2.0666	1.8933	1.7353	1.5977	1.5184
.005		2.7440	2.4113	2.2270	2.0230	1.8400	1.6809	1.5897
.100		1.6308	1.5218	1.4575	1.3826	1.3100	1.2418	1.1991
.050		1.8783	1.7166	1.6233	1.5164	1.4146	1.3206	1.2626
.025	200	2.1130	1.8996	1.7780	1.6403	1.5108	1.3927	1.3205
.010		2.4106	2.1294	1.9713	1.7941	1.6295	1.4811	1.3912
.005		2.6292	2.2970	2.1116	1.9051	1.7147	1.5442	1.4416

REFERENCES FOR APPENDICES

Kendall, Maurice G., and Alan Stuart (1969). *The Advanced Theory of Statistics, Vol. 1.* 3rd ed. Hafner, New York.

The RAND Corporation (1955). *A Million Random Digits with 100,000 Normal Deviates.* The Free Press (Macmillan), New York.

Zehna, P. W., and D. R. Barr (1970). *Tables of Common Probability Distributions.* United States Naval Postgraduate School, Monterey, Calif.

Answers to Selected Exercises

CHAPTER 2

2.1 (a) ordinal (b) ratio (c) nominal (d) ordinal (e) interval

2.2 discrete: (a), (d), (e); continuous: (b), (c), (f), (g)

2.8 (b)

Midpoint of class interval	31	34	37	40	43	46
Relative frequency	.005	.065	.258	.495	.169	.008

2.11 (a) 3   **2.22** (b), (e)   **2.25** (a) $n' - n$ (b) $100\% \times (n' - n)/n$ (e) b'/n'

CHAPTER 3

3.1 (a) 25 (b) 4 (c) 6.25   **3.6** (a) Mode ($= 2$) is most descriptive: 9 of the 12 families have 2 children.

3.7 Number of children: mean $= 3.5$, median $= 3$ **3.9** $13,000

3.11 (a)

Midpoint of class interval (100's)	50	75	100	125	150	175	200
Number of families		5	17	17	7	2	0	1

(b) $9,338 (c) $9,118 (d) $8,750 (e) $9,365 (f) $9,000

3.13 4, 9, 5; 6   **3.15** (b) 23 hours per week   **3.17** 143.0 lb

3.19 206.67, 180, 190, 206 lb   **3.21** 73.27, 73, 75, 76 inches

3.23

MACHINE	MEAN	MEDIAN
A	339	340
B	338.67	339
C	348.5	348

3.28 1960: 72.5; 1970: 64.1

3.32 (a) $\sum_{i=1}^{12} x_i$ (b) $\sum_{i=7}^{12} x_i$ (c) $\sum_{i=6}^{8} x_i$ **3.34** (a) $\sum_{i=1}^{50} x_{2i}/50$ (b) $\sum_{i=1}^{50} x_{2i-1}/50$

3.37 \$13,529

CHAPTER 4

4.2 +4   **4.3** 2   **4.4** 37.6, 41.1, 3.5 inches   **4.7** 3.67   **4.9** (a) 4.3, 3.7, 2.5

4.11 205, 46   **4.12** 30.5, 52 lb   **4.15** \$2896   **4.16** \$2911

4.20 males: $\bar{x} = 2.90$ kg, $s^2 = 0.22$; females: $\bar{x} = 2.36$ kg, $s^2 = 0.08$

4.21 2.54 inches   **4.25** Both equal 2665.   **4.31** 10, 6, 5.36, 5.16

CHAPTER 5

5.1 (a)

1	0	1
0	1	1
1	1	2

0	1	1
1	0	1
1	1	2

(b)

1	1	2
0	1	1
1	2	3

0	2	2
1	0	1
1	2	3

5.3

0	4	4
4	0	4
4	4	8

1	3	4
3	1	4
4	4	8

2	2	4
2	2	4
4	4	8

3	1	4
1	3	4
4	4	8

4	0	4
0	4	4
4	4	8

5.6 (a) .84 (b) .62

5.7 (a)

	No	YES	TOTAL
Low	32 (58%)	23 (42%)	55 (100%)
High	64 (33%)	128 (67%)	192 (100%)
Total	96	151	247

(b)

	No	YES	TOTAL
Low	83 (49%)	87 (51%)	170 (100%)
High	13 (17%)	64 (83%)	77 (100%)
Total	96	151	247

5.8 (a)(i)

	F	S	J
Favor	60.0%	66.7%	37.5%
Opposed	40.0%	33.3%	62.5%
Total	100.0%	100.0%	100.0%

(ii)

	F	S	J	TOTAL
Favor	42.9%	28.6%	28.6%	100.1%
Opposed	31.6%	15.8%	52.6%	100.0%

(b)(i)

	LOWER CLASSES	JUNIORS	TOTAL
Favor	75	30	105
Opposed	45	50	95
Total	120	80	200

(ii)

	LOWER CLASSES	JUNIORS
Favor	62.5%	37.5%
Opposed	37.5%	62.5%
Total	100.0%	100.0%

(iii)

	LOWER CLASSES	JUNIORS	TOTAL
Favor	71.4%	28.6%	100.0%
Opposed	47.4%	52.6%	100.0%

(c)(i)

	FRESHMEN	SOPHOMORES	TOTAL
Favor	45	30	75
Opposed	30	15	45
Total	75	45	120

(ii)

	FRESHMEN	SOPHOMORES	TOTAL
Favor	60.0%	40.0%	100.0%
Opposed	66.7%	33.3%	100.0%

5.10

	T	C	OS	DK	TOTAL
Squirrels	6.0%	12.0%	18.1%	63.9%	100.0%
Bears	9.2%	11.1%	23.5%	56.2%	100.0%
Cats	11.6%	14.5%	21.7%	52.2%	100.0%

5.13 (a) (i) 65.5% (ii) 21.9% (iii) 41.7% (iv) 32.6% (v) 58.3% (vi) 16.5%

(b) *Percent Voting Democratic, by Age and Productivity*

		PRODUCTIVITY		
		Low	*Medium*	*High*
	Less than 41	69%	79%	79%
AGE	*41–50*	50%	63%	72%
	More than 50	42%	50%	54%

5.25 (a) 30%, 48% (b)

		RESPONDENT'S LEVEL	
		Low	*High*
TRUE	*Low*	19.0%	52.1%
RATE	*High*	28.0%	64.1%

5.27

		NUMBER OF MAGAZINES READ					
		0	*1*	*2*	*3*	*4*	*Total*
	0	29%	33%	30%	6%	2%	100%
NUMBER OF	1	27%	33%	30%	8%	2%	100%
CHILDREN	2	25%	35%	30%	7%	3%	100%
	3 or more	25%	31%	31%	8%	5%	100%
All nos. of children		27%	33%	30%	7%	3%	100%

5.29 0.92, 1.49, 1.62

5.36

d:	−8	−7	−6	−5	−4	−3	−2	−1	0	+1	+2	+3	+4
f:	1	0	2	1	7	8	15	14	35	6	3	2	2

5.37 (a) −0.990 (b) 4.137

CHAPTER 6

6.2 (a) a person is either wealthy or a college graduate
(b) a person is a wealthy college graduate
(c) a person is either wealthy or a college graduate, but not both
(d) a person is not a college graduate
(e) a person is not wealthy
(f) a person is neither wealthy nor a college graduate

6.4 (a) $\Pr(C)$ (b) $\Pr(D)$

6.6 (a) .6 (b) .1 (c) .4

6.8 (a) independent (b) dependent (c) dependent

6.10 (a) .0121 (b) .0118

6.12 .0965 **6.15** 1/3 **6.17** 1/6 **6.19** .06

6.21 (a) $\Pr(D|S) = .8$ (b) $\Pr(D|\bar{S}) = .05$

6.23 (a) .064 (b) .124

6.25 brown: 3/4; blue: 1/4

6.26 1/3 **6.28** 1/2, 1/4, 1/8, . . . **6.31** .84 **6.32** .81

6.35 .5625

CHAPTER 7

	(2)	(3)		
7.3 (a)	.9192	.9452	**7.5** .3125	**7.7** .2009
(b)	.1587	.2119		
(c)	.2743	.3446		
(d)	.1935	.2898		

7.8 .9377 **7.11** .3214

7.13 x:	0	1	2	3
prob.:	.0156	.1406	.4219	.4219

7.15 (a) .4505 (b) .0256 (c) .4032 (d) .8413

7.17 (a) .933 (b) .242 (c) .533 (d) .050 (e) .049 (f) 650, 93.32 percentile

7.19 (a) .9332 (b) .0606 (c) .0062

7.23 F: $-\infty$ to -1.04
  D: -1.04 to $-.39$
  C: $-.39$ to $+.39$
  B: $+.39$ to 1.04
  A: 1.04 to ∞

7.27 .3553   **7.28** (a) .2422 (b) .4844

7.29 .8397

7.37 $\{a,b\}$, $\{a,c\}$, $\{a,d\}$, $\{b,c\}$, $\{b,d\}$, $\{c,d\}$

CHAPTER 8

8.1 ($10,104, $10,496)   **8.3** ($10,136, $10,464)

8.5 (158.4 lb, 163.6 lb)   **8.7** (86.1, 93.9)

8.9 (a) (208.4, 219.6)   **8.13** Expected attendance is 42.

8.17 1374 hours

8.20 (48 hours, 52 hours)

8.30 (a) 1/16 (b) 1/16 (c) 1/8 (d) 7/8

8.32 $1 - 1/2^{n-1}$ **8.34** $0 < 1/(n+2) \leq \bar{p} \leq (n+1)/(n+2) < 1$, MSE($\bar{p}$) < MSE($\hat{p}$) if $\frac{1}{2} - \frac{1}{2}\sqrt{(n+1)/(2n+1)} < p < \frac{1}{2} + \frac{1}{2}\sqrt{(n+1)/(2n+1)}$

8.35 (a) N is *at least* as large as $x_{(n)}$.

CHAPTER 9

9.2 $z = -1.68 < -1.645$, so reject H.

9.4 $t = .38 < 1.38 = t_9(.10)$, so accept H.

9.5 reject   **9.10** (a) reject (b) reject

9.12 (a) Let μ be the mean tail length of offspring of rat mothers kept on a rotating turntable. Test $H\!: \mu \geq 4.0$ cm against Alternative: $\mu < 4.0$ cm.

9.14 reject   **9.17** reject (Value of standard normal test statistic is 7.35, very large indeed!)

CHAPTER 10

10.1 Confidence interval is $(-25, 39)$, which includes 0; hence accept the hypothesis.

10.3 (b) 26,581 (c) 64 (d) $s_p^2 = 415$, $s_p = 20.38$

10.6 reject   **10.11** (a) reject (b) accept (c) reject

10.12 reject   **10.14** 429,000

10.20 Accept hypothesis of equality of variances.

CHAPTER 11

11.1 (a) 5.992 (b) 0.103 (c) 9.488 (d) 0.711

11.3 (a) (55.2, 118) (b) accept   **11.5** (4.88, 17.2)

11.7 (1.76, 2.68)   **11.9** 1.67

11.11 (a) 2.3479 (b) 1.9367 (c) 4.1028 (d) 19.396

11.13 $F = 2 < 3.4185 = F_{20,10}(.025)$, so accept.

11.15 $F = 1.17 < 1.81 = F_{38,30}(.05)$, so accept.

11.17 $s/\bar{x}$; males: 0.161; females: 0.080

CHAPTER 12

12.1 Chi-square statistic $= 9 > 3.84 = \chi_1^2(.05)$, so reject.

12.3 Chi-square statistic $= 2.116 < 2.706 = \chi_1^2(.10)$, so accept.

12.5 Chi-square statistic $= 8.98 < 9.210 = \chi_2^2(.01)$, so accept.

12.7 Chi-square statistic $= 12.07 > 11.07 = \chi_5^2(.05)$, so reject.

12.9 Chi-square statistic $= 8.84 < 11.07 = \chi_5^2(.05)$, so accept.

12.11 (b) 6.45 (c) reject

12.13 Chi-square statistic $= 4.81$ without continuity correction $(.025 < P < .05)$ and 3.21 with continuity correction $(.05 < P < .10)$.

12.15 Chi-square statistic $= 25.0$ $(P < .005)$.

12.17 (c) 0.04, 0.03, 0.31; no significant differences.

12.19 (a) Chi-square statistic $= 3.77$ (not significant)
(b) Chi-square statistic $= 2.32$ (not significant)

12.21 Chi-square statistic $= 25.42$ $(P < .005)$

12.23 (a) Chi-square statistic $= 40.45$ $(P < .005)$ (b) $\phi = -.6819$

12.25

BIRTH ORDER	2	3–4	≥ 5
(a) ϕ	.05038	.04152	.1711
(b) Chi-square	0.421	0.195	2.518

12.27 (b) Democrats: chi-square statistic $= 158.8$ $(P < .005)$
Other: chi-square statistic $= 70.86$ $(P < .005)$

12.29 (a) 0.458 (b) 0.408

12.31 (a) 0 (b) .025 (c) .0139 (d) .298 (e) There is some slight association; this was detected by the coefficient based on ordering.

12.33 $\phi_{RB} = .50$, $\phi_{RC} = .42$, $\phi_{CB} = .26$

12.35 (d) 390.17 $(P < .005)$ **12.37** 408.1 $(P < .005)$ **12.43** $+0.06$

CHAPTER 13

13.1 $F = 9.58$; reject   **13.3** $F = 1.63$ $(P > .20)$ **13.5** $H = 172$; accept

13.7 (a) $F = 61.0$; reject (b) 0   **13.9** $W = 6$ $(P < .025)$

13.11

SOURCE OF VARIATION	F	P
Persons (random effect)	2.72	$.025 < P < .05$
Cans	20.3	$P < .005$
Interaction	0.728	$P > .20$

CHAPTER 14

14.1

	1	2	4	5	7	8	Total
1	1	0	1	0	0	0	2
3	1	1	0	1	1	0	4
5	0	0	1	1	1	1	4
Total	2	1	2	2	2	1	10

14.3 (a)

		WEIGHT					
		$23.5-$ 28.5	$28.5-$ 33.5	$33.5-$ 38.5	$38.5-$ 43.5	$43.5-$ 48.5	$48.5-$ 53.5	*Total*
	32.5–35.5	1	2	0	0	0	0	3
	35.5–38.5	0	2	7	1	0	0	10
HEIGHT	38.5–41.5	0	1	9	8	0	0	18
	41.5–44.5	0	0	1	5	3	0	9
	44.5–47.5	0	0	0	0	0	1	1
	Total	1	5	17	14	3	1	41

(b)

Height:	32.5–35.5	35.5–38.5	38.5–41.5	41.5–44.5	44.5–47.5	
Frequency:	3	10	18	9	1	

Weight:	23.5–28.5	28.5–33.5	33.5–38.5	38.5–43.5	43.5–48.5	48.5–53.5
Frequency:	1	5	17	14	3	1

(c) The mode in the height distribution is 40 inches, the midpoint of the interval 38.5–41.5. The mode of the weight distribution is 36 pounds, the midpoint of the interval 33.5–38.5.

(d)

		WEIGHT					
	23.5–28.5	28.5–33.5	33.5–38.5	38.5–43.5	43.5–48.5	48.5–53.5	Total
32.5–35.5	.02	.05	.00	.00	.00	.00	.07
35.5–38.5	.00	.05	.17	.02	.00	.00	.24
HEIGHT 38.5–41.5	.00	.02	.22	.20	.00	.00	.44
41.5–44.5	.00	.00	.02	.12	.07	.00	.21
44.5–47.5	.00	.00	.00	.00	.00	.02	.02
Total	.02	.12	.41	.34	.07	.02	.98

14.5 $\hat{y} = 4 - 0.2x$

14.7 *CD:* grade independent of hours of study. *EF:* positive correlation between grade and hours of study; slope is about 5 grade points per hour of study.

14.9 (a) +   (b) −   (c) 0   (d) +   (e) +

14.11 (a) 2.5 (b) 1.581 (c) 2.5 (d) 1.581 (e) −2.0 (f) −.8

14.13 (b) $s_x = 1, s_y = 3, s_{xy} = 3, r = 1$ **14.15** .615   **14.17** (b) .849

14.19 (b) $\hat{y} = 26.6 + .60x$ (c) 5.60 (d) (0.16, 1.04) (e) reject (f) The residual sum of squares is 54.5% of the total sum of squares. Note, however, that $s_{y \cdot x} = 5.60\%$, not much smaller than $s_y = 7.38\%$.

14.21 .7202   **14.23** (a) 5 (b) 4 (c) 2.8   **14.25** $b = -0.000429, a = 1.00$

CHAPTER 15

15.1 (a) $b_1 = 1.5$, $b_2 = 0.75$, $a = -5.5$ (b) $F = 18.5 > 6.3547 = F_{2,30}(.005) > F_{2,37}(.005)$; reject.

15.3 (c) $b_1 = -0.1330$, $b_2 = -0.03533$, $a = 11.08$ (d) $s_{y \cdot 12}^2 = 1.3027$ (e) $R_{y \cdot 12}^2 = .892$, $R_{y \cdot 12} = .944$

15.5 (0.4676, 10.77)   **15.7** $F = 71$ $(P < .005)$ **15.8** (a) .579

15.11 The reduction in the residual sum of squares must be greater than the original mean square.

CHAPTER 16

16.4 $\sigma_1 N_1 = \sigma_2 N_2$ **16.5** (a) (i) 1.26 (ii) 2.28 (c) 2.530

16.7 (a) Situation A: {4, 10, 8} is the only possible sample. Situation B: {10, 10, 10}, {8, 8, 8}, and {4, 4, 4} are the only possible samples.

(b) Situation A: standard error is 0. Situation B: standard error is 2.494.

INDEX

[A dagger (†) indicates a reference to an example used in the text or in an exercise.]